Wave Mechanics of Crystalline Solids

WAVE MECHANICS OF CRYSTALLINE SOLIDS

R. A. SMITH

C.B.E., F.R.S.

Director, Centre for Materials Science and Engineering,
Massachusetts Institute of Technology

LONDON

CHAPMAN AND HALL LTD

11 New Fetter Lane, E.C.4

First published 1961
Reprinted 1963 and 1967

Reprinted lithographically in Great Britain
by John Dickens & Co. Ltd., Northampton
Type set by Spottiswoode, Ballantyne & Co. Ltd.,
London and Colchester
SBN 412 06630 0

Preface

THE phenomenal increase of research effort concentrated on the physics of solids in the past decade has led to a much fuller understanding of electronic conduction in crystals and of the processes associated with emission and absorption of radiation by solids. The great technological interest in semiconductors has stimulated theoretical research into their electronic structure and it is now clear that these substances as well as metals and insulators are of the first importance in understanding the basic physics of the solid state. Unfortunately the underlying theory both of the forces which bind solids into their crystalline form and of the motion of charge carriers through the solids is by no means simple, and is frequently beyond the experience and training of experimental research workers in this field. The number of such workers is now very great and it was in an attempt to make some of the basic theory of crystalline solids more readily available to experimental physicists that this book came to be written. Its aim is to treat the fundamental theory of wave motion in solids, both that of the lattice, and of the electrons, in as elementary a way as is possible. Much of the material has been given as lectures in the Physics Department of the Royal Radar Establishment, Malvern, to an audience consisting mainly of experimental physicists who had either some previous experience of quantum mechanics as applied to atomic problems at the university, or had attended an introductory course covering this aspect of the subject. I am of the opinion as a result of giving these lectures that the subject can readily be taught to experimental physicists without an extensive training in mathematical physics and that their experimental work will benefit from the attempt to learn it. This requires some effort on the part of the beginner but if the subject is taken logically, step by step, there need be no major difficulties.

A considerable number of elementary and advanced treatments of quantum mechanics is now readily available and no attempt will be made in this book to deal with the fundamental concepts of the quantum theory or to derive the well-known results of the application of this theory to the structure of atoms or molecules. The book is entirely concerned with the application of the theory to the understanding of the electrical, thermal, and optical properties of crystalline solids. The reader will therefore be expected to have a knowledge of elementary quantum mechanics including the properties of wave functions. Some knowledge will be assumed of the wave-mechanical treatment of simple atomic spectra including those of hydrogen-like atoms and ions; such fundamental concepts as Heisenberg's Uncertainty Principle and Pauli's Exclusion Principle will also be assumed to be known, together with the part played by the latter in determining the form of the spectrum of the helium atom and in the binding of simple molecules.

We begin with a simple treatment of the fundamental principles of various types of wave motion including the concepts of phase velocity and group velocity. We then deal with the non-relativistic motion of free electrons under various conditions and with the formation of wave packets from plane waves. These ideas are used to treat the Sommerfeld model of a metal in which the electrons are regarded as essentially free. From this it becomes clear that we must take the periodic structure of the atomic lattice of a solid into account. A treatment is then given of the vibrational motion of the lattice, leading to the main part of the discussion which is concerned with the motion of electrons in the periodic field of the lattice. There follows an account of the interaction between the electrons and the lattice vibrations, and of the effect of impurities in an otherwise perfect crystal.

The main object of this treatment is to provide a theoretical basis for various parameters used in the phenomenological description of the motion of charge carriers in solids, such as effective mass, mobility, lifetime, relaxation time, carrier density, etc. Such a theoretical basis should lead to a better understanding of the variation of these quantities with such things as impurity content, temperature, etc. Most of these subjects have already been given detailed quantum-mechanical treatment both in original papers and in books, but these are frequently beyond the experience of all but theoretical physicists specializing in this field. There are, for example, standard works such as *The Theory of the Properties of Metals and Alloys*, by N. F. Mott and H. Jones (Oxford University Press, 1936), *The Modern Theory of Solids*, by F. Seitz (McGraw-Hill, 1940), *The Theory of Metals* (2nd Ed.), by A. H. Wilson (Cambridge University Press, 1953), and *The Quantum Theory of Solids*, by R. E. Peierls (Oxford University Press, 1955). To all of these I am more indebted than I can possibly say, and a good deal of such knowledge of the subject as I possess has been gleaned from them. There are, in addition, a number of excellent review articles on special topics to which I have made reference in the text, and also some more recent advanced treatments such as in *Electrons and Phonons*, by J. M. Ziman (Oxford University Press, 1960). It is hoped that this book may serve as an introduction to these more advanced treatments and may encourage more experimental workers in the field to explore still further the extensive theoretical work which has already been done.

There are, moreover, still many outstanding problems to be solved in which a great deal of careful experimentation will be required. In conducting such experiments the investigator should be helped by a better understanding of the fundamental atomic processes which underly the phenomena being observed.

In preparing the material for publication I have extended it somewhat in scope and have tried to bring it reasonably up to date. I have also had in mind that this should serve as a theoretical supplement to my book *Semiconductors* (Cambridge University Press, 1959). I have been greatly helped both by colleagues who have read and commented on the manuscript and also by those to whom the original lectures were given. By asking searching questions they have helped me to clarify many points in the exposition.

R. A. S.

Malvern
1 *June* 1961

Preface to the Second Impression

In this second printing the opportunity has been taken to correct a number of errors which occurred in the first printing, and also to make one or two small changes in order to clarify the text. In making the necessary corrections and changes, I have been greatly helped by some of my students who worked through the book during a graduate course which I taught at the Massachusetts Institute of Technology in the Spring Semester 1963. To my students and to some of my colleagues I should like to acknowledge my thanks for their help in this matter.

R. A. S.

MIT
12 *June* 1963

Contents

List of Symbols

(a) Fundamental constants, etc.

$\boldsymbol{e}$	magnitude of electronic charge	$= 1{\cdot}602 \times 10^{-19}$ C.
$\boldsymbol{m}$	mass of electron	$= 9{\cdot}107 \times 10^{-31}$ kg.
$\boldsymbol{h}$	Planck's constant	$= 6{\cdot}624 \times 10^{-34}$ joule sec.
$\boldsymbol{\hbar}$	$= \boldsymbol{h}/2\pi$	$= 1{\cdot}054 \times 10^{-34}$ joule sec.
$\boldsymbol{k}$	Boltzmann's constant	$= 1{\cdot}380 \times 10^{-23}$ joule/°C.
ϵ_0	permittivity of free space	$= 8{\cdot}854 \times 10^{-12}$ F/m.
μ_0	permeability of free space	$= 1{\cdot}257 \times 10^{-6}$ H/m.
$\boldsymbol{c}$	$= (\epsilon_0 \mu_0)^{-1/2}$ velocity of light in free space	$= 2{\cdot}998 \times 10^8$ m/sec.
$\boldsymbol{\beta}$	$= \boldsymbol{e}\hbar/2\boldsymbol{m}$ Bohr magneton	$= 9{\cdot}273 \times 10^{-24}$ A m^2.
a_0	$(= \boldsymbol{h}^2 \epsilon_0/\pi \boldsymbol{m}\boldsymbol{e}^2)$ Bohr radius	$= 5{\cdot}32 \times 10^{-11}$ m.
$\boldsymbol{N}$	Avogadro's number	$= 6{\cdot}025 \times 10^{23}$ per mole.
$\boldsymbol{R}$	$= \boldsymbol{k}\boldsymbol{N}$ gas constant.	
Ω	= volume per atom.	
$\boldsymbol{V}$	= volume of one mole.	

(b) Commonly used symbols

ν	frequency.
ω	$= 2\pi\nu$, angular frequency.
λ	wavelength.
k or K	$= 2\pi/\lambda$, magnitude of wave vector.
$\mathbf{k}$ or $\mathbf{K}$	wave vector*.
$\mathbf{P}$	$= \hbar\boldsymbol{k}$, crystal momentum.
P	magnitude of crystal momentum.
$\mathbf{p}$	true momentum.
E	energy.
ϵ	permittivity.
μ	permeability.
n	$= (\epsilon\mu/\epsilon_0\mu_0)^{1/2}$, refractive index.
R	Hall constant.
κ	thermal conductivity.
$\mathscr{P}$	thermoelectric power.
$\boldsymbol{\mathscr{E}}$	electric field.
$\mathscr{E}$	magnitude of electric field.
$\mathbf{B}$	magnetic flux.
B	magnitude of magnetic flux.

* The wave vector **k** need not be confused with Boltzmann's constant since the former is a vector and the latter is a scalar quantity which nearly always occurs as $T\boldsymbol{k}$ but sometimes as $\boldsymbol{k}/\mathbf{e}$. Boltzmann's constant $\boldsymbol{k}$ is set in italic type and the wave vector **k** in roman type.

m_e	effective mass of electron.
m_h	effective mass of hole.
ω_c	$= eB/m_e$ or eB/m_h, cyclotron frequency.
μ_e	= electron mobility.
μ_h	= hole mobility.
eV	electron volt ($= 1{\cdot}602 \times 10^{-19}$ joule).

1

Wave Motion in a Homogeneous Medium

1.1 Non-dispersive medium

THE simplest form of partial differential equation describing wave motion in one dimension has the well-known form

$$\frac{\partial^2 \psi}{\partial x^2} = \frac{1}{c^2}\frac{\partial^2 \psi}{\partial t^2} \tag{1}$$

ψ might represent, for example, the displacement of a uniform stretched string, or the electric field in free space, the quantity c being a constant velocity appropriate to the problem. The striking feature of equation (1) is that it has solutions of the form

$$\psi = f(x-ct) \tag{2}$$

or

$$\psi = g(x+ct) \tag{3}$$

where f and g are arbitrary functions, as may readily be verified by substitution; indeed any linear combination of these is also a solution. The solution (2) represents an arbitrary disturbance which is propagated *without change of form* in the positive direction of the x-axis, with velocity c, as may readily be seen by making the substitution $x' = x-ct$. The origin of x' has moved at time t to $x = ct$ and then $\psi = f(x')$, i.e. ψ is the same in the x'-coordinate at time t as it was in the x-coordinate at $t = 0$. The solution (3) represents an arbitrary disturbance propagated towards the negative direction of the x-axis.

If at $t = 0$ we have $\psi = f(x)$ and $\dot{\psi} = 0$ then clearly

$$\psi = \tfrac{1}{2}f(x-ct)+\tfrac{1}{2}f(x+ct) \tag{4}$$

The solutions (2) and (3) are of particular interest when f and g are sinusoidal in form, i.e.

$$\psi = A\sin k(x \pm ct) \tag{5}$$

The quantity k is related to the wavelength λ by the equation $k = 2\pi/\lambda$ since, for a fixed value of t, the disturbance repeats itself whenever x changes by a multiple of λ. The frequency ν is given by $\nu = c/\lambda$ since, for a fixed value of x, the disturbance repeats itself in time ν^{-1} equal to λ/c.

A related quantity $\omega = 2\pi\nu$, called the angular frequency, is also frequently used. It is common to pair k with ω and λ with c or ν and to write (5) in one of the alternative forms

$$\psi = A \sin(kx \pm \omega t) \tag{5a}$$

$$= A \sin\frac{2\pi}{\lambda}(x \pm ct) \tag{5b}$$

$$= A \sin 2\pi\left(\frac{x}{\lambda} \pm \nu t\right) \tag{5c}$$

The quantity c is called the phase velocity and is given by the equivalent expressions

$$c = \lambda\nu = \omega/k \tag{6}$$

Combinations of solutions of the form (5) may be used to fit boundary conditions appropriate to the problem in hand; in particular, a standing wave of the form

$$\begin{aligned}\psi &= A\sin(kx+\omega t)+A\sin(kx-\omega t)\\ &= 2A\sin kx\cos\omega t\end{aligned} \tag{7}$$

may be constructed.

The form (7) is the appropriate one to use when the boundary conditions are such that $\psi = 0$ at $x = 0$ and at $x = L$. In order to satisfy these k is no longer arbitrary but must have one of the values k_n given by

$$k_n = n\pi/L \tag{8}$$

where n is an integer. We have here an example of a condition which we shall frequently meet, in which a parameter appearing in a solution of a differential equation is restricted by boundary conditions to have certain discrete values. Such values are frequently known as 'allowed' values or 'eigenvalues'. The corresponding allowed values of the wavelength λ_n and frequency ν_n are given by

$$\lambda_n = 2L/n \tag{9}$$

$$\nu_n = \omega_n/2\pi = nc/2L \tag{10}$$

It is interesting, at this stage, to see what is the general form of solution of equation (1) harmonic in the time. We may note first that solutions of the type (5a) may also be written in the exponential form

$$\psi = A\,e^{ikx}\,e^{\pm i\omega t} \tag{5d}$$

This form is frequently more convenient mathematically*. The type (5a)

* If the solution we require is a real function given by Re $\psi(x,t)$ we need only use for $\psi(x,t)$ solutions of the form

$$\psi(x,t) = A\,e^{ikx-i\omega t} \tag{a}$$

where A is complex, and k can take both positive and negative values. Solutions of the form

$$\phi(x,t) = B\,e^{ik+i\omega t} \tag{b}$$

can give the same real part and so contribute nothing new. If we adopt solutions of the form (a) then $k > 0$ corresponds to propagation in the positive x-direction with can of course be formed by making linear combinations of two solutions of type (5d).

Let us therefore seek a solution of equation (1) of the form

$$\psi = f(x)\,\mathrm{e}^{-i\omega t} \tag{11}$$

The equation to be satisfied by $f(x)$ is then easily seen to be

$$\frac{\mathrm{d}^2 f}{\mathrm{d}x^2}+\frac{\omega^2}{c^2}f = 0 \tag{12}$$

or

$$\frac{\mathrm{d}^2 f}{\mathrm{d}x^2}+\frac{4\pi^2}{\lambda^2}f = \frac{\mathrm{d}^2 f}{\mathrm{d}x^2}+k^2 f = 0 \tag{12a}$$

Solutions of the form (5d), or those derivable from them, are therefore seen to be the only appropriate ones. It follows that it must be possible to construct more general solutions of the form (2) and (3) from solutions of the form (5d) by summing over all the appropriate values of k.

Two situations arise: when the values of k are restricted by boundary conditions as in equation (8) to a series of discrete values, the appropriate solution is in the form of a sum over all the allowed values of k; when k is unrestricted clearly we must use an integral.

If we restrict ourselves to waves travelling in the positive x-direction the most general form of solution when k is unrestricted is then

$$\psi(x,t) = \int_{-\infty}^{\infty} A(k)\,\mathrm{e}^{ik(x-ct)}\,\mathrm{d}k \tag{13}$$

$$= \int_{-\infty}^{0} A(k)\,\mathrm{e}^{i(kx+\omega t)}\,\mathrm{d}k+\int_{0}^{\infty} A(k)\,\mathrm{e}^{i(kx-\omega t)}\,\mathrm{d}k \tag{14}$$

If at $t = 0$ we have $\psi(x,0) = f(x)$ we then have

$$f(x) = \int_{-\infty}^{\infty} A(k)\,\mathrm{e}^{ikx}\,\mathrm{d}x \tag{15}$$

$A(k)$ is then just the Fourier transform of the function $f(x)$ and is given by*

$$A(k) = \frac{1}{2\pi}\int_{-\infty}^{\infty} f(x)\,\mathrm{e}^{-ikx}\,\mathrm{d}x \tag{16}$$

$\omega = kc$, and $k < 0$ to propagation in the negative x-direction with $\omega = -kc$. Alternatively we may adopt solutions of the type (b) and this is usually done in electromagnetic theory. In quantum theory, on the other hand, the solution we require for the wave function ψ is of the form (a) with negative time exponent, and solutions of the type (b) correspond to the complex conjugate ψ^* (see § 1.2.1). For the most general form of complex solution we must use a linear combination

$$\chi(x,t) = A\,\mathrm{e}^{i(kx-\omega t)}+B\,\mathrm{e}^{i(kx+\omega t)} \tag{c}$$

A and B both being complex and k taking both positive and negative values. The wavelength λ is then equal to $2\pi/|k|$.

* This result is proved in most books on advanced integral calculus or on Fourier series and integrals.

If we put $x' = x - ct$ in (13) it will be seen at once on comparison with (15) that

$$\psi(x,t) = f(x') = f(x-ct)$$

Thus the solution (13) is equivalent to (2) with $A(k)$ given by (16).

We may note one or two interesting examples. Suppose that initially the disturbance consists of a single isolated 'pulse'. We shall take for $f(x)$ the function

$$f(x) = e^{-x^2/a^2} \tag{17}$$

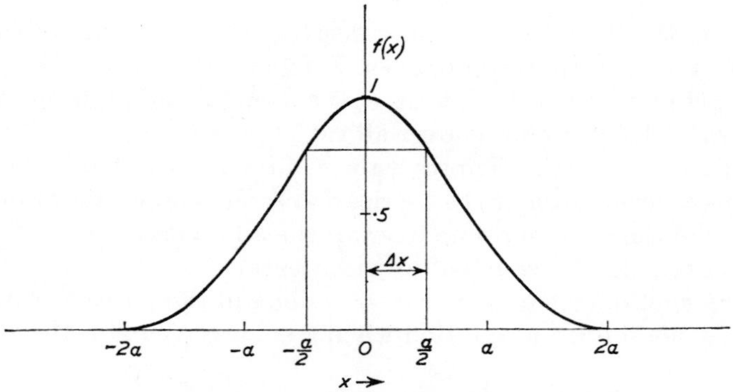

FIG. 1.1. A function $f(x)$ representing a single pulse

to represent such a 'pulse' since this leads to an integral for $A(k)$ which may be evaluated at once (see Fig. 1.1). We have

$$A(k) = \frac{1}{2\pi}\int_{-\infty}^{\infty} e^{-x^2/a^2 - ikx}\,dx$$

If we make the substitution $z = x + \frac{1}{2}ia^2 k$ we get

$$A(k) = \frac{e^{-a^2k^2/4}}{2\pi}\int_{-\infty}^{\infty} e^{-z^2/a^2}\,dz$$

$$= a\,e^{-a^2k^2/4}/2\pi^{1/2} \tag{18}$$

The function $A(k)$ is shown in Fig. 1.2. We may note here a general result: if a is a measure of the extent of the 'pulse', and the range of k for which $A(k)$ is appreciable is of the order a^{-1}. The solution for $t > 0$ consists of the pulse, unchanged in shape, moving in the positive x-direction with velocity c.

Another interesting example is a localised group of waves, or 'wave packet'; in this case we have

$$f(x) = g(x)\cos k_0 x \tag{19}$$

where $g(x)$ is appreciable only over a limited range of the variable x, which may, however, contain many wavelengths $2\pi/k_0$; the form of such

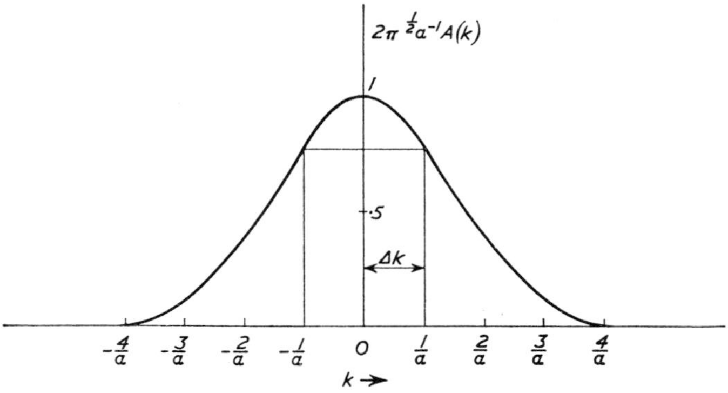

FIG 1.2. The function $A(k)$ corresponding to the function $f(x)$ shown in Fig. 1.1

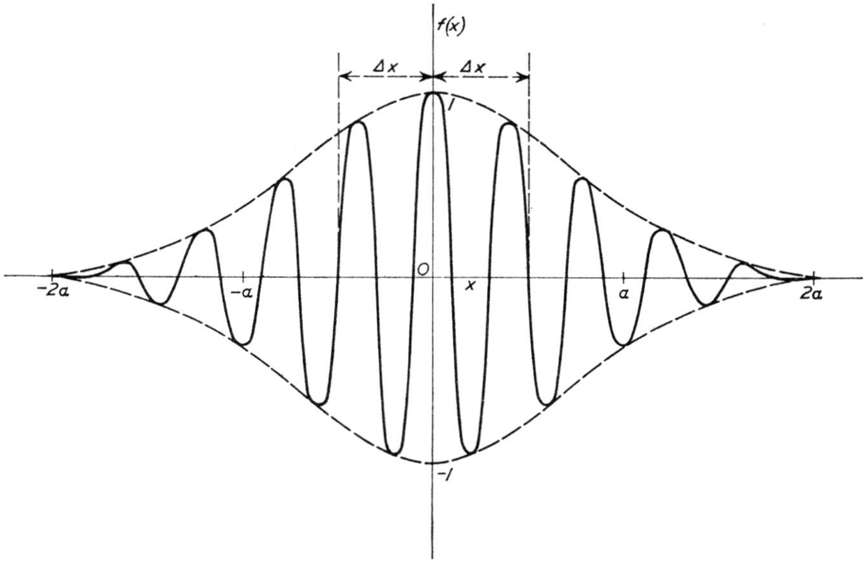

FIG. 1.3. A function $f(x)$ representing a wave packet

a function is shown in Fig. 1.3. For mathematical convenience it is generally desirable to use the complex form

$$f(x) = g(x)\,e^{ik_0 x} \tag{19a}$$

from which (19) may be obtained by taking the real part. If we take the function (17) for $g(x)$ we have

$$f(x) = \exp[-x^2/a^2 + ik_0 x] \tag{20}$$

The solution for $t > 0$ is simply

$$y(x,t) = g(x-ct)\,e^{ik_0(x-ct)}$$

There is no relative motion of the waves and envelope, the whole group moving in the positive x-direction with velocity c. We shall see that this condition is only possible when c is a constant, independent of k. It will readily be verified by proceeding as above that in this case

$$A(k) = \frac{a}{2\pi^{1/2}}\,e^{-a^2(k-k_0)^2/4} \tag{21}$$

$A(k)$ has the same form as for the function $f(x)$ given by (17) but is now centred on $k = k_0$ (see Fig. 1.4). This is a quite general result for any

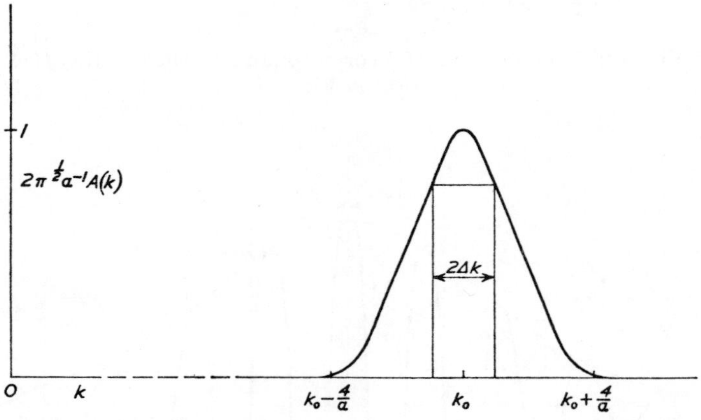

FIG. 1.4. The function $A(k)$ corresponding to the function $f(x)$ shown in Fig. 1.3

form of $g(x)$ as the reader will readily be able to prove by making a change of variable in the expression for $A(k)$.

Exercise Show that, if the boundary conditions are $\psi = 0$ at $x = 0$ and at $x = L$, the general solution such that $\psi = f(x)$ at $t = 0$ while $\dot{\psi} = 0$, may be built up by means of the series

$$\psi = \sum_{n=1}^{\infty} A_n \sin k_n x \cos \omega_n t$$

k_n and ω_n being given by (8) and (10), and that A_n is given by the integral

$$A_n = \frac{2}{L}\int_0^L f(x)\sin k_n x\,dx$$

(Note that $\int_0^L \sin k_n x \sin k_m x\,dx = 0$ if $k_m \neq k_n$.)

1.1.1 Three-dimensional form of the wave equation

The equation for wave propagation in a non-dispersive medium in three dimensions, corresponding to equation (1), is

$$\nabla^2 \psi = \frac{\partial^2 \psi}{\partial x^2} + \frac{\partial^2 \psi}{\partial y^2} + \frac{\partial^2 \psi}{\partial z^2} = \frac{1}{c^2}\frac{\partial^2 \psi}{\partial t^2} \qquad (22)$$

A functional solution of a very general type is also available for this equation, namely

$$\psi = f(lx + my + nz - ct) \qquad (23)$$

where l, m, n are direction cosines defining a direction in space. That (23) is a solution will readily be verified on substitution in (22) and recalling that for direction cosines $l^2 + m^2 + n^2 = 1$. We note that ψ has the same value at all points on any one of the planes

$$lx + my + nz = a \qquad (24)$$

a being a constant. Such waves are therefore called plane waves, and the planes given by equation (24) for various values of a are equi-phase surfaces, the direction of propagation being normal to them.

For sinusoidal waves the solution (23) may be written in complex form as

$$\psi = A\,\mathrm{e}^{ik(lx+my+nz) - i\omega t} \qquad (25)$$

or introducing the wave vector $\mathbf{k}$, whose components are (kl, km, kn), we may write (25) in the form

$$\psi = A\,\mathrm{e}^{i\mathbf{k}.\mathbf{r} - i\omega t}$$

The general solution (23) again may be built up by using (triple) integrals of the form

$$\psi(\mathbf{r}, t) = \int_{-\infty}^{\infty} A(\mathbf{k})\,\mathrm{e}^{i\mathbf{k}.\mathbf{r} - i\omega t}\,\mathrm{d}\mathbf{k} \qquad (26)$$

or expressed in Cartesian coordinates we have

$$\psi(x, y, z, t) = \int_{-\infty}^{\infty}\int_{-\infty}^{\infty}\int_{-\infty}^{\infty} A(\xi, \eta, \zeta)^{i(\xi x + \eta y + \zeta z - \omega t)}\,\mathrm{d}\xi\,\mathrm{d}\eta\,\mathrm{d}\zeta \qquad (26a)$$

The form of the function $A(\xi, \eta, \zeta)$ may be found as before by means of a triple Fourier integral.

Exercise Show that when we have spherical symmetry there exists a solution of equation (22) of the form

$$\psi = \frac{A}{r} f(r - ct)$$

where f is an arbitrary function.

1.2 Dispersive medium

The equation (1) is unique in being the only second-order partial differential equation to have solutions of the forms (2) and (3) which represent waves that propagate without change of shape. Many of the equations of mathematical physics may also, however, be reduced to the derived form (12a) when solutions which vary periodically with the time are sought. Only for equation (1), however, has k the simple form ω/c, where c is a constant. In general we may write the derived equation in the form

$$\nabla^2 \phi + \frac{\omega^2}{\mathscr{V}^2}\phi = 0 \tag{27}$$

where $\mathscr{V}$ is a function of ω. In one dimension the required solution will be of the form $\psi = \phi(x)\,\mathrm{e}^{-i\omega t}$, with $\phi(x)$ satisfying the equation

$$\frac{\mathrm{d}^2\phi}{\mathrm{d}x^2} + \frac{\omega^2}{\mathscr{V}^2}\phi = 0 \tag{28}$$

By considering a sinusoidal solution for a *given* value of ω we see that $\mathscr{V}$ is again the phase velocity but that, in general, it varies with ω. Indeed for a *given* value of ω all the considerations of § 1.1 apply, and we have the possibility of having pure *sinusoidal* waves propagated without change of shape; it is when we combine such waves with different values of ω, and hence with different velocities of propagation, that differences arise. This will become clearer when we treat in detail one specific example.

1.2.1 Electron waves in free space

As we shall be much concerned with the motion of electrons, we take as example the wave equation for free electrons. If we take the potential energy as constant, and therefore zero, since its absolute value is arbitrary, Schrödinger's time-dependent wave equation in one dimension may be written in the form, using the usual notation

$$\frac{\partial^2 \Psi}{\partial x^2} = -\frac{4\pi i \boldsymbol{m}}{\boldsymbol{h}}\frac{\partial \Psi}{\partial t} \tag{29}$$

$\boldsymbol{m}$ being the mass of the electron and $\boldsymbol{h}$ Planck's constant. $|\Psi|^2\,\mathrm{d}V$ gives the probability of finding an electron in the volume $\mathrm{d}V$. If we seek a solution of frequency ν corresponding to energy E, then we have $E = \boldsymbol{h}\nu$ and we may take for Ψ the form

$$\Psi = f(x)\,\mathrm{e}^{-2\pi i\nu t} = f(x)\,\mathrm{e}^{-2\pi iEt/\boldsymbol{h}} \tag{30}$$

The equation satisfied by $f(x)$ is then

$$\frac{\mathrm{d}^2 f}{\mathrm{d}x^2} + \frac{8\pi^2 \boldsymbol{m}}{\boldsymbol{h}^2}Ef = 0 \tag{31}$$

In terms of ω this may be written as

$$\frac{d^2f}{dx^2}+\frac{4\pi \boldsymbol{m}\omega}{\boldsymbol{h}}f = 0 \tag{32}$$

Comparing (32) with (28) we see that $\mathscr{V}^2 = \boldsymbol{h}\omega/4\pi\boldsymbol{m}$, i.e. $\mathscr{V}$ is a function of ω. Various other relationships of interest are

$$k^2 = 4\pi\boldsymbol{m}\omega/\boldsymbol{h} = \omega^2/\mathscr{V}^2 \tag{33}$$

and

$$k^2 = \frac{8\pi^2\boldsymbol{m}E}{\boldsymbol{h}^2} = \frac{2\boldsymbol{m}E}{\hbar^2} \tag{34}$$

where $\hbar = \boldsymbol{h}/2\pi$. If we write $E = \frac{1}{2}\boldsymbol{m}v^2$, where v is the electron's velocity, we have

$$k = \boldsymbol{m}v/\hbar \tag{35}$$

so that the wavelength λ, is given by

$$\lambda = 2\pi/k = \boldsymbol{h}/\boldsymbol{m}v \tag{36}$$

The angular frequency ω is then given by

$$\omega = 2\pi\nu = 2\pi E/\boldsymbol{h} = \pi\boldsymbol{m}v^2/\boldsymbol{h} = \tfrac{1}{2}\boldsymbol{m}v^2/\hbar \tag{37}$$

The phase velocity $\mathscr{V}$ may then be expressed as

$$\mathscr{V} = \omega/k = \tfrac{1}{2}v \tag{38}$$

The phase velocity is thus seen to be equal to *half* the velocity of the electron.

Let p represent the momentum of the electron; we then have

$$p = \boldsymbol{m}v = \hbar k = \boldsymbol{h}/\lambda \tag{39}$$

Thus the quantity k is simply a constant multiple of the momentum and is frequently used in its place. Again we have

$$E = \tfrac{1}{2}\boldsymbol{m}v^2 = p^2/2\boldsymbol{m} = \hbar^2k^2/2\boldsymbol{m} \tag{40}$$

A solution of the form

$$\Psi = A\,e^{i(kx-\omega t)} \qquad (k > 0) \tag{41}$$

represents a stream of electrons of uniform average density travelling with velocity v in the positive x-direction. At first sight it seems curious that the phase velocity $\mathscr{V}$ is not equal to the velocity of the electrons. The reason for this, as we shall see, lies in the fact that $\mathscr{V}$ is a function of ω. The charge density ρ associated with the wave function (41) is given by

$$\rho = -\boldsymbol{e}\Psi\Psi^\star = -\boldsymbol{e}|A|^2 \tag{42}$$

where $-\boldsymbol{e}$ is the electronic charge and $\Psi^\star$ is the complex conjugate of Ψ, and so is constant over all space. This wave function therefore tells us nothing about the position of an individual electron. To obtain such information we must form a localised wave function in the form of a wave packet of the type (19a) by superposition of functions like (41) with different values of k. Before considering such wave functions let us calculate the current density given by the wave function (41). In quantum mechanics the current density I is given for one-dimensional flow by the expression

$$I = \frac{-e\hbar}{2mi}\left[\Psi^{\star}\frac{\partial\Psi}{\partial x} - \Psi\frac{\partial\Psi^{\star}}{\partial x}\right] \tag{43}$$

Thus for the wave function (41) we have

$$I = -e|A|^2 k\hbar/m = -e|A|^2 v = \rho v \tag{44}$$

This is just what we should expect from a uniform stream of particles having a density $|A|^2$ per unit volume and moving with velocity v.

If we wish to normalise the wave function (41) to represent N particles crossing unit area per unit time we must have $|A|^2 v = N$ so that

$$\Psi = N^{1/2} v^{-1/2} \mathrm{e}^{i(kx-\omega t)} \tag{45}$$

1.3 Group velocity

If we wish to have a wave function representing a single particle we must use a wave packet of the form

$$\Psi(x,t) = \int_{-\infty}^{\infty} A(k)\,\mathrm{e}^{i(kx-\omega t)}\,\mathrm{d}k \tag{46}$$

In order to see how this moves we proceed as in § 1.1 but we must now remember that ω/k is a function of k. Let us suppose the wave packet to be similar in form to (19a) consisting of waves having a mean value k_0 and modulated by a slowly varying function. The spread in the values of k, and therefore of the momentum p is a requirement of the Uncertainty Principle, and is necessary in order to localise the position of the electron (see § 1.4).

We shall suppose that $A(k)$ is appreciable only for values of k near $k = k_0$. Thus we may write $k = k_0 + k'$ and shall assume k' to be so small that k'^2 may be neglected. Then we may write

$$\omega(k) = \omega_0 + k'\frac{\mathrm{d}\omega}{\mathrm{d}k_0} \tag{47}$$

where ω_0 and $\mathrm{d}\omega/\mathrm{d}k_0$ are the values of ω and $\mathrm{d}\omega/\mathrm{d}k$ at $k = k_0$. Inserting this expression for ω in (46) we have

$$\Psi(x,t) = \int_{-\infty}^{\infty} A(k_0+k')\exp\left(ik_0 x + ik'x - i\omega_0 t - i\frac{\mathrm{d}\omega}{\mathrm{d}k_0}k't\right)\mathrm{d}k' \tag{48}$$

If we write $A(k_0+k') = B(k')$ we see that $B(k')$ is appreciable only for small values of k'. Thus we may write, retaining the infinite limits for convenience, since this makes no difference,

$$\Psi(x,t) = \mathrm{e}^{i(k_0 x-\omega_0 t} \int_{-\infty}^{\infty} B(k')\exp\left[ik'\left(x-\frac{\mathrm{d}\omega}{\mathrm{d}k_0}t\right)\right]\mathrm{d}k' \tag{49}$$

When $t = 0$, we have

$$\Psi(x,0) = \mathrm{e}^{ikx_0}f(x) \tag{50}$$

where
$$f(x) = \int_{-\infty}^{\infty} B(k')\,\mathrm{e}^{ik'x}\,\mathrm{d}k'$$

Since $B(k')$ is appreciable only for small values of k', we see from the general properties of Fourier transforms that $f(x)$ is a slowly varying function of x. (Compare equations (17) and (18).) At $t = 0$ Ψ has the form of the type of wave packet we desire. Since $f(x)$ may be restricted to a range of values of x of the order of $1/k'_m$ where k'_m is the value of $|k'|$ beyond which $B(k')$ is inappreciable, Ψ may be normalised to unity. The condition is

$$1 = \int_{-\infty}^{\infty} \Psi\Psi^{\star}\,\mathrm{d}x = \int_{-\infty}^{\infty} |f(x)|^2\,\mathrm{d}x \tag{51}$$

We may now write down the form of Ψ at later values of the time t. Comparing (49) and (50), we have,

$$\Psi(x,t) = \mathrm{e}^{i(k_0 x-\omega_0 t)}f\left(x-\frac{\mathrm{d}\omega}{\mathrm{d}k_0}t\right) \tag{52}$$

We note that Ψ remains normalised to unity. Thus we see on comparing (50) and (52) that the individual waves travel with velocity $\mathscr{V}$ equal to ω_0/k_0, i.e. with the mean phase velocity, but that the whole wave packet moves with velocity U given by

$$U = \frac{\mathrm{d}\omega}{\mathrm{d}k_0} \tag{53}$$

(see Fig. 1.5). The velocity U with which the group of waves moves is called the group velocity.

For an electron wave we have from (53), (37) and (40)

$$U = \frac{\mathrm{d}\omega}{\mathrm{d}k_0} = \frac{\hbar k_0}{m} = v \tag{54}$$

The group velocity is therefore equal to the particle velocity corresponding to k_0, the mean value of k; this is reasonable in that the group of waves moves along with the particle. Individual waves, whose phases have no physical meaning in this case, move more slowly and pass back in turn through the wave packet as it advances.

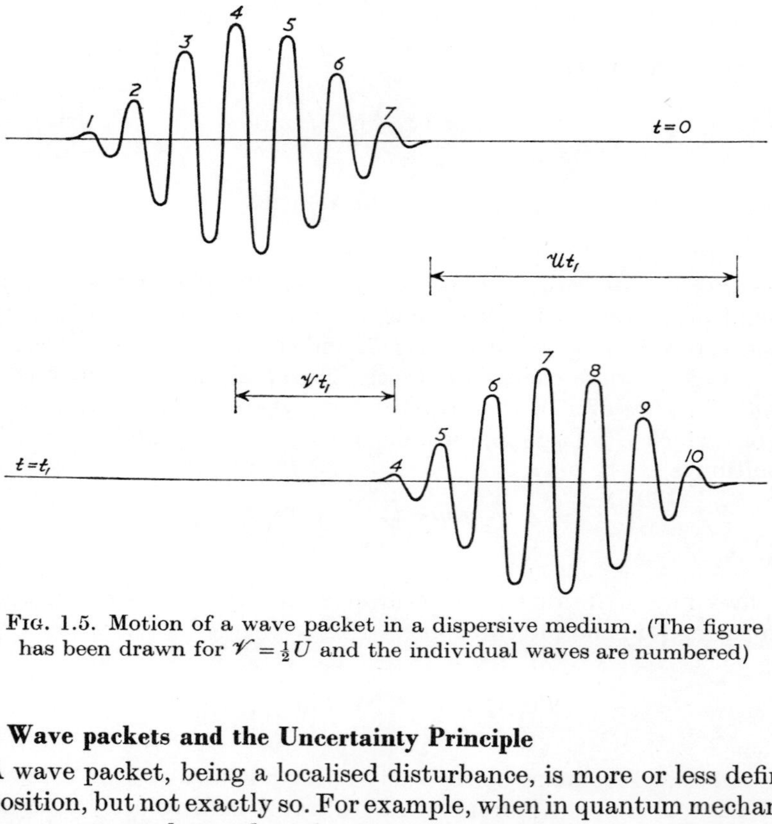

FIG. 1.5. Motion of a wave packet in a dispersive medium. (The figure has been drawn for $\mathscr{V} = \frac{1}{2}U$ and the individual waves are numbered)

1.4 Wave packets and the Uncertainty Principle

A wave packet, being a localised disturbance, is more or less defined in position, but not exactly so. For example, when in quantum mechanics we use a wave packet to describe the position of an electron there is some uncertainty as to the actual position of the electron. Associated with this uncertainty there is a spread in the values of k required to form the wave packet, and hence in the momentum, the precise expression of which is Heisenberg's Uncertainty Principle, one of the fundamental principles of quantum mechanics. It is of interest, however, to formulate the general uncertainty relationship between the spread in values of k, and hence of λ, and in the positional coordinates, for a wave packet. To do this we must define more precisely what we mean by the 'position' of a wave packet and its 'extent'.

We shall use the notation of quantum mechanics for convenience, since the analysis will then apply immediately to electron waves. We define the mean value $\bar{x}$, or 'centre of gravity' of the wave packet by the relationship

$$\bar{x} = \int_{-\infty}^{\infty} x|\Psi|^2\,\mathrm{d}x \Big/ \int_{-\infty}^{\infty} |\Psi|^2\,\mathrm{d}x \tag{55}$$

If Ψ is normalised we have

$$\int_{-\infty}^{\infty} |\Psi|^2\,\mathrm{d}x = 1 \tag{56}$$

The uncertainty Δx in x we define by means of the relationship

$$\Delta x^2 = \overline{(x-\bar{x})^2} = \int_{-\infty}^{\infty} (x-\bar{x})^2\,|\Psi|^2\,\mathrm{d}x \Big/ \int_{-\infty}^{\infty} |\Psi|^2\,\mathrm{d}x \tag{57}$$

Similarly we define $\bar{k}$ the mean value of k by means of the equation

$$\bar{k} = \int_{-\infty}^{\infty} k|A(k)|^2\,\mathrm{d}k \Big/ \int_{-\infty}^{\infty} |A(k)|^2\,\mathrm{d}k \tag{58}$$

and Δk, the uncertainty in k, by means of the equation

$$\Delta k^2 = \overline{(k-\bar{k})^2} = \int_{-\infty}^{\infty} (k-\bar{k})^2\,|A(k)|^2\,\mathrm{d}k \Big/ \int_{-\infty}^{\infty} |A(k)|^2\,\mathrm{d}k \tag{59}$$

The expressions (58) and (59) may be simplified by noting that if Ψ is normalised* we have also

$$2\pi \int_{-\infty}^{\infty} |A(k)|^2\,\mathrm{d}k = 1 \tag{60}$$

As an example let us consider the function Ψ given by the equation (20) with $A(k)$ given by (21), for which we have clearly $\bar{x} = 0$, $\bar{k} = k_0$. We also have

$$\Delta x^2 = \left(\frac{2}{\pi}\right)^{1/2} \frac{1}{a} \int_{-\infty}^{\infty} x^2\,\mathrm{e}^{-x^2/a^2}\,\mathrm{d}x = \frac{a^2}{4} \tag{61}$$

Similarly we have

$$\Delta k^2 = 1/a^2 \tag{62}$$

(see Figs. 1.1, 1.2). We thus obtain the important relationship

$$2\Delta x \Delta k = 1 \tag{63}$$

* This result follows from a theorem proved in most books on Fourier transforms

$$\frac{1}{2\pi} \int_{-\infty}^{\infty} |f(x)|^2\mathrm{d}x = \int_{-\infty}^{\infty} |A(k)|^2\mathrm{d}k$$

We shall show that, as the wave packet proceeds, Δx increases so that (63) gives a minimum value for the product $\Delta x \Delta k$. The exact Gaussian form for the envelope has in fact this property of giving a minimum value of $\Delta x \Delta k$.* We shall therefore write equation (63) in the form

$$2\Delta x \Delta k \geqslant 1 \tag{64}$$

This is a very general relationship.

For electromagnetic waves in free space, for example, $\Delta k = 2\pi \Delta f/c$, so that we have

$$4\pi \Delta x \Delta f \geqslant c \tag{65}$$

The frequency band width (as defined by Δf) of a pulse of radiation lasting a time ΔT equal to $\Delta x/c$ is therefore at least as great as $(1/4\pi\Delta T)$—a result well known to radio engineers. For electron waves we have, expressing k in terms of the momentum p by means of equation (39),

$$\Delta x \Delta p \geqslant \hbar/2 \tag{66}$$

This is the well-known Heisenberg Uncertainty Relationship.

We may also derive another well-known relationship between the uncertainty of energy ΔE and of time Δt. The uncertainty in time of arrival of a wave packet at a point Δt may be reasonably defined as equal to $\Delta x/v$. Also $\Delta E = \boldsymbol{m} v \Delta v$ so that

$$\Delta t \Delta E = \boldsymbol{m} \Delta x \Delta v = \Delta x \Delta p$$

Hence we have

$$\Delta t \Delta E \geqslant \hbar/2 \tag{67}$$

1.4.1 Dispersion of a wave packet

It is now of interest to see what happens to the wave packet at later values of the time when we have a dispersive medium. We consider as an example electron waves obeying the dispersion law given by equation (33). We have, starting with the wave packet (20) at $t = 0$, and inserting the value given by (21) for $A(k)$,

$$\Psi(x,t) = \frac{a}{2\pi^{1/2}} \int_{-\infty}^{\infty} \exp\left[-\frac{a^2}{4}(k-k_0)^2 + ikx - i\omega t\right] \mathrm{d}k \tag{68}$$

The exponent may be written, using (33), as

$$-\frac{a^2}{4}(k-k_0)^2 + ikx - \frac{ihk^2 t}{4\pi \boldsymbol{m}}$$

If we now make the change of variable $k' = k - k_0$ this becomes

$$-k'^2\left[\frac{a^2}{4} + \frac{iht}{4\pi \boldsymbol{m}}\right] + ik'\left(x - \frac{hk_0}{2\pi \boldsymbol{m}}t\right) + ik_0 x - \frac{ihtk_0^2}{4\pi \boldsymbol{m}}$$

* D. Ter Haar and W. M. Nicol, *Nature*, **175**, 1046 (1955).

The integral is then of the same form as we had above and may be evaluated at once to give

$$\Psi(x,t) = a\left(a^2+\frac{iht}{\pi m}\right)^{-1/2} \exp\left[-(x-v_0 t)^2\Big/\left(a^2+\frac{iht}{\pi m}\right)+ik_0(x-\tfrac{1}{2}v_0 t)\right] \tag{69}$$

where $v_0 = hk_0/2\pi m$.

We see that the packet moves with velocity v_0, this being the group velocity and equal to $d\omega/dk_0$, and the individual waves have phase velocity $\frac{1}{2}v_0$. We note also that the pulse spreads, the value of Δx being given at time t by

$$\Delta x^2 = \frac{1}{4a^2}\left[a^4+\frac{h^2 t^2}{\pi^2 m^2}\right] \tag{70}$$

For large values of t we have

$$\Delta x \to \frac{ht}{2\pi m}\cdot\frac{1}{a} = \frac{ht}{2\pi m}\Delta k = t\Delta v \tag{71}$$

The rate of spread is therefore that expected from the uncertainty in the velocity. The factor before the exponential in (69) is just a normalising factor. There is also a phase-shift term in (69) arising from the fact that the faster components of v tend to move to the front of the wave packet as it proceeds.

Exercise Show by expanding $\omega(k)$ about $k = k_0$ as far as second-order terms, i.e.

$$\omega(k) = \omega_0 + (k-k_0)\frac{d\omega}{dk_0} + \tfrac{1}{2}(k-k_0)^2\frac{d^2\omega}{dk_0^2}$$

and then evaluating the integrals in the equation (68) that the general expression for the rate of spread of the wave packet is $(d^2\omega/dk_0^2)\,\Delta k$. Note that this is zero for a non-dispersive medium.

1.5 Wave packets in three dimensions

The above considerations may all readily be extended to waves in three dimensions in a dispersive medium. The general form of the derived equation corresponding to solutions periodic in the time will now be given by (27). In an isotropic medium the phase velocity $\mathscr{V}$ will be a function of the wavelength λ and of the magnitude k of the vector $\mathbf{k}$ and not of its individual components. In general, however, this will not necessarily be so and we may regard ω as a function of the three components $k_x = kl$, $k_y = km$, $k_z = kn$ of the vector $\mathbf{k}$, (l, m, n) being the direction cosines of the direction of propagation. Solutions may be built up in exactly the same way as in equation (46) using the triple Fourier integral.

Suppose we are concerned with a wave packet whose propagation vector $\mathbf{k}$ is centred round the vector $\mathbf{k}_0$ whose components are (k_{x0}, k_{y0}, k_{z0}) then if

$$k_x = k_{x0}+\xi, \qquad k_y = k_{y0}+\eta, \qquad k_z = k_{z0}+\zeta$$

we may express ω, when ξ, η, ζ are small, in the form

$$\omega = \omega_0+\xi\frac{\partial\omega}{\partial k_{x0}}+\eta\frac{\partial\omega}{\partial k_{y0}}+\zeta\frac{\partial\omega}{\partial k_{z0}}$$

the quantities $\partial\omega/\partial k_{x0}$, etc., being the values of $\partial\omega/\partial k_x$, etc., when $\omega=\omega_0$. We then have

$$\Psi(x,y,z,t) = \iiint A(k_x,k_y,k_z)\,\mathrm{e}^{i(xk_x+yk_y+zk_z-\omega t)}\,\mathrm{d}k_x\,\mathrm{d}k_y\,\mathrm{d}k_z \tag{72}$$

and we may write this as before (at $t=0$) in the form

$$\Psi(x,y,z,0) = f(x,y,z)\,\mathrm{e}^{i\mathbf{k}_0.\mathbf{r}}$$

where

$$f(x,y,z) = \iiint B(\xi,\eta,\zeta)\,\mathrm{e}^{i(\xi x+\eta y+\zeta z)}\,\mathrm{d}\xi\,\mathrm{d}\eta\,\mathrm{d}\zeta$$

and we have written $B(\xi,\eta,\zeta)$ for $A(k_{x0}+\xi, k_{y0}+\eta, k_{z0}+\zeta)$. If the function B is appreciable only for small values of ξ, η, ζ, as we have assumed, $f(x,y,z)$ will be slowly varying compared with $\exp(i\mathbf{k}_0.\mathbf{r})$. The function Ψ therefore represents a three-dimensional form of wave packet similar to that already discussed for one dimension. For values of $t>0$ we have then

$$\Psi(x,y,z,t) = \exp(i\mathbf{k}_0.\mathbf{r}-i\omega_0 t)\iiint B(\xi,\eta,\zeta)\times$$

$$\times\exp\left[i\xi\left(x-\frac{\partial\omega}{\partial k_{x0}}t\right)+i\eta\left(y-\frac{\partial\omega}{\partial k_{y0}}t\right)+i\zeta\left(z-\frac{\partial\omega}{\partial k_{z0}}t\right)\right]\mathrm{d}\xi\,\mathrm{d}\eta\,\mathrm{d}\zeta$$

$$= \exp(i\mathbf{k}_0.\mathbf{r}.-i\omega_0 t)\,f\left(x-\frac{\partial\omega}{\partial k_{x0}}t,\, y-\frac{\partial\omega}{\partial k_{y0}}t,\, z-\frac{\partial\omega}{\partial k_{z0}}t\right) \tag{73}$$

Thus we see that the wave packet moves as a whole with vector velocity $\mathbf{U}$ whose components are $\partial\omega/\partial k_{x0}$, $\partial\omega/\partial k_{y0}$, $\partial\omega/\partial k_{z0}$. Thus we may write

$$\mathbf{U} = \nabla_k.\omega \tag{74}$$

This is the three-dimensional form of equation (53) for the group velocity.

1.5.1 Application to electron waves

For free electrons, the derived equation for wave functions harmonic in the time (energy E constant) is the well-known Schrödinger equation (with $V=0$)

$$\nabla^2\psi+\frac{8\pi^2 m}{h^2}E\psi = 0 \tag{75}$$

the full wave function Ψ being given by

$$\Psi = \psi(x, y, z)\,\mathrm{e}^{-i\omega t} \tag{76}$$

with $\omega = 2\pi E/h$.

Wave functions of the form

$$\psi = A\,\mathrm{e}^{i\mathbf{k}.\mathbf{r}} \tag{77}$$

correspond to a uniform stream of electrons moving with velocity **v** which has the same direction as the vector **k**. The value of E is given by

$$E = \frac{\hbar^2}{2m}(k_x^2 + k_y^2 + k_z^2) = \tfrac{1}{2}mv^2 \tag{78}$$

so that

$$\omega = \frac{\hbar}{2m}(k_x^2 + k_y^2 + k_z^2) = \frac{\hbar k^2}{2m} \tag{79}$$

Thus for the group velocity **U** we have

$$\mathbf{U} = \nabla_k \omega = \hbar^{-1}\nabla_k E = \hbar\mathbf{k}/m \tag{80}$$

We thus identify $\hbar\mathbf{k}/m$ as the velocity vector **v** and $\hbar\mathbf{k}$ as the momentum vector **p**.

In later chapters we shall have occasion to discuss problems for which the energy E is given in terms of the wave vector **k** by expressions more complex than (78). However, the group velocity and hence the particle velocity vector **v** is still given by the equation (74) and hence, using the relationship between E and ω, by

$$\mathbf{v} = \hbar^{-1}\nabla_k E \tag{81}$$

1.6 Vector and scalar waves

The Schrödinger wave function Ψ for particles in non-relativistic quantum mechanics is a scalar quantity, i.e. has no direction, and the corresponding waves are known as scalar waves; another example of such waves is sound waves in air. Electromagnetic waves are, however, vector waves, since the electric force must be specified in direction as well as in magnitude. For the complete description of electromagnetic wave motion in free space it turns out that two transverse components must be given and so we have the two directions of polarisation. For waves on an elastic string there are the two components of the lateral displacement wave, at right angles; these are transverse waves. There is in addition a compressional wave in which the displacement is in the direction of motion and so is a longitudinal wave. Similarly there are three components of the elastic waves in a solid; for an isotropic solid they may be resolved into two transverse shear components and one compressional component. In crystalline solids this simple resolution can be made only for certain special directions in the crystal having a high degree of symmetry. We shall later have to consider such waves when we come to deal with the vibrations of the crystalline lattice.

1.7 Inhomogeneous media

So far, we have considered derived equations of the form

$$\nabla^2\psi + k^2\psi = 0 \tag{82}$$

obtained for solutions periodic in the time, only when k is independent of the space coordinates. It is in these circumstances that the circular and exponential functions form an appropriate class of functions for building up more general solutions; generally, these are by far the simplest equations to deal with. However, there are many other equations for which k^2 is a function of (x, y, z) which may be solved in terms of analytical functions, and from these functions general solutions may be built up in the same way. Indeed if we have a complete set of such functions, any arbitrary continuous function may be expanded in an infinite series (or integral) just as we have done for the functions like $\exp(i\mathbf{k}.\mathbf{r})$. It is a well-known theorem in quantum mechanics that if the functions ψ_n are solutions of Schrödinger's equation

$$\nabla^2\psi + \frac{8\pi^2 \boldsymbol{m}}{\boldsymbol{h}^2}[E-V]\psi = 0 \tag{83}$$

for which $E = E_n$ then any solution Ψ of the full wave equation satisfying the boundary conditions may be expanded in the form*

$$\Psi(x,t) = \sum A_n \psi_n \mathrm{e}^{-2\pi i E_n t/\boldsymbol{h}} \tag{84}$$

V is the potential energy and will, in general, be a function of the space coordinates, so that, since k is related to V through the equation

$$k^2 = \frac{8\pi^2 \boldsymbol{m}}{\boldsymbol{h}^2}[E-V] \tag{85}$$

k^2 is also a function of the space coordinates.

Using such expansions, the required solutions may, in principle, be found in exactly the same way as we have done for the simpler case when $V = 0$. We shall later have occasion to use this procedure. The equation (82) may be regarded as the equation of wave motion in a medium whose refractive index is equal to $(E-V)^{1/2}/E^{1/2}$.

An example of a series of functions, other than the simple exponentials, which form a complete set of functions is given by the Hermite Polynomials, which are solutions of the wave equation for a linear harmonic oscillator. The derived equation in this case is

$$\frac{\mathrm{d}^2\psi}{\mathrm{d}x^2} + \frac{8\pi^2 \boldsymbol{m}}{\boldsymbol{h}^2}[E-Ax^2]\psi = 0 \tag{86}$$

The solutions $H_n(x)$ corresponding to E_n form an orthogonal set of functions and may be used to form a wave packet. An example of how this is done is given by L. I. Schiff†.

* L. I. Schiff, *Quantum Mechanics* (2nd Ed.). McGraw-Hill (1955), p. 48.
† L. I. Schiff, *loc. cit.*, p. 67.

2

Free-electron Theory of Metals

2.1 Average potential

THE first application of wave mechanics to discuss the motion of electrons in solids was that of A. Sommerfeld* who proposed a model to describe the motion of electrons in metals. This model, though simple and crude, enabled some of the most puzzling features concerned with the electrons to be clarified, and led naturally to more sophisticated models; we shall accordingly begin with a consideration of the Sommerfeld model.

From the high electrical conductivity of metals, and also from their high reflectivity, it may be deduced that they contain a large number of essentially free electrons. Various estimates of the number of free electrons have been made, and will be discussed later, and these all show that the number is of the same order as the number of atoms in the metal. If we assume that each atom gives up one electron which is free to move through the metal, the remaining positive ions may be considered as remaining fixed at the lattice sites of the crystal of which the metal is formed. Each free electron therefore interacts with the positive ions and with the other electrons. Now it is clearly very difficult to take all these interactions into account in detail, just as it is for a complex atom with a much smaller number of electrons. The interaction of one electron with all the others may, however, be averaged in wave mechanics as was done by W. Hartree† in calculating the average potential for an electron in an atom. Taken over a volume of the crystal containing many positive ion cores this potential will tend to average out to a constant value except very near to each ion core; Sommerfeld proposed to take as a first approximation a constant potential inside the metal. Near the surface of the metal an electron will experience a strong force which tends to keep it within the metal; this Sommerfeld proposed to represent by a step in potential. The potential in which an electron is supposed to move is shown in one dimension in Fig. 2.1; we may take the potential energy of an electron as zero outside the metal and as $-W$ inside.

The problem is thus reduced to studying the motion of a *single* electron in the above potential. The model is not quite so crude as may appear at first sight; its main defect is that it does not take into account the rather

* A. Sommerfeld, *Z. Phys.*, **47**, 1 (1928).
† W. Hartree, *Proc. Camb. Phil. Soc.*, **24**, 89 (1928).

large attractive forces acting on an electron when it is very close to one of the lattice sites. The interaction with the other electrons and with the ions is represented by an average which is independent of the time and therefore cannot take account of collisions between the electrons. In spite of these deficiencies the model gives some interesting results, as we shall see, but also fails in some important aspects, as we should expect. Another point about the interaction between electrons which we must not forget is that these are not to be regarded as quite independent, being subject to the Pauli Principle which forbids more than one electron to occupy the same stationary state. We shall now consider the problem in wave mechanics associated with the potential shown in Fig. 2.1.

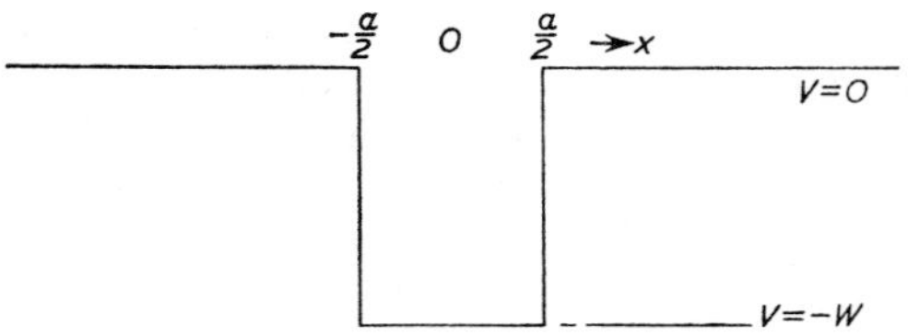

FIG. 2.1. Potential function for Sommerfeld's model

2.2 Energy levels of an electron in a deep one-dimensional potential well

We shall first of all consider the problem of the motion of an electron in a one-dimensional potential well such as that shown in Fig. 2.1. If a is the width of the well and W its depth we shall have for the wave equation for the electron

$$\left.\begin{aligned} \frac{d^2\psi}{dx^2}+\frac{2mE}{\hbar^2}\psi &= 0 \qquad |x| > a/2 \\ \frac{d^2\psi}{dx^2}+\frac{2m}{\hbar^2}[E+W]\psi &= 0 \quad |x| \leqslant a/2 \end{aligned}\right\} \tag{1}$$

Suppose we now write

$$\left.\begin{aligned} E &= -W+E_k \\ E_k &= \hbar^2 k^2/2m \\ W &= \hbar^2\gamma^2/2m \\ \alpha^2 &= \gamma^2-k^2 \end{aligned}\right\} \tag{2}$$

We may then write the equations (1) in the form

$$\left.\begin{aligned} \frac{d^2\psi}{dx^2}-\alpha^2\psi &= 0 \quad |x| > a/2 \\ \frac{d^2\psi}{dx^2}+k^2\psi &= 0 \quad |x| < a/2 \end{aligned}\right\} \tag{3}$$

E_k is then the kinetic energy of the electron, the zero of the energy E being taken as that of an electron at rest *outside* the well.

We have two types of solution to consider, those for which $E_k > W$ and hence α is imaginary, and those for which $E_k < W$ and hence α is real. We shall consider only the latter, which correspond to electrons bound in the potential well. The solutions with $E_k > W$ correspond to electrons which move outside the well, with too much energy to be bound in it. From the definition of α we may clearly take it either as positive or negative when real; to be definite we shall therefore assume that $\alpha > 0$.

For $x > a/2$ the appropriate solution is of the form

$$\psi = A\,\mathrm{e}^{-\alpha x} \tag{4}$$

the solution with positive exponent being inadmissible since it becomes infinite as x tends to infinity. The solution (4) tends rapidly to zero for $x > a/2$ if αa is large; this will be so if a is large unless E_k is nearly equal to W. If we remember that

$$\frac{2\boldsymbol{m}}{\hbar^2} W_H = \frac{1}{a_0^2} \tag{5}$$

where W_H is the ionisation energy of a hydrogen atom (13·5 eV) and a_0 is the so-called Bohr radius ($0{\cdot}53 \times 10^{-8}$ cm), we may write

$$\gamma a = \frac{a}{a_0}\left(\frac{W}{W_H}\right)^{1/2} \tag{6}$$

For a metal, the potential well turns out to have a depth of the order of 10 eV, so for a crystal 1 cm long $\gamma a \simeq 2 \times 10^8$. In this case we may therefore neglect the wave function for $x > a/2$ and use the simpler boundary conditions

$$\psi = 0 \quad |x| > a/2 \tag{7}$$

We shall examine this limiting condition more carefully in § 2.9. We have thus reduced the problem to finding solutions of the second of equations (3) with the boundary condition (7).

From the physical symmetry of the problem we must have $|\psi|^2$ a symmetrical function of x. Thus we may either have $\psi(x) = \psi(-x)$ or $\psi(x) = -\psi(-x)$; the former type of solution is called symmetrical and the latter antisymmetrical. The appropriate symmetrical form of solution is

$$\psi_s(x) = \left(\frac{2}{a}\right)^{1/2} \cos kx \tag{8}$$

the multiplying constant being chosen to normalise the wave function by means of the condition

$$\int_{-a/2}^{a/2} |\psi_s|^2\,\mathrm{d}x = 1$$

In order that we may have $\psi_s(a/2) = \psi_s(-a/2) = 0$ we must have

$$\frac{ka}{2} = (r+\tfrac{1}{2})\pi \tag{9}$$

where r is an integer.

The antisymmetrical form of solution is given by

$$\psi_a = \left(\frac{2}{a}\right)^{1/2} \sin kx \tag{10}$$

In order to have $\psi(a/2) = -\psi(-a/2) = 0$ we must have

$$\frac{ka}{2} = r\pi \tag{11}$$

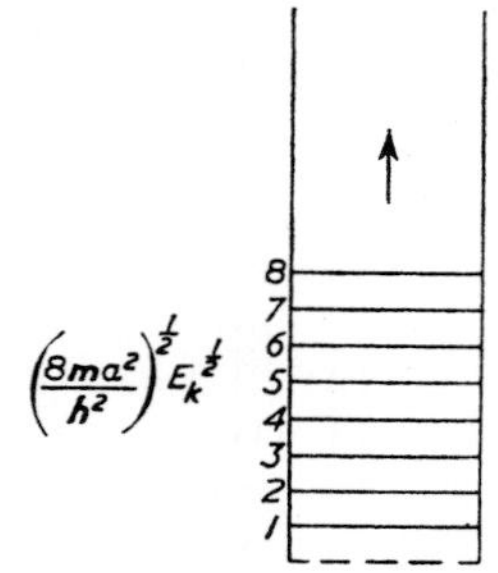

FIG. 2.2. Energy levels in potential well

where again r is an integer. The conditions (9) and (11) may be combined in a single equation

$$ka = n\pi \tag{12}$$

where n is an integer. When n is odd we have a symmetrical wave function and when n is even an antisymmetrical one. The allowed values of the kinetic energy are then given by

$$E_k = n^2 \boldsymbol{h}^2/8\boldsymbol{m}a^2 \tag{13}$$

$$= n^2\pi^2 W_H a_0^2/a^2 \tag{14}$$

We may note that the condition (9) for a state described by a symmetrical wave function is simply that we have an odd integral number of half wavelengths in the length a. The condition (12) for a state described by an antisymmetrical wave function indicates that in this case we have an even integral number of half wavelengths, i.e. an integral number of wavelengths, in the length a.

For the values of W and a considered above, E_k is very much smaller than W unless n is very large; thus when $n = 1$ we have $E_k \simeq 2{\cdot}5 \times 10^{-15}$ eV. For a crystal of finite size the allowed values of E_k therefore lie very close together, and for many purposes may be regarded as forming a continuum. The allowed values E_n of the total energy E are given by

$$E_n = -W+E_k = -W+n^2\boldsymbol{h}^2/8\boldsymbol{m}a^2 \tag{15}$$

Some of the lower energy levels are shown in Fig. 2.2 in which $E_k^{1/2}$ (and not E_k) is plotted vertically.

2.3 The Sommerfeld model

For an actual crystal we must, of course, solve a problem in three dimensions. Suppose then we consider a piece of metal in the form of a cube of side a; we choose this form for simplicity of calculation, but the results are quite general (see exercise below). We now have a potential given by

$$\begin{aligned} V(x,y,z) &= -W \quad 0 \leqslant x,y,z \leqslant a \\ &= 0 \text{ otherwise} \end{aligned}$$

We have now transferred the origin of coordinates to one corner of the cube. We shall assume that, as before, we may use as boundary conditions the condition that ψ vanishes at the surface of the cube. The wave equation which we now have to solve is

$$\frac{\partial^2\psi}{\partial x^2}+\frac{\partial^2\psi}{\partial y^2}+\frac{\partial^2\psi}{\partial z^2}+k^2\psi = 0 \tag{16}$$

where the kinetic energy E_k is again given by $E_k = \boldsymbol{h}^2k^2/8\pi^2\boldsymbol{m}$. We may write the solutions in symmetrical and antisymmetrical forms as before but the analysis is rather simpler with our new choice of origin. The appropriate form of solution which makes ψ vanish when $x = 0$, $y = 0$, or $z = 0$ is

$$\psi = A\sin\frac{\pi lx}{a}\sin\frac{\pi my}{a}\sin\frac{\pi nz}{a} \tag{17}$$

when in order to satisfy equation (16) we must have

$$a^2k^2 = \pi^2(l^2+m^2+n^2) \tag{18}$$

In order that ψ may also vanish when $x = a$, $y = a$, or $z = a$ we must have l, m, and n integers, which may be taken to be positive.

The allowed values of E_k are then given by

$$E_k = \boldsymbol{h}^2(l^2+m^2+n^2)/8\boldsymbol{m}a^2 \tag{19}$$

This should be compared with equation (13) for the one-dimensional problem. As before, the spacing of the energy levels is extremely small and we may regard them for many purposes as forming a continuum, when the metal has macroscopic dimensions.

Now suppose l, m, n are *large* integers, which they will be unless E_k is very small, and we enquire how many allowed values of k correspond to energies less than a given value E_{k0} of E_k. If we write

$$E_{k0} = \boldsymbol{h}^2r_0^2/8\boldsymbol{m}a^2 \tag{20}$$

this is equivalent to asking how many integral points lie inside the positive octant of a sphere of radius r_0, with centre of the origin, i.e. such that

$$l^2+m^2+n^2 < r_0^2 \tag{21}$$

For large values of l, m, n this is simply equal to the volume of the octant or $\pi r_0^3/6$. The required number of values $M(E_{k0})$ is therefore given by

$$M(E_{k0}) = \pi r_0^3/6 = \pi(8\boldsymbol{m}a^2 E_{k0}/\boldsymbol{h}^2)^{3/2}/6 \tag{22}$$

Another quantity of interest is the number of energy levels $N(E)\,\mathrm{d}E$ lying between E and $E+\mathrm{d}E$. This is clearly equal to the number of allowed values of E_k lying between E_k and $E_k+\mathrm{d}E$, and may be obtained from (22). We have

$$N(E)\,\mathrm{d}E = M(E_k+\mathrm{d}E)-M(E_k) = \frac{\mathrm{d}M(E_k)}{\mathrm{d}E_k}\mathrm{d}E$$

Hence

$$N(E)\,\mathrm{d}E = \frac{\pi a^3}{4}\left(\frac{8\boldsymbol{m}}{\boldsymbol{h}^2}\right)^{3/2} E_k^{1/2}\,\mathrm{d}E \tag{23}$$

$$= \frac{\pi V}{4}\left(\frac{8\boldsymbol{m}}{\boldsymbol{h}^2}\right)^{3/2} E_k^{1/2}\,\mathrm{d}E = \frac{V(2\boldsymbol{m})^{3/2}E_k^{1/2}\,\mathrm{d}E}{4\pi^2\hbar^3} \tag{24}$$

where V is the volume of the crystal.

There is another very instructive way of writing this result. To obtain it we must first consider the three-dimensional form of a plane wave which is an extension of the form given in equation (45) of Chapter 1. Suppose we have the wave function

$$\psi = A\,\mathrm{e}^{i(k_x x+k_y y+k_z z-\omega t)} \tag{25}$$

which, as we have seen (§ 1.5.1) represents a stream of electrons travelling in a direction parallel to the vector **k** whose components are k_x, k_y, k_z. The wave function (17) is clearly a combination of such wave functions with

$$k_x = \pm\frac{\pi l}{a}, \qquad k_y = \pm\frac{\pi m}{a}, \qquad k_z = \pm\frac{\pi n}{a}$$

The magnitude of the vector **k** is just that given by the quantity k defined in equation (18), and the magnitude of the corresponding momentum p is $\hbar k$ (see § 1.5.1). The kinetic energy may be written in terms of k in the form

$$E_k = \hbar^2k^2/2\boldsymbol{m} \tag{26}$$

$$= \tfrac{1}{2}p^2/\boldsymbol{m} \tag{27}$$

so that

$$N(E)\,\mathrm{d}E = Vk^2\,\mathrm{d}k/2\pi^2 \tag{28}$$

$$= 4\pi Vp^2\,\mathrm{d}p/\boldsymbol{h}^3 \tag{29}$$

Now $4\pi p^2\,dp$ is just the volume of momentum space corresponding to the energy interval dE. This result then shows that the number of energy levels is proportional to the volume V and that for each volume $\boldsymbol{h}^3$ of momentum space we should allocate one level. We have derived the above result with a rather special model but it may be shown to be quite generally true.

Exercise Discuss the energy levels of a crystal in the form of a rectangular parallelopiped with sides a, b, c, and derive an equation for $N(E)\,dE$ similar to that given by (29).

2.4 Distribution of the free electrons among the allowed energy levels in the state of lowest energy

Having found the allowed energy levels in a metal according to the Sommerfeld model we must now consider how the electrons are distributed among them. According to the Pauli Principle only one electron may occupy each level. So far, we have, neglected the electron's spin, which may take two values, so that we may have *two* electrons in each of the levels we have found. Let us then consider what happens at a very low temperature; we may assume that in this condition the electrons will try to occupy as low-lying levels as possible. Suppose we have N electrons in all, then we must fill up the lowest $N/2$ levels with the electrons. This is a very different picture from that given by the classical theory of an electron gas, in which the mean kinetic energy of an electron at temperature T is $3/2\ \boldsymbol{k}T$. The kinetic energy therefore would tend to zero as T tends to zero, and this would indicate that all the electrons would try to crowd into the lowest level; this, however, is prevented by the Pauli Principle.

We may readily calculate the maximum value to which the levels are filled by writing down the condition that the lowest $N/2$ are occupied. If we write E_m for this maximum energy it is obtained at once from equation (22) by equating $M(E_m)$ to $N/2$. Thus we have

$$N/2 = \pi V(8\boldsymbol{m}E_m)^{3/2}/6\boldsymbol{h}^3 \tag{30}$$

If we write $n = N/V$, so that n is equal to the concentration of free electrons in the metal, we have

$$E_m = \left(\frac{3}{\pi}\right)^{2/3}\left(\frac{\boldsymbol{h}^2}{8\boldsymbol{m}}\right) n^{2/3} \tag{31}$$

If we assume that we have n_d free electrons per atom and d^3 is equal to the volume per atom, equation (30) may be written as

$$E_m = \left(\frac{3}{\pi}\right)^{2/3}\left(\frac{\boldsymbol{h}^2}{8\boldsymbol{m}d^2}\right) n_d^{2/3} \tag{31a}$$

In terms of the ionisation energy W_H of a hydrogen atom (13·5 eV) we have

$$E_m/W_H = (3n_d/\pi)^{2/3}\,\pi^2 a_0^2/d^2 \tag{31b}$$

We note that E_m does not depend on the actual volume of the crystal but only on the concentration of free electrons. If we assume that $n_d = 1$, i.e. that each atom provides one free electron we may calculate the values of E_m for various metals, obtaining the quantity d from X-ray data on the crystal structure. The calculated values are shown in Table 2.1.

Table 2.1

Values of E_m the maximum kinetic energy of an electron in the lowest energy state of a metal assuming one free electron per atom ($n = d^{-3}$ is equal to the number of atoms per unit volume).

Metal	n (cm^{-3})	d (cm)	E_m (eV)
Li	$4{\cdot}80 \times 10^{22}$	$2{\cdot}74 \times 10^{-8}$	4·75
Na	$2{\cdot}60 \times 10^{22}$	$3{\cdot}36 \times 10^{-8}$	3·15
K	$1{\cdot}37 \times 10^{22}$	$4{\cdot}17 \times 10^{-8}$	2·1
Cu	$8{\cdot}70 \times 10^{22}$	$2{\cdot}26 \times 10^{-8}$	7·2
Ag	$5{\cdot}95 \times 10^{22}$	$2{\cdot}56 \times 10^{-8}$	5·5

From the values given in Table 2.1 it will be seen that even at the absolute zero of temperature the majority of the electrons have very considerable kinetic energy; this is very different from the classical picture. We shall later (§ 2.7) discuss the distribution of electrons between the available energy levels at any temperature. In Fig. 2.3 we show the distribution function giving the number of electrons with energy between E and $E + \mathrm{d}E$ for the lowest electronic energy state of the metal. The number $\phi(E)\,\mathrm{d}E$ is equal to $2N(E_k)\,\mathrm{d}E$ where $N(E_k)$ is given by equation (24), when $E_k \leqslant E_m$, and is equal to zero when $E_k > E_m$. The maximum value ϕ_m of $\phi(E)$ is obtained by inserting the value for E_m in equation (24) and is given by

$$\phi_m = \frac{4\pi V m}{h^2 d}\left(\frac{3n_d}{\pi}\right)^{1/3} \tag{32}$$

Clearly we have

$$\int_0^\infty \phi(E)\,\mathrm{d}E = 2\int_0^{E_m} N(E)\,\mathrm{d}E = N \tag{33}$$

It is of interest to calculate the average kinetic energy $\overline{E_k}$ per electron. It is given by

$$\begin{aligned}\overline{E_k} &= \int_0^\infty E_k\phi(E)\,\mathrm{d}E \bigg/ \int_0^\infty \phi(E)\,\mathrm{d}E \\ &= \int_0^{E_m} E^{3/2}\,\mathrm{d}E \bigg/ \int_0^{E_m} E^{1/2}\,\mathrm{d}E \\ &= \tfrac{3}{5}E_m \end{aligned} \tag{34}$$

Thus we see that even in the lowest energy state the average kinetic energy is quite large.

Because of the approximation of neglecting the wave function outside the metal, the depth W of the potential well has not so far figured in our calculation apart from its occurrence in the expression for the total energy (referred to an electron at rest outside the metal). The question now arises as to what is the magnitude of E_m with respect to W. The maximum energy of an electron $E_{max.}$ is given by

$$E_{max.} = E_m - W \tag{35}$$

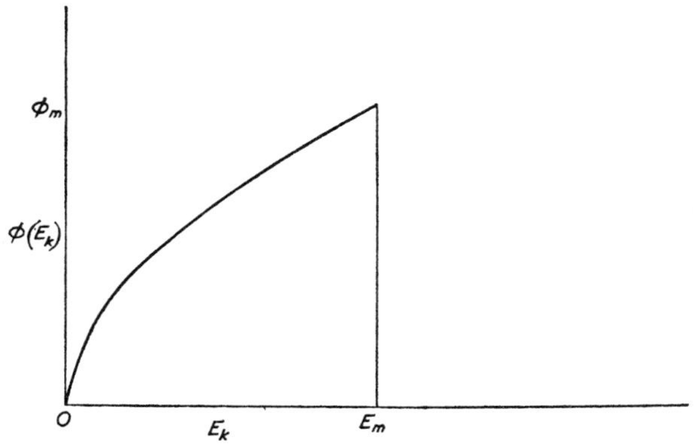

FIG. 2.3. Number of electrons $\phi(E_k)\,dE_k$ with kinetic energies between E_k and $E_k + dE_k$ in the lowest electronic energy state of a metal. ϕ_m is given by equation (32)

Clearly this must be negative otherwise the electron would not stay within the metal. The quantity ϕ given by

$$e\phi = W - E_m \tag{36}$$

where e is the magnitude of the electronic charge, is called the work function of the metal and represents the smallest energy required to extract an electron from the metal at $T = 0$. In principle, ϕ may be determined from an observation of the photoelectric effect and, as we shall see later, from the variation of thermionic emission with temperature. Surface effects have a marked influence on the measured values of ϕ, which do not always represent the quantity defined by equation (36). Values of ϕ which are thought to represent the pure substance fairly well are, however, available for a number of metals, and some are given in Table 2.2. Knowing ϕ, we may use the values of E_m given in Table 2.1 to calculate W; these are also included in Table 2.2.

The values of W obtained in this way are reasonable; experimental confirmation is scanty but the value of the average potential inside a metal may be obtained from experiments on the diffraction of slow electrons by thin metal films. An electron whose energy is $E(>0)$ outside the metal will have average wavelength $\lambda_m = 2\pi/k$ inside the metal where $k^2 = 8\pi^2 \boldsymbol{m}(E+W)/\boldsymbol{h}^2$. Values of W of the right order have been obtained by observation of λ_m. In particular, we note that the quantity αa (cf. equation (2)) which determines how much of the wave functions appear outside the crystal, is large unless an electron is excited by an amount nearly equal to $\boldsymbol{e}\phi$ above the highest occupied level in the lowest energy state of the metal; such excitation is, as we shall see, very improbable except at high temperatures. The approximation we have made of letting the wave function ψ equal zero at the boundary of the metal is therefore reasonable.

Table 2.2

Work function ϕ and depth W of potential well for some metals

Metal	ϕ (V)	W (eV)
Na	2·3	5·45
K	2·7	4·8
Ag	4·8	10·3

2.5 Travelling electron waves in a metal

Wave functions such as those given in equations (8), (10) or (17) represent standing waves and so carry no electric current (see § 1.5.1). In order to represent a current we must use travelling waves of the form

$$\psi = A\,\mathrm{e}^{ikx} \tag{37}$$

or in three dimensions

$$\psi = A\,\mathrm{e}^{i(\mathbf{k}.\mathbf{r})} \tag{37a}$$

Such waves, however, imply an infinite medium for their propagation. There is available, however, a mathematical trick, first proposed by M. Born, whereby travelling waves may be used for a finite crystal. Consider first a one-dimensional problem; suppose that we have a linear crystal of length a and we repeat this length indefinitely to make up an infinite crystal. Moreover, we postulate that the solutions of the wave equation for each section of length a are identical. The boundary condition which must be satisfied in this case is then

$$\psi(x) = \psi(x+sa) \tag{38}$$

where s is any integer. This situation may be approximately achieved physically if we consider bending the crystal into the form of a closed ring. These boundary conditions are called cyclic or periodic boundary conditions. Since we may take a to be any length we please it is clear that properties which are found not to depend on a may be considered as characteristic of the material of the crystal.

Applying the boundary condition (38) to the wave function (37) we see that the only allowed values of k are given by

$$k = \pm 2n\pi/a \tag{39}$$

where n is a positive integer. Comparing this with equation (12) we note that the allowed values of k are now spaced twice as far apart as before, but that each is doubly degenerate, the energy having the same value for positive and negative values of k. These now correspond to different solutions. As we have already seen, when a refers to a length of macroscopic dimensions, the allowed energy levels are very close together. We have therefore just the same number of energy levels in an interval dE, large compared with their separation, as we had before, and the expressions for E_m, etc., may be taken over from the previous analysis. The wave function (37) now represents a stream of electrons flowing in the positive direction of x when k is positive and in the negative direction of x when k is negative (see § 1.5.1).

In three dimensions the cyclic boundary conditions have no simple physical interpretation since we cannot imagine a three-dimensional crystal being bent round on itself in all three dimensions. The same device is, however, useful in allowing us to have wave functions representing travelling waves. Suppose we have a cube of side a which is repeated in three dimensions to make up an infinite crystal, we may have solutions of the form

$$\psi = A\,e^{i(2\pi lx/a + 2\pi my/a + 2\pi ny/a)} \tag{40}$$

so that the vector $\mathbf{k}$ has components given by

$$k_x = 2\pi l/a, \qquad k_y = 2\pi m/a, \qquad k_z = 2\pi n/a \tag{41}$$

In order to satisfy the periodic boundary conditions we must have

$$\psi(x, y, z) = \psi(x+ra, y+sa, z+ta)$$

r, s, t being any integers so that l, m, n must also be integers. The allowed values of the kinetic energy E_k are now given by

$$E_k = \boldsymbol{h}^2(l^2+m^2+n^2)/2\boldsymbol{m}a^2 \tag{42}$$

where now l, m, n may take either positive or negative integral values. We now wish to calculate $M(E_m)$ the number of energy levels with E_k less than E_m. Let us write

$$E_m = \boldsymbol{h}^2 r_0^2/2\boldsymbol{m}a^2 \tag{43}$$

The number of integral points such that

$$l^2+m^2+n^2 < r_0^2$$

is now equal to the volume of the whole of a sphere of radius r_0. Hence the number $M(E_m)$ is given by

$$M(E_m) = \tfrac{4}{3}\pi r_0^3 = \pi a^3(8\boldsymbol{m}E_m)^{3/2}/6\boldsymbol{h}^3 \tag{44}$$

This is exactly the same value as we had for the standing-wave solution (cf. equation (22)). The calculation of the maximum value of the kinetic energy for the lowest energy state is then as before, and gives the same value (equation (31)).

Clearly the state of lowest energy of the metal carries no current since each level will be occupied by as many electrons with positive values of **k** as with negative values. When we have a current we have an excited state in which some electrons are raised above the lowest level by the electric field; except at the absolute zero of temperature, some electrons will in any case be excited thermally and we must now determine the distribution of such electrons at any temperature.

2.6 Statistical mechanics—basic equations

When we have determined the possible values which the energy of any system can take, there remains the problem of determining what is the probability of finding it in any particular state, corresponding to one of the possible energy values, when it is in thermal equilibrium with other systems or with its surroundings. Closely related to this problem is the following; suppose we have a large number of identical systems in thermal equilibrium, how are the systems distributed among the various possible states? These problems are dealt with by means of statistical mechanics. We do not propose to derive the basic formulae of this subject and we shall simply discuss one or two elementary principles and quote three of the basic formulae. These are all we shall require for our study of the equilibrium distribution of electrons and atomic vibrations in solids. (The subject of statistical mechanics is treated extensively in a number of textbooks; a short and elegant account is given by E. Schrödinger, *Statistical Thermodynamics*, Cambridge, 1946.)

The theory of statistical mechanics was well worked out before the advent of wave mechanics and the new quantum theory, but these brought to the subject some quite fundamental changes; we shall try to indicate briefly how this came about. Suppose we consider a single system which may be an atom, a complex molecule or even a macroscopic crystal. It will have certain states which may be described by an energy E; for the moment we shall suppose that to each energy E there corresponds a single state of the system. In classical theory the values of E form a continuum, all values from a lower limit upwards being permissible. In the quantum theory, on the other hand, over certain ranges of the energy, only discrete values are allowed. For example if the system in question is a hydrogen atom we have a series of discrete values extending from $-W_H$ to zero, and then a continuum stretching from zero to infinity, provided we choose the zero of energy as that of the ionised atom with the electron at rest at infinite separation. If this system were treated classically the energy would extend in a continuum from $+\infty$ to $-\infty$. This brings us to the question of the choice of the zero point from which to measure energy. Although the kinetic energy of a system has a natural zero point, namely that in which all the velocities relative to the co-ordinate system in use are zero, the potential energy always contains an arbitrary constant. Thus the total energy always has an arbitrary constant, and any general formulation of the theory of distribution of systems among energy states must take this into account, so as to be independent of the choice of energy zero. We should thus expect that only *differences* in energy would enter into the theory.

Now suppose we consider two states of a single system having energies $E_1, E_2, (E_2 > E_1)$. Let the probabilities of the system being in these states be P_1 and P_2, respectively; then clearly the ratio P_2/P_1 is a function of (E_2-E_1) and not of E_1 and E_2 independently. This function has some general properties which are almost intuitively clear; the higher the value of E_2 relative to E_1 the smaller will be the ratio P_2/P_1. This is simply a statement of the tendency of the system to occupy the lowest levels, a tendency which is opposed by thermal agitation which increases with the temperature, so that we expect the ratio of P_2/P_1 to increase as the absolute temperature T increases. The most fundamental of all the formulae of statistical mechanics is in fact

$$P_2/P_1 = \exp[(E_1-E_2)/\boldsymbol{k}T] \tag{45}$$

where $\boldsymbol{k}$ is a constant, known as Boltzmann's constant, and has the value $1{\cdot}38 \times 10^{-23}$ joule/°C. (A useful approximate value to remember is that when $T = 290$°K $\boldsymbol{k}T \simeq 1/40$ eV.) This result, known as Boltzmann's law, will be seen to have the properties discussed above. It holds equally well in classical and in quantum theory with the proviso that in the latter E_1 and E_2 are allowed values of the energy; if E is not an allowed value then we have $P(E) = 0$. If the states corresponding to E_1 and E_2 have degeneracies g_1 and g_2 we clearly have

$$P_2/P_1 = (g_2/g_1)\exp[(E_1-E_2)/\boldsymbol{k}T] \tag{46}$$

since each state may be regarded as g neighbouring states and we should sum over these to get the total probability of occupation.

When the states of the system form a denumerable set having energies $E_1, E_2 \ldots E_m \ldots$ it is easy to write down an expression for the actual value of the various probabilities of occupation $P_1, P_2 \ldots P_m \ldots$. We may write

$$P_r = A\,\mathrm{e}^{-E_r/\boldsymbol{k}T} \tag{47}$$

We must then have

$$\sum_r P_r = 1 \tag{48}$$

so that

$$A \sum_r \mathrm{e}^{-E_r/\boldsymbol{k}T} = 1 \tag{49}$$

Hence we have

$$P_r = \mathrm{e}^{-E_r/kT} / \sum_s \mathrm{e}^{-E_s/kT} \tag{50}$$

We note that equation (50) involves only energy differences since any constant value may be added to E_r and to each of the quantities E_s without affecting the value of P_r. We shall apply equation (50) in Chapter 3 to discuss the distribution of energy among the vibrational states of the atoms of a crystal (see § 3.5).

We now come to the second main problem of statistical mechanics, to determine the distribution of available states among a large number N of identical systems. It might be thought at first sight that this is just the same problem as finding in turn the probability of each system being in the state having each of the allowed energies E_r. In classical theory this is so, the fact that we may have discrete values of the energy making no essential difference to the *statistics*. Two new ideas emerge, however, in the quantum theory. The first and most subtle is that we must not count as distinct states those in which a pair of *identical* systems are interchanged—in other words we cannot *label* identical systems, and this has a fundamental effect on the statistical calculation. The second new idea is that some types of system, for example electrons, are subject to Pauli's Exclusion Principle, which allows not more than one system to occupy any state. This may modify the distribution of the systems between the various states in a drastic fashion, and we have already had an example of this for the free electrons in a metal. On the basis of classical statistics we should expect that the electrons would all crowd into the lowest state as $T \to 0$, but because of the Exclusion Principle they are distributed among all allowed states having energies up to a maximum value E_m.

If we adopt the classical statistics we simply argue that we may treat the systems as independent, and hence if $P(E)$ is the probability that a system is in a state corresponding to energy E, the 'expectation' number, giving the expected number of systems in that state when the total number is N, is just $NP(E)$. If $N(E)\,\mathrm{d}E$ is the number of *available* states between E and $E+\mathrm{d}E$ then the number of *occupied* states $\phi(E)\,\mathrm{d}E$ is according to classical statistics given by

$$\phi(E)\,\mathrm{d}E = AN(E)\,\mathrm{e}^{-E/kT}\,\mathrm{d}E \tag{51}$$

The constant A is determined by the condition that

$$\int \phi(E)\,\mathrm{d}E = N \tag{52}$$

the integral being taken over all available states.

For the free electrons in a metal we have seen that $N(E) = BE^{1/2}$, where B is a constant, provided we now take as the zero of our energy scale that of an electron at rest *inside* the metal so that $E = E_k$. We then have

$$\phi(E)\,\mathrm{d}E = NE^{1/2}\,\mathrm{e}^{-E/kT}\,\mathrm{d}E \bigg/ \int_0^\infty E^{1/2}\,\mathrm{e}^{-E/kT}\,\mathrm{d}E \tag{53}$$

We have taken the upper limit of the integral as infinity although this is not strictly true, since the expression used for $N(E)$ only holds for $E < W$. The exponential $\exp(-W/kT)$ is, however, very small for all ordinary temperatures ($W/kT \sim 200$ for $T = 300°\mathrm{K}$).

The integral in the denominator of (53) has the value $\frac{1}{2}\pi^{1/2}(kT)^{3/2}$ so we have

$$\phi(E)\,\mathrm{d}E = \frac{2N}{\pi^{1/2}\,kT}\left(\frac{E}{kT}\right)^{1/2} \mathrm{e}^{-E/kT}\,\mathrm{d}E \tag{54}$$

This shows that, as $T \to 0$, $\phi(E) \to 0$ except when $E \to 0$, as stated above.

When we take account of Pauli's Exclusion Principle we should expect to find in the limit when $T \to 0$ the result we have already obtained in § 2.4, namely that

$$\begin{aligned} \phi(E) &= 2N(E) \quad & E \leqslant E_m \\ &= 0 & E > E_m \end{aligned}$$

When this is done, and we also take account of the fact that electrons cannot be labelled, we find that this is so.

Before giving the formula for $P(E)$ under these conditions there are some general remarks about its form which we might make. The classical formula for $P(E)$ indicates that the lower a state lies the more probable it is that it will be occupied. In fact if we have one single energy level at a great depth below all the others this would contain nearly all the systems. The quantum modification for electrons must clearly exclude such a possibility. We should again expect $P(E)$ to decrease monotonically as E/kT increases, but we should now expect that for values of E well below a certain value E_F the probability of occupation would tend to the constant value of unity. For values of E much greater than E_F we might expect to obtain the classical result since we should expect the energy levels to be sparsely populated and the Exclusion Principle to be unimportant.

If we choose the energy E_F to be such that $P(E) > \frac{1}{2}$ when $E < E_F$ and $P(E) < \frac{1}{2}$ when $E > E_F$, the function $P(E)$ given by the equation

$$P(E) = \frac{1}{\mathrm{e}^{(E-E_F)/kT} + 1} \tag{55}$$

has all the above properties. It is in fact the function given by statistical mechanics for systems or particles which are subject to the Exclusion Principle. Such statistics are known as Fermi–Dirac statistics and the function

$$F(x) = \frac{1}{\mathrm{e}^x + 1} \tag{56}$$

is known as the Fermi–Dirac function and is shown in Fig. 2.4. The probability of occupation of a degenerate energy level is then given by

$$P(E) = gF[(E-E_F)/\boldsymbol{k}T] \tag{57}$$

where g is the degree of degeneracy. The energy level corresponding to E_F is generally known as the Fermi level and may or may not be one of the allowed levels of the system. It is generally a slowly varying function

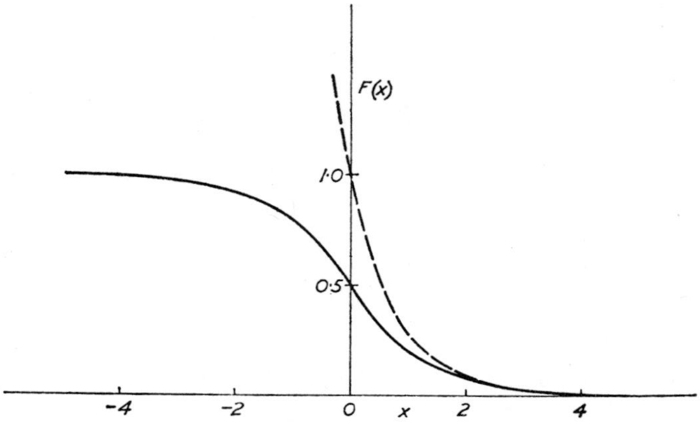

FIG. 2.4. The Fermi–Dirac function $F(x)$. (The classical approximation is shown dashed)

of T and is determined by expressing the fact that the total number of systems is equal to N. Thus we have

$$N = \int_{-\infty}^{\infty} gN(E)\, F[(E-E_F)/\boldsymbol{k}T]\,\mathrm{d}E \tag{58}$$

when $N(E)$ can be represented as a continuous function. For discrete energy levels this will be replaced by a sum of the form

$$N = \sum_r g_r\, F[(E_r-E_F)/\boldsymbol{k}T] \tag{58a}$$

and for some problems both a sum and an integral may have to be included.

For the problem we have already considered, that of the free electrons in a metal, we see that we get the same probability distribution when $T \to 0$ as before provided we take $E_F = E_m$. Thus we have

$$\begin{aligned} P(E) &\to 0 \quad E > E_m \\ &\to 1 \quad E < E_m \end{aligned}$$

The form of the function $P(E)$ for this case is shown in Fig. 2.5 for $T = 0$ and also for a finite value of T. It will be seen that only electrons with energies near the Fermi level are affected by the increase in temperature unless this is very large indeed. We shall defer to § 2.7 a detailed discussion of the problem when $T \neq 0$.

In Fig. 2.4 we also show the classical function e^{-x}; it will be seen that only when $x > 2$ is this a reasonable approximation. Thus when

$$(E - E_F)/\boldsymbol{k}T > 2$$

no appreciable error will be made by using the simpler classical formula.

To complete the discussion we must now consider systems which are not subject to the Exclusion Principle; examples of such are photons, α-particles, etc. For such systems the probability $P(E)$ is expressed in

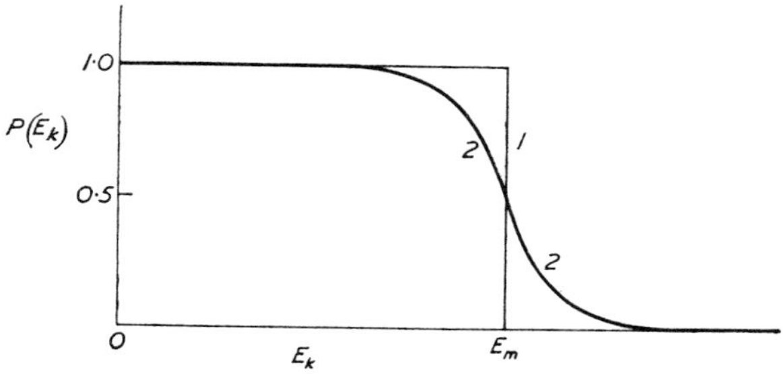

FIG. 2.5. Probability of occupation for electrons in a metal according to Fermi–Dirac statistics, (1) $T = 0$, (2) $T > 0$

terms of a function $B(x)$ which at first sight appears to be only slightly different from $F(x)$, namely

$$B(x) = \frac{1}{e^x - 1} \tag{59}$$

These statistics are known as Einstein–Bose statistics and the function $B(x)$ as the Einstein–Bose function; it is shown in Fig. 2.6. We now have

$$P(E) = B[(E - E_B)/\boldsymbol{k}T] \tag{60}$$

the quantity E_B is again being determined by the condition that the total number of particles in the system should have a given value. When $(E - E_B)/\boldsymbol{k}T \gg 1$ we see that $P(E)$ again tends the classical value. This is also shown for comparison in Fig. 2.6. We should note that in this case that, as $E \to E_B$, $P(E) \to \infty$, so that we are only concerned with values of the energy greater than E_B. This type of function is thus applicable only to quantized systems with a well-defined lower limit to the energy. Let the lowest energy level have the value E_0; then if $N(E)\,dE$ is the density of levels between E and $E + dE$, and we have N systems in all, we have

$$N = \int_{E_0}^{\infty} \frac{N(E)\,dE}{e^{(E - E_B)/\boldsymbol{k}T} - 1} \tag{61}$$

For small values of x, $B(x) \to x^{-1}$ so the integral diverges near $E = E_B$; thus we see that we must have $E_B < E_0$. As $T \to 0$ the function $P(E)$ tends to zero except when $E \to E_B$. It is easier to see what is happening when we have a discrete set of energy levels, the lowest having $E = E_0$; we have then

$$N = \sum_r \frac{g_r}{e^{(E_r - E_B)/kT} - 1} \tag{62}$$

g_r being the degree of degeneracy of the level with energy E_r, and we assume that the energies E_r are arranged in an increasing sequence of which the lowest is E_0. As $T \to 0$ all the terms tend to zero except that with $E_r = E_0$.

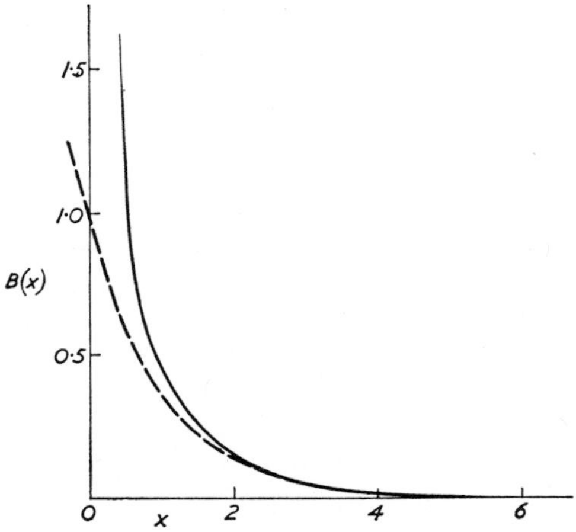

FIG. 2.6. The Einstein–Bose function $B(x)$. (The classical approximation is shown dashed)

We then have approximately

$$N = \frac{g_0}{e^{(E_0 - E_B)/kT} - 1} \tag{63}$$

so that

$$E_B = E_0 - \boldsymbol{k}T \ln\left(1 + \frac{g_0}{N}\right) \tag{64}$$

For large values of N we have

$$E_B = E_0 - \boldsymbol{k}Tg_0/N \tag{65}$$

so that E_B is very nearly equal to E_0 and, in particular, if N is infinite $E_B = E_0$, a result which must hold even when T does not tend to zero. Apart from the first term, the sum (62) is clearly finite, since the series will converge rapidly when $E_B \neq E_r$. The infinity must therefore come from the first term and this will only be so if $E_B = E_0$.

As pointed out above, the energy contains an arbitrary constant and it is frequently convenient to take the lowest energy level as the zero of energy, i.e. to take $E_0 = 0$. With this choice we have, when N is infinite,

$$P(E) = B(E/\boldsymbol{k}T) = \frac{1}{e^{E/\boldsymbol{k}T} - 1} \tag{66}$$

One of the most interesting applications of this result is to the statistical theory of radiation. On the photon picture of radiation we regard it as consisting of particles which can either be in a ground state with zero energy, and are not observable in this state, or in one of a series of higher energy states corresponding to a set of allowed frequencies ν_s. If the radiation is contained in a cavity these are just the resonant frequencies of the cavity. The energy of the state with frequency ν_s is $\boldsymbol{h}\nu_s$, so that we have

$$P(E_s) = \frac{1}{e^{\boldsymbol{h}\nu_s/\boldsymbol{k}T} - 1} \tag{67}$$

The mean energy $\bar{E}_s$ associated with the frequency ν_s is therefore given by

$$\bar{E}_s = \frac{\boldsymbol{h}\nu_s}{e^{\boldsymbol{h}\nu_s/\boldsymbol{k}T} - 1} \tag{68}$$

We shall show that for a cavity of volume V the number of allowed frequencies between ν and $\nu + d\nu$ is equal to $8\pi V\nu^2 d\nu/\boldsymbol{c}^3$. The energy per unit volume $U(\nu)\,d\nu$ of radiation with frequencies in the range ν to $\nu + d\nu$ is thus given by

$$U(\nu)\,d\nu = \frac{8\pi\boldsymbol{h}}{\boldsymbol{c}^3}\frac{\nu^3\,d\nu}{e^{\boldsymbol{h}\nu/\boldsymbol{k}T} - 1} \tag{69}$$

This is Planck's well-known formula for the energy density in black-body radiation. We shall later see how similar formulae may be applied to the vibrations of the atoms of a crystal.

The formula for the number of allowed frequencies lying between ν and $\nu + d\nu$ is derived in exactly the same way as for equation (24). If we have a perfectly reflecting cavity with walls in the form of a cube of side a, the electric field of the possible standing waves in the cavity is given by a wave function of the form (17) and the allowed wavelength $\lambda = 2\pi/k$ by equation (18). The only difference now is that the dispersion law relating to the frequency ν and k is $\nu = k\boldsymbol{c}/2\pi$. If we now write $a^2 k_0^2/\pi^2 = r_0^2$ the number $M(\nu_0)$ of allowed values for ν less than a value ν_0 is given by the volume of the octant of a sphere of radius r_0 as before; thus we have

$$M(\nu_0) = \frac{\pi}{6}r_0^3 = a^3 k_0^3/6\pi^2 = 4\pi V\nu_0^3/3\boldsymbol{c}^3 \tag{70}$$

where $V = a^3$ is equal to the volume of the cavity.

The number of allowed values with frequencies between ν and $\nu+\mathrm{d}\nu$ is thus given by

$$\frac{\mathrm{d}M}{\mathrm{d}\nu}\mathrm{d}\nu = 4\pi V\nu^2\,\mathrm{d}\nu/\boldsymbol{c}^3$$

We must remember, however, that each allowed mode is degenerate and corresponds to two directions of polarisation so that we have

$$N(\nu)\,\mathrm{d}\nu = 8\pi V\nu^2\,\mathrm{d}\nu/\boldsymbol{c}^3 \tag{71}$$

2.7 Distribution of free electrons among the allowed levels at absolute temperature T

We have discussed in § 2.4 the distribution of the free electrons in the Sommerfeld model for a metal when they occupy the lowest possible states permitted by the Exclusion Principle; this is the lowest energy state of the metal, and will be that state to which the metal tends as the temperature approaches absolute zero*. We are now in a position to calculate the average distribution of electrons among the allowed levels for any value of the absolute temperature T. From the discussion of § 2.6 it will be seen that the average number of electrons $\phi(E)\,\mathrm{d}E$ occupying levels corresponding to a kinetic energy E lying between E and $E+\mathrm{d}E$ is given by

$$\phi(E)\,\mathrm{d}E = 2N(E)\,F[(E-E_F)/\boldsymbol{k}T]\,\mathrm{d}E \tag{72}$$

where $N(E)$ is given by (24), the factor 2 being included to take account of the two possible values of the spin of the electron. The energy E_F corresponding to the Fermi level is then obtained from the equation

$$N = 2\int_0^\infty N(E)\,F[(E-E_F)/\boldsymbol{k}T]\,\mathrm{d}E \tag{73}$$

where N is the total number of free electrons. As the value of the absolute temperature T tends to zero we have simply as before

$$N = 2\int_0^{E_m} N(E)\,\mathrm{d}E \tag{74}$$

Equation (73) then becomes

$$\int_0^{E_m} N(E)\,\mathrm{d}E = \int_0^\infty N(E)\,F[(E-E_F)/\boldsymbol{k}T]\,\mathrm{d}E \tag{75}$$

This gives an equation for E_F as a function of T.

The mean electronic energy $\bar{E}$ of the metal is given by

$$\bar{E} = 2\int_0^\infty N(E)\,F[(E-E_F)/\boldsymbol{k}T]\,E\,\mathrm{d}E \tag{76}$$

* There are complications arising from the phenomenon of superconductivity but we ignore them for the moment.

In order to evaluate integrals of the form (75) and (76) we make use of an important property of the Fermi–Dirac function $F(x)$; its derivative is appreciable only for a small range of values of x near $x = 0$. To make use of this property we transform the integrals by means of an integration by parts. Suppose we have to evaluate the integral

$$I = \int_0^\infty F[(E-E_F)/\boldsymbol{k}T]\frac{\mathrm{d}}{\mathrm{d}E}[G(E)]\,\mathrm{d}E \tag{77}$$

where the function $G(E) \to 0$ as $E \to 0$. The integrals in (75) and (76) are of this form with $G(E) = \frac{2}{3}CE^{3/2}$ and $G(E) = \frac{2}{5}CE^{5/2}$, respectively, with

$$C = V(2\boldsymbol{m})^{3/2}/4\pi^2\hbar^3$$

On integrating (77) by parts we have, since $G(0) = 0$,

$$I = -\int_0^\infty G(E)\frac{\mathrm{d}}{\mathrm{d}E}F[(E-E_F)/\boldsymbol{k}T]\,\mathrm{d}E \tag{78}$$

In (78) let us put $x = (E-E_F)/\boldsymbol{k}T$; we have then

$$I = -\int_{-E_F/\boldsymbol{k}T}^\infty G(E_F+\boldsymbol{k}Tx)\frac{\mathrm{d}F(x)}{\mathrm{d}x}\,\mathrm{d}x \tag{79}$$

Since $\mathrm{d}F(x)/\mathrm{d}x$ tends rapidly to zero for large negative values of x we may replace the lower limit in the integral in (79) by $-\infty$ provided $\boldsymbol{k}T \ll E_F$; this condition we shall now assume. Also, we may expand the function $G(E_F+\boldsymbol{k}Tx)$ in power of x obtaining

$$I = -\int_{-\infty}^\infty \left[G(E_F)+\boldsymbol{k}TxG'(E_F)+\frac{\boldsymbol{k}^2T^2x^2}{2}G''(E_F)+\ldots\right]\frac{\mathrm{d}F(x)}{\mathrm{d}x}\,\mathrm{d}x \tag{80}$$

Now we have

$$\int_{-\infty}^\infty \frac{\mathrm{d}F(x)}{\mathrm{d}x}\,\mathrm{d}x = [F(x)]_{-\infty}^{\infty} = -1$$

$$\int_{-\infty}^\infty x\frac{\mathrm{d}F(x)}{\mathrm{d}x}\,\mathrm{d}x = -\int_{-\infty}^\infty \frac{xe^x\,\mathrm{d}x}{(e^x+1)^2} = 0$$

$$\int_{-\infty}^\infty x^2\frac{\mathrm{d}F(x)}{\mathrm{d}x}\,\mathrm{d}x = -\int_{-\infty}^\infty \frac{x^2e^x\,\mathrm{d}x}{(e^x+1)^2} = -\frac{\pi^2}{3}$$

so that

$$I = G(E_F)+\frac{\pi^2\boldsymbol{k}^2T^2}{6}G''(E_F)+\ldots \tag{81}$$

If we have $G(E) = \frac{2}{3}E^{3/2}$ we have

$$I_{1/2} = \int_0^\infty E^{1/2}\, F[(E-E_F)/\boldsymbol{k}T]\,\mathrm{d}E$$

$$= \tfrac{2}{3}E_F^{3/2}\left[1+\frac{\pi^2}{8}\left(\frac{\boldsymbol{k}T}{E_F}\right)^2+\ldots\right] \tag{82}$$

Inserting this value for $I_{1/2}$ in equation (75) we have

$$E_m^{3/2} = E_F^{3/2}\left[1+\frac{\pi^2}{8}\left(\frac{\boldsymbol{k}T}{E_F}\right)^2+\ldots\right] \tag{83}$$

Since we have assumed $\boldsymbol{k}T \ll E_F$ this equation may be inverted to give

$$E_F = E_m\left[1-\frac{\pi^2}{12}\left(\frac{\boldsymbol{k}T}{E_m}\right)^2+\ldots\right] \tag{84}$$

At room temperature $\boldsymbol{k}T \simeq \frac{1}{40}$ eV whereas in § 2.4 we have seen that E_m is of the order of 5 eV. Thus at ordinary temperatures the Fermi level

Table 2.3

Degeneracy temperature T_0

Metal	Li	Na	Cu	Ag
T_0 (°K)	$5{\cdot}5\times10^4$	$3{\cdot}6\times10^4$	$8{\cdot}0\times10^4$	$6{\cdot}5\times10^4$

lies very close the energy E_m and departs from it only to the second order in $\boldsymbol{k}T/E_m$; for many purposes we may therefore take $E_F = E_m$. In this condition the assembly of free electrons is said to be degenerate; the condition will hold provided $T \ll T_0$ where $\boldsymbol{k}T_0 = E_m$. The temperature T_0 is called the degeneracy temperature; values of T_0 may be calculated from Table 2.1 and a few values are given in Table 2.3. Thus we see that at all temperatures below the melting point the free electrons in these metals are in the degenerate condition.

If we take $G(E) = \frac{2}{5}E^{5/2}$ we obtain

$$I_{3/2} = \int_0^\infty E^{3/2}\, F[(E-E_F)/\boldsymbol{k}T]\,\mathrm{d}E \tag{85}$$

$$= \tfrac{2}{5}E_F^{5/2}\left[1+\frac{5\pi^2}{8}\left(\frac{\boldsymbol{k}T}{E_F}\right)^2+\ldots\right] \tag{86}$$

This enables us to evaluate the integral in (76). We have for the mean energy $\bar{E}$

$$\bar{E} = \frac{4C}{5}E_F^{5/2}\left[1+\frac{5\pi^2}{8}\left(\frac{\boldsymbol{k}T}{E_F}\right)^2+\ldots\right] \tag{87}$$

On substituting for E_F from (84) we have

$$\bar{E} = \frac{3NE_m}{5}\left[1+\frac{5\pi^2}{12}\left(\frac{\boldsymbol{k}T}{E_m}\right)^2+\ldots\right] \tag{88}$$

since $N = 4CE_m^{3/2}/3$ and we may replace E_F by E_m in the small term of order $(\boldsymbol{k}T/E_F)^2$. The first term in (88) is just the average kinetic energy in the lowest energy state corresponding to $T = 0$ (cf. equation (34)). The second term may be regarded as giving approximately the 'thermal' energy E_T of the electrons. Neglecting powers of $(\boldsymbol{k}T/E_m)$ higher than the second this is given by

$$E_T = \left(\frac{3N\boldsymbol{k}T}{2}\right)\left(\frac{\pi^2\,\boldsymbol{k}T}{6E_m}\right) \tag{89}$$

This should be compared with the classical value

$$E_T = 3N\boldsymbol{k}T/2 \tag{90}$$

and will be seen to be very much smaller. This was the main new result of Sommerfeld's treatment of the free electrons in a metal. It had always been difficult to understand why the free electrons in a metal do not contribute an amount $\frac{3}{2}\boldsymbol{R}$ to the atomic heat as they should do if equation (90) were correct; it had long been known from experiment that they make no such contribution. If we assume that we have one free electron per atom and that $N = \boldsymbol{N}$, the number of atoms in a mole of the material, we see from equation (89) that the contribution of the electrons to the atomic heat C_v is given by

$$C_v = \frac{\mathrm{d}\bar{E}}{\mathrm{d}T} = \frac{3\boldsymbol{R}}{2}\left(\frac{\pi^2\,\boldsymbol{k}T}{3E_m}\right) \tag{91}$$

since $\boldsymbol{R} = \boldsymbol{N}\boldsymbol{k}$. This contribution is observable only at low temperatures, since otherwise it is much less than that due to the atomic vibrations, which, as we shall see later, tends to zero as T^3 as T tends to zero. The linear term has been observed for a number of metals and agrees fairly well in magnitude with that given by (91). If we insert the value of E_m from equation (30) in (91) we have

$$C_v = \frac{4\pi^3\,\boldsymbol{m}\boldsymbol{k}^2\,\boldsymbol{V}}{3\boldsymbol{h}^2}\left(\frac{3n}{\pi}\right)^{1/3} T \tag{92}$$

where n is the number of free electrons per unit volume and $\boldsymbol{V}$ is the atomic volume (i.e. the volume of one mole). If we assume one free electron per atom all the quantities in (92) are known, and an experimental check of the theory is possible. For the transition metals there may be some doubts as to how many free electrons we should assume per atom and hence on the interpretation of equation (92). For the alkali metals, which have only one valence electron outside a closed shell, there seems little doubt that we are justified in assuming one electron per atom in the

Sommerfeld model. Recent accurate measurements of the electronic specific heat by D. H. Parkinson* and his colleagues have shown that in this case equation (92) is obeyed provided we replace the electronic mass $\boldsymbol{m}$ by a quantity $m^\star$ which is generally referred to as the 'effective electronic mass'. The magnitude of $m^\star$ for the alkali metals does not differ greatly from $\boldsymbol{m}$ being 1·22 $\boldsymbol{m}$ for Na, 1·1 $\boldsymbol{m}$ for K and 2·3 $\boldsymbol{m}$ for Li. We shall see later that the replacement of $\boldsymbol{m}$ by $m^\star$ takes account, to some extent, of the interaction of the electrons with the atomic cores for which we have only included an average value. It represents a slightly better approximation than that obtained with the simple Sommerfeld model.

The physical reason for the small electronic specific heat is now quite clear. Most of the electrons cannot readily be excited into higher energy states as the temperature is raised since there are no *empty* neighbouring states with allowed values of the energy into which they can go. It is only the electrons whose energies lie near the Fermi level that can find neighbouring *unoccupied* states. It is, in fact, misleading to label all the electrons under discussion as 'free', since most of them cannot change their state without a big change in energy. We might interpret equation (89) as showing that the number of really free electrons is $\pi^2 N\boldsymbol{k}T/6E_m$ and not N. This is not unreasonable since it is only those with energies in a range of approximately $\pm \boldsymbol{k}T$ from the Fermi level that can be regarded as free. We shall retain for the moment the term 'free' to refer to electrons not bound to individual atoms. We shall later see that even this concept has to be abandoned.

2.7.1 The classical limit

Although it is quite clear from the above discussion that the condition $T \gg T_0$ cannot refer to any real metal in the solid state, it is of interest to consider this condition, which (as we shall see) corresponds to the classical form of the probability distribution. In contrast to our previous assumption that $E_F/\boldsymbol{k}T$ was large and positive let us now assume that $E_F/\boldsymbol{k}T$ is large and negative. We may now expand the integrand in (73) in powers of $\exp E_F/\boldsymbol{k}T$ and write equation (75) in the form

$$\left.\begin{aligned} \tfrac{2}{3}E_m^{3/2} &= \mathrm{e}^{E_F/\boldsymbol{k}T}\int_0^\infty E^{1/2}(\mathrm{e}^{-E/\boldsymbol{k}T}+\mathrm{e}^{E_F/\boldsymbol{k}T}\,\mathrm{e}^{-2E/\boldsymbol{k}T}+\ldots)\,\mathrm{d}E \\ &= \frac{\pi^{1/2}}{2}(\boldsymbol{k}T)^{3/2}\,e^{E_F/\boldsymbol{k}T}[1+2^{-3/2}\,\mathrm{e}^{E_F/\boldsymbol{k}T}+\ldots] \end{aligned}\right\} \qquad (93)$$

Taking only the first term in the expansion we have, on writing, as before $T_0 = E_m/\boldsymbol{k}$,

$$\mathrm{e}^{E_F/\boldsymbol{k}T} = \frac{4}{3\pi^{1/2}}\left(\frac{T_0}{T}\right)^{3/2} \qquad (94)$$

* D. H. Parkinson, *Rep. Progr. Phys.*, **21**, 226 (1958).

Thus we see that, if $T \gg T_0$, $E_F/\boldsymbol{k}T$ must be large and negative, and our approximation is justified. Under this condition, since $E > 0$ the probability distribution $P(E)$ may be written in the form

$$P(E) = \exp[(E_F - E)/\boldsymbol{k}T]$$
$$= \frac{4}{3\pi^{1/2}}\left(\frac{T_0}{T}\right)^{3/2} e^{-E/\boldsymbol{k}T} \tag{95}$$

On substituting for T_0 this becomes

$$P(E) = \frac{n\boldsymbol{h}^3 e^{-E/\boldsymbol{k}T}}{2(2\pi \boldsymbol{m}\boldsymbol{k}T)^{3/2}} \tag{96}$$

This is just the classical distribution function. We might have assumed this form and determined E_F from the condition

$$N = 2 e^{E_F/\boldsymbol{k}T} \int_0^\infty N(E) e^{-E/\boldsymbol{k}T} dE \tag{97}$$

It is easily verified that this just leads to equation (94). It follows at once that we have, under these conditions, the mean electronic energy given by the expression

$$\bar{E} = N \int_0^\infty E^{3/2} e^{-E/\boldsymbol{k}T} dE \bigg/ \int_0^\infty E^{1/2} e^{-E/\boldsymbol{k}T} dE$$
$$= 3N\boldsymbol{k}T/2 \tag{98}$$

This is the well-known classical value.

2.8 Thermionic emission

We now return to the condition $E_F/\boldsymbol{k}T \gg 1$ which holds for a normal metal. Suppose that the energy E is so much greater than E_F that $(E - E_F)/\boldsymbol{k}T \gg 1$. We may again write approximately

$$P(E) = e^{(E_F - E)/\boldsymbol{k}T}$$

These high-energy electrons therefore have a classical distribution function. Suppose that for E we take the value W corresponding to an electron which can just get out of the potential well; then we have

$$P(E) = e^{-(W - E_F)/\boldsymbol{k}T} = e^{-e\phi/\boldsymbol{k}T} \tag{99}$$

where ϕ is the work function (see equation (36)). From this we may calculate the number of electrons with kinetic energy greater than W; this will clearly be proportional to $\exp[-e\phi/\boldsymbol{k}T]$. The condition that $E > W$ is a necessary condition for emission of electrons from the metal but not a sufficient one. The required condition for emission from the face $x = 0$ is $\frac{1}{2}\boldsymbol{m}v_x^2 > W$, and we may readily calculate the number of electrons satisfying this condition and hence the emitted current per unit area. We shall not give details of this calculation*, but it will be clear that the current i must be given by an expression of the form

$$i = g(T)\,e^{-e\phi/kT} \tag{100}$$

where $g(T)$ is a function of T which varies slowly compared with $\exp[e\phi/kT]$. A classical treatment would give a different function $g(T)$ and would replace $e\phi$ by the depth W of the potential well. As we have seen, the treatment based on the quantum theory gives the more reasonable interpretation of the work function.

Exercises (1) Show that the energy E_F corresponding to the Fermi level may be expressed in the form

$$E_F = E_m - \frac{\pi^2 k^2 T^2}{6}\frac{d}{dE}\{\ln N(E)\}_{E=E_m} + \ldots$$

where $N(E)\,dE$ is the number of allowed energy levels between E and $E+dE$ and E_m is given by the equation

$$N = 2\int_0^{E_m} N(E)\,dE$$

(Take for $G(E)$ in equation (77) the function

$$G(E) = \int_0^E N(E)\,dE)$$

(2) Hence show that the mean electronic energy $\overline{E}$ may be expressed in the form

$$\overline{E} = \overline{E}_0 + \frac{\pi^2 k^2 T^2}{3} N(E_m) + \ldots$$

where

$$\overline{E}_0 = N\int_0^{E_m} EN(E)\,dE \Big/ \int_0^{E_m} N(E)\,dE$$

(Take for $G(E)$ in equation (77) the function

$$G(E) = \int_0^E EN(E)\,dE)$$

(3) Verify that when $N(E) = 2\pi V(2m)^{3/2}E^{1/2}/h^3$ we get the same formulae as before for E_m, E_F, $\overline{E}$ and C_v.

2.9 Energy levels for a potential well

In our treatment in § 2.2 of the motion of a particle in a potential well of depth W and width a we assumed that $\gamma a \gg 1$, where γ is given in terms of W by the equation $\gamma^2 = 8\pi^2 mW/h^2$. This condition holds for the Sommerfeld model of a metal, as we have seen, but does not necessarily

* See, for example, M. Born, *Atomic Physics* (4th Ed.). Blackie, London (1946), p. 232.

hold when the width a is of atomic dimensions. Later we shall use such a potential well to represent crudely the attractive potential of each of the individual ion cores in a crystal, which we have smoothed out in the 'free-electron' model. We must therefore find the wave functions and corresponding energy levels for the motion of an electron in a potential well without the restriction $\gamma a \gg 1$.

As we have seen in § 2.2, the symmetry of the problem demands that $|\psi(x)|^2$ be a symmetrical function of x and that this may be achieved either by making $\psi(x)$ symmetrical or antisymmetrical. Let us first consider the symmetrical form of the solution of equation (1). It is

$$\left.\begin{aligned} \psi_s(x) &= A\,\mathrm{e}^{-\alpha x} && x > a/2 \\ &= B\cos\beta x && -a/2 \leqslant x \leqslant a/2 \\ &= A\,\mathrm{e}^{\alpha x} && x < a/2 \end{aligned}\right\} \tag{101}$$

where $\alpha^2 = -(8\pi^2 \boldsymbol{m}/\boldsymbol{h}^2)E$ and $\beta^2 = (8\pi^2\boldsymbol{m}/\boldsymbol{h}^2)(E+W)$. We have already specified that $\alpha > 0$ and clearly we may also take $\beta > 0$. The boundary conditions which ensure continuity of $|\psi|^2$ and also continuity of current are that ψ and $\partial\psi/\partial x$ should be continuous at $x = a/2$ and at $x = -a/2$ (see § 1.5.1). If we satisfy these conditions at $x = a/2$ the symmetry condition ensures that they will be satisfied at $x = -a/2$. We have then

$$\left.\begin{aligned} B\cos\tfrac{1}{2}\beta a &= A\,\mathrm{e}^{-\frac{1}{2}\alpha a} \\ \beta B\sin\tfrac{1}{2}\beta a &= \alpha A\,\mathrm{e}^{-\frac{1}{2}\alpha a} \end{aligned}\right\} \tag{102}$$

We have an allowed solution only when

$$\beta a\tan\tfrac{1}{2}\beta a = \alpha a \tag{103}$$

On inserting the values for β and α in terms of the energy E we get an equation for the allowed values of E. (The zero of E here corresponds to the energy of an electron at rest *outside* the well.)

The antisymmetrical form of solution is given by

$$\left.\begin{aligned} \psi_a(x) &= A\,\mathrm{e}^{-\alpha x} && x > a/2 \\ &= B\sin\beta x && -a/2 \leqslant x \leqslant a/2 \\ &= -A\,\mathrm{e}^{\alpha x} && x < a/2 \end{aligned}\right\} \tag{104}$$

In this case we have

$$\left.\begin{aligned} B\sin\tfrac{1}{2}\beta a &= A\,\mathrm{e}^{-\frac{1}{2}\alpha a} \\ \beta B\cos\tfrac{1}{2}\beta a &= -\alpha A\,\mathrm{e}^{-\frac{1}{2}\alpha a} \end{aligned}\right\} \tag{105}$$

giving

$$\beta a\cot\tfrac{1}{2}\beta a = -\alpha a \tag{106}$$

Equations (103) and (106) may be combined into a single equation giving all the allowed energy levels, namely

$$\tan\beta a = 2\alpha\beta/(\beta^2-\alpha^2) \tag{107}$$

These equations taken along with the equation

$$\alpha^2+\beta^2 = \gamma^2 \tag{108}$$

enable us to determine α and hence the value of E, when $E < 0$.

There is a simple graphical method which enables the form of the solution to be readily seen. Suppose we make a plot of αa as a function of βa, remembering that we are only concerned with positive values of α and β. The relationship (108) represents an arc of a circle of radius γa. The relationships (103) and (106) may also readily be plotted, the general form being shown in Fig. 2.7. (The relationship (106) is shown dashed

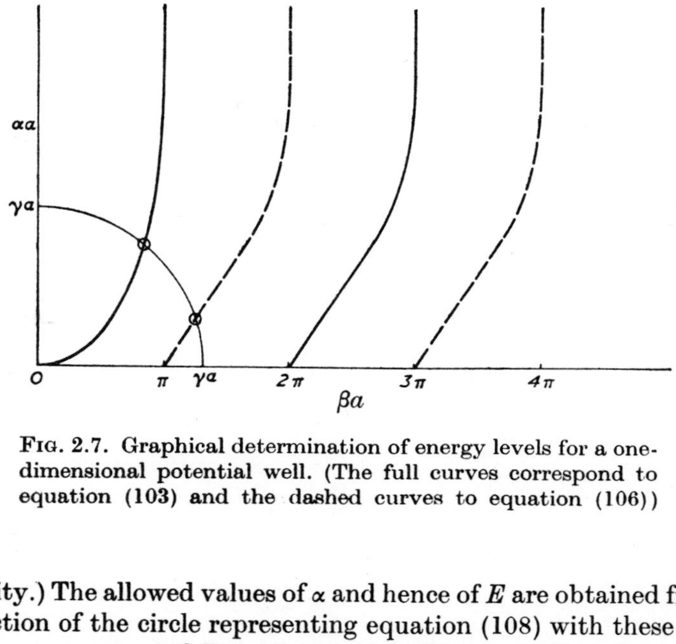

FIG. 2.7. Graphical determination of energy levels for a one-dimensional potential well. (The full curves correspond to equation (103) and the dashed curves to equation (106))

for clarity.) The allowed values of α and hence of E are obtained from the intersection of the circle representing equation (108) with these curves. Several consequences follow at once (see Fig. 2.7). If we have $\gamma a < \pi$ there is only one allowed value and it corresponds to a symmetrical solution. If $\pi < \gamma a < 2\pi$, we have one symmetrical and one antisymmetrical solution, and so on; in particular, if $(n-1)\pi < \gamma a < n\pi$ we have n allowed energy levels altogether, with $E < 0$. When n is a large integer the circle of radius γa clearly makes its intersections (except for one or two near $\beta a = \gamma a$) at values of βa approximately equal to $n\pi$, where n is an integer, odd integers corresponding to symmetrical solutions and even integers to antisymmetrical solutions. This is just the result we had before when we neglected the wave function for $|x| > a/2$ and justifies our previous simplified treatment.

The symmetrical and antisymmetrical wave functions corresponding to the two lowest states are shown in Fig. 2.8. There is in addition, a continuum of allowed values of E with $E > 0$, but we shall not be concerned with these at present.

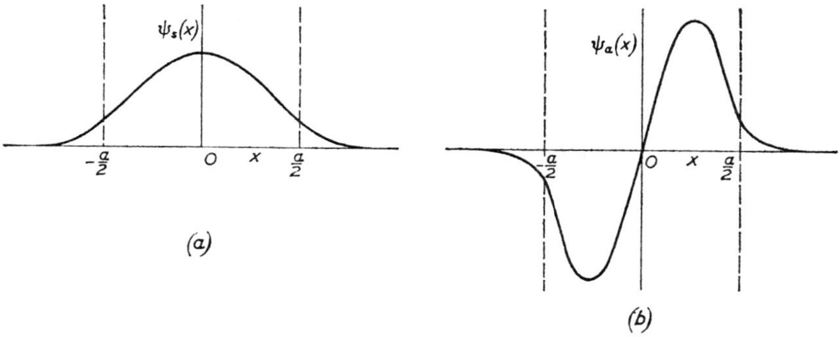

FIG. 2.8. Wave functions for the two lowest states in a one-dimensional potential well; (a) lowest state ψ_s, (b) second lowest state, ψ_a

2.9.1 Potential well in three dimensions

Let us now consider the three-dimensional form of the above problem. In this case we have a spherical potential 'well' given by

$$\left.\begin{aligned} V &= -W \quad r \leqslant a/2 \\ &= 0 \qquad r > a/2 \end{aligned}\right\} \tag{109}$$

We shall consider only spherically symmetrical solutions. (It is always found that these correspond to the lowest energy levels.) The wave equation in this case is given by

$$\left.\begin{aligned} \frac{d^2}{dr^2}(r\psi)-\alpha^2(r\psi) &= 0 \quad r > a/2 \\ \frac{d^2}{dr^2}(r\psi)+\beta^2(r\psi) &= 0 \quad r \leqslant a/2 \end{aligned}\right\} \tag{110}$$

where α and β have the same meanings as before.

The appropriate solutions are

$$\left.\begin{aligned} \psi &= \frac{A}{r}e^{-\alpha r} \qquad r > a/2 \\ &= \frac{B}{r}\sin\beta r \quad r \leqslant a/2 \end{aligned}\right\} \tag{111}$$

Here there cannot be a 'cosine' type of solution since, for this, $\psi \to \infty$ as $r \to 0$. The boundary conditions, ψ and $\partial\psi/\partial r$ continuous at $r = a/2$ are equivalent to $r\psi$ and $\partial(r\psi)/\partial r$ continuous at $r = a/2$. The analysis therefore proceeds as already carried out for the *antisymmetrical* one-dimensional solutions, and the allowed energy levels correspond to the intersections with the dashed curves in Fig. 2.7. In particular, we note that if $\gamma a < \pi$ there is now no allowed solution at all, and this is the outstanding difference between the three-dimensional and one-dimensional problems. In the former, unless we have the well sufficiently deep or wide that $\gamma a > \pi$ the particle will not stay in the well. Physically we are generally concerned with three-dimensional wells and must be careful not to be misled by the curious feature of a one-dimensional well which always permits a solution corresponding to a bound particle. Let us consider some numerical values; if the particle is an electron and a has the value 10^{-8} cm, we may write the condition $\gamma a > \pi$ in the form (cf. equation (6))

$$\gamma^2 a^2 = \frac{W}{W_H} \cdot \frac{a^2}{a_0^2} > \pi^2$$

or approximately $W > 2{\cdot}5\ W_H \simeq 34$ eV. If this condition is not satisfied we have only solutions for which $E > 0$ corresponding to electrons which are not bound in the potential well.

2.10 Transmission of electrons through a potential barrier

The analytical form of the solutions of the problems discussed in § 2.9 for $E > 0$ may be readily found; they are not of very great practical interest and correspond to partial reflection of electrons by a potential well. Of greater practical interest is the corresponding problem when the potential energy of the electron is increased instead of being lowered in the 'well'; in this case we have what may be termed a potential barrier, and we have clearly no bound states with $E < 0$. If the potential is given by (see Fig. 2.9):

$$\left.\begin{aligned} V &= W \quad 0 \leqslant x \leqslant a \\ &= 0 \quad \text{elsewhere} \end{aligned}\right\} \tag{112}$$

with $W > 0$, the interesting condition is $0 < E < W$. In this condition the electron would have insufficient kinetic energy to surmount the barrier and, on classical theory, would certainly be reflected from it. We shall see, however, that wave mechanics indicates that there is a finite probability that a particle may pass through the barrier.

The wave equation to be solved is

$$\left.\begin{aligned} \frac{d^2\psi}{dx^2} + k^2\psi &= 0 \quad x < 0 \\ \frac{d^2\psi}{dx^2} - \alpha^2\psi &= 0 \quad 0 \leqslant x \leqslant a \\ \frac{d^2\psi}{dx^2} + k^2\psi &= 0 \quad a < x \end{aligned}\right\} \tag{113}$$

where now we have

$$\left.\begin{aligned} k^2 &= 2\boldsymbol{m}E/\hbar^2 \\ \alpha^2 &= 2\boldsymbol{m}[W-E]/\hbar^2 \end{aligned}\right\} \tag{114}$$

We can no longer invoke symmetry considerations; the problem is now physically unsymmetrical since we assume electrons incident on the barrier from the left-hand side only. Thus we have for the appropriate solution

$$\left.\begin{aligned}\psi &= A\,e^{ikx}+B\,e^{-ikx} && x<0\\ &= C\,e^{\alpha x}+D\,e^{-\alpha x} && 0\leqslant x\leqslant a\\ &= F\,e^{ikx} && x>a\end{aligned}\right\} \tag{115}$$

The first term in the solution for $x < 0$ represents the incident stream of electrons and A is determined by the magnitude of the current. The constants B, C, D, F are determined by the four boundary conditions which ensure the continuity of ψ and $\partial\psi/\partial x$ at $x = 0$ and at $x = a$. The form of solution for $x > a$ shows that there is a finite current of electrons beyond the potential barrier. The transmission coefficient t which gives the ratio

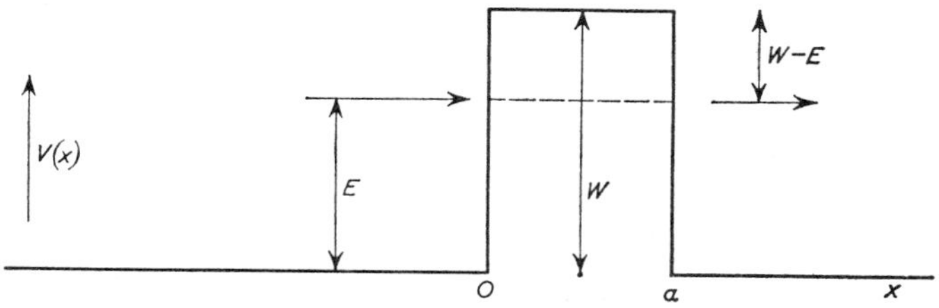

FIG. 2.9. Reflection of electrons by a potential barrier.

of transmitted current to incident current is equal to $|F/A|^2$. On applying the boundary conditions we find, after some algebra, that we have

$$\frac{F}{A} = \frac{4ik\alpha\,e^{-ika}}{e^{-\alpha a}(\alpha+ik)^2-e^{\alpha a}(\alpha-ik)^2} \tag{116}$$

If we have $e^{-\alpha a} \ll 1$ we may simplify this expression to obtain

$$t = |F/A|^2 = \frac{16k^2\alpha^2}{(\alpha^2+k^2)^2}\,e^{-2\alpha a} \tag{117}$$

It will thus be seen that, unless αa is large, the potential barrier will be quite transparent to electrons even though on classical theory they cannot surmount it. This phenomenon is the equivalent of the transmission of light through a thin metal film; it has very important consequences, as we shall see, for the theory of electrons in solids.

The factor $e^{-\alpha a}$ corresponds to the attenuation of the wave function inside the barrier; the factor $16k^2\alpha^2/(\alpha^2+k^2)^2$ in equation (117) corresponds to the reflection of the particle waves at the sharp discontinuity at $x = 0$. The reflection at $x = a$ is included by means of the first term in the denominator of the right-hand side of equation (116) and in our approximation we have neglected it; its effect is small when $e^{-\alpha a}$ is small.

2.11 Slowly varying potential—approximate solution

In the problem discussed above, the potential changes very rapidly at certain values of x. When, however, the change in potential is gradual we may use an approximate method of solution. If we write the wave equation in the form

$$\frac{d^2\psi}{dx^2} - g^2(x)\psi = 0 \tag{118}$$

where $$g^2(x) = 2m[V(x)-E]/\hbar^2 \tag{119}$$

and assume a solution of the form

$$\psi(x) = Ah(x)\exp[S(x)] \tag{120}$$

it may be shown that an approximate solution is given by

$$S(x) = \pm \int^x g(x)\,dx \tag{121}$$

$$h(x) = A[g(x)]^{-1/2} \tag{122}$$

This type of solution, originally discussed by H. Jeffreys*, has been used extensively in quantum mechanics and is frequently referred to as the Wentzel–Brillouin–Kramers (W.B.K.) type of solution. A discussion of this method of solution and conditions for its applicability will be found in most elementary textbooks on quantum mechanics.†

When $E < V(x)$ we have a region which cannot be penetrated by a 'classical' electron; the function $S(x)$ is then real and we have an exponential type of solution. When $E > V(x)$, $S(x)$ is imaginary and we have an oscillatory type of solution. The problem of fitting these solutions across the points where $E = V(x)$ is a little difficult, but we shall not require to perform this operation to arrive at an approximate expression for the transmission of a slowly varying potential barrier, such as that shown in Fig. 2.10. Between the points P and Q, $g(x)$ is real. As in the problem of the constant-potential barrier, when the transmission coefficient is small, only one of the exponential solutions is of importance inside the barrier (the other corresponds to a small reflection at the distant wall). The approximate solution inside the barrier is then

$$\psi = Ag(x)^{-1/2}\exp\left[-\int_{x_1}^{x} g(x)\,.\,dx\right] \tag{123}$$

x_1 and x_2 being the coordinates of the points P and Q where $E = V(x)$. If we consider two points P' Q' near P and Q inside the barrier (see Fig. 2.10), with coordinates x_1', x_2' such that $g(x_1') = g(x_2')$ then we have

$$\psi_{Q'}/\psi_{P'} = \exp\left[-\int_{x_1'}^{x_2'} g(x)\,dx\right] \tag{124}$$

* H. Jeffreys, *Proc. Lond. Math. Soc.*, **23**, 428 (1924).

† See, for example, L. I. Schiff, *Quantum Mechanics* (2nd Ed.). McGraw-Hill (1955), p. 184.

$\psi(x)$ is not varying rapidly with x near $x = x_1$, and $x = x_2$ since there we have $\mathrm{d}^2\psi/\mathrm{d}x^2 \simeq 0$. We may thus write approximately

$$\psi_Q/\psi_P = \exp\left[-\int_{x_1}^{x_2} g(x)\,\mathrm{d}x\right] \tag{125}$$

this gives the attenuation of the wave function inside the barrier. Since the function $V(x)$ varies slowly, the reflection at the front of the barrier will now be small and we have approximately

$$t = |\psi_Q/\psi_P|^2 = \exp\left[-2\int_{x_1}^{x_2} g(x)\,\mathrm{d}x\right] \tag{126}$$

Thus we have

$$-\ln t = \frac{2}{\hbar}\int_{x_1}^{x_2} [2\boldsymbol{m}(V-E)]^{1/2}\,\mathrm{d}x \tag{127}$$

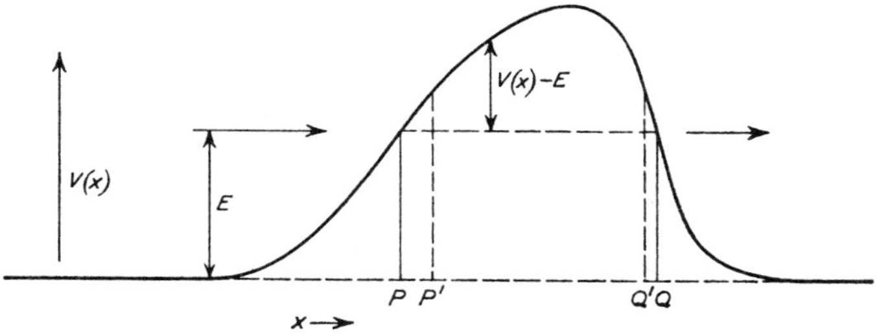

FIG. 2.10. Transmission through a slowly varying potential barrier

When V is constant between $x = x_1$ and $x = x_2$ this reduces to

$$-\ln t = 2\alpha(x_2 - x_1) \tag{128}$$

where $\alpha^2 = 2\boldsymbol{m}(V-E)/\hbar^2$. Apart from the reflection factor, which is generally unimportant, this is the same result as we had before (equation (117)).

2.12 Field emission of electrons from a cold metal

The analysis of § 2.11 has been applied to the emission of electrons from a metal under the influence of a strong electric field; in contrast to thermionic emission this may take place when the metal is cold. The strong electric field is usually obtained by using a sharp point, or edge and the form of the potential function near the surface of the metal is shown in Fig. 2.11. If we take $x = 0$ at the surface, the potential energy outside the metal is given by $V = -e\mathscr{E}x$. We may define the width l of the barrier for an electron at the top of the Fermi distribution by the relationship $\phi = \mathscr{E}l$, where ϕ is the work function. The transmission coefficient t of the triangular barrier is obtained from equation (127) in the form

$$-\ln t = 2\beta \int_0 \left(1 - \frac{x}{l}\right)^{1/2} \mathrm{d}x \qquad (129)$$

where $\beta^2 = 2\boldsymbol{me}\phi/\hbar^2$.

Thus we have

$$\ln t = -\tfrac{4}{3}\beta l \qquad (130)$$

$$= -8\pi(2\boldsymbol{me})^{1/2}\phi^{3/2}/3\boldsymbol{h}\mathscr{E} \qquad (131)$$

In order to calculate the current emitted under the influence of a field $\mathscr{E}$ we must integrate over a narrow band of energies on either side of the Fermi level; this calculation has been carried out by R. H. Fowler and

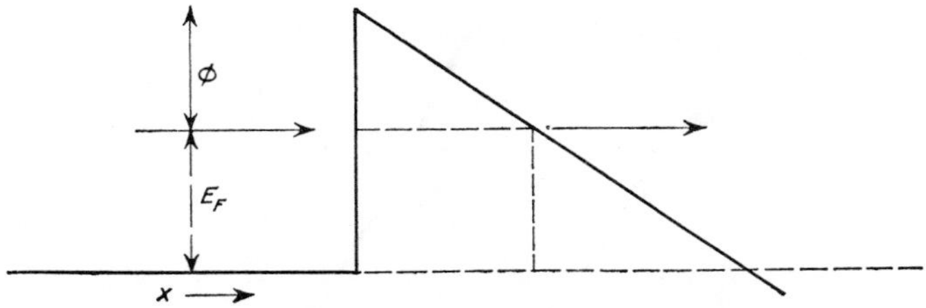

FIG. 2.11. Potential function near the surface of a metal subject to a strong electric field

W. Nordheim* and the reader is referred to their paper for details. It is not difficult to see, however, that the current will be given by an expression of the form

$$i = tI(\mathscr{E})$$

$$= I(\mathscr{E})\exp[-8\pi(2\boldsymbol{me})^{1/2}\phi^{3/2}/3\boldsymbol{h}\mathscr{E}] \qquad (132)$$

where $I(\mathscr{E})$ is a slowly varying function of $\mathscr{E}$. Such cold emission has been observed, and although experimental difficulties do not permit an exact check of equation (132), the rapid variation with $\mathscr{E}$ as predicted is confirmed. There is also a refinement which we have omitted—namely the attractive image force acting on the emitted electron. The general agreement with theoretical prediction is, however, good enough to give a striking confirmation of the ideas on which equation (132) is based, and in particular, of the wave-mechanical picture of the electron; the penetration of the potential barrier is essentially a wave phenomenon which cannot take place on the classical theory of the electron. This so-called 'tunnel' effect accounts also for many other observed phenomena, such as the ready passage of electrons from one piece of metal to another in spite of the fact that each is always coated with a thin insulating oxide film when exposed to the atmosphere. As we shall see in the next section, the 'tunnel' effect has far-reaching implications in the theory of crystalline solids.

* R. H. Fowler and W. Nordheim, *Proc. Roy. Soc.*, A **119**, 173 (1928).

2.13 Potential barriers in a crystal

In our discussion of the average potential to be used in the Sommerfeld model (see § 2.3) no *detailed* account was taken of the attractive potential near each atomic core, or of the other electrons presumably held in this potential. The true potential in which an electron finds itself must be of the form shown in Fig. 2.12, which refers to a line of atoms in the crystal. For a free electron, it has been assumed that no bound state exists in the potential well existing at each atomic core and the average potential in which it finds itself as it moves through the crystal has been taken to replace the true form. It is tacitly assumed that the other electrons are firmly bound in the potential wells at each atomic core and only contribute to the average potential by screening off the nuclear field. From our knowledge of the energy levels of heavy atoms, we may assume that the first few electrons will be bound in states having a depth of a few electron-volts. This energy will increase as we proceed towards deeper levels, the

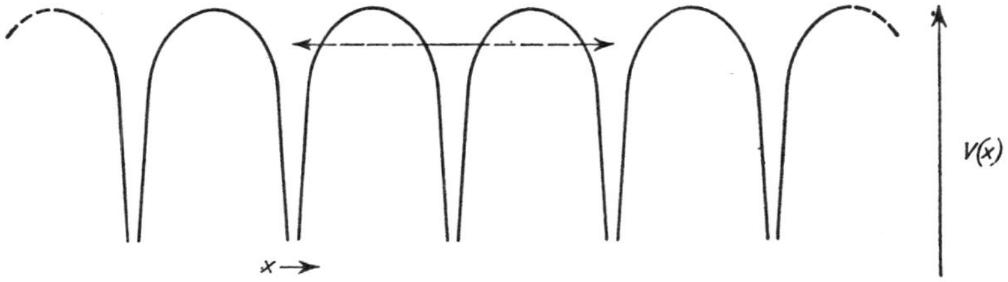

FIG. 2.12. Form of the actual potential in a crystal

innermost electrons, which give rise to X-ray levels, having binding energies of the order of 10^5 eV.

For these electrons there will be a potential barrier separating them from neighbouring atoms of width equal to the lattice spacing (see Fig. 2.12) and of height approximately equal to the binding energy W_B. It is of interest to enquire what is the transmission coefficient of such a barrier. If we take the separation as 5×10^{-8} cm and consider an equivalent 'rectangular' barrier we have

$$\left.\begin{aligned} \alpha a &= \left(\frac{W_B}{W_H}\right)^{1/2} \cdot \frac{a}{a_0} \\ &\simeq 10(W_B/W_H)^{1/2} \end{aligned}\right\} \tag{133}$$

If we take $W_B = \frac{1}{4}W_H \simeq 3{\cdot}3$ eV we see that αa has a value of about 5. The barrier is thus not very transparent and the electron has a chance of only about $\exp(-10)$ or 10^{-4} of penetrating at each approach to the barrier. If, however, as a rather crude picture of what is happening we envisage the electron as oscillating inside its potential well with wavelength of the order of $2a_0$ and hence with velocity $\boldsymbol{h}/2\boldsymbol{m}a_0$ it will collide $\boldsymbol{h}/2\boldsymbol{m}a_0^2$ times per second with the barrier, i.e. about 10^{17} times. Thus an electron would not be kept bound by such a barrier for longer than about 10^{-13} sec. Even if $W_B = 50$ eV the electron would be bound for only about 0·1 sec. It seems therefore pointless to talk about the outer electrons being bound to individual atoms at all when the atoms are crowded together with spacings of a few times10^{-8} cm in a crystal. Such considerations led F. Bloch* to the view that we should not regard *any* of the electrons in a solid as bound but that they should *all* be regarded as 'free', i.e. as not belonging to any atom in particular. We should then consider their motion in the periodic field of the array of bare atomic nuclei, and should start building up the energy states of the metal as a whole, filling the lowest levels with electrons just as we build up a heavy atom. These electrons move in the fields of *all* the nuclei, and as for atoms, the effect of the other electrons on any given electron would be obtained as an average. This would seem at first sight to be in violent contrast to the Sommerfeld model of a metal, but, as we shall see this is not really so, but we have to reinterpret what we mean by a 'bound electron'. To discuss this problem we need to be able to solve the wave equation for the motion of an electron in a periodic potential; this will be dealt with in Chapter 4. We may remark at this stage that the solution of this problem leads not only to a better understanding of metals but also of crystalline solids which are not metals, and for which, according to Sommerfeld's model, all the electrons would be 'bound'.

In an absolutely fundamental treatment of the problem we should, of course, not have to assume that the atomic nuclei are arranged in a crystalline order; we should be able to treat any arrangement and to show that the particular crystalline arrangement is that having the least energy, but this is too ambitious a programme for the methods at present available. It is, in principle, possible to consider one or two alternative crystalline forms but in very few cases is the theory good enough to pick out the one with the lowest energy; we may also vary the lattice spacing and find the value giving minimum energy. The method of treatment now generally adopted is less ambitious and is as follows: we assume the atomic nuclei to be fixed at the lattice sites of a particular crystalline form, and try to find one-electron wave functions for the motion of the electrons in the field of the nuclei and the averaged field of all the other electrons. In practice we shall find that it is a good approximation to regard all except the valence electrons as bound to individual atoms and in their atomic states. We shall find that the approximation of regarding the atomic nuclei as fixed at the lattice points is inadequate to explain many interesting phenomena, which arise from their departure from these points in executing small oscillations under thermal agitation. We shall therefore have to study these oscillations and then consider their effect on the electronic motion.

* F. Bloch, *Z. Phys.*, **52**, 555 (1928).

3

Lattice Vibrations

3.1 Normal modes of oscillation

WE have seen in the previous chapter how, for an adequate description of solids, we have to take the periodic arrangement of atoms in the crystal into account. We shall first study the small oscillations of the atoms of the crystal about their equilibrium configuration as a dynamical problem. We shall then consider the effect of a perfect stationary crystal lattice* on the wave motion of electrons in the solid, and finally the interaction of the electron waves with the atomic vibrations.

Many of the fundamental properties of a three-dimensional crystal lattice are also possessed by a linear lattice. The latter presents a much less formidable mathematical problem and we shall accordingly start with a study of the properties of such a lattice.

Suppose that we have a one-dimensional crystal consisting of a number of similar particles equally spaced along a line (Fig. 3.1); the equilibrium positions of the particles then form a linear lattice. We shall assume to begin with that only nearest neighbours interact and that these exert an attractive force F on each other; we shall also assume that for small lateral displacements the force may be taken as of constant magnitude and that it acts in the direction of the nearest neighbour. Let y_r be the displacement of the rth particle from its equilibrium position. For the moment we shall assume that such displacements are at right angles to the line of the lattice, in one plane, and are very much less than the lattice spacing a. Under these conditions the equation of motion of the rth particle of the crystal is

$$m\ddot{y}_r = \frac{F}{a}(y_{r+1}-y_r)-\frac{F}{a}(y_r-y_{r-1}) \tag{1}$$

where m is the mass of each particle.

* The term 'lattice' used in its strict mathematical sense means a regular periodic array of points. The term is frequently used in the literature on the properties of crystalline solids to refer to the arrangement of atoms making up a crystal. We shall generally refer to a regular arrangement of particles which, in equilibrium, would occupy the points of a lattice simply as a crystal. When the particles of such a crystal vibrate about their positions of equilibrium it should be noted that at any time their positions do not form a lattice in the strict sense. Since, however, the vibrations are executed about the points of a lattice we shall refer to them as 'lattice vibrations'. When we refer to a crystal lattice we mean the lattice giving the equilibrium positions of the particles of the crystal.

If we write $\alpha = F/ma$ this becomes

$$\ddot{y}_r - \alpha y_{r-1} + 2\alpha y_r - \alpha y_{r+1} = 0 \tag{2}$$

If we have $n+2$ particles in the crystal numbered $0, 1, \ldots n+1$, the equations for the displacements y_0 and y_{n+1} will be modified by the conditions holding at the two ends of the crystal. In particular, if the end points are fixed we have simply

$$y_0 = y_{n+1} = 0 \tag{3}$$

In this case we have n equations of the form (2) for the n particles free to move, and we may regard the crystal under these conditions as a dynamical system having n degrees of freedom described by coordinates y_1, $y_2 \ldots y_n$.

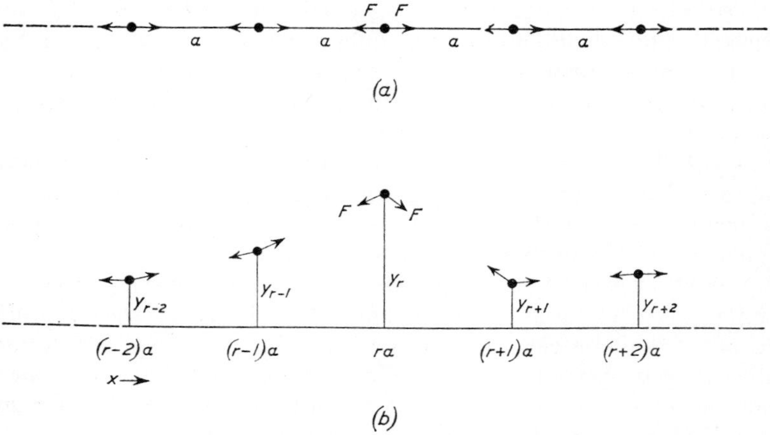

FIG. 3.1. One-dimensional crystal consisting of equally spaced identical particles

The usual procedure in dealing with such a system, when n is small, is to assume solutions of the form

$$y_r = Y_r e^{-i\omega t} \tag{4}$$

the quantities Y_r being constants which we have to determine. When we substitute these in the series of equations (2) we obtain a determinantal equation of degree n in ω^2. This method of solution is dealt with in most textbooks on particle dynamics; from the n solutions $\omega_1, \omega_2 \ldots \omega_n$ of the determinantal equation we obtain the so-called normal periods of oscillation. A solution corresponding to one value ω_s of ω will be of the form

$$y_r = Y_{rs} e^{-i\omega_s t} \tag{5}$$

the constant Y_{rs} being appropriate to the particular value ω_s of ω. The motion corresponding to this particular solution will be periodic, with period $T_s = 2\pi/\omega_s$, and the ratio of the amplitudes of the displacements of the various particles will be given by

$$|y_1|:|y_2|\ldots:|y_r|:\ldots:|y_n| = Y_{1s}:Y_{2s}\ldots:Y_{rs}:\ldots:Y_{ns} \tag{6}$$

These ratios are obtained from equations (2) on substitution of the solution (5), and such a solution is known as a normal mode. There are n such normal modes corresponding to the n solutions $\omega_1, \omega_2 \ldots \omega_n$.

The general solution consists of a linear combination of the normal modes and is in the form

$$y_r = \sum_s Y_{rs}\,\mathrm{e}^{-i\omega_s t} \tag{7}$$

The quantities Y_{rs}, for one value of r, may be obtained from the initial conditions and the others are then obtained from equation (6). Note that each quantity Y_{rs} is a complex number and so effectively contains two unknown constants; there are therefore $2n$ conditions to determine the $2n$ constants obtained from the $2n$ values of y_r, $\dot{y}_r$ at $t = 0$.

While this treatment is readily applied when n is a small number, the solution of the determinantal equation becomes tedious when n is large, and we shall adopt a different procedure. Before doing this, however, we shall discuss the question of finding a more convenient set of coordinates to describe the system. The coordinates y_r do not, in general, vary periodically, since they contain several terms like $\mathrm{e}^{i\omega_s t}$ with different values of ω_s. It would be very convenient if we had a set of coordinates $Q_1, Q_2 \ldots Q_n$ each of which varied periodically with one of the periods $T_1, T_2 \ldots T_n$. Such a set can always be found for a dynamical system whose kinetic and potential energies T, V may both be described, respectively, as quadratic functions of the velocities and the coordinates. The crystal under discussion is such a system, under the conditions assumed, since we have

$$T = \tfrac{1}{2}m \sum_{r=1}^{n} \dot{y}_r^2 \tag{8}$$

$$V = Fa^{-1} \sum_{r=1}^{n} y_r(y_r - y_{r+1}) \tag{9}$$

The form (8) for T is obvious; equation (9) for V may readily be obtained by evaluating to the second order the total increased distance between the particles when displaced, and multiplying by the force F. The problem of finding the coordinates $Q_1, Q_2 \ldots Q_n$ is also dealt with in most books on particle dynamics and we shall simply outline the principles.

3.2 Normal coordinates

Suppose a dynamical system may be described by the n coordinates $q_1, q_2 \ldots q_n$ and that the kinetic energy T and potential energy V are given by

$$T = \tfrac{1}{2} \sum \sum a_{rs}\dot{q}_r\dot{q}_s \tag{10}$$

$$V = \tfrac{1}{2} \sum \sum b_{rs}q_r q_s \tag{11}$$

the quantities a_{rs}, b_{rs} being constants.

Now it is a well-known result of the theory of small oscillations, or indeed of the algebraic study of quadratic forms, that there exists a linear transformation of coordinates of the form

$$Q_r = \Sigma A_{rs} q_s \tag{12}$$

which turns *both* (8) and (9) into a sum of squares. Making this transformation we then have

$$T = \tfrac{1}{2} \sum_{s=1}^{n} c_s \dot{Q}_s^2 \tag{13}$$

$$V = \tfrac{1}{2} \sum_{s=1}^{n} d_s Q_s^2 \tag{14}$$

where c_s and d_s are constants. These are just what we should obtain for the kinetic and potential energies of a set of n *independent* harmonic oscillators each of mass c_s and having restoring force $-d_s Q_s$. The equations of motion in terms of the coordinates Q_s are then

$$\ddot{Q}_s + \omega_s^2 Q_s = 0; \quad s = 1 \dots n \tag{15}$$

where $\omega_s^2 = d_s/c_s$. The coordinates Q_s therefore vary harmonically with periods $T_s = 2\pi/\omega_s$, and are normal coordinates of the system. There are exactly n such coordinates and periods so long as all the values of ω_s are different. When two or more are equal the system is said to be degenerate. The general solution in terms of the coordinates q_r is thus seen to be a linear combination of normal modes; this follows since each q_r is expressible as a linear combination of the coordinates $Q_1 \dots Q_n$, as will be seen by inverting equations (12). A normal mode is a solution involving only *one* of the quantities Q_s, and this is just what we have in equation (5).

Many examples of the determination of the normal periods, modes and coordinates of various dynamical systems are given in textbooks on particle dynamics. The reader will, however, find it instructive to consider the following example.

Exercise A linear crystal of the type described above consists of five equally spaced particles of equal mass, the end ones being fixed in position. Show that the equations of motion of the three free particles, whose lateral displacements are y_1, y_2, y_3 are

$$\begin{aligned} \ddot{y}_1 + 2\alpha y_1 - \alpha y_2 &= 0 \\ \ddot{y}_2 - \alpha y_1 + 2\alpha y_2 - \alpha y_3 &= 0 \\ \ddot{y}_3 - \alpha y_2 + 2\alpha y_3 &= 0 \end{aligned}$$

with $\alpha = F/ma$.

Hence show that the normal periods T_1, T_2, T_3 are given by $2\pi/\omega_1, 2\pi/\omega_2, 2\pi/\omega_3$ where $\omega_1, \omega_2, \omega_3$ are the roots of the determinantal equation

$$\begin{vmatrix} (2\alpha - \omega^2) & -\alpha & 0 \\ -\alpha & (2\alpha - \omega^2) & -\alpha \\ 0 & -\alpha & (2\alpha - \omega^2) \end{vmatrix} = 0$$

This equation has roots given by

$$\omega_1^2 = (2-\sqrt{2})\alpha$$
$$\omega_2^2 = 2\alpha$$
$$\omega_3^2 = (2+\sqrt{2})\alpha$$

Show that the normal modes corresponding to ω_1, ω_2, ω_3 are given by

$$\left.\begin{aligned} y_1 &= \frac{A_1}{\sqrt{2}}\mathrm{e}^{-i\omega_1 t} \\ y_2 &= A_1\mathrm{e}^{-i\omega_1 t} \\ y_3 &= \frac{A_1}{\sqrt{2}}\mathrm{e}^{-i\omega_1 t} \end{aligned}\right\}$$

$$\left.\begin{aligned} y_1 &= A_2\mathrm{e}^{-i\omega_2 t} \\ y_2 &= 0 \\ y_3 &= -A_2\mathrm{e}^{-i\omega_2 t} \end{aligned}\right\}$$

$$\left.\begin{aligned} y_1 &= \frac{A_3}{\sqrt{2}}\mathrm{e}^{-i\omega_3 t} \\ y_2 &= -A_3\mathrm{e}^{-i\omega_3 t} \\ y_3 &= \frac{A_3}{\sqrt{2}}\mathrm{e}^{-i\omega_3 t} \end{aligned}\right\}$$

Sketch out the form of the normal modes. Show that the kinetic and potential energies of the system are given by

$$T = \tfrac{1}{2}m(\dot{y}_1^2+\dot{y}_2^2+\dot{y}_3^2)$$
$$V = \tfrac{1}{2}\frac{F}{a}(2y_1^2+2y_2^2+2y_3^2-2y_1y_2-2y_2y_3)$$

If we now make the transformation

$$y_1 = \tfrac{1}{2}Q_1+\frac{1}{\sqrt{2}}Q_2+\tfrac{1}{2}Q_3$$
$$y_2 = \frac{1}{\sqrt{2}}Q_1-\frac{1}{\sqrt{2}}Q_3$$
$$y_3 = \tfrac{1}{2}Q_1-\frac{1}{\sqrt{2}}Q_2+\tfrac{1}{2}Q_3$$

we obtain

$$T = \tfrac{1}{2}m(\dot{Q}_1^2+\dot{Q}_2^2+\dot{Q}_3^2)$$
$$V = \tfrac{1}{2}m(\omega_1^2Q_1^2+\omega_2^2Q_2^2+\omega_3^2Q_3^2)$$

Hence Q_1, Q_2, Q_3 are normal coordinates. Show by inverting the above equations for y_1, y_2, y_3 that

$$Q_1 = \tfrac{1}{2}y_1+\frac{1}{\sqrt{2}}y_2+\tfrac{1}{2}y_3$$
$$Q_2 = \frac{1}{\sqrt{2}}y_1-\frac{1}{\sqrt{2}}y_3$$
$$Q_3 = \tfrac{1}{2}y_1-\frac{1}{\sqrt{2}}y_2+\tfrac{1}{2}y_3$$

3.3 The infinite linear lattice

It is frequently easier to solve a more general problem without restrictions than to deal with a number of special cases, and this turns out to be so for the linear lattice as described in § 3.1. We shall accordingly first discuss the motion of an infinite linear crystal and obtain solutions for crystals containing finite numbers of particles by application of suitable restrictive boundary conditions. This method of approach not only leads to a simplification of the analysis as compared with the more elementary method outlined in § 3.1, but also gives an excellent example of wave motion in a dispersive medium.

The equations for the transverse oscillations of an infinite linear crystal of the type described in § 3.1, are of the form

$$\ddot{y}_r + 2\alpha y_r - \alpha y_{r-1} - \alpha y_{r+1} = 0 \tag{16}$$

with $\alpha = F/ma$ and y_r the lateral displacement of the rth particle (see § 3.1); there is now an infinite number of such equations.

The method we now adopt to solve this infinite set of equations is to see whether solutions in the form of travelling waves can be found, similar to those discussed in Chapter 1. Here, however, we have no continuous medium in which the waves move, but only a set of discrete particles, so that these waves have no direct *physical* significance except at the positions of the particles, and are known as 'mathematical' or 'descriptive' waves. We shall see that, not only can such waves be found to describe the motion, but this can be done in an infinite variety of ways.

Let us measure distance along the crystal by means of a coordinate x, the lattice points being given by $x = ra$, where r takes on all positive and negative integral values. We shall try to find a function of the form

$$y = \phi(x, t) \tag{17}$$

which for *all* values of t and *the values of x corresponding to lattice points* gives the motion of the particles of the crystal; if such a function can be found the displacements of the particles y_r may be expressed in the form

$$y_r = \phi(ra, t) \tag{18}$$

The variation with t we shall assume to be harmonic, and we shall seek solutions of the form

$$y(x, t) = \psi(x)\, \mathrm{e}^{-i\omega t} \tag{19}$$

Proceeding as in Chapter 1, we first of all see if we may use simple solutions in the form of plane waves, and then build up more general solutions by means of linear combinations. We have therefore to determine whether or not we may use a function of the form

$$y(x, t) = A\, \mathrm{e}^{ikx - i\omega t} \tag{20}$$

to generate solutions of the infinite set of equations (16). The values of y_r would be given in this case by

$$y_r = y(ra, t) = A\, \mathrm{e}^{ikra - i\omega t} \tag{21}$$

That the quantities y_r given by (21) do in fact form a solution of equation (16) when k is given as a particular function of ω, may be seen at once by substitution, since the particular form of (21) leads to an equation for k in terms of ω, independent of r. We have for all values of r

$$-\omega^2+2\alpha-\alpha\,e^{-ika}-\alpha\,e^{ika} = 0$$

or

$$\omega^2 = 2\alpha(1-\cos ka) \tag{22}$$

Equation (22) shows that for any positive value of k we may also have a solution with $-k$, corresponding to a wave travelling in the opposite direction. Remembering this we may then write (22) in the form

$$\omega = 2\alpha^{1/2}\left|\sin\frac{ka}{2}\right| \tag{23}$$

The form (23) giving ω as a function of k is shown in Fig. 3.2. It will be seen that ω is a periodic function of k with period $2\pi/a$ in k. Thus if, for

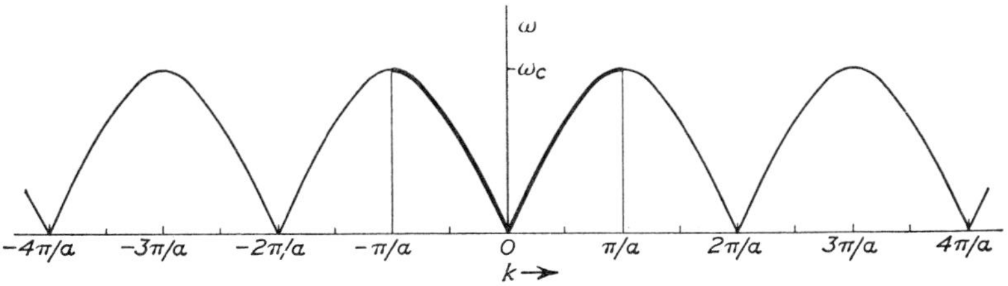

FIG. 3.2. Dispersion law for linear crystal containing one type of particle ($\omega_c = 2\alpha^{1/2}$, $\alpha = F/ma$)

a given value of ω we have a solution k_0 such that $|k_0| \leqslant \pi/a$ then equally good solutions are given by $k = k_n$, where

$$k_n = k_0+2n\pi/a$$

and n is any positive or negative integer. We shall show that the solution k_n gives exactly the same motion of the particles of the crystal as k_0. We have if $k = k_0$

$$y_r = A\,e^{ik_0ra-i\omega t} \tag{24}$$

If $k = k_n$ we have

$$\begin{aligned} y_r &= A\,e^{i(k_0+2n\pi/a)ra-i\omega t} \\ &= A\,e^{ik_0ra-i\omega t} \end{aligned}$$

We thus have exactly the same motion for the particles of the crystal although the form of the descriptive waves is different; the solutions for various values of n are therefore *exactly equivalent*, and it is usual to restrict k so that $|k| \leqslant \pi/a$. With this restriction there are thus just two values of k, i.e. $\pm k_0$ for which we have a solution. For a given value of ω the general solution is therefore of the form

$$y = A\,e^{ikx-i\omega t}+B\,e^{-ikx-i\omega t} \tag{25}$$

giving

$$y_r = (A\,e^{ikra}+B\,e^{-ikra})\,e^{-i\omega t} \tag{26}$$

This solution corresponds to waves travelling to the right and to the left in a medium whose dispersion law is given by the equation (23). If we were to cut slots so as to observe the motion in this medium only at the points $x = ra$, the motion we should see would be just that of the particles of the crystal. This representation of the motion in terms of 'mathematical' or 'descriptive' waves turns out to be most convenient.

To obtain the general motion we proceed as in Chapter 1, forming linear combinations of solutions for a band of values of ω, the dispersion law being given by (23). For a finite lattice, however, only certain discrete values of ω are allowed, as we shall see, and we have to sum over these to obtain the general solution. We note from equation (23) that real values of k are obtained only for values of ω less than ω_c where

$$\omega_c = 2\alpha^{1/2} = 2(F/ma)^{1/2} \tag{27}$$

This shows that waves corresponding to frequencies higher than

$$\nu_c = (F/ma)^{1/2}/\pi$$

are not propagated along the crystal; the frequency ν_c is called the upper cut-off frequency. This does not mean that there are no solutions for higher values of the frequency. If we write $\beta = ik$ equation (22) may be written in the form

$$\sinh\frac{\beta a}{2} = \pm\,\omega/2\alpha^{1/2} \tag{28}$$

and the corresponding solutions are of the form

$$y_r = A\,\mathrm{e}^{\pm\beta ra - i\omega t} \tag{29}$$

These correspond to exponentially attenuated waves, and will have to be included in a general solution.

3.3.1 Finite linear crystal

When we have a finite number of particles in the crystal we may obtain the solution from that for the infinite crystal by applying appropriate boundary conditions. These will generally be determined by the conditions at the extremities. For example, suppose we have a crystal consisting of $N+2$ particles, the end particles (labelled 0 and $N+1$) being fixed; the boundary conditions are then $y_0 = 0$ and $y_{N+1} = 0$ for all values of t. In terms of the wave function $y(x, t)$ of equation (25) we have

$$\left.\begin{aligned} y &= 0, \qquad x = 0 \\ y &= 0, \qquad x = (N+1)a \end{aligned}\right\} \tag{30}$$

The ratio of the constants A/B required in order to fulfill the first of these conditions is such that the required solution may be written in the form

$$y = A\sin kx\,.\,\mathrm{e}^{-i\omega t} \tag{31}$$

The second will also be satisfied provided we have

$$\sin (N+1)\, ka = 0$$

i.e. if

$$k = k_s = \frac{s\pi}{(N+1)a} = \frac{s\pi}{L} \qquad (32)$$

where s is an integer, and L is the total length of the crystal.

From the form of (31) it is clear that negative values of k, in this case, give the same solution as positive values, and, as we have seen, values of k greater than π/a lead to no new solutions. The only values of s which lead to *independent* solutions are therefore

$$s = 1, 2, \ldots N$$

($s = 0$ and $s = N+1$ give $y_r = 0$ for all values of r). We have therefore just N allowed values ω_s of the angular frequency ω given by equation (23) for the n allowed values of k. (Note that values of k_n given by (32) which are greater than π/a just lead to the same values of ω_s.) The allowed values of the frequency are, of course, just the normal frequencies discussed in § 3.1. We have then

$$\omega_s = \omega_c \sin \frac{s\pi}{2(N+1)}, \quad s = 1, 2, \ldots N \qquad (33)$$

where $\omega_c = 2\alpha^{1/2}$. The solutions y_{rs} obtained for $\omega = \omega_s$ are given by

$$y_{rs} = A_s \sin\left(\frac{rs\pi}{N+1}\right) e^{-i\omega_s t}$$

and correspond to the normal modes. The general solution is obtained by forming a linear combination of all the allowed solutions in the form

$$y_r = \sum_s A_s \sin\left(\frac{rs\pi}{N+1}\right) e^{-i\omega_s t} \qquad (34)$$

The physical interpretation of equation (32) is simply that in order to have both ends fixed, we must have an integral number of half wavelengths of the 'descriptive' waves in the length L of the crystal. Thus if we write $\lambda = 2\pi/k$ we have for the allowed values, $\lambda_s = 2L/s$, s being an integer.

For the particular case $N = 3$ we have,

$$\omega_1 = \omega_c \sin \frac{\pi}{8} = \alpha^{1/2}[2-\sqrt{2}]^{1/2}$$

$$\omega_2 = \omega_c \sin \frac{\pi}{4} = 2^{1/2}\alpha^{1/2}$$

$$\omega_3 = \omega_c \sin \frac{3\pi}{8} = \alpha^{1/2}[2+\sqrt{2}]^{1/2}$$

These are just the values obtained by the elementary method (see exercise in § 3.1). It will readily be verified that the values of y_1, y_2, y_3 given by (34) are also the same as those obtained in the exercise in § 3.1.

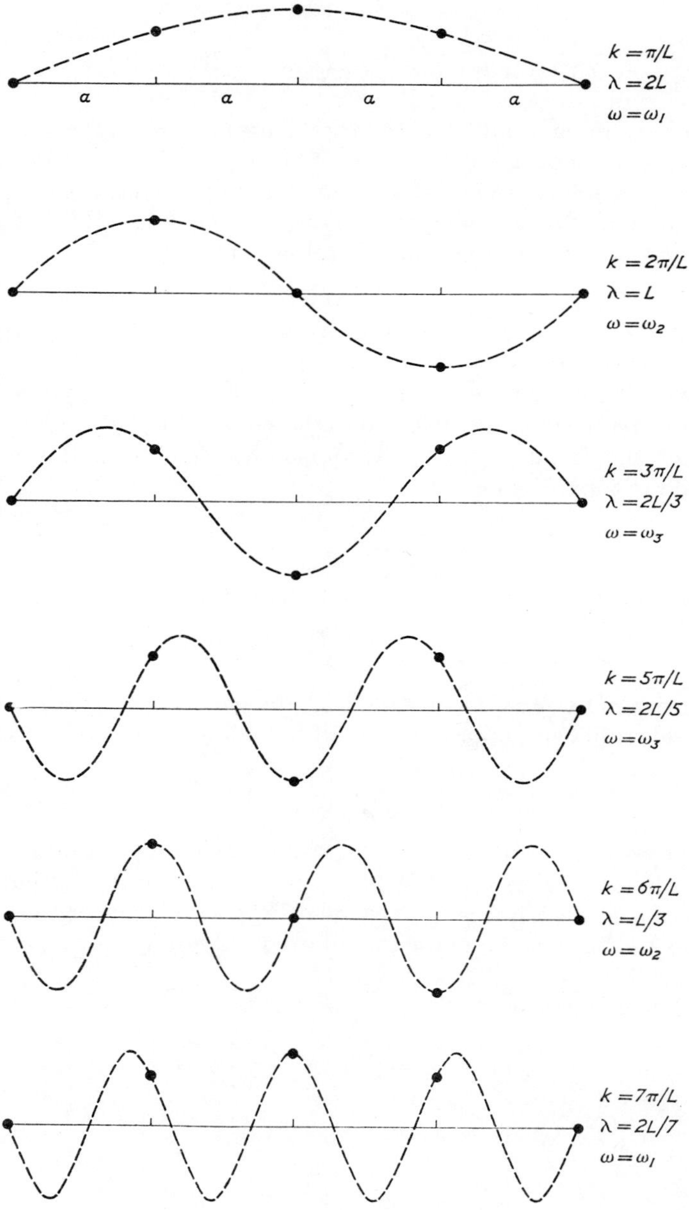

FIG. 3.3. Descriptive wave forms for the normal modes of a crystal having three free particles

The form of the 'descriptive' waves corresponding to the solutions with $k = \pi/L, 2\pi/L, 3\pi/L$ ($\lambda = 2L, L, 2L/3$) are shown in Fig. 3.3, and also those corresponding to some higher values of k; that for $k = 5\pi/L$ gives the same displacements of the particles as when $k = 3\pi/L$, and similarly $k = 6\pi/L$ corresponds to $k = 2\pi/L$, and $k = 7\pi/L$ to $k = \pi/L$. The allowed values ω_s of ω are shown in Fig. 3.4 together with the corresponding values of k_s of k.

As the number of particles in the crystal becomes very large the allowed values of k, and corresponding values of ω, become very closely spaced and for many purposes may be regarded as forming a continuous distribution. Since values of k greater in magnitude than π/a lead to no new solutions, the ω/k relationship is frequently shown only for the

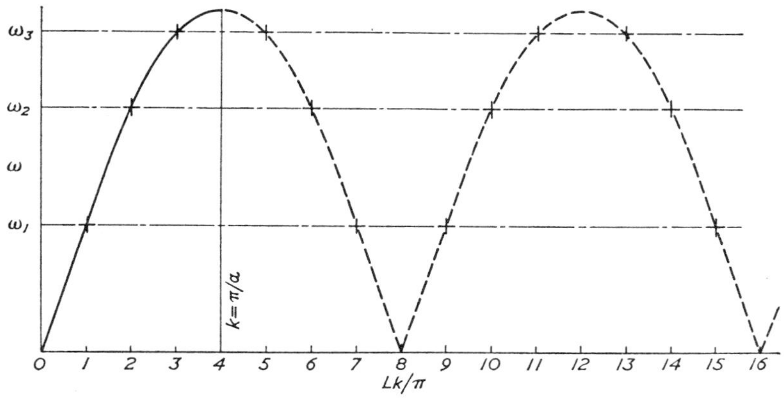

FIG. 3.4. Allowed values of ω and k for a crystal having three free particles

interval $0 \leqslant k \leqslant \pi/a$, as in Fig. 3.5. Such a representation of the dispersion relationship is generally referred to as a 'reduced' representation whereas that shown in Fig. 3.2 is called a 'full' representation; we shall use both forms later. When using the reduced representation we must remember that certain negative values of k in the range $-\pi/a \leqslant k < 0$ may also give allowed and different solutions from those obtained with positive values of k. The latter refer to waves travelling to the right and the former to waves travelling to the left, and when we have solutions like (31) both are included with equal amplitude.

We note that as $k \rightarrow 0$ ($\lambda \rightarrow \infty$), corresponding to low values of the frequency, equation (23) becomes

$$\omega/k = a\alpha^{1/2} = (Fa/m)^{1/2} = (F/\rho)^{1/2} \tag{35}$$

where ρ is the average linear density of the crystal, which therefore behaves like a continuous string of linear density ρ stretched with tension

F, the phase velocity $\mathscr{V}$ (and hence also the group velocity) being a constant $\mathscr{V}_0$ equal to $(F/\rho)^{1/2}$. As the frequency increases so that $\omega \to \omega_c$ we have $k \to \pi/a$. The phase velocity $\mathscr{V}$ therefore tends to the value $2\mathscr{V}_0/\pi$ whereas the group velocity tends to zero since $d\omega/dk$ tends to zero (see Fig. 3.5). The value of k equal to π/a corresponds to a standing

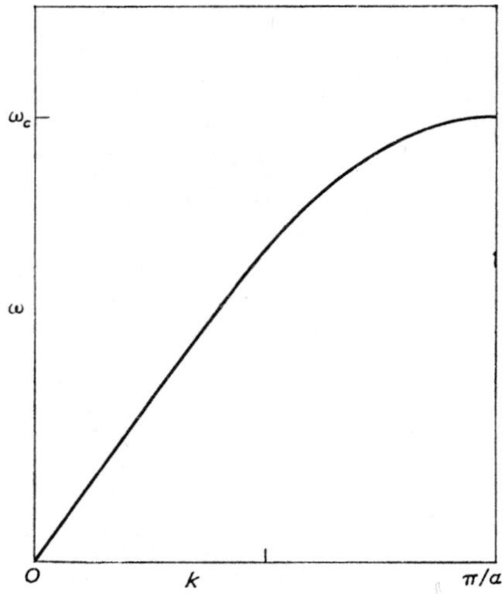

FIG. 3.5. Reduced representation of the dispersion curve for a linear crystal with one type of particle

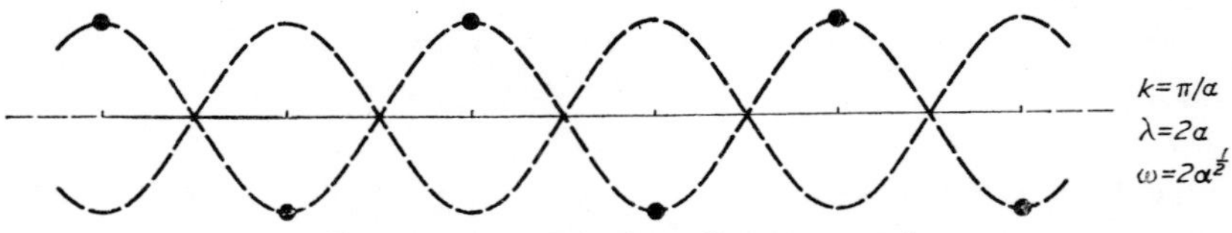

FIG. 3.6. π-mode for a linear crystal

wave and not to a progressive wave, as may be seen from equation (21). We have then

$$y_r = A\,e^{i\pi r}.e^{-i\omega t}$$
$$= A(-1)^r e^{-i\omega t} \tag{36}$$

and all the particles have the same amplitude of oscillation but neighbouring particles differ in phase by π; this mode, called the π-mode, is illustrated in Fig. 3.6. Clearly the physical motion of the particles does not in this case correspond to a progressive motion along the crystal. The 'descriptive' waves may either be travelling waves with wavelength $2a$ to the right or to the left or standing waves obtained by combining these; the resulting motion of the particles is the same in either case.

3.3.2 Periodic boundary conditions

The solutions obtained for a crystal with fixed terminations consist entirely of standing waves. When N is large this is not always convenient, and a mathematical trick similar to that already discussed in § 2.5 is frequently used to obtain travelling waves with simple boundary conditions for finite values of N. We assume that we have indeed an infinite lattice but that it repeats itself in groups of N points. The length of each group will be $L = Na$, and this condition may be approximately realised physically if the crystal is bent to form a closed ring containing N

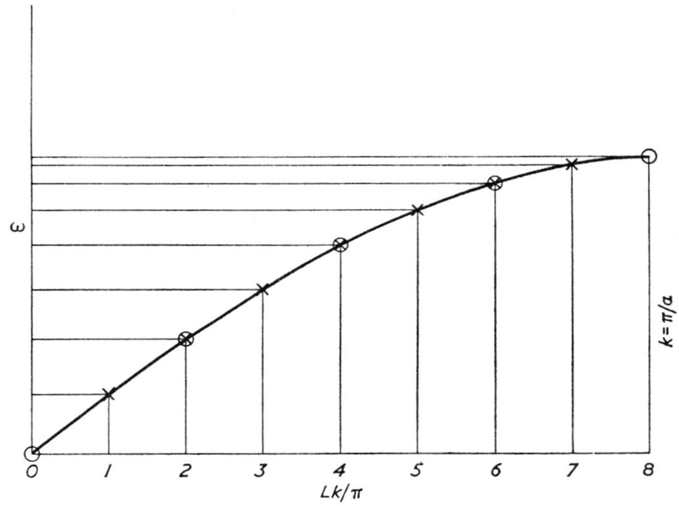

FIG. 3.7. Allowed values of ω and k for a linear crystal with fixed ends having seven equal free particles and for one having eight equal free particles with periodic boundary conditions; the length L of each is $8a$. (×, fixed ends, ○, periodic conditions)

particles. The appropriate solutions must then repeat exactly at intervals L along the lattice, which means that the function $y(x, t)$ must be periodic in x with period L. If we have a solution of the form (20) corresponding to a wave travelling in the positive direction of x we shall obtain the required periodicity provided $kL = 2s\pi$, since $e^{2\pi si} = 1$, if s is an integer. This gives for the allowed values k_s of k

$$k_s = \frac{2s\pi}{Na} \tag{37}$$

For large values of N the allowed values of k_s are spaced twice as far apart as for the lattice with fixed ends, and correspond now to the physical condition $\lambda_s = 2\pi/k_s = L/s$, i.e. there are s *full* waves in the interval L. At first sight we seem to get only about half the number of allowed values compared with the condition in which the ends are fixed. We must note, however, that in this case negative values of k give different modes since we are dealing with travelling waves, and each normal mode is therefore degenerate. An exception occurs when N is even and we have a π-mode allowed, corresponding to $k = \pi/a$ with $s = N/2$; this is a single mode the value $k = -\pi/a$ giving the same motion, since here, as we have seen, we really have a standing wave. From the considerations of § 3.1 we should again expect exactly N modes. When N is even we have $(N-2)/2$ values of s given by (37) for $0 < k \leqslant \pi/a$, each to be counted twice, and one for $k = \pi/a$, giving $2N-1$ in all. The remaining mode corresponds to $k = 0$, which gives an allowed motion in this case, all the particles oscillating in phase. Thus we again have N normal modes, as expected. When N is odd we have $\frac{1}{2}(N-1)$ double modes and that for $k = 0$, giving N in all. For large values of N the density of allowed values of k per unit interval of ω is thus the same as for a crystal with fixed ends. In Fig. 3.7 are shown the allowed values of k and ω for a crystal of length $L = 8a$ having seven particles with fixed ends, and for one with eight particles, also of length $L = 8a$ with periodic boundary conditions.

3.3.3 Longitudinal and other modes

So far, we have restricted the motion of the particles to a plane, and have considered only transverse motion. If the restriction to a plane is removed, transverse motion of each particle will be possible in two dimensions. If the force between two particles is symmetrical about the line joining them, transverse motion restricted to a plane at right angles to that which we have already considered will lead to exactly the same types of normal modes and to the same allowed frequencies for a finite crystal. It may readily be shown that, to the first order, the equations for motion in the two planes are independent, so their solutions may be combined linearly to give the more general motion, each mode being degenerate, owing to the possibility of the two-dimensional transverse motion. If the force is not symmetrical, two planes may generally still be found in which the motion to the first order is independent and we have two types of waves which may be superimposed to give the general motion. In this case we shall have two ω/k curves corresponding to the two types of motion giving two values of ω for each value of k.

In addition we must also consider longitudinal displacements of the particles of the crystal. These may readily be shown, for small displacements, to be independent of the transverse displacements to the first order, and we shall treat them assuming we have no transverse displacement. The restoring force on a particle moved from its position of equilibrium now depends on the spatial variation of the force F. If z_r is the displacement of the rth particle and F is now regarded as a function of the separation we have for the net force on the rth particle

$$F(a+z_{r+1}-z_r)-F(a+z_r-z_{r-1})$$

For small displacements this is equal to

$$(z_{r+1}+z_{r-1}-2z_r)\,F'$$

where F' is the derivative of F with respect to the linear displacement, evaluated at the equilibrium position. The equations of motion for the longitudinal motion of the particles are then

$$\ddot{z}_r-\alpha' z_{r-1}+2\alpha' z_r-\alpha' z_{r+1} = 0 \tag{38}$$

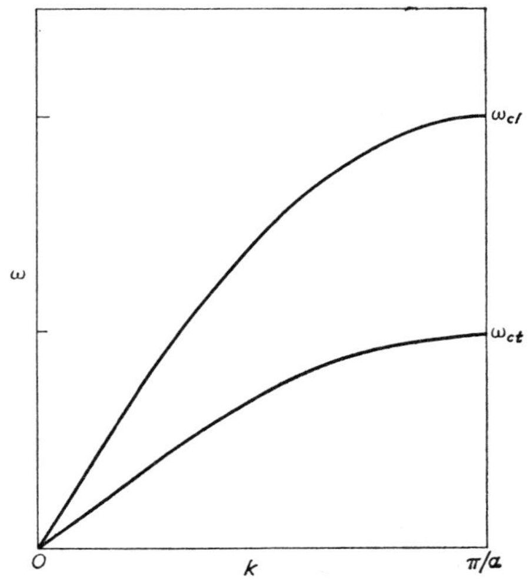

FIG. 3.8. Dispersion curves for a linear crystal, showing longitudinal and transverse modes

with $\alpha' = F'/m$. This is of the same form as equation (2) and the previous analysis also applies to the longitudinal oscillations. A different dispersion law is now found corresponding to the different constant α' which occurs in (38). For an isotropic linear crystal we have then two dispersion curves as shown in Fig. 3.8; for a non-isotropic crystal we shall have three. If the crystal were not under tension, for example if the ends were free, only the longitudinal mode would be significant.

3.3.4 Normal coordinates for a linear crystal

It is of interest to determine the normal coordinates for the linear crystal discussed in § 3.3.1. We shall consider a crystal with a finite number $N+2$ of particles, the end particles being fixed. We have then N

normal modes whose frequencies $\nu_s = 2\pi\omega_s$ are obtained from equation (33), the displacements y_r of the particles being given by equation (34). From § 3.2 we see that the normal coordinates Q_s corresponding to frequency ν_s may be expressed in the form

$$Q_s = B_s e^{-i\omega_s t} \tag{39}$$

where B_s is a constant. Thus we see that we may express the displacements y_r in terms of the normal coordinates Q_s by equations of the form

$$y_r = \sum_s C_s \left[\sin\left(\frac{rs\pi}{N+1}\right)\right] Q_s \tag{40}$$

The constants C_s are not determined since Q_s multiplied by any constant is also a normal coordinate. It is, however, mathematically convenient to choose the constants C_s so that the sum of the squares of the constants of the linear transformation connecting the coordinates y_r and Q_s is equal to unity; such a transformation is said to be 'normalised'. It may readily be shown that

$$\sum_{s=1}^{N} \sin^2\left(\frac{rs\pi}{N+1}\right) = \tfrac{1}{2}(N+1) \tag{41}$$

so that if we take $C_s = 2^{1/2}/(N+1)^{1/2}$ for all values of s we have

$$y_r = \left(\frac{2}{N+1}\right)^{1/2} \sum_{s=1}^{n} \left[\sin\left(\frac{rs\pi}{N+1}\right)\right] Q_s \tag{42}$$

It will also readily be verified that if t is an integer not equal to s

$$\sum_{r=1}^{N} \sin\left(\frac{rs\pi}{N+1}\right)\sin\left(\frac{rt\pi}{N+1}\right) = 0 \tag{43}$$

If we arrange the coefficients of the transformation in a square array of elements A_{rs}, known as the matrix of the transformation, we have

$$\left.\begin{aligned} \sum_r A_{rs} A_{rt} &= \delta_{ts} \\ \sum_s A_{rs} A_{us} &= \delta_{ru} \end{aligned}\right\} \tag{44}$$

where $\delta_{ts} = 1$ when $t = s$ and is equal to zero otherwise. Such a transformation is known as a normal, orthogonal transformation. The transformation from y_r to Q_s is always orthogonal and we have also made it normal by our choice of C_s. The advantage of such a transformation is that it may be used at once to express the normal coordinates Q_s in terms of the displacements y_r. It will be readily seen from the symmetry of the quantities

$$A_{rs} = \left(\frac{2}{N+1}\right)^{1/2} \sin\left(\frac{rs\pi}{N+1}\right) \tag{45}$$

that the transformation may be read both ways, so that we have the normal coordinates Q_s given by the equations

$$\begin{aligned} Q_s &= \sum_{r=1}^{N} A_{rs} y_r \\ &= \left(\frac{2}{N+1}\right)^{1/2} \sum_{r=1}^{N} \sin\left(\frac{rs\pi}{N+1}\right) y_r \end{aligned} \tag{46}$$

Exercise (1) Verify that the expression (46) gives the normal coordinates for the example with $N = 3$ discussed in § 3.2.

(2) Verify that the kinetic energy T of the oscillations of the particles may be written as (see § 3.2)

$$T = \tfrac{1}{2}m \sum_{s=1}^{N} \dot{Q}_s^2 \tag{47}$$

and also that the potential energy V may be expressed as

$$V = \tfrac{1}{2}m \sum_{s=1}^{N} \omega_s^2 Q_s^2 \tag{48}$$

The equations of motion may be then expressed in the form

$$\ddot{Q}_s + \omega_s^2 Q_s = 0 \quad (s = 1, 2, \ldots N) \tag{49}$$

The crystal therefore behaves like N independent harmonic oscillators whose frequencies are $\omega_1, \omega_2 \ldots \omega_N$. The motion associated with each oscillator is a normal mode.

(3) Show that when the motion is subject to periodic boundary conditions (repeating after N lattice points) the displacements of the N particles of a repeating group may be expressed, if N is odd, in the form

$$y_r = \sum_{-(N-1)/2}^{(N-1)2} A_s e^{2\pi i r s/N} e^{-i\omega_s t}$$

where the values of ω_s are those given in § 3.3.2.

This may be written in the form

$$y_r = B_0 + \sum_{s=1}^{(N-1)/2} \left[C_s \cos\frac{2\pi rs}{N} + D_s \sin\frac{2\pi rs}{N}\right] e^{-i\omega_s t}$$

We have thus a degenerate system having two normal coordinates associated with each value of ω_s. The normal coordinates are therefore not, uniquely defined. By writing

$$\begin{aligned} Q_0 &= a_0^{-1} B_0 \\ Q_s &= a_s^{-1} C_s e^{-i\omega_s t} \quad 1 \leqslant s \leqslant (N-1)/2 \\ Q_{N-s} &= a_s^{-1} D_s e^{-i\omega_s t} \quad s > (N-1)/2 \end{aligned}$$

show that y_r may be expressed in the form

$$y_r = a_0 Q_0 + \sum_{s=1}^{(N-1)/2} a_s \cos\frac{2\pi rs}{N} Q_s + \sum_{s=1}^{(N-1)/2} a_s \sin\frac{2\pi rs}{N} Q_{N-s}$$

Determine the constants a_0, a_s to make the transformation from y_r to Q_s a 'normal' one. Hence show that the relationship between y_r and a set of normal coordinates Q_s having this property is given by

$$y_r = \frac{1}{N^{1/2}}Q_0 + \frac{2^{1/2}}{N^{1/2}}\sum_{s=1}^{(N-1)/2} Q_s \cos\frac{2\pi rs}{N} + \frac{2^{1/2}}{N^{1/2}}\sum_{s=1}^{(N-1)/2} Q_{N-s}\sin\frac{2\pi rs}{N}$$

(Use the relationships

$$\left.\sum_{s=1}^{N}\cos^2\frac{2\pi rs}{N} = \sum_{s=1}^{N}\sin^2\frac{2\pi rs}{N} = \frac{N}{2}\right)$$

Verify that the transformation is also orthogonal and hence show that the normal coordinates Q_s may be expressed in terms of the displacements y_r by the equations

$$Q_0 = \frac{1}{N^{1/2}}\sum_{r=1}^{N} y_r$$

$$Q_s = \frac{2^{1/2}}{N^{1/2}}\sum_{r=1}^{N}\cos\frac{2\pi rs}{N}y_r \qquad 1 \leqslant s \leqslant \frac{N-1}{2}$$

$$= \frac{2^{1/2}}{N^{1/2}}\sum_{r=1}^{N}\sin\frac{2\pi r(N-s)}{N}y_r \qquad \frac{N+1}{2} \leqslant s \leqslant N-1$$

(4) For $N = 3$ show that the normal coordinates are given by the transformation

	Q_1	Q_2	Q_3
y_1	$1/\sqrt{3}$	$-1/\sqrt{6}$	$1/\sqrt{2}$
y_2	$1/\sqrt{3}$	$-1/\sqrt{6}$	$-1/\sqrt{2}$
y_3	$1/\sqrt{3}$	$\sqrt{2}/\sqrt{3}$	0

Check this by solving the problem by the elementary method of § 3.1.

(5) For example (4) show that the kinetic energy T and the potential energy V are given by

$$T = \tfrac{1}{2}m[\dot{Q}_1^2 + \dot{Q}_2^2 + \dot{Q}_3^2]$$

$$V = \tfrac{1}{2}\frac{F}{a}[Q_2^2 + Q_3^2]$$

(6) For the general case verify that

$$T = \tfrac{1}{2}m[\dot{Q}_0^2 + \dot{Q}_1^2 \ldots \dot{Q}_{N-1}^2]$$

$$V = \tfrac{1}{2}m\sum_{s=1}^{(N-1)/2}\omega_s^2[Q_s^2 + Q_{N-s}^2]$$

3.4 Quantization of the lattice vibrations

So far, we have treated the motions of the particles of the crystal according to classical mechanics; for atomic crystals this is inadequate except for certain limiting conditions. To treat the problem of the linear crystal by means of quantum mechanics we should form the N-dimensional Schrödinger equation in terms of the displacements y_r (assuming we restrict the motion, for the moment to transverse motion in a plane); this leads to a series of coupled second-order differential equations which are difficult to solve. The situation is greatly simplified if we use instead the normal coordinates Q_s, and this was indeed the main reason for introducing them. In terms of the normal coordinates the kinetic and potential energies are such that the Schrödinger equation may be separated into N equations, each involving only one coordinate Q_s. This is just equivalent to the statement that the crystal may be treated as N independent harmonic oscillators. The wave equation for a single oscillator of frequency $\omega/2\pi$ is*

$$\frac{\partial^2 \psi}{\partial x^2}+\frac{8\pi^2 m}{\boldsymbol{h}^2}[E-\tfrac{1}{2}m\omega^2 x^2]\psi = 0 \tag{50}$$

For the crystal the wave equation is then

$$\sum_{s=1}^{N}\left\{\frac{\partial^2 \psi}{\partial Q_s^2}+\frac{8\pi^2 m}{\boldsymbol{h}^2}[E-\tfrac{1}{2}m\omega_s^2 Q_s^2]\psi\right\} = 0 \tag{51}$$

This is equivalent to N equations of the form

$$\frac{\partial^2 \psi_s}{\partial Q_s^2}+\frac{8\pi^2 m}{\boldsymbol{h}^2}[E_s-\tfrac{1}{2}m\omega_s^2 Q_s^2]\psi_s = 0 \tag{52}$$

with

$$E = \sum_{s=1}^{N} E_s \tag{53}$$

and

$$\psi = \prod_{s=1}^{N} \psi_s(Q_s) \tag{54}$$

The solutions of equation (52) may be expressed in terms of the well known Hermitian Polynomials*, but we shall not require the actual expression for ψ_s for the moment. The allowed energy values for equation (52) are*

$$E_s = (n_s+\tfrac{1}{2})\,\boldsymbol{h}\nu_s \tag{55}$$

where n_s is an integer, and $\nu_s = \omega_s/2\pi$.

The energy $\boldsymbol{h}\nu_s$ may be regarded as a quantum of energy associated with a frequency ν_s. The solution $\psi_s(Q_s)$, corresponding to E_s, represents a state having n_s quanta of energy in the sth normal mode, in excess of the energy of the lowest state. The total energy of the system is obtained by summing over all the normal modes and we have

$$E = \boldsymbol{h}\sum_{s=1}^{N} n_s\nu_s+\tfrac{1}{2}\boldsymbol{h}\sum_{s=1}^{N}\nu_s \tag{56}$$

* See any textbook on elementary quantum mechanics.

The lowest allowed energy E_0 is obtained when $n_1 = n_2 = \ldots n_N = 0$ and is given by

$$E_0 = \tfrac{1}{2}\boldsymbol{h} \sum_{s=1}^{N} \nu_s \tag{57}$$

This energy is called the zero-point energy of the crystal, and corresponds to the lowest vibrational energy state, which the system would have in equilibrium at the absolute zero of temperature. This is a well-known feature in which quantum mechanics differs from classical mechanics for which the zero-point energy is equal to zero.

Inserting the value of ω_s from equation (33) we have, writing n to represent the set of integers $n_1, n_2, \ldots n_N$,

$$E_n = \boldsymbol{h}\nu_c \sum_{s=1}^{N} n_s \sin\left[\frac{s\pi}{2(N+1)}\right] + E_0 \tag{58}$$

with

$$E_0 = \frac{\boldsymbol{h}\nu_c}{2} \sum_{s=1}^{N} \sin\left[\frac{s\pi}{2(N+1)}\right] \tag{59}$$

ν_c being the cut-off frequency of the vibrational spectrum. The sum (59) may be evaluated to give

$$E_0 = \frac{\boldsymbol{h}\nu_c}{4} \frac{\cos\left[\dfrac{\pi}{4(N+1)}\right] - \cos\left[\dfrac{(2N+1)\pi}{4(N+1)}\right]}{\sin\left[\dfrac{\pi}{4(N+1)}\right]}$$

When N is large we have simply

$$E_0 = N\boldsymbol{h}\nu_c/\pi \tag{60}$$

When we no longer restrict the motion of the particles to transverse motion in a plane we have to add the other transverse modes and the longitudinal modes. For an isotropic crystal we have then

$$E_n = \boldsymbol{h}\nu_{ct} \sum_{s=1}^{N} n_s \sin\left[\frac{s\pi}{2(N+1)}\right] + \boldsymbol{h}\nu_{cl} \sum_{s=1}^{N} m_s \sin\left[\frac{s\pi}{2(N+1)}\right] + E_0 \tag{61}$$

where

$$E_0 = \boldsymbol{h}(\nu_{ct} + \tfrac{1}{2}\nu_{cl}) \sum_{s=1}^{N} \sin\left[\frac{s\pi}{2(N+1)}\right] \tag{62}$$

and n_s, m_s are integers; ν_{ct} is the high-frequency cut-off for transverse oscillations and ν_{cl} for longitudinal oscillations. For large values of N we have

$$E_0 = N\boldsymbol{h}(2\nu_{ct} + \nu_{cl})/\pi \tag{63}$$

3.4.1 Quantization of travelling waves

An exactly similar analysis may be carried out using the periodic boundary conditions; the only change is that the appropriate form for ω_s must be inserted. When N is large this makes no appreciable difference to the value of the total energy.

In terms of the normal coordinates Q_s for periodic boundary conditions introduced in the exercises at the end of § 3.3, the particle displacements are expressed, in general, as a sum over all the normal modes. Now suppose that only one of the normal coordinates Q_s is different from zero. If $1 \leqslant s \leqslant (N-1)/2$ (N odd), we have the displacements y_r given by

$$y_r = 2^{1/2} N^{-1/2} Q_s \cos k_s ra \tag{64}$$

where $k_s = 2\pi s/Na$. Q_s expressed as a function of the time has the form $A_s \mathrm{e}^{-i\omega_s t}$, A_s being a real constant and $\omega_s/2\pi$ the frequency associated with k_s. If we derive the quantities y_r from a 'descriptive' wave given by the wave function $y_s(x,t)$ we have

$$y_s(x,t) = A_s' \cos k_s x \,\mathrm{e}^{-i\omega_s t} \tag{65}$$

where $A_s' = 2^{1/2} N^{-1/2} A_s$; this is clearly a standing wave. Similarly if all the quantities except one in the range $(N+1)/2 \leqslant s \leqslant N-1$ are zero we have

$$y_r = 2^{1/2} N^{-1/2} Q_{N-s} \sin k_s ra \tag{66}$$

where $Q_{N-s} = B_s \mathrm{e}^{-i\omega_s t}$. This again may be derived from a wave function of the form

$$y_s'(x,t) = B_s' \sin k_s x \,\mathrm{e}^{-i\omega_s t} \tag{67}$$

where $B_s' = 2^{1/2} N^{-1/2} B_s$; this is also a standing wave. Finally when all the quantities Q_s are zero except that for $s = 0$ we have

$$y_r = N^{-1/2} Q_0 \tag{68}$$

This mode simply gives the position of the centre of gravity of the system, and apart from noting that it counts in enumerating the modes we may omit it from our discussion.

The modes described by y_s and y_s' are clearly degenerate; in enumerating the quantities k_s in this description we need not be concerned with negative values since these give no new types of motion; this is characteristic of standing waves, as we have already seen. From the two degenerate modes we may form others by linear combination and, in particular, may form travelling waves. If we take $A_s = B_s$ and write

$$y_s'' = y_s + i y_s' \tag{69}$$

We have

$$y_s''(x,t) = A_s' \mathrm{e}^{i(k_s x - \omega_s t)} \tag{70}$$

which is a travelling wave having a frequency $\omega_s/2\pi$ and propagation constant k_s. We may now also have negative values of k_s and these give different modes. Instead, therefore, of counting k_s from $s = 1$ to $s = (N-1)/2$, each value being taken twice (once for y_s and once for y_s') we may count k_s from $s = -(N-1)/2$ to $(N-1)/2$, giving N values again when we also include $s = 0$.

If we now express the particle displacements arising from the wave function y_s'' in terms of the normal coordinates Q_s, Q_{N-s}, we at once run into a difficulty. For the particle displacements we have

$$y_r = A_s' \mathrm{e}^{ik_s ra - i\omega_s t} \tag{71}$$

There are now two normal coordinates Q_s and Q_{N-s} corresponding to frequency ν_s and in terms of these we may express the displacements y_r in the form (cf. equations (64) and (66))

$$y_r = 2^{1/2} N^{-1/2}[(Q_s \cos k_s ra + Q_{N-s} \sin k_s ra)] \tag{72}$$

This may also be written in the form

$$y_r = 2^{-1/2} N^{-1/2}[(Q_s - iQ_{N-s}) \mathrm{e}^{ik_s ra} + (Q_s + iQ_{N-s}) \mathrm{e}^{-ik_s ra}] \tag{73}$$

The first of the terms in the above expression represents a travelling wave to the right like (71) and the second a travelling wave to the left. If therefore we were to introduce two new quantities

$$\left.\begin{aligned} q_s &= (Q_s - iQ_{N-s})/\sqrt{2} \\ q_{-s} &= (Q_s + iQ_{N-s})/\sqrt{2} \end{aligned}\right\} \tag{74}$$

these would represent travelling waves in the sense that if all the quantities q_s except one were zero, and all the quantities q_{-s} were also zero, we should have a travelling wave like that representing the motion given by (71). The difficulty is that the quantities q_s as described above appear to be complex, and as the coordinates Q_s are real, we could not, at first sight, express the real coordinates y_r in terms of q_s; also if $q_s \neq 0$ then $q_{-s} \neq 0$. If however, we express the equations (74) rather differently and write

$$\left.\begin{aligned} q_s &= (Q_s + \omega_s^{-1} \dot{Q}_{N-s})/\sqrt{2} \\ q_{-s} &= (Q_s - \omega_s^{-1} \dot{Q}_{N-s})/\sqrt{2} \end{aligned}\right\} \tag{75}$$

(the factors $\sqrt{2}$ are just convenient normalising constants), the quantities q_s are real if Q_s and $\dot{Q}_{N-s}$ are real. In fact when we write $Q_s = A\,\mathrm{e}^{-i\omega t}$ we introduce a complex quantity, but it must be remembered throughout that this is really just a shorthand notation and we *mean the real part*; the transformation (75) takes care of this automatically, and we may also have $q_{-s} = 0$ when $q_s \neq 0$.

The quantities q_s are a combination of what in classical mechanics are called generalised coordinates and velocities; the velocities could also be expressed in terms of momenta $p_s = m\dot{Q}_s$. Transformations involving both coordinates and momenta are known as contact transformations. We need not pursue the matter here other than to note that we may express the total energy E of the system in the form

$$E = T + V = \tfrac{1}{2}m\dot{Q}_0^2 + \tfrac{1}{2}m \sum_{s=1}^{(N-1)/2} (\dot{Q}_s^2 + \dot{Q}_{N-s}^2 + \omega_s^2 Q_s^2 + \omega_s^2 Q_{N-s}^2) \tag{76}$$

Now we have

$$\left.\begin{aligned} \dot{q}_s &= (\dot{Q}_s + \omega_s^{-1}\ddot{Q}_{N-s})/\sqrt{2} \\ &= (\dot{Q}_s - \omega_s Q_{N-s})/\sqrt{2} \end{aligned}\right\} \tag{77}$$

since Q_{N-s} satisfies the equation

$$\ddot{Q}_{N-s} + \omega^2 Q_{N-s} = 0$$

Similarly

$$\dot{q}_{-s} = (\dot{Q}_s + \omega_s Q_{N-s})/\sqrt{2} \tag{78}$$

Substituting in the expression for the total energy we obtain

$$E = \tfrac{1}{2}m \sum_{s=-(N-1)/2}^{(N-1)/2} (\dot{q}_s^2 + \omega_s^2 q_s^2) \tag{79}$$

if we write $Q_0 = q_0$. Thus the quantities q_s may also be taken as describing the motion of N independent simple harmonic oscillators, and are thus a set of normal coordinates. We now have degeneracy, except for the trivial case $s = 0$, between the modes with $+s$ and $-s$. The quantum mechanical wave equation with its associated energy levels, is just the same in terms of the coordinates q_s as we had before in terms of the coordinates Q_s. It simply remains to show formally that the quantities q_s represent travelling waves. Substituting from equations (74) in equation (72) and summing over all the values of s we have the following expressions for the coordinates y_r in terms of the quantities q_s,

$$y_r = N^{-1/2} \sum_{s=-(N-1)/2}^{(N-1)/2} q_s \, e^{ik_s ra} \tag{80}$$

If all the quantities q_s except one are zero and this has the form

$$q_s = A_s e^{-i\omega_s t}$$

we have, writing $A_s'' = N^{-1/2} A_s$,

$$y_r = A_s'' \, e^{i(k_s ra - \omega_s t)} \tag{81}$$

which is derived from a travelling wave of the form

$$y_s''(x, t) = A_s'' \, e^{i(k_s x - \omega_s t)}$$

Finally if we wish to express the new normal coordinates q_s in terms of the displacements y_r we have, on substituting for Q_s, Q_{N-s} from § 3.3,

$$q_s = N^{-1/2} \sum_{r=1}^{N} y_r e^{-ikra} \tag{82}$$

The elements of the transformation matrix from q_s to y_r are therefore the same as for y_r to q_s but with i replaced by $-i$. Thus if we write equation (80) in the form

$$y_r = \sum a_{rs} q_s$$

we have

$$q_s = \sum a_{rs}^{\star} y_r$$

where $a_{rs}^{\star}$ is the complex conjugate of a_{rs}.

Exercise (1) Show that for the above transformation we have

$$\left.\begin{aligned}\sum_r a_{rs} a_{rt}^{\star} &= \delta_{st} \\ \sum_s a_{rs} a_{us}^{\star} &= \delta_{ru}\end{aligned}\right\} \tag{83}$$

This is a generalisation of the normal orthogonal transformation when the elements of the transformation matrix are complex.

(2) Show that similar equations may be found relating the quantities y_r and q_s when N is even.

Since the form of the energy function E in terms of the coordinates q_s is the same as for the coordinates Q_s the energy levels obtained by quantization are exactly the same as we had before. The energy associated with a travelling wave with propagation constant k_s and corresponding frequency $\nu_s = \omega_s/2\pi$ is then just $(n_s + \frac{1}{2})h\nu_s$ where n_s is an integer. For many purposes, as we shall see later, the travelling waves are more convenient to use, and the quanta of energy $h\nu_s$ associated with these waves are known as phonons. When a state corresponding to any one of the travelling waves has its energy increased by one quantum, we say that a phonon is emitted; when the energy is decreased by one quantum a phonon is said to be absorbed. This is analogous to the emission and absorption of photons or light quanta (see § 3.10).

3.5. Distribution of energy among the normal modes

We may now calculate for any given temperature the relative probability of excitation of the various normal modes of the crystal and thus obtain the mean energy of each mode. From this we may then obtain the heat capacity of the crystal corresponding to the lattice vibrations. Each normal mode may be regarded as a single independent dynamical system to which Boltzmann's law may be applied (see § 2.6). Thus if E_{sn} is the energy of the nth excited state of the sth mode, the probability of its being excited is proportional to $\exp[-E_{sn}/kT]$. The actual probability of excitation P_{sn} is then given by

$$P_{sn} = \mathrm{e}^{-E_{sn}/kT} \Big/ \sum_{n=0}^{\infty} \mathrm{e}^{-E_{sn}/kT} \tag{84}$$

where the sum in the denominator is introduced so as to ensure that

$$\sum_{n=0}^{\infty} P_{sn} = 1$$

The expression (84) may clearly also be written in the form

$$P_{sn} = \mathrm{e}^{-(E_{sn}-E_{s0})/kT} \Big/ \sum_{n=0}^{\infty} \mathrm{e}^{-(E_{sn}-E_{s0})/kT} \tag{85}$$

Inserting the value of E_{sn} from equation (55) we have,

$$P_{sn} = \mathrm{e}^{-nh\nu_s/kT} \Big/ \sum_{n=0}^{\infty} \mathrm{e}^{-nh\nu_s/kT} \tag{86}$$

Now if $|x| < 1$ we have

$$y(x) = \sum_{n=0}^{\infty} x^n = 1/(1-x)$$

Hence
$$P_{sn} = e^{-nh\nu_s/kT}(1-e^{-h\nu_s/kT}) \tag{87}$$

The mean energy $\bar{E}_s$ associated with the sth mode given by

$$\left.\begin{aligned}\bar{E}_s &= \sum_{n=0}^{\infty} E_{sn} P_{sn} \\ &= E_0 + (1-e^{-h\nu_s/kT})\, h\nu_s \sum_{n=1}^{\infty} n\, e^{-nh\nu_s/kT}\end{aligned}\right\} \tag{88}$$

Now we have

$$\sum_{n=0}^{\infty} nx^n = x\frac{dy}{dx} = x/(1-x)^2$$

Hence we have

$$\bar{E}_s - E_0 = \frac{h\nu_s}{e^{h\nu_s/kT}-1} = h\nu_s B(h\nu_s) \tag{89}$$

The function

$$B(E/kT) = \frac{1}{e^{E/kT}-1}$$

is the Einstein–Bose function for the occupational probability of a state of energy E by one of an assembly of identical particles obeying Einstein–Bose statistics (see § 2.6 equation (59)). The significance of this result is further discussed in § 3.10.

The total vibrational energy of the crystal is obtained by summing (89) over values of s corresponding to the N normal modes. We have for the mean energy W in excess of the zero-point energy

$$W = \sum_{s=1}^{N} \frac{h\nu_s}{e^{h\nu_s/kT}-1} \tag{90}$$

when N is very large the sum may be replaced by an integral. Thus we have

$$W = \int_0^{\nu_c} \frac{h\nu n(\nu)\, d\nu}{e^{h\nu/kT}-1} \tag{91}$$

where $n(\nu)\,d\nu$ is the number of normal modes having frequencies between ν and $\nu + d\nu$, and ν_c is the highest allowed value of ν_s. From equation (33) we have, writing $\nu_c = \omega_c/2\pi$,

$$\nu_s = \nu_c \sin\frac{s\pi}{2(N+1)} \tag{92}$$

For large values of N the interval between successive allowed values of ν is obtained by differentiation with respect to s and is given by

$$\left.\begin{aligned} \Delta\nu &= \frac{\nu_c \pi}{2(N+1)} \cos \frac{s\pi}{2(N+1)} \\ &= \frac{\pi\nu_c}{2(N+1)}\left[1-\left(\frac{\nu}{\nu_c}\right)^2\right]^{1/2} \end{aligned}\right\} \tag{93}$$

In an interval $d\nu$ we therefore have $d\nu/\Delta\nu$ allowed states. Thus we have

$$n(\nu)\,d\nu = \frac{2(N+1)}{\pi\nu_c}\left[1-\left(\frac{\nu}{\nu_c}\right)^2\right]^{-1/2} d\nu \tag{94}$$

Inserting this value in (91) we have for large values of N

$$W = \frac{2hN}{\pi}\int_0^{\nu_c} \frac{\nu\,d\nu}{[e^{h\nu/kT}-1][\nu_c^2-\nu^2]^{1/2}} \tag{95}$$

The heat capacity $\mathscr{C}$ of the crystal is given by

$$\mathscr{C} = \frac{dW}{dT} \tag{96}$$

The expression (95) is very similar to one which we shall later obtain for a crystalline solid. Unfortunately the integral (95) cannot be evaluated by simple analytical methods, and we shall later discuss various approximations which have been made. For large values of T the expression (95) reduces to

$$W = \frac{2N}{\pi}\int_0^{\nu_c} \frac{kT\,d\nu}{[\nu_c^2-\nu^2]^{1/2}} = NkT \tag{97}$$

This is just the classical expression for the mean energy content of a dynamical system having N degrees of freedom. (N.B. the next term in the approximation gives $W = E_0$ and cancels the zero-point energy.)

So far, we have restricted the motion to transverse oscillations in a plane. When we remove this restriction and also include longitudinal oscillations we shall have to add to (95) the contributions from these. If the transverse oscillations in two planes at right angles are degenerate we shall have

$$W = \frac{4hN}{\pi}\int_0^{\nu_{ct}} \frac{\nu\,d\nu}{[e^{h\nu/kT}-1][\nu_{ct}^2-\nu^2]^{1/2}} + \frac{2hN}{\pi}\int_0^{\nu_{cl}} \frac{\nu\,d\nu}{[e^{h\nu/kT}-1][\nu_{cl}^2-\nu^2]^{1/2}} \tag{98}$$

for large values of T this reduces to

$$W = 3NkT \tag{99}$$

which again is the well-known classical value.

3.6 Linear crystal with two types of particle

The crystal which we have considered, so far, has all its particles of the same mass; we now deal with a linear crystal containing two types of particle. These we shall suppose to be spaced at equal distances apart and alternate particles to have masses M and m (see Fig. 3.9). We shall suppose that $M > m$. The force system we shall take to be as before and shall consider first transverse oscillations in a plane.

We shall use the coordinate x to describe the position along the crystal, as before; the particles of mass M we shall take to be at the points

$$x = (2r+1)a \qquad r = 0, 1, 2, \ldots (N-1)$$

in equilibrium, and those of mass m at the points

$$x = 2ra \qquad r = 0, 1, 2, \ldots N$$

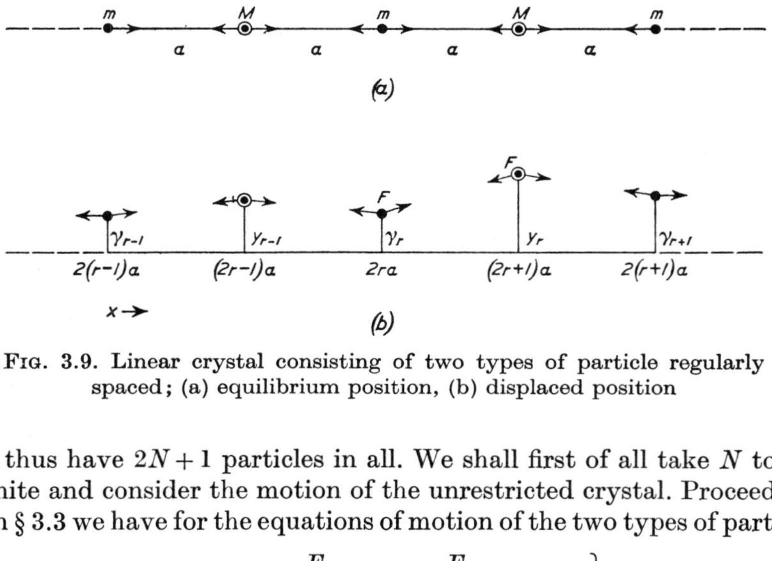

FIG. 3.9. Linear crystal consisting of two types of particle regularly spaced; (a) equilibrium position, (b) displaced position

We thus have $2N+1$ particles in all. We shall first of all take N to be infinite and consider the motion of the unrestricted crystal. Proceeding as in § 3.3 we have for the equations of motion of the two types of particle

$$\left.\begin{aligned} M\ddot{y}_r &= -\frac{F}{a}(y_r - z_r) - \frac{F}{a}(y_r - z_{r+1}) \\ m\ddot{z}_r &= -\frac{F}{a}(z_r - y_{r-1}) - \frac{F}{a}(z_r - y_r) \end{aligned}\right\} \tag{100}$$

where y_r, z_r, represent the displacements of the particles of mass M and m, respectively. If we write $\alpha = F/Ma$, $\beta = F/ma$ these become

$$\left.\begin{aligned} \ddot{y}_r &= -\alpha(2y_r - z_r - z_{r+1}) \\ \ddot{z}_r &= -\beta(2z_r - y_r - y_{r-1}) \end{aligned}\right\} \tag{101}$$

If we try to solve these equations by proceeding as in § 3.3 we find that we cannot do this by means of a single wave function $y(x, t)$ which gives the displacement of all the particles on insertion of the appropriate values of x. We find, however, that by using *two* separate wave functions, $y(x,t)$ and $z(x,t)$ to describe, respectively, the motion of the particles of mass M and those of mass m, a solution may be found which is very similar in form to that found for the simpler lattice already discussed. We assume that the functions $y(x,t)$ and $z(x,t)$ may be expressed in the form

$$\left.\begin{aligned} y(x,t) &= A\,e^{ikx-i\omega t} \\ z(x,t) &= B\,e^{ik'x-i\omega t} \end{aligned}\right\} \tag{102}$$

Clearly we take the *same* value for ω in both, since we are looking for solutions periodic in the time, but for the moment we may suppose k and k' to be different. We now substitute the expressions (102) in the equations (101), and we obtain

$$\omega^2 A\,e^{ik(2r+1)a} = \alpha(2A\,e^{ik(2r+1)a} - B\,e^{2ik'ra} - B\,e^{2ik'(r+1)a}) \tag{103}$$

$$\omega^2 B\,e^{2ikra} = \beta(2B\,e^{2ik'ra} - B\,e^{ik(2r+1)a} - B\,e^{ik(2r-1)a}) \tag{104}$$

It will at once be clear that we cannot reduce these to simple equations *independent* of r unless $k = k'$. When this is so, however, we obtain

$$\left.\begin{aligned} \omega^2 A &= \alpha(2A - B\,e^{-ika} - B\,e^{ika}) \\ \omega^2 B &= \beta(2B - A\,e^{-ika} - A\,e^{ika}) \end{aligned}\right\} \tag{105}$$

These equations may be further simplified and written in the form

$$\left.\begin{aligned} \omega^2 A &= 2\alpha(A - B\cos ka) \\ \omega^2 B &= 2\beta(B - A\cos ka) \end{aligned}\right\} \tag{106}$$

We thus have two homogeneous equations from which we may obtain k as a function of ω and also the ratio A/B. The equations (106) have a non-trivial solution only if

$$\begin{vmatrix} \omega^2 - 2\alpha & 2\alpha\cos ka \\ 2\beta\cos ka & \omega^2 - 2\beta \end{vmatrix} = 0 \tag{107}$$

or

$$\omega^4 - 2(\alpha+\beta)\,\omega^2 + 4\alpha\beta\sin^2 ka = 0 \tag{108}$$

We have now two possible solutions for ω in terms of k,

$$\omega_1^2 = (\alpha+\beta) - \{(\alpha+\beta)^2 - 4\alpha\beta\sin^2 ka\}^{1/2} \tag{109}$$

and

$$\omega_2^2 = (\alpha+\beta) + \{(\alpha+\beta)^2 - 4\alpha\beta\sin^2 ka\}^{1/2} \tag{110}$$

Let us first consider the solution giving ω_1; when k is small we have

$$\omega_1^2 \simeq \frac{2\alpha\beta k^2 a^2}{(\alpha+\beta)} \tag{111}$$

Thus the phase velocity $\mathscr{V}_1 = \omega_1/k$ is given by

$$\mathscr{V}_1 = (Fa/M^\star)^{1/2} \tag{112}$$

where

$$M^\star = \tfrac{1}{2}(M+m)$$

This is the same as for a crystal with one type of particle whose mass is equal to the arithmetic mean of the masses of the two particles. For small values of k (long waves) this solution behaves like that for a continuous string of line density ρ under tension F, where ρ has the value

$$\rho = (M+m)/2a \tag{113}$$

and is thus equal to the average line density of the crystal. The group velocity $\mathscr{U}_1$ for small values of k is equal to $\mathscr{V}_1$ and there is no dispersion for long waves. In view of this behaviour the waves occurring in the solution corresponding to ω_1 are generally known as acoustic waves and the ω_1/k curve as the acoustic branch of the dispersion curve; the corresponding normal modes of motion of the particles are known as the acoustic modes. As k increases, ω_1 reaches a maximum

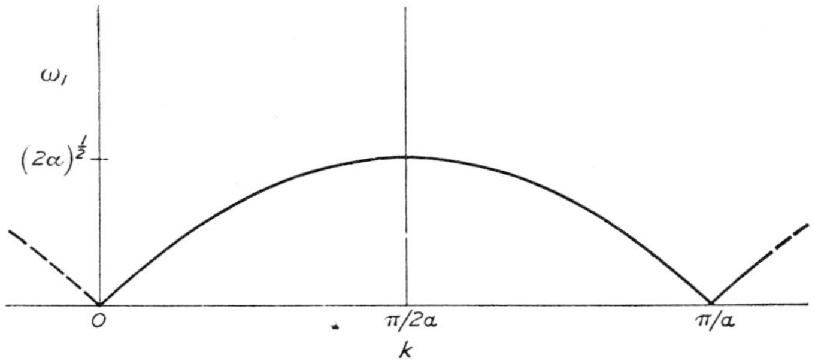

FIG. 3.10. The frequency ω_1 corresponding to the acoustic branch of the dispersion curve for a linear crystal with two types of equally spaced particles ($\alpha = F/Ma$)

value $2^{1/2}\alpha^{1/2}$ when $k = \pi/2a$, provided $M > m$. This is equal to the high-frequency cut-off for a crystal consisting of particles of mass M spaced $2a$ apart. The form of the function giving ω_1 in terms of k is shown in Fig. 3.10. It will be seen that it is periodic and that the period is now π/a, as compared with $2\pi/a$ for a crystal consisting of equal particles spaced a distance a apart. For the two-particle crystal $2a$ is the repetition period of the crystal. We must now consider the second type of solution given by equation (110) for ω_2. For small values of k we now have

$$\omega_2^2 \simeq 2(\alpha+\beta) \tag{114}$$

so that ω_2 is independent of k. We thus see that for small values of k the phase velocity $\mathscr{V}_2 = \omega_2/k$ tends to infinity and the group velocity $d\omega_2/dk$ tends to zero as k tends to zero. As k increases, ω_2 begins to decrease, reaching a minimum value of $2^{1/2}\beta^{1/2}$ when $k = \pi/2a$; this is equal to the high-frequency cut-off for a crystal consisting of equal particles of mass m spaced $2a$ apart. We note that if $M > m$ this frequency is higher than the high-frequency cut-off for ω_1 so that the curves for ω_1 and ω_2 do not cross. We also note that when $k = \pi/2a$ the group velocities for both the solutions ω_1 and ω_2 are zero. The form of the ω_2/k curve is shown in Fig. 3.11; for reasons which will appear later the normal modes corresponding to the solution ω_2 are called optical modes, and this branch of

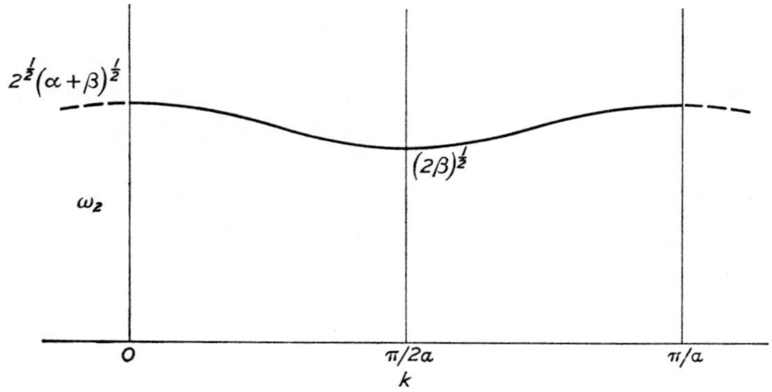

FIG. 3.11. The frequency ω_2 corresponding to the optical branch of the dispersion curve for a linear crystal with two types of equally spaced particles ($\alpha = F/Ma$, $\beta = F/ma$, $M > m$)

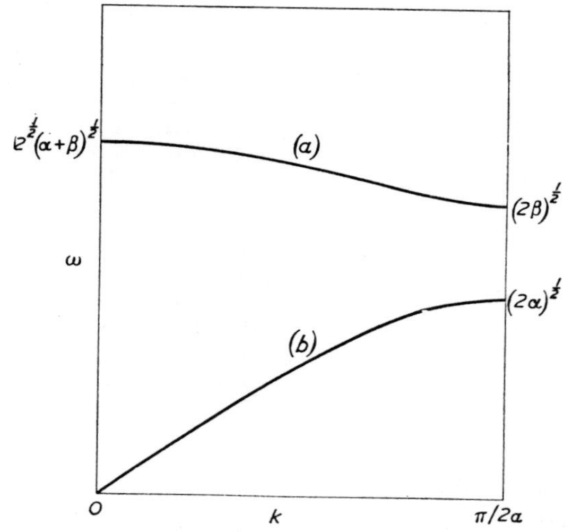

FIG. 3.12. Dispersion curves for a crystal with two types of equally spaced particles in the reduced representation; (a) optical branch, (b) acoustic branch ($\alpha = F/Ma$, $\beta = F/ma$, $M > m$)

the dispersion curve is called the optical branch. As before, because of the periodicity of the ω/k curves, we may use a reduced representation, all the useful information being contained in the interval $0 \leqslant k \leqslant \pi/2a$. This is shown for the full dispersion curve (both branches) in Fig. 3.12.

It is interesting to see how the dispersion curves for a crystal having two types of particle go over into that for one type as $M \to m$; clearly we must use a representation for the interval $0 \leqslant k \leqslant \pi/a$ for this (see Fig. 3.13). The appropriate curves for crystals having equally spaced particles of equal masses m and M are also shown in Fig. 3.13. It will be seen that as $M \to m$ the acoustic branch tends to the curve for one type of particle for $0 \leqslant k \leqslant \pi/2a$ and the optical branch to this curve for $\pi/2a \leqslant k \leqslant \pi/a$. The gap between the acoustic and optical branches at $k = \pi/2a$ tends to zero as $M \to m$.

As we shall see, it is a general result that the ω/k curves for a periodic lattice are periodic with period π/d where d is the periodic distance of the

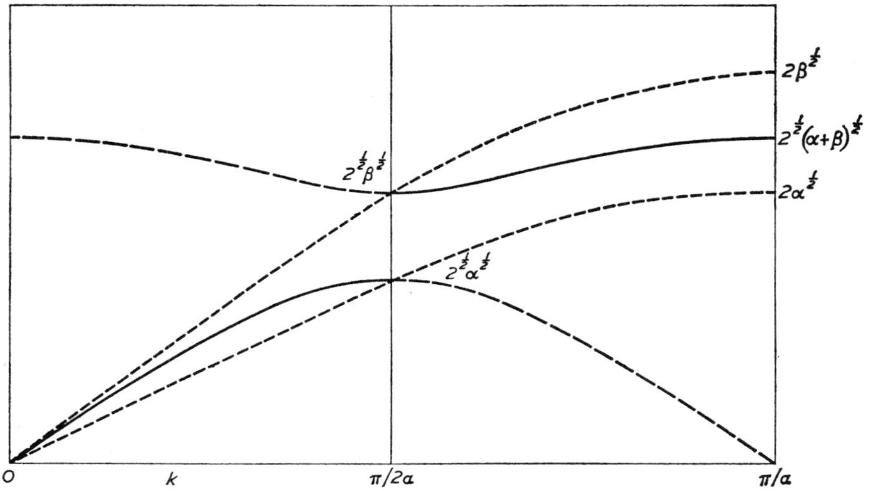

FIG. 3.13. Transition to a crystal with one type of particle ($\alpha = F/Ma$, $\beta = F/ma$, $M > m$)

crystal lattice. The effect of the unequal masses is to reduce the periodicity of the ω/k curves from π/a to $\pi/2a$ and to introduce a break at $k = \pi/2a$. This leads naturally to a representation in the reduced form using two branches. The representation of the ω/k curves for the lattice with two types of particle shown by the full line in Fig. 3.13 is an example of one which is sometimes used instead of the reduced representation. In it the lowest branch is shown for $0 \leqslant k \leqslant \pi/d$, the next branch for $\pi/d \leqslant k \leqslant 2\pi/d$, and so on. This representation is less compact than the reduced representation but for some purposes is more instructive; it is known as the 'expanded' representation. It is the ω/k curve in the expanded representation that tends to the single curve when we have one branch only.

We must now return to examine the actual form of the motions of the particles corresponding to the various types of solution. To complete the solution we have still to determine the ratio of the two constants A and B occurring in equations (106); we have

$$A_1/B_1 = -2\alpha \cos ka/(\omega_1^2 - 2\alpha) \tag{115}$$

$$A_2/B_2 = -2\alpha \cos ka/(\omega_2^2 - 2\alpha) \tag{116}$$

ω_1 and ω_2 being given as functions of k by equations (109) and (110). This completes the formal solution, the absolute value of one of the constants A or B being determined by the initial conditions. So far, the value of k is unrestricted but will be determined by the boundary conditions imposed on the crystal as before.

Let us now examine the form of the solutions corresponding to various limiting values of k.

(a) *Acoustic modes,* $k \to 0$ $(\lambda \to \infty)$

On substituting in equation (115) the value of ω_1 given by equation (111) for small values of k, we obtain when $k \to 0$, $A = B$. Thus we have

$$y_r = z_r = A\,e^{i(kx_r - \omega t)} \tag{117}$$

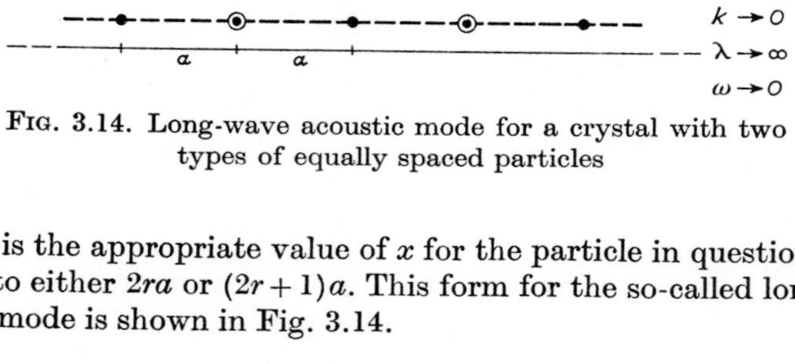

FIG. 3.14. Long-wave acoustic mode for a crystal with two types of equally spaced particles

where x_r is the appropriate value of x for the particle in question, i.e. x_r is equal to either $2ra$ or $(2r+1)a$. This form for the so-called long-wave acoustic mode is shown in Fig. 3.14.

(b) *Modes for* $k \to \pi/2a$ $(\lambda \to 4a)$

On substituting in equation (115) $\omega_1 = (2\alpha)^{1/2}$ we find that as $k \to \pi/2a$ $B_1 \to 0$. For this mode we have motion of the particles of mass M only, those of mass m remaining at rest. Thus we have

$$\left.\begin{aligned} y(x,t) &= A\,e^{2\pi ix/a - i\omega t} \\ z(x,t) &= 0 \end{aligned}\right\} \tag{118}$$

giving

$$\left.\begin{aligned} y_r &= (-1)^r A\,e^{-i\omega t} \\ z_r &= 0 \end{aligned}\right\} \tag{119}$$

The displacements of the particles of mass M are therefore at all times equal in magnitude but alternate particles differ by π in phase. This mode is therefore just the π-mode for a crystal consisting of particles of mass M spaced $2a$ apart, and is shown in Fig. 3.15.

Similarly for the optical branch, when $k \to \pi/2a$, $A_2 \to 0$ and we have a motion in which all the particles of mass M are at rest while those of mass m move in the π-mode for a crystal consisting of particles of mass

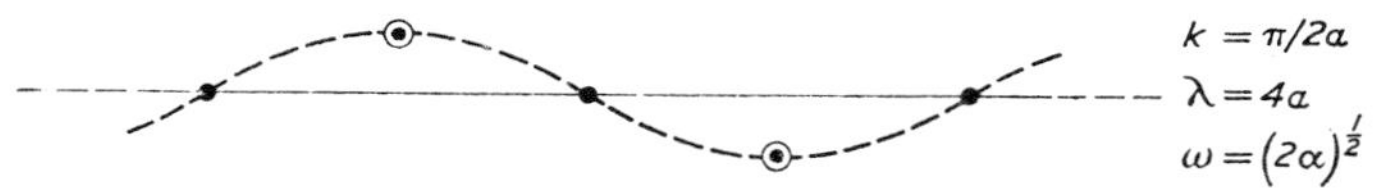

FIG. 3.15. Mode corresponding to $k = \pi/2a$ in which particles of mass m are at rest

m spaced $2a$ apart (Fig. 3.16). The physical distinction between optical and acoustic modes is irrelevant for this case and has only significance for small values of k. We see, however, that one mode is derived from the optical branch of the dispersion curve and the other from the acoustical branch.

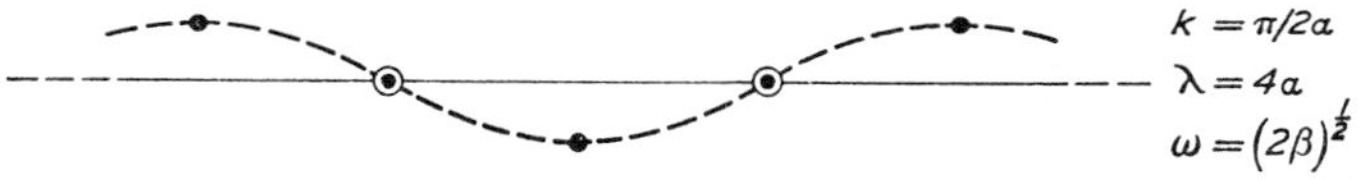

FIG. 3.16. Mode corresponding to $k = \pi/2a$ in which particles of mass M are at rest

(c) *Optical modes,* $k \to 0$ $(\lambda \to \infty)$

On substituting in equation (116) for ω_2 as given by equation (110) we have, when $k \to 0$, $MA_2 \to -mB_2$. Thus we have, when $k = 0$,

$$\left.\begin{aligned} y_r &= A\,e^{-i\omega_0 t} \\ z_r &= -(M/m)\,A\,e^{-i\omega_0 t} \end{aligned}\right\} \qquad (120)$$

where ω_0 is given by (114) and is the limiting frequency of the long-wave optical modes, i.e. ω_0 is given by

$$\omega_0^2 = \frac{2F}{a}\left(\frac{1}{M}+\frac{1}{m}\right) \qquad (121)$$

In this mode the particles of mass M all move together, as do those of mass m, there being a phase difference π between the two motions, i.e. the two types of particle move in opposite directions. The relative amplitudes of the two types of particle are such that the centre of gravity of the system remains at rest; the form of the motion is shown in Fig. 3.17; this looks physically like a π-mode and indeed goes over into a π-mode when $m \to M$, as will be seen from Fig. 3.13. We note that $k \to 0$ is equivalent to $k \to \pi/a$ for this case; the mode is, however, best described as a 'long-wave' mode since the waves corresponding to *each* type of particle become infinitely long as $k \to 0$. If the two types of particle are indistinguishable we may use only *one* type of wave to describe the motion of the particles and the wavelength tends to $2a$ for this type of motion, corresponding to $k \to \pi/a$; this is shown by the dotted curve in Fig. 3.17.

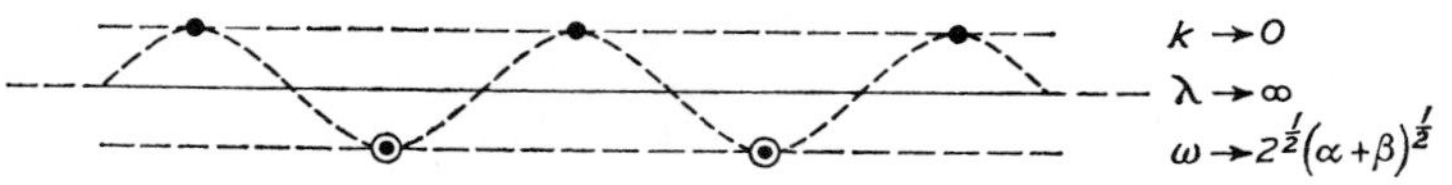

FIG. 3.17. Optical mode corresponding to $k=0$ ($\lambda=\infty$)

3.6.1 Excitation of the optical modes by electromagnetic waves

If we suppose the particles of the linear crystal discussed above to be charged they will be subject to external forces in the presence of electromagnetic waves. We shall assume that the particles of mass M carry a charge e and those of mass m a charge $-e$. In a constant electric field at right angles to the axis of the crystal the particles will be displaced in opposite directions in static equilibrium. If we have an even number of particles the total external force acting on the crystal will be zero.

Let us now consider what occurs when we have a plane electromagnetic wave propagated in the x-direction along the crystal. If the wavelength is much longer than a, the distance between adjacent particles, these will again be urged in opposite directions and we shall consider only this condition which generally holds for the interaction of infra-red radiation with the ions of an ionic crystal; for example if $\lambda = 100\,\mu$ and $a = 10^{-8}$ cm we have $\lambda/a = 10^6$. The problem of the interaction with a linear crystal has many features in common with that of the interaction with a three-dimensional crystal, as we shall see later, and is somewhat simpler to treat mathematically.

Clearly only the long-wave modes will interact strongly with the electromagnetic waves as otherwise the exciting force will average out; again, only the long-wave *optical* modes will interact strongly since for an acoustic mode, in which the displacement of neighbouring particles is in the same direction, the exciting force will be very small. Also, the interaction of a vibrational mode with the electromagnetic field will be small unless the phase velocity of the waves describing the mode is comparable with the phase velocity of the electromagnetic waves; only the long-wave optical modes have such a high phase velocity. We need therefore consider only the long-wave optical modes, and it is for this reason that they are so named, being those modes which interact strongly with electromagnetic waves in the form of light or infra-red radiation.

To study completely the interaction of the electromagnetic waves with the particles of the crystal we should first of all evaluate the forces F_r, F

acting on the particles; we should then write the equations of motion in the form (cf. equations (101)):

$$\left.\begin{aligned}\ddot{y}_r+\alpha(2y_r-z_r-z_{r+1}) &= F_r/M \\ \ddot{z}_r+\beta(2z_r-y_r-y_{r-1}) &= F'_r/m\end{aligned}\right\} \tag{122}$$

These form a coupled set of equations which cannot readily be solved exactly in terms of the coordinates y_r, z_r. If, however, we transform to normal coordinates Q_s, the equations (122) may be transformed into the set

$$\ddot{Q}_s+\omega_s^2 Q_s = G_s \tag{123}$$

(see exercise (1) below). If we express the electromagnetic wave in the form

$$\mathscr{E} = \mathscr{E}_0\,\mathrm{e}^{i(k_0x-\omega t)}$$

where $\mathscr{E}$ is the electric field at distance x along the crystal, and $k_0 = 2\pi/\lambda$, we have

$$\left.\begin{aligned}F_r &= e\mathscr{E}_0\,\mathrm{e}^{ik_0(2r+1)a-i\omega t} \\ F'_r &= -e\mathscr{E}_0\,\mathrm{e}^{2ik_0ra-i\omega t}\end{aligned}\right\} \tag{124}$$

The quantities G_s will then have the form

$$G_s = A_s\,\mathrm{e}^{-i\omega t} \tag{125}$$

the constants A_s being calculated by means of equations (124) and the transformation giving Q_s in terms of the displacements y_r.

The solution of equation (123) for the forced oscillations is

$$Q_s = \frac{A_s\,\mathrm{e}^{-i\omega t}}{(\omega_s^2-\omega^2)} \tag{126}$$

Thus we see that we have a resonant condition whenever ω is equal to ω_s, one of the normal frequencies. This completes the formal solution if the normal coordinates Q_s are known, since the displacements y_r can then be expressed in terms of them. In this case the transformation matrix for the normal coordinates is rather complicated and we shall instead give an approximate treatment, since it may be shown, that the quantities A_s are only large when s corresponds to a long-wave optical mode (see exercise below).

In the approximate treatment we neglect the variation in phase ka of the electromagnetic wave between neighbouring particles. Thus we replace equations (124) by

$$\left.\begin{aligned}F_r &= e\mathscr{E}_0\,\mathrm{e}^{-i(\omega t+\epsilon_r)} \\ F'_r &= -e\mathscr{E}_0\,\mathrm{e}^{-i(\omega t+\epsilon_r)}\end{aligned}\right\} \tag{127}$$

where $\epsilon_r = 2rk_0a$; also we assume that only long-wave optical modes are appreciably excited so that we may take as an approximation (cf. equations (120))

$$My_r = My_{r-1} = -mz_r = -mz_{r+1}$$

With these approximations, equations (122) become

$$\left.\begin{aligned} \ddot{y}_r + 2\alpha\frac{(M+m)}{m}y_r &= \frac{e\mathscr{E}_0}{M}\mathrm{e}^{-i(\omega t+\epsilon_r)} \\ \ddot{z}_r + 2\beta\frac{M+m}{M}z_r &= -\frac{e\mathscr{E}_0}{m}\mathrm{e}^{-i(\omega t+\epsilon_r)} \end{aligned}\right\} \tag{128}$$

Inserting the values for α, β, these may be written in the form

$$\left.\begin{aligned} \ddot{y}_r + \omega_0^2 y_r &= \frac{e\mathscr{E}_0}{M}\mathrm{e}^{-i(\omega t+\epsilon_r)} \\ \ddot{z}_r + \omega_0^2 z_r &= -\frac{e\mathscr{E}_0}{m}\mathrm{e}^{-i(\omega t+\epsilon_r)} \end{aligned}\right\} \tag{129}$$

Thus we have

$$|y_r| = \frac{e\mathscr{E}_0}{M|\omega^2-\omega_0^2|} \tag{130}$$

$$|z_r| = \frac{e\mathscr{E}_0}{m|\omega^2-\omega_0^2|} \tag{131}$$

where $\omega_0/2\pi$ is the frequency corresponding to the upper limit of the optical branch; we thus get a resonant condition when $\omega = \omega_0$. We shall see later that this frequency corresponds to that giving strong absorption and reflection of infra-red radiation from crystals—the so-called *reststrahl* frequency.

If we define a quantity P_m which is the electric polarisation in unit field, per pair of particles, we have

$$\left.\begin{aligned} P_m &= e(y_r - z_r)/\mathscr{E} \\ &= e^2\left(\frac{1}{M}+\frac{1}{m}\right)\Big/(\omega_0^2-\omega^2) \end{aligned}\right\} \tag{132}$$

As $\omega \to 0$ we have for P_{m0} the static polarisation

$$P_{m0} = e^2\left(\frac{1}{M}+\frac{1}{m}\right)\Big/\omega_0^2 \tag{133}$$

Exercise (1) Show that if the $2N$ normal coordinates Q_s are related to the displacements y_r, z_r ($r = 1, 2, \ldots N$) by the normal, orthogonal transformation

$$Q_s = \sum_r A_{rs} y_r + \sum B_{rs} z_s$$

then the quantities G_s in equation (123) are given by

$$G_s = \sum_r M^{-1} A_{rs} F_r + \sum m^{-1} B_{rs} F'_r$$

(Perhaps the easiest way to see this is to multiply the equations (122) in proper order by A_{rs}, B_{rs} and add. The first term clearly gives $\ddot{Q}_s$, the rest of the right-hand side must give $\omega_s^2 Q_s$ as can be seen by putting $F_r = F'_r = 0$. The right-hand side then gives the above result.)

(2) Consider a rather artificial example in the form of a linear crystal having particles of equal mass but with charges of opposite sign. The normal coordinates Q_s for this system have been determined in § 3.3.4 for periodic boundary conditions. Show that in this case

$$G_0 = 0$$

$$G_s = \frac{2^{1/2} e\mathscr{E} \mathrm{e}^{-i\omega t}}{N^{1/2} m} \sum_{r=1}^{N} (-1)^r \cos k_s ra \, \mathrm{e}^{ik_0 ra}$$

$$G_{N-s} = \frac{2^{1/2} e\mathscr{E} \mathrm{e}^{-i\omega t}}{N^{1/2} m} \sum_{r=1}^{N} (-1)^r \sin k_s ra \, \mathrm{e}^{ik_0 ra}$$

Hence show that for small values of k_s

$$|G_s| = |G_{N-s}| = 2^{1/2} N^{1/2} e\mathscr{E} ka/m(N^2 k_0^2 a^2 - 4\pi^2 s'^2)$$

where $s' = N/2 - s$.

G_s, G_{N-s} are thus only appreciable for small values of s'. $Q_{N/2}$ corresponds to the π-mode and for small values of s', $Q_{N/2-s'}$ to neighbouring modes. These correspond to the optical modes of the two-particle crystal.

3.6.2 Distribution of energy between the different modes

The mean energy of the normal modes for a two-particle crystal may be expressed in terms of the allowed frequencies ν_s as for a crystal having only one type of particle. For each allowed value of k we have now two allowed values of ν, which we shall denote by ν_{sa}, ν_{so}, one corresponding to the acoustic branch and one to the optical branch. If we have a crystal consisting of $2N$ particles spaced at equal distances apart, N being of mass M and N of mass m, and if we impose periodic boundary conditions, the allowed values k_s of the propagation constant will be given by (N even)

$$k_s = \frac{s\pi}{Na} \qquad -\frac{N}{2}+1 \leqslant s \leqslant \frac{N}{2}$$

The corresponding values of ν_{sa}, ν_{so} are then determined from the dispersion curves given by equations (109), (110) (see also Fig. 3.12). The quantization of the normal modes is carried out exactly as before, and the probability of any particular mode being excited at temperature T is proportional to $\exp[-\boldsymbol{h}\nu/\boldsymbol{k}T]$. Thus at low temperatures only the long-wave acoustic modes are excited, the optical modes being in their lowest state in equilibrium, since $\boldsymbol{h}\nu_{so} \gg \boldsymbol{k}T$ for all values of s. By comparison with equation (91) it will be seen that the mean energy W in excess of the zero-point energy is given by

$$W = \int_0^{\nu_{ma}} \frac{\boldsymbol{h}\nu n_a(\nu)\,\mathrm{d}\nu}{\mathrm{e}^{\boldsymbol{h}\nu/\boldsymbol{k}T} - 1} + \int_{\nu_{mo}}^{\nu_o} \frac{\boldsymbol{h}\nu n_o(\nu)\,\mathrm{d}\nu}{\mathrm{e}^{\boldsymbol{h}\nu/\boldsymbol{k}T} - 1} \tag{134}$$

ν_{ma} being the upper frequency limit of the acoustic modes, ν_o that of the optical modes, and ν_{mo} the lower frequency limit of the optical branch. The density of allowed frequencies $n_a(\nu)$ and $n_o(\nu)$ for the acoustic and optical branches are obtained from equations (109), (110). For large values of T, i.e. such that $\boldsymbol{k}T \gg \boldsymbol{h}\nu_o$, we have

$$W = \boldsymbol{k}T \int_0^{\nu_{ma}} n_a(\nu)\,\mathrm{d}\nu + \boldsymbol{k}T \int_{\nu_{mo}}^{\nu_o} n_o(\nu)\,\mathrm{d}\nu$$

$$= 2\boldsymbol{k}TN \tag{135}$$

since

$$\int_0^{\nu_{ma}} n_a(\nu)\,\mathrm{d}\nu = \int_{\nu'_{mo}}^{\nu_o} n_o(\nu)\,\mathrm{d}\nu = N$$

the integrals representing, respectively, the number of allowed frequencies in the acoustic and optical branches.

When we include the two types of transverse modes and the longitudinal modes we shall have four branches if the transverse modes at right angles are degenerate and six if they are not. The corresponding integrals will have to be evaluated over all the branches. For large values of T we shall then have

$$W = 6\boldsymbol{k}TN \tag{136}$$

3.7 Vibrations of a linear crystal with any number of particles per unit cell

In § 3.6 we have discussed the motion of a linear crystal having equally spaced particles of different mass. We need not have taken the spacing between consecutive particles as equal; we could, for example, have taken a mass M at $x = a$ and a mass m at $x = d$ $(a < d \neq 2a)$ and made up the crystal by repeating this unit, with period d. By writing down the equations of motion it will readily be verified that we again obtain an equation for the frequency $\omega/2\pi$ in terms of the propagation constant k quadratic in ω^2 giving two branches of the ω/k curves in the reduced representation, and periodic with respect to k of period $2\pi/d$; the reduced representation extends from $k = 0$ to $k = \pi/d$ as before.

This analysis may be extended to a linear crystal made up of equal groups each containing any number n of particles; for a crystal containing N such groups we should have Nn particles. We may again define the equilibrium position of the particles in terms of a linear lattice; such a lattice we may take as consisting of the points specified by the coordinates $x = rd$, with the particles in equilibrium at the points $x = rd + a_s$, a_s taking n values corresponding to the n particles of a group. The aggregate of points occupied by the particles in equilibrium is referred to as a *lattice with a basis*; the points $x = rd$ are called a *simple* lattice. It is usual to take $a_1 = 0$ so that in equilibrium a particle is at each point of the simple lattice. Let y_{rs} be the displacement of the sth particle in the rth group; we have now for the equations of motion, for transverse displacements,

m_s being the mass of the sth particle,

$$m_s \ddot{y}_{rs} + \frac{F(y_{rs} - y_{r,s-1})}{(a_s - a_{s-1})} + \frac{F(y_{rs} - y_{r,s+1})}{(a_{s+1} - a_s)} = 0 \quad (137)$$

provided $s \neq 1$ or $s \neq n$. For the first and last particles of the group we have $(a_1 = 0)$,

$$m_1 \ddot{y}_{r1} + \frac{F(y_{r1} - y_{r-1,n})}{(d - a_n)} + \frac{F(y_{r1} - y_{r2})}{a_2} = 0 \quad (138)$$

$$m_n \ddot{y}_{rn} = \frac{F(y_{rn} - y_{r,n-1})}{a_n - a_{n-1}} + \frac{F(y_{rn} - y_{r+1,1})}{d - a_n} = 0 \quad (139)$$

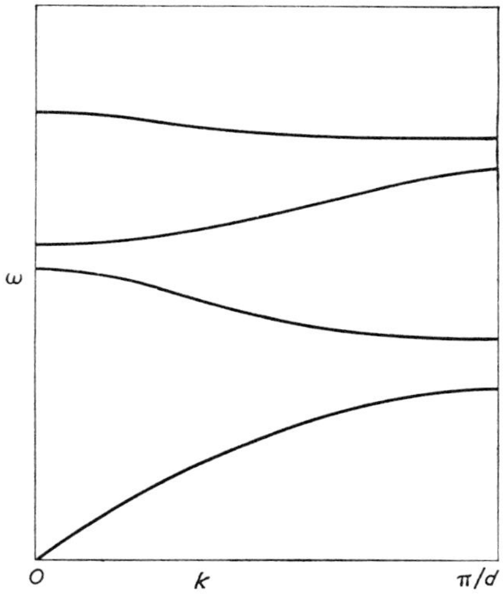

FIG. 3.18. ω/k curves in the reduced representation for a linear crystal containing periodic groups of four particles (planar transverse modes)

These n equations may be solved by means of n wave functions of the form

$$y_s(x, t) = A_s e^{ikx - i\omega t} \quad (140)$$

the value of the displacement of the sth particle in the rth group being given by $y_s(rd + a_s, t)$. On substituting in the equations (137)–(139) we obtain a set of equations independent of r as before, namely

$$-m_1 \omega^2 A_1 + \frac{F(A_1 - A_n e^{-ikd + ika_n})}{(d - a_n)} + \frac{F(A_1 - A_2 e^{ika_2})}{a_2} = 0 \quad (141)$$

$$-m_s \omega^2 A_s + \frac{F(A_s e^{ika_s} - A_{s-1} e^{ika_{s-1}})}{(a_s - a_{s-1})} + \frac{F(A_s e^{ika_s} - A_{s+1} e^{ika_{s+1}})}{(a_{s+1} - a_s)} = 0 \quad (142)$$

$s = 2, \ldots, n-1.$

$$-m_n \omega^2 A_n + \frac{F(A_n e^{ika_n} - A_{n-1} e^{ika_{n-1}})}{(a_n - a_{n-1})} + \frac{F(A_n e^{ika_n} - A_1 e^{ikd})}{d - a_n} = 0 \quad (143)$$

We have thus n linear equations for the n quantities A_s, and we have a non-trivial solution only if the determinant of the coefficients is equal to zero. This gives an algebraic equation of degree n in ω^2 and gives n solutions for ω^2 in terms of k, so that we have now n branches for the ω/k curves; for each branch we may determine the ratios $A:A_2:\ldots:A_n$ and so complete the solution. As before, only one of the solutions is such that $\omega \to 0$ as $k \to 0$. Again it will be seen that ω is an even function of k being unchanged by replacing k by $-k$; ω is also a periodic function of k of period $2\pi/d$ since it will readily be seen that the value of the determinant is unchanged by replacing k by $k \pm 2\pi l d$ where l is an integer. The form of the ω/k curves in the reduced representation is shown for $n = 4$ in

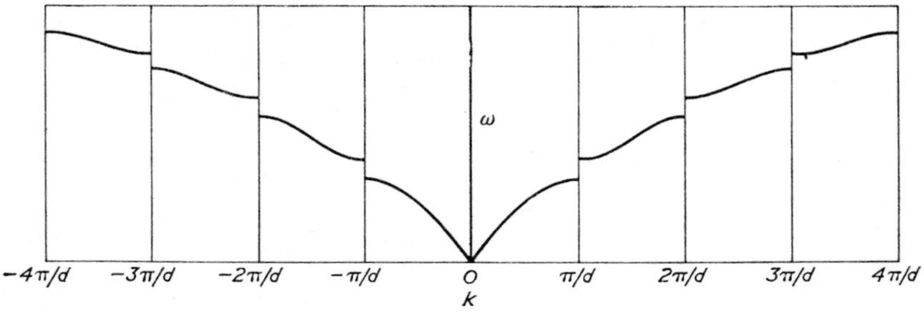

FIG. 3.19. ω/k curves in the expanded representation for a linear crystal containing periodic groups of four particles (planar transverse modes)

Fig. 3.18 and in the expanded representation in Fig. 3.19, where we have also included negative values of k.

3.8 Passage to continuous distributions

The solution for the problem of a crystal made up of identical groups of n particles may be expressed by means of n wave functions, $y_s(x,t)$, as given by equation (140). The displacement of y_{rs} of the sth particle in the rth group is then given by the value of the function $y_s(x,t)$ when $x = rd + a_s$, i.e. we have

$$y_{rs} = y_s(rd + a_s, t), \quad s = 1, 2 \ldots n \tag{144}$$

$$= A_s \, e^{ik(rd+a_s)-i\omega t} \tag{144a}$$

We may, however, combine all the n wave functions y_s in a single wave function $y(x,t)$ defined as follows: the function $y(x,t)$ has the value y_{rs} at each of the points $x = rd + a_s$ and at other points is only defined so as to be continuous. The function $y(x,t)$ may therefore be taken to have the form

$$y(x,t) = A(x) \, e^{ikx - i\omega t} \tag{145}$$

the function $A(x)$ being such that

$$A(rd+a_s) = A_s \tag{146}$$

for all values of the integer r. For the points at which the function $A(x)$ is exactly defined it is therefore a periodic function of period d, i.e.

$$A(x+rd) = A(x)$$

where r is any integer.

Now suppose we increase the number of particles indefinitely so that we have effectively a periodic line density $\rho(x)$. The function $A(x)$ then becomes defined for all values of x and the function $y(x,t)$ gives the displacement; it is given by a function of the form (145), where $A(x)$ is a periodic function of period d. For a continuous line subject to tension F and line density $\rho(x)$ the displacement satisfies the differential equation

$$\rho(x)\frac{\partial^2 y}{\partial t^2} = F\frac{\partial^2 y}{\partial x^2} \tag{147}$$

For a solution periodic in the time this may be written in the form $y = Y(x)\mathrm{e}^{-i\omega t}$ where $Y(x)$ is a solution of the equation

$$\frac{\mathrm{d}^2 Y}{\mathrm{d}x^2}+f(x)\, Y = 0 \tag{148}$$

and $f(x) = \omega^2\rho(x)/F$; $f(x)$ is a periodic function of period d, since $\rho(x)$ is periodic. The solution in this case has therefore the form

$$Y(x) = A(x)\,\mathrm{e}^{ikx} \tag{149}$$

where $A(x)$ is a periodic function of period d; this is a most important result. The demonstration we have given can hardly be called a *proof* but gives some insight as to how this type of solution comes about. In pure mathematics the theorem is generally referred to as Floquet's theorem and as applied to wave mechanics as Bloch's theorem. (For a more rigorous mathematical proof of this theorem see § 4.9.) We shall apply it in the next Chapter to the problem of the one-dimensional motion of electrons in a periodic potential.

The solution (149) may also be written in an alternative form which is frequently useful; suppose we are able to solve the equation (148) for the region $0 \leqslant x \leqslant d$ and the solution is given by

$$Y(x) = \phi_0(x), \quad 0 \leqslant x \leqslant d \tag{150}$$

On comparing this with the solution (149) we must have

$$A(x) = \phi_0(x)\,\mathrm{e}^{-ikx} \tag{151}$$

The general solution may be written in the form

$$Y(x) = A(x'+nd)\,\mathrm{e}^{i(knd+kx')}$$

where $x' = x-nd$. Since $A(x)$ is periodic this may also be written in the form

$$Y(x) = A(x')\,\mathrm{e}^{ikx'}\,\mathrm{e}^{iknd} \tag{152}$$

Now if $nd \leqslant x \leqslant (n+1)d$ we also have $0 \leqslant x' \leqslant d$ and $A(x')$ may be expressed in terms of the solution $\phi_0(x)$ which is only defined for the range of the variable $0 \leqslant x \leqslant d$. We have then

$$\left.\begin{aligned} Y(x) &= \phi_0(x')\,\mathrm{e}^{iknd} \\ &= \mathrm{e}^{iknd}\,\phi_0(x-nd), \quad nd \leqslant x \leqslant (n+1)\,d \end{aligned}\right\} \tag{153}$$

The boundary conditions required to ensure the continuity of the solution between the various regions determine the propagation constant k in terms of ω. Since the problem is equivalent to that of an infinite number of particles in a group which is repeated periodically we now have an infinite number of branches in the representation of the ω/k curves.

It is interesting to show now that if the value of k is replaced by $k \pm 2\pi r/d$ (where r is an integer) the form of the solution is unchanged. If we substitute in the solution (149) we have

$$Y(x) = A(x)\,\mathrm{e}^{ikx \pm 2\pi irx/d} = B(x)\,\mathrm{e}^{ikx} \tag{154}$$

where $B(x) = A(x)\exp(\pm 2\pi inx/d)$, so that $B(x)$ is also a periodic function of period d and the new value of k is just as good as the former. Again it will be seen that the form of the solution (153) is unchanged by such a substitution. Thus we are again justified in restricting k to a region for which $|k| \leqslant \pi/d$. As we shall see later, however, when we have chosen a value k the values $k \pm 2\pi n/d$ can be given a *physical* interpretation, when we are dealing with differential equations for wave motion in a continuous periodic medium.

In passing from a discontinuous distribution to a continuous one we could easily have let a finite number of the masses remain finite. This would correspond to the problem of dealing with a periodic medium with a finite number of periodic discontinuities. We shall consider some problems of this kind in the next Chapter.

3.9 Inclusion of forces from beyond nearest neighbours

In the problems we have discussed so far in the present Chapter we have taken account of forces only between adjacent particles in the crystal. It is, however, fairly straightforward to take into account forces arising from a few more particles other than nearest neighbours, but as the number of contributing particles increases, the analysis becomes rapidly more complex. The subject has been treated for example by L. Brillouin*, in the case of a linear crystal having only one type of particle. It may be shown that the ω/k curves may still be represented for each type of transverse or longitudinal motion as a number of branches, the number being equal to the number of different types of particle making up the crystal. If the ω/k curves are shown in the expanded representation each value of k still corresponds to a single value of ω.

* L. Brillouin, *Wave Propagation in Periodic Structures* (2nd Ed.). Dover (1953), p. 33.

When only interactions between nearest neighbours are taken into account the converse is also true, to each value of ω there is only one value of k in the expanded representation. When interactions between more distant particles are taken into account this is no longer true and a given value of ω may correspond to more than one value of k in the expanded representation. Indeed if the M nearest particles on either side are included, corresponding to forces of range Ma, there may be as many as M values of k for a given value of ω; all of these may not, however, be real. Moreover, the ω/k curves are no longer monotonic increasing functions of k in the expanded representation; the maximum and minimum values of ω may not now even occur at the edges of the interval $0 \leqslant k \leqslant \pi/d$ in the reduced representation. Apart from this most of the results which we have already obtained still hold.

Exercise (1) Show that quite generally the potential energy V of a linear crystal with one kind of particle, neglecting terms higher than those of the second order in the displacements y_r from the positions of equilibrium, may be written in the form

$$V = V_0 - \tfrac{1}{2} \sum_{r,r'} a_{rr'} y_r y_{r'}$$

where $a_{rr'}$ depends only on the difference $(r-r')$, $a_{rr'} = a_{r'r}$ and $\sum_r a_{rr'} = 0$.

Hence show that the equations of motion may be written in the form

$$m\ddot{y}_r + \sum_{r'} a_{rr'} y_{r'} = 0$$

(2) Using the results of Exercise 1 show that, for a solution periodic in the time with frequency $\omega/2\pi$, the equations of motion may be reduced to the form

$$m\omega^2 = \sum_{s=1}^{N} a_s \cos(ska)$$

by assuming a solution of the form

$$y_r = B\mathrm{e}^{ikra - i\omega}$$

where $a_s = a_{rr'}$ when $|r-r'| = s$.

Hence show that $\omega^2(k) = \omega^2(-k)$ and also that ω^2 is a periodic function of k of period $2\pi/a$.

3.10 Photons and phonons

The dual wave and particle character of electromagnetic radiation is well known. On the one hand we may represent a travelling light wave in free space by means of a wave function describing the variation of electric and magnetic field, characterised by a wavelength λ and corresponding propagation constant $k = 2\pi/\lambda$, and also by a velocity of propagation $\boldsymbol{c}$; the dispersion relationship is simply $\omega = k\boldsymbol{c}$. On the other hand we may consider the radiation as consisting of particles (photons) having energy $\boldsymbol{h}\nu$, momentum $\boldsymbol{h}\nu/\boldsymbol{c}$ and hence mass $\boldsymbol{h}\nu/\boldsymbol{c}^2$. The relations between mass and energy follow from the fundamental equation connecting mass m and energy E, $E = mc^2$, derived from the theory of relativity. The momentum equation follows from the fact that the particle velocity must also be c, since this is also the group velocity and hence the velocity with which energy is propagated. Thus we have

$$p = \text{momentum} = \text{mass} \times \text{velocity} = \frac{h\nu}{c^2} \times c = \frac{h\nu}{c}$$

The relationship between momentum p and wavelength λ is

$$p = h/\lambda = \hbar k \tag{155}$$

This is the same as for material particles like electrons (see § 1.2.1).

In dealing with the statistical properties of radiation we have used the particle model and treated a cavity containing radiation as an assembly of identical particles (photons) which may take the set of energies 0, $h\nu_1$, $h\nu_2$ ... where ν_1, ν_2 ... are the frequencies of the allowed resonant modes of the cavity. We could, however, have treated the normal modes of the cavity in exactly the same way as we have done for the normal modes of a linear crystal. We should have a doubly degenerate mode for each allowed value of k, corresponding to the two possible states of polarisation of the radiation. Each mode can be represented as before as a harmonic oscillator, which can have energy $(n + \frac{1}{2}) h\nu_s$ where n is an integer. Treating each mode by *classical* statistics (since we have a single system), and taking our energy zero as that in which all the oscillators are in their ground state, we get exactly the same result as that given by equation (89) for the lattice vibrations

$$\bar{E}_s = \frac{h\nu_s}{e^{h\nu_s/kT} - 1} \tag{156}$$

We then sum as before over all values of ν_s, but we now have an infinite number of modes and not a finite number as for the crystal. This part of the calculation is exactly as before (see § 2.5) and we get Planck's law for the energy density of radiation. Thus we see that the particle and wave picture lead to the same results provided the normal modes in the wave picture are quantized. We may note that it is essential to use Einstein–Bose statistics for the photons and not Fermi–Dirac statistics if the two methods of approach are to give the same results. Accordingly we conclude that photons do not obey the Exclusion Principle and we can have more than one photon in the same state. The average number $\bar{n}_s$ in a state with frequency ν_s is given by

$$\bar{n}_s = \frac{1}{e^{h\nu_s/kT} - 1} \tag{157}$$

as follows from equation (156). We see that $\bar{n}_s$ is always less than unity if $h\nu_s > kT$, but this does not imply that we cannot have more than one photon with energy $h\nu_s$ at any one time, since sometimes we shall have none; $\overline{n}_s$ simply gives the average number. Clearly if we have n photons of energy $h\nu_s$ at any one time this is equivalent to the sth normal mode being excited to an energy $(n+\frac{1}{2})h\nu_s$. When a photon passes into a state with zero energy and so disappears from observation the energy of the sth mode drops from $(n+\frac{1}{2})h\nu_s$ to $(n-\frac{1}{2})h\nu_s$ $(n \geqslant 1)$ and when all the photons of energy $h\nu_s$ have disappeared the sth mode is in its lowest state with zero-point energy $\frac{1}{2}h\nu_s$. There is a slight philosophical difficulty about the fact that we appear to have an infinite zero-point energy, but in fact this is academic since we never measure this energy in radiation processes; it is only *changes* in energy that are significant.

An exactly similar dual description is available for the lattice vibrations. We have calculated the mean energy using the quantized waves and obtained equation (89). This could have been also obtained by considering the lattice vibrations as an assembly of particles whose allowed energies are 0, $h\nu_1$, $h\nu_2$ ... $h\nu_N$ obeying Einstein–Bose statistics. The momentum duality is, however, not yet clear; we shall also see later (§ 10.9.2) that when an electron interacts with lattice vibrations and its momentum is changed from p_1 to p_2 then

$$p_1 = p_2 + \hbar k_s \tag{158}$$

therefore, in a sense, the quantity $\hbar k_s$ acts as a momentum. If we write $p_s = \hbar k_s$ we have $p_s = h/\lambda_s$ which is the usual relationship between momentum and wavelength. We must note, however, that the quantity p_s cannot be a momentum in the strict physical sense since $k_s' = k_s + 2\pi n/d$ is an equally good propagation vector, where n is an integer and d the lattice spacing. This would correspond to a 'momentum' $p_s + nh/d$. The 'momentum' p_s is therefore not uniquely defined, but apart from this has many properties similar to those of real momentum and is frequently referred to as the 'phonon momentum'.

4

Motion of Electrons in a One-dimensional Periodic Potential

4.1 Motion in a periodic potential

WE have already discussed some aspects of the mathematical techniques required for a study of the motion of particles in a periodic potential in § 3.8. Let the potential energy $V(x)$ be a periodic function of x of period d such that

$$V(x) = V(x+rd) \tag{1}$$

where r is any integer. The wave equation for one-dimensional motion of electrons of energy E in the potential field defined by $V(x)$ is

$$\frac{\mathrm{d}^2\psi}{\mathrm{d}x^2}+\frac{8\pi^2 \boldsymbol{m}}{\boldsymbol{h}^2}[E-V(x)]\psi = 0 \tag{2}$$

This is an equation of the form

$$\frac{\mathrm{d}^2\psi}{\mathrm{d}x^2}+f(x)\psi = 0 \tag{3}$$

discussed in § 3.8, with

$$f(x) = \frac{8\pi^2 \boldsymbol{m}}{\boldsymbol{h}^2}[E-V(x)]$$

The function $f(x)$ is also periodic with period d since E is a constant. Solutions of equation (2) may therefore be written in the form

$$\psi(x) = A(x,k)\,\mathrm{e}^{ikx} \tag{4}$$

where $A(x,k)$ is a periodic function of period d and k is a constant to be determined by the boundary conditions (see § 3.8 equation (149)). Alternatively, if we know the solution $\psi_0(x)$ for the interval $0 \leqslant x \leqslant d$ we may express the solution for any other value of x in the form (see § 3.8 equation (153))

$$\psi(x) = \mathrm{e}^{ikrd}\psi_0(x-rd), \quad rd \leqslant x \leqslant (r+1)d \tag{5}$$

This form of solution is particularly useful when we may readily find the solution for the interval $0 \leqslant x \leqslant d$ apart from certain constants which have to be determined so as to make it fit to the solutions for neighbouring intervals.

A particularly simple example of this kind of problem occurs when $V(x) = 0$ except at the points $x = rd$ where it is discontinuous and causes discontinuities in the derivative of the wave function. We shall begin by discussing this problem which, although somewhat artificial, so far as application to the motion of electrons in solids is concerned, provides us with a simple solution which brings out some very important features common to all such problems.

4.2 One-dimensional motion of electrons in an infinite array of deep narrow potential wells

Suppose we represent the atomic cores in a crystal by potential wells of the form discussed in § 2.9 and we consider the one-dimensional motion of an electron in such a potential. The potential energy is shown in Fig. 4.1. We shall take each well to be of depth W and of width a, and the centre of each to be separated by distance d from the centres of its neighbours. Thus each well may be regarded as being separated from its

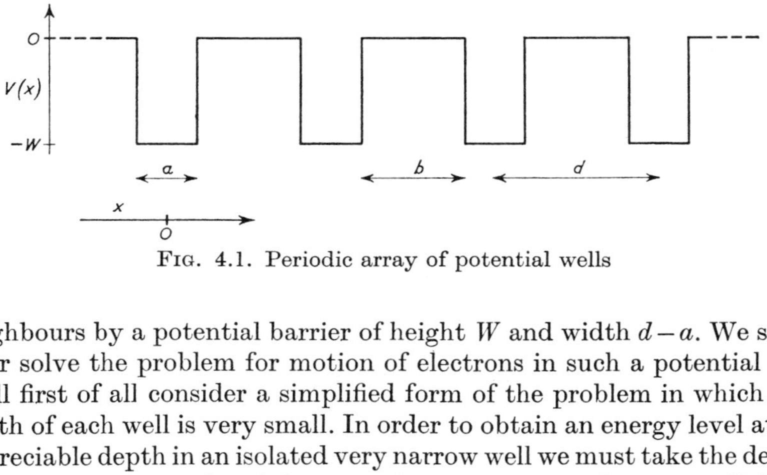

FIG. 4.1. Periodic array of potential wells

neighbours by a potential barrier of height W and width $d-a$. We shall later solve the problem for motion of electrons in such a potential but shall first of all consider a simplified form of the problem in which the width of each well is very small. In order to obtain an energy level at an appreciable depth in an isolated very narrow well we must take the depth to be very great, so that it is convenient to measure the energy E from the *top* of the wells. This form of potential is shown in Fig. 4.2. As the width of each well tends to zero we see that an electron in an energy level at *depth* E_0 in one well is separated from a corresponding position in a neighbouring well by a potential barrier of height E_0 and width d. Since such a barrier is somewhat transparent (see § 2.13) we cannot regard the electron as belonging to any one well; this will also be apparent from the mathematical form of the solution which we shall obtain.

We consider first an isolated well and let us write, as before

$$\gamma^2 = 2mW/\hbar^2$$

We shall assume that $W \to \infty$ and $a \to 0$ in such a way that Wa remains finite but that $Wa^2 \to 0$; thus we have $\gamma a \to 0$. In this case we have only one energy level (see § 2.9) but we can adjust its position to any value we please by choosing the limiting value of aW. We have for the normalised wave function outside the well (§ 2.9 equations (101))

$$\left.\begin{aligned}\psi(x) &= \alpha_0^{1/2}\,\mathrm{e}^{-\alpha_0 x} \quad x > 0\\ &= \alpha_0^{1/2}\,\mathrm{e}^{\alpha_0 x} \quad x < 0\end{aligned}\right\} \tag{6}$$

the energy E being given by

$$E = -E_0 = -\hbar^2\alpha_0^2/2m \tag{7}$$

Thus we see that ψ is continuous at $x = 0$ but that $\partial\psi/\partial x$ has a discontinuity given by

$$\frac{\partial\psi_+}{\partial x} - \frac{\partial\psi_-}{\partial x} = -2\alpha_0\psi \tag{8}$$

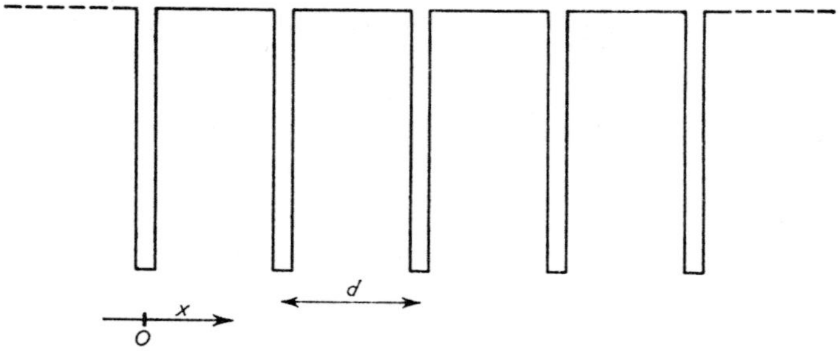

FIG. 4.2. Array of very deep narrow potential wells

where ψ_+ and ψ_- refer, respectively, to the values of ψ for $x > 0$ and $x < 0$. Mathematically the limiting form of the potential well may be represented by this boundary condition. The limiting form of the wave function is shown in Fig. 4.3. It is fairly easy to see that the condition holds also for each well in the array.

First let us consider the continuity of ψ; before we let a tend to zero we note that ψ is continuous at $x = \pm a/2$. The change in ψ is therefore that obtained in crossing the well. If E is the energy, we have inside the well

$$\psi = C\,\mathrm{e}^{i\beta x} + D\,\mathrm{e}^{-i\beta x} \tag{9}$$

where $\beta^2 = 2m[W+E]/\hbar^2$. Now $\beta^2 a^2 \to 0$ since $Wa^2 \to 0$ and $Ea^2 \to 0$ as $a \to 0$. Thus, since $\beta a \to 0$ the change in ψ tends to zero as will be seen from equation (9).

On integrating equation (2) across the well we have, since ψ changes very little,

$$\begin{aligned}\frac{\partial\psi_+}{\partial x} - \frac{\partial\psi_-}{\partial x} &= -2m[E+W]\psi a/\hbar^2\\ &\to -2mWa\psi/\hbar^2\end{aligned} \tag{10}$$

as $a \to 0$. Since $\gamma a \to 0$ we see from Fig. 2.7 that $\beta a \to 0$ for an allowed level and equation (103) of § 2.9 simplifies to the form

$$\alpha_0 = \tfrac{1}{2}\beta^2 a = mWa/\hbar^2$$

since $Ea^2 \to 0$ as $a \to 0$, so that equation (10) reduces to equation (8).

We may therefore reduce the problem of the motion of an electron in a series of narrow deep potential wells to one of periodic discontinuous boundary conditions. We have to solve the equation

$$\frac{d^2\psi}{dx^2} + \frac{2mE}{\hbar^2}\psi = 0 \tag{11}$$

with the condition

$$\left.\begin{aligned} \psi_{r+1}(x) &= \psi_r(x) & x &= (r+1)d \\ \frac{\partial \psi_{r+1}}{\partial x} - \frac{\partial \psi_r}{\partial x} &= -2\alpha_0 \psi_r & x &= (r+1)d \end{aligned}\right\} \tag{12}$$

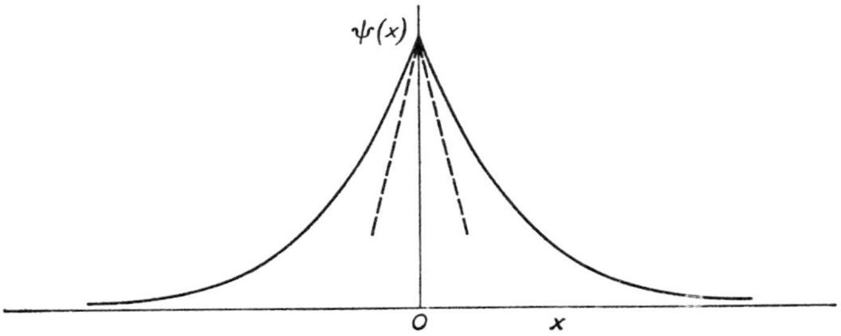

FIG. 4.3. Limiting form of the wave function for an electron in a deep narrow potential well

$\psi_r(x)$ being the solution for $rd \leqslant x \leqslant (r+1)d$, and $-E_0 = -\hbar^2\alpha_0^2/2m$ being the energy of an electron in a single isolated well.

The form of solution is slightly different according as the energy E is positive or negative, i.e. according as the energy lies above the top of the potential wells or not. The latter condition would clearly correspond to a bound electron at infinite separation of the wells and we shall consider it first. The general solution of equation (11) has the form

$$\psi(x) = A\,e^{\alpha x} + B\,e^{-\alpha x} \tag{13}$$

with $\alpha^2 = -2mE/\hbar^2$, and we may take this as $\psi_0(x)$, the solution for $0 \leqslant x \leqslant d$. From equation (5) we see that

$$\psi(x) = e^{ikrd}[A\,e^{\alpha(x-rd)} + B\,e^{-\alpha(x-rd)}] \tag{14}$$

when $rd \leqslant x \leqslant (r+1)d$.

The solution (14) contains the constant k as well as the constants A, B. The absolute magnitudes of A and B are determined by the normalising condition so that we really have only two constants to determine. This is done by means of the boundary conditions (12), the form of solution ensuring that we get a set of equations independent of r.

For the interval $0 \leqslant x \leqslant d$ we have $\psi(x)$ given by (13); for $d \leqslant x \leqslant 2d$ we have

$$\psi(x) = \psi_1(x) = e^{ikd}[A\,e^{\alpha(x-d)} + B\,e^{-\alpha(x-d)}] \tag{15}$$

Applying the conditions (12) at $x = d$ we have

$$A\,e^{\alpha d} + B\,e^{-\alpha d} = e^{ikd}(A+B) \tag{16}$$

$$\alpha A\,e^{\alpha d} - \alpha B\,e^{-\alpha d} - \alpha A\,e^{ikd} + \alpha B\,e^{ikd} = 2\alpha_0\,e^{ikd}(A+B) \tag{17}$$

It will readily be verified that exactly the same two equations are obtained by applying the boundary conditions (12) at $x = rd$ where r is any integer. The constants A and B may now be eliminated from equations (16) and (17) and we obtain the equation

$$\cos kd = \cosh \alpha d - \frac{\alpha_0}{\alpha} \sinh \alpha d \tag{18}$$

Equation (18) enables us to determine α and hence E as a function of k. From the form of equation (18) it will be seen that α, and hence E, is an even periodic function of k of period $2\pi/d$. Because of this periodic relationship between E and k we may again use a reduced representation (see § 3.3) restricting the value of k to the range $0 \leqslant k \leqslant \pi/d$. It is important to note that this is exactly the same range as for the propagation constant connected with the vibrations of a linear crystal with equal particles spaced a distance d apart.

The quantity k may be regarded as a propagation constant, but the electron waves which are propagated along the array are not *plane* waves but have amplitude varying with x as shown by the form (4) of the solution; we shall later find the form of the amplitude function $A(x)$. Since the angular frequency ω is connected with the energy E by means of the relationship $E = \hbar\omega$ equation (18) may be regarded as giving the ω/k dispersion curve for these waves, but for this type of problem it is more usual to plot the corresponding E/k curves.

In order to see the form of this relationship let us first suppose that $\alpha_0 d$ is so large that $\exp(-\alpha_0 d)$ is very small. Clearly, for large values of α_0, we must have $\alpha \to \alpha_0$ if k is to be real, and we may then write equation (18) in the approximate form

$$\alpha = \alpha_0 + 2\alpha_0\,e^{-\alpha_0 d} \cos kd \tag{19}$$

The energy E is then given in terms of k by the equation

$$E = -E_0 - 4E_0\,e^{-\alpha_0 d} \cos kd \tag{20}$$

For large values of d the value of E tends to $-E_0$ the value for an isolated potential well. As d is decreased, however, the energy values are spread into a band of breadth $8E_0\exp(-\alpha_0 d)$ as shown in Fig. 4.4. For small values of k equation (20) may be written in the form

$$E = -E_0 - 4E_0\,e^{-\alpha_0 d} + 2E_0 k^2 d^2 e^{-\alpha_0 d} \qquad (20a)$$

If we now shift the zero of energy to the bottom of the band this has the same form as the energy of free particles whose mass $m^\star$ is given by the equation

$$m^\star = \frac{h^2 e^{\alpha_0 d}}{16\pi^2 d^2 E_0} \qquad (21)$$

For large values of d, $m^\star$ is very large, but if d is of atomic dimensions and $\alpha_0 d \simeq 1$ $m^\star$ is of the same order of magnitude as the electronic mass.

We thus see that the effect of the regular array of potential wells is to broaden the single energy level associated with an isolated well into a

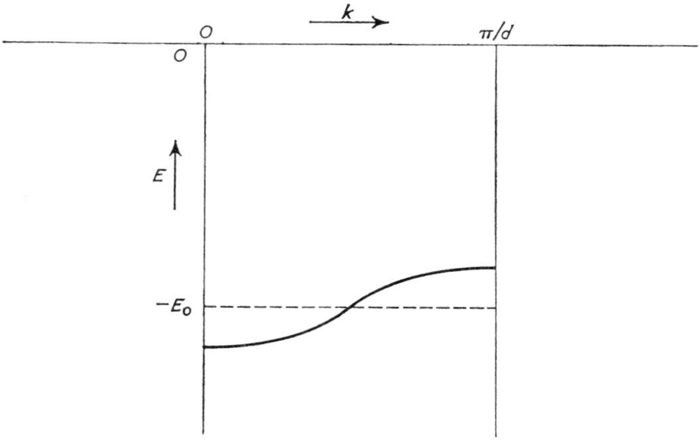

FIG. 4.4. Lowest band of allowed energy levels arising from a 'bound' state

band. For energies in this band the electron waves are propagated freely along the array and an individual electron cannot be regarded as belonging to any single well. When the separation is large, however, the band becomes very narrow and the energy tends to that for an isolated well.

Let us now determine the wave function $\psi(x)$. From equation (16) we have

$$\frac{A}{B} = -\frac{e^{-\alpha d} - e^{ikd}}{e^{\alpha d} - e^{ikd}} \qquad (22)$$

and when $\alpha d \gg 1$ we have approximately

$$A = B\,e^{-\alpha d + ikd} \simeq B\,e^{-\alpha_0 d + ikd} \qquad (23)$$

so that

$$\psi(x) = B[e^{-\alpha_0(d-x)+ikd} + e^{-\alpha_0 x}] \quad 0 \leqslant x \leqslant d \qquad (24)$$

$$= B\,e^{ikd}[e^{-\alpha_0(2d-x)+ikd} + e^{-\alpha_0(x-d)}] \quad d \leqslant x \leqslant 2d \qquad (25)$$

etc.

Thus unless x is nearly equal to rd, where r is an integer, $\psi(x)$ is small. For an isolated well at $x = rd$ we have $\psi(x) = B_0 \exp[-\alpha_0 |x-rd|]$ (cf. equation (6)). When x is nearly equal to rd, $\psi(x)$ is approximately equal to the wave function for an isolated well apart from factors of the form $\exp(ikrd)$, and the constant B. Thus, apart from a factor B^2, $|\psi(x)|^2$ is approximately the same as for an isolated well, near each well of the array, and is small elsewhere. If the wave function $\psi(x)$ refers to a single electron the constant B will be determined by the normalising condition that $\int |\psi(x)|^2 \mathrm{d}x = 1$. Setting aside for the moment the problem of

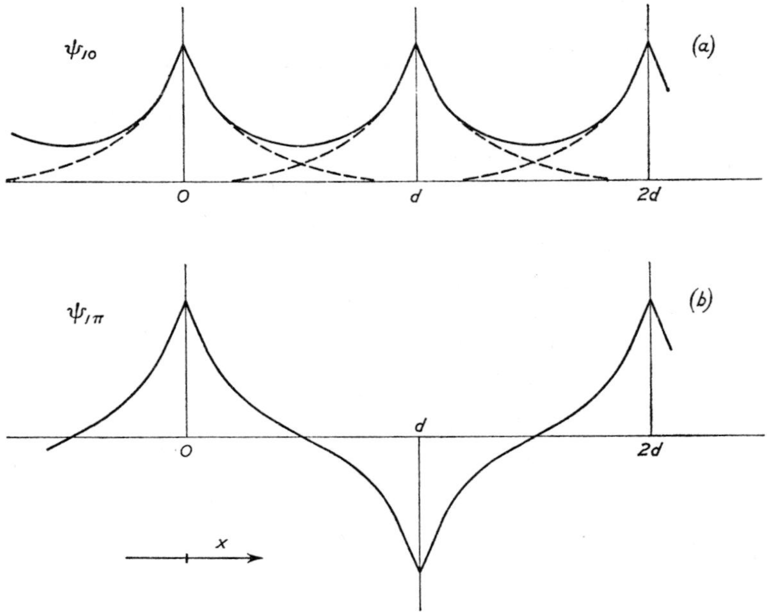

FIG. 4.5. Wave function $\psi_{10}(x)$ ($k=0$) and $\psi_{1\pi}$ ($k=\pi/d$) for the potential of Fig. 4.2. (The dashed curve shows $\psi_0(x)$, the wave function for each well regarded as isolated)

boundary conditions for a finite array, suppose we have N wells; then clearly we have approximately $B^2 \simeq N^{-1} B_0^2$ where B_0 is the normalising factor of the wave function for a single isolated well. If the wave function is normalised to represent N electrons we have $B \simeq B_0$. The wave function near each well, apart from phase factors of the form $\exp[ikrd]$, is then nearly the same as for a single electron in each well, regarded as isolated, and the charge distribution given by $|\psi|^2$ is approximately the same. The interpretation for a single electron is simply that it is most likely to be near one of the N wells, but which cannot be specified.

The form of the wave function $\psi(x)$ is shown for $k = 0$ (denoted by ψ_{10})

in Fig. 4.5 when $\alpha d \gg 1$ together with ψ_0, the wave function for each well regarded as isolated. In Fig. 4.5 is also shown the wave function $\psi(x)$ for $k = \pi/d$ (denoted by $\psi_{1\pi}$). In this case the wave function changes sign as we pass from one well to the next (but $|\psi(x)|^2$ does not differ much from $|\psi_0(x)|^2$); also $\psi_{1\pi}(x)$ has the value zero half-way between the wells.

When the value of $\alpha_0 d$ is not large compared with unity, the approximate equation (20) giving the allowed energy values no longer holds, and we must use the exact form given by equation (18). As $\alpha_0 d$ decreases from a large value we may note from equation (20) that the allowed energy band becomes broader. To find the limits of the band we plot the function f appearing on right-hand side of equation (18) as a function of αd; such

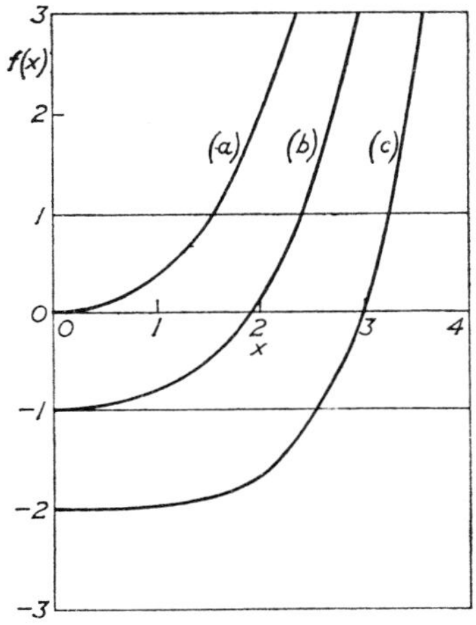

FIG. 4.6. The function
$f(x) = \cosh x - (x_0 \sinh x)/x$,
(a) $x_0 = 1$, (b) $x_0 = 2$, (c) $x_0 = 3$

a plot is shown in Fig. 4.6. We may note that $f(x)$ is a monotonic increasing function of x whose value at $x = 0$ is equal to $1 - \alpha_0 d$. Equation (18) shows that we have a real value of k only when $|f(\alpha d)| \leqslant 1$; the allowed values of the energy therefore come only from values of α corresponding to the sections of the curves of Fig. 4.6 whose ordinates lie between $+1$ and -1. The intersections of these curves with the lines corresponding to $f(x) = \pm 1$ then give us the values of α, and hence of E, corresponding to the band edges. We have plotted the curves corresponding to values of $\alpha_0 d$ equal to 3, 2, 1. We note that when $\alpha_0 d > 2$ the curves intersect both the lines $f(x) = \pm 1$ and the band edges both occur for values of α corresponding to $E < 0$. For example, for $\alpha_0 d = 3$ we have the band edges at $\alpha d = 2{\cdot}6$ and $\alpha d = 3{\cdot}25$ corresponding, respectively, to $E = -0{\cdot}755\, E_0$ and $E = -1{\cdot}18\, E_0$; these correspond, respectively, to $k = \pi/d$ and $k = 0$.

We note that the band includes the value $E = -E_0$; this must *always be so* since, when $\alpha = \alpha_0$, we have

$$\cos kd = e^{-\alpha_0 d} \tag{26}$$

so that $|\cos kd| < 1$.

For any value of α in the allowed band we may find from Fig. 4.6 the value of $\cos kd$ and hence of k; thus we may plot the form of the band. A good approximation to the form may be obtained by calculating the values of E corresponding to the band edges as above and obtaining the point corresponding to $E = E_0$ from (26). From the form of equation (20) we see that the E/k curves are parallel to the k-axis at the band edges; thus the form of the band may be readily drawn to a good approximation. The actual form as calculated for $\alpha_0 d = 3$ is shown in Fig. 4.8. (It corresponds to the lowest band shown in the figure.)

When $\alpha_0 d = 2$ the upper edge of the band rises to $E = 0$ corresponding to $\alpha = 0$, with $k = \pi/d$, as will be seen from Fig. 4.6, and equation (18); this is a limiting condition. The lower edge of the band occurs when $k = 0$ and $\alpha d = 2{\cdot}37$, corresponding to $E = -1{\cdot}40\ E_0$ so that the band has now a width equal to $1{\cdot}40\ E_0$ as compared with $0{\cdot}405\ E_0$ when $\alpha_0 d = 3$ (see Fig. 4.8).

When $\alpha_0 d < 2$ the value of $f(\alpha d)$ corresponding to $\alpha = 0$ has a value lying between -1 and 1 so that $E = 0$ is an allowed energy. This energy does not occur at a band edge since the lowest band, as we shall see, now extends continuously to positive values of E. When $\alpha_0 d = 1$ the bottom edge of the band, with $k = 0$, has $\alpha d = 1{\cdot}57$ and so corresponds to $E = -2{\cdot}45\ E_0$. As $\alpha_0 d \to 0$ corresponding, for a fixed value of d, to small values of E_0, more and more of the lowest allowed band is forced into the region corresponding to $E > 0$ (see Fig. 4.8). When $\alpha_0 d = 0$ we have free electrons and negative values of E have no significance.

We must now consider solutions of equation (11) corresponding to positive values of the energy E. Let us now write

$$E = \hbar^2 k_0^2/2\boldsymbol{m} \tag{27}$$

The analysis now proceeds exactly as before if we replace α by ik_0. The wave function $\psi(x)$ is given by

$$\psi(x) = A' e^{ik_0 x} + B' e^{-ik_0 x} \tag{28}$$

when $0 \leqslant x \leqslant d$ and by

$$\psi(x) = [A' e^{ik_0(x-d)} + B' e^{-ik_0(x-d)}] e^{ikd} \tag{29}$$

when $d \leqslant x \leqslant 2d$.

As before, the equation giving the propagation constant k in terms of k_0 and hence of E, corresponding to (18), may be found by applying the conditions for continuity of ψ and $\partial\psi/\partial x$ at $x = d$ and is

$$\cos kd = \cos k_0 d - \alpha_0 d \sin k_0 d/k_0 d \tag{30}$$

As $\alpha_0 d \to 0$ we have $k_0 \to k$ so that $E \to \hbar^2 k^2/2m$ as for free electrons; for other values of $\alpha_0 d$ we must solve equation (30) to obtain k_0, and hence E, as a function of k. If we write the right-hand side of equation (30) as $g(k_0 d)$ we see that we only have real values of k corresponding to allowed values of E, provided $-1 \leqslant g(k_0 d) \leqslant 1$; the function $g(x)$ is plotted in Fig. 4.7 for the three values of $\alpha_0 d$ previously discussed, namely $\alpha_0 d = 1$, 2, 3. We note that, for a range of values of k_0 just less than $2\pi/d$, $g(k_0 d)$ is greater than 1; this implies that values of the energy corresponding to these values of k_0 are not permitted. From the form of $g(x)$ it will be seen that $g(2n\pi) = 1$ for all integral values of n and, for values of $k_0 d$ just less than $2n\pi$, $g(k_0 d) > 1$; thus we have a whole series of *forbidden* energy bands whose upper limits are given by $k_0 = 2n\pi/d$. Similarly we have $g[(2n+1)\pi] = -1$ and $g(k_0 d) < -1$ when k_0 is just less than $(2n+1)\pi/d$; we have therefore another series of forbidden energy bands whose upper

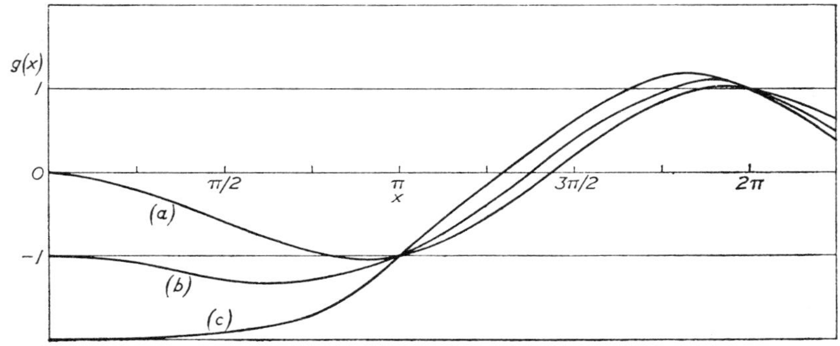

FIG. 4.7. The function $g(x) = \cos x - (x_0 \sin x)/x$, (a) $x_0 = 1$, (b) $x_0 = 2$, (c) $x_0 = 3$

edges occur at $k_0 = (2n+1)\pi/d$. When $E > 0$ we might have expected the electrons to be essentially free and to take on any value of the kinetic energy as for an isolated well; the periodicity of the array, however, introduces this feature of forbidden energy bands. These are quite analogous to the forbidden frequency bands for propagation of oscillations along a periodic array of particles as discussed in Chapter 3. Here energy and frequency are effectively the same thing and it is not therefore surprising that this feature should again appear; it is the most remarkable new effect introduced by the wave-mechanical treatment. For large values of n and hence of E the forbidden bands become very narrow and the electrons may be regarded as substantially free. The first few forbidden bands are, however, of appreciable width, unless $\alpha_0 d$ is very small, and impose an important restriction on the values which the energy may take when electrons are ejected from the lowest band which they will generally occupy. This restriction of energies to certain bands is, as we shall see, a characteristic feature of motion in a periodic potential and not just a peculiarity of the particular potential we have chosen for this detailed analysis. This potential has the advantage of providing an analytical solution which enables us to determine the forbidden energy bands exactly and to compute the energy in the allowed bands as a function of the propagation constant k. We shall later discuss in more detail the physical significance of the propagation constant k, but in the meantime, an analogy with the corresponding quantity for free electrons (see § 1.2.1) we may refer to the quantity $\hbar k$ as the momentum of the electron in the periodic array. It is frequently called the *crystal momentum* to distinguish it from the momentum of a perfectly free electron. That

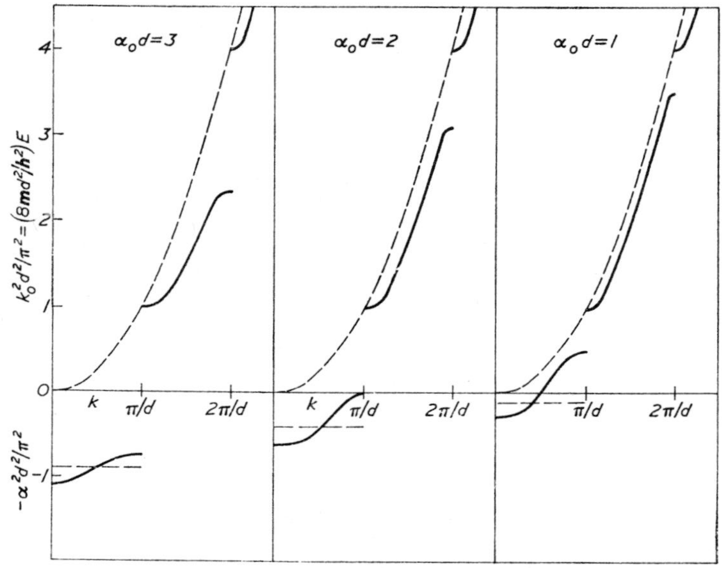

FIG. 4.8. Energy bands in the expanded representation ($\alpha_0 d = 1, 2, 3$)

there is some distinction is clear from the fact that k is not uniquely defined, the quantity $k + 2n\pi/d$ being equally good as a propagation constant (see § 3.8).

The upper edges of the allowed bands in each case must be obtained from Fig. 4.7, or from the conditions that when $k = 0$ we have

$$k_0/\alpha_0 = -\cot \tfrac{1}{2}k_0 d \tag{31}$$

and when $k = \pi/d$ we have

$$k_0/\alpha_0 = \tan \tfrac{1}{2}k_0 d \tag{32}$$

When $\alpha_0 d > 2$ the energies corresponding to $0 < k_0 d < \pi$ are forbidden, the first allowed band corresponding to $E > 0$ beginning with $k_0 = \pi/d$; this point lies on the E/k curve for free electrons since $k = k_0$. In Fig. 4.8 we have used an expanded representation so as to bring out the relationship with the curve for free electrons. The energy bands could just as well have been shown in a reduced representation; this is shown for $\alpha_0 d = 3$ in Fig. 4.9. For $\alpha_0 d = 2$ we have seen that the lowest band ($E \leqslant 0$) just reaches to $E = 0$; the region immediately above is forbidden. When $\alpha_0 d < 2$ however, the region just above $E = 0$ becomes allowed and joins on to complete the band for $E < 0$; the value $E = 0$ occurs within the band, and the upper edge again occurs for $k = \pi/d$. For $\alpha_0 d = 1$ this edge occurs when $k_0 d = 0{\cdot}745\,\pi$; a forbidden zone then comes between this value and the bottom of the next band at $k_0 d = \pi$.

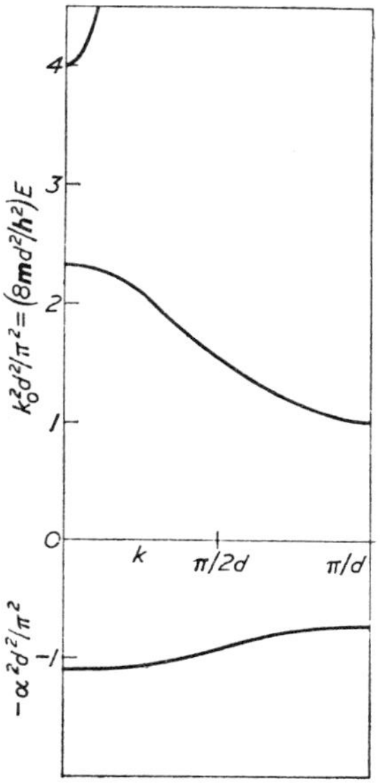

FIG. 4.9. Energy bands in the reduced representation ($\alpha_0 d = 3$)

From Fig. 4.8 it will be seen that, as $\alpha_0 d$ decreases, the lower band gradually moves up into the region for which $E > 0$. Under these conditions the distinction between states corresponding to 'free' and 'bound' electrons loses its meaning, and we can only refer to states belonging to a particular band of energies. In the normal state, an electron will be in the lowest band which it can occupy, and then higher bands will correspond to excited states. We shall now see how these ideas may be applied to discuss the motion of electrons in a crystal.

4.3 Application to the motion of electrons in a crystalline solid

The problem we have discussed in § 4.2 may be applied to throw some light on the difficulties outlined in § 2.13 concerning the motion of electrons in crystalline solids. It was shown that there appeared to be no point in associating an electron, even in a deep level, with a particular atom, since it would quickly leak through the potential barrier dividing it from a neighbouring atom. The array of potential wells discussed in § 4.2 is a crude but not too unrealistic one-dimensional representation of the narrow deep potential wells created by the nuclear attractive fields of the atoms of the crystal. It is most useful for the discussion of deep atomic levels. Suppose such a level is at depth E_0 in an isolated atom; if we express E_0 in terms of the ionisation energy of a hydrogen atom W_H, we have, in the notation of § 4.2,

$$\alpha_0^2 d^2 = E_0 d^2 / W_H a_0^2$$

If we take $d = 10a_0 \simeq 5 \times 10^{-8}$ cm, and $E_0 = 100 W_H$, we have $\alpha_0 d = 100$. We thus see that we shall have large values of $\alpha_0 d$ for all but the valence electrons of the isolated atoms. Because of the presence of the neighbouring atoms, the single energy level of the isolated atom is spread into a band, but with such large values of $\alpha_0 d$ the band is a very narrow one. The wave function is spread over all the atoms of the crystal and not localised on any particular one, and this is just as it should be, since owing to the tunnel effect we cannot attribute an electron for very long to any particular atom of the crystal (see § 2.13). Near each atom, however, the form of the wave function is approximately the same, apart from a normalising constant, as that for an isolated atom when $\alpha_0 d$ is large. If therefore we superimpose just the right number of independent wave functions to give the right number of electrons we get a result that does not differ appreciably from what we would get by treating the individual atoms of the crystal as isolated. Thus we are justified in doing this for the deep-lying levels.

To calculate the number of energy levels in the band which we must use to represent the crystal, we have to consider more carefully the boundary conditions. There are difficulties associated with the boundaries of finite crystals, as we have seen in § 2.5, and it is convenient to use travelling waves, although the same results may be obtained by assuming, for example that the wave function $\psi(x)$ vanishes at the boundary; we therefore use periodic boundary conditions. Suppose we have an infinite number of potential wells but that the system repeats itself exactly in groups of N, i.e. in a length L where $L = Nd$. The condition that $\psi(x)$, in the form given by equation (4), should be periodic with period L is that we must have

$$k = 2\pi r/L = 2\pi r/Nd \tag{33}$$

where r is an integer; since we restrict k to the range $-\pi/d < k \leqslant \pi/d$ this leads to exactly N values. *Thus within each energy band there are exactly* N *discrete energy levels which are permitted.* When N is large these lie very close together so that for most purposes we may regard the energy band as continuous; it is sometimes, however, necessary to remember that each band is made up of a large number of discrete levels. Let us assume that each atom (well) in the array contributes one electron; we then have N electrons in separate states in the band, each contributing to the charge density by addition of the appropriate value of $e|\psi|^2$. When $\alpha_0 d$ is large we thus get, as we have seen, approximately the same charge density as for N isolated atoms. The wave-mechanical description using travelling waves is more logical, but practically leads to the same energy levels and charge density for the deep-lying states as we should get by regarding the atoms as isolated.

Suppose we now wish to know to what extent the energy bands are occupied by electrons. Remembering that we may put two electrons, because of their spin, on each level, we see that with one electron per atomic level we should just fill half the allowed levels of a band—the lowest half of the lowest band first. All deep-lying states have, however, two electrons with opposite spin in each, and for these we should assume two electrons per atom. These then fill *completely* the allowed levels in the band which arises from this (double) state. One very important consequence is that the electrons in the deep atomic levels, although free to move through the crystalline lattice, cannot take part in conduction. There are no neighbouring states to which they can go without a large energy change, and to each state with a certain value of k there exists a filled state with the value $-k$, so that the full band carries no net current. This is why these electrons, although in a sense 'free', may be left out of account in conduction phenomena, and why the Sommerfeld model of a metal, which ignores them, is not a bad approximation; we shall return to this point later (see § 4.13).

While the simple problem we have discussed gives a fair description of what happens to a deep-lying atomic level in a crystal, it is clearly inadequate to discuss the behaviour of the conduction electrons. For these, $\alpha_0 d$ is not large, and also the potential well has width comparable with the atomic spacing. Moreover, the deep narrow well gives only a single level which must be pre-selected. A much better approximation is provided by considering an array of potential wells of finite width and depth as shown in Fig. 4.1, and we shall later discuss this problem for which an exact solution is also possible, as shown by R. de L. Krönig and W. J. Penney* who first used it to discuss conduction in metals. This involves some slightly more complex mathematics. The simpler problem already discussed brings out some very important results which will also be found for this potential. In particular, we shall again find allowed energy bands separated by regions of forbidden energy.

4.4 Normalisation of the wave function

Before leaving the problem of the array of deep narrow potential wells, let us see how the wave function for this type of problem may be normalised. This is only possible when the boundary conditions have been

* R. de L. Krönig and W. J. Penney, *Proc. Roy. Soc.*, **A130**, 499 (1930).

specified, and we shall carry out the normalisation procedure for the periodic boundary conditions discussed in § 4.3. For a perfectly free electron, subject to the periodic boundary conditions, we have

$$\psi(x) = A\,e^{ikx} \tag{34}$$

with $k = 2\pi r/L = 2\pi r/Nd$, r being an integer. If we assume that we have one electron for each period L we have

$$\int_0^L |\psi|^2\,dx = 1 = LA^2 \tag{35}$$

so that the normalised wave function is

$$\psi = L^{-1/2}\,e^{ikx} = N^{-1/2}\,d^{-1/2}\,e^{ikx} \tag{36}$$

Alternatively we may normalise $\psi(x)$ to represent a unit current of particles, i.e. one particle crossing unit area in unit time. The particle current is $v|\psi|^2$ where v is the electron's velocity given by $v = \hbar k/\boldsymbol{m}$. Thus if $\psi(x)$ represents unit particle current we have

$$\begin{aligned}\psi(x) &= v^{-1/2}\,e^{ikx}\\ &= (\boldsymbol{m}/\hbar k)^{1/2}\,e^{ikx}\end{aligned} \tag{37}$$

Let us now consider the wave function for an electron in the periodic potential. From equations (13) and (22) we have when $E < 0$

$$\psi(x) = C[e^{-\alpha x}(1-e^{(ik-\alpha)d}) + e^{ikd}.e^{-\alpha(d-x)}(1-e^{-(ik+\alpha)d})] \quad 0 \leqslant x \leqslant d \tag{38}$$

where C is a constant. For large values of αd

$$\psi(x) \simeq C[e^{-\alpha x} + e^{ikd}\,e^{-\alpha(d-x)}] \quad 0 \leqslant x \leqslant d \tag{39}$$

which is equivalent to equation (24). Equation (38) may be expressed in the rather more elegant form

$$\psi(x) = D[\sinh\alpha(d-x) + e^{ikd}\sinh\alpha x] \quad 0 \leqslant x \leqslant d \tag{40}$$

where D also is a constant. We have

$$\left.\begin{aligned}|\psi(x)|^2 = D^2[\sinh^2\alpha x + 2\cos kd\sinh\alpha x\sinh\alpha(d-x) + \sinh^2\alpha(d-x)]\\ 0 \leqslant x \leqslant d\end{aligned}\right\} \tag{41}$$

The normalising condition is given by

$$1 = \int_0^L |\psi(x)|^2\,dx = \int_0^L |A(x)|^2\,dx \tag{42}$$

using equation (4) for ψ. Because $A(x)$ is periodic with period d, equation (42) may also be written in the form

$$1 = N\int_0^d |A(x)|^2\,dx = N\int_0^d |\psi(x)|^2\,dx \tag{43}$$

We may then insert the form given by equation (41) for $|\psi(x)|^2$ and carry out the integrations to obtain the constant D. We thus obtain for the normalised wave function the expression

$$\left.\begin{aligned}\psi_k(x) &= \frac{\alpha^{1/2}}{N^{1/2}}\frac{\sinh\alpha(d-x)+e^{ikd}\sinh\alpha x}{\{(\sinh\alpha d\cosh\alpha d-\alpha d)+\cos kd(\alpha d\cosh\alpha d-\sinh\alpha d)\}^{1/2}}\\ &\qquad 0\leqslant x\leqslant d\end{aligned}\right\}\tag{44}$$

It will readily be verified that, when $e^{-\alpha d}$ is small, the expression (44) gives, as $\alpha\to\alpha_0$, for small values of x

$$\psi\to\alpha_0^{1/2}N^{-1/2}e^{-\alpha_0 x}\tag{45}$$

which is the normalised wave function for a single well multiplied by $N^{-1/2}$. Of particular interest are the wave functions when $k=0$ and $k=\pi/d$. When $k=0$ we have

$$\psi_{10}(x)=\frac{2^{1/2}\alpha^{1/2}N^{-1/2}\cosh\alpha\left(\dfrac{d}{2}-x\right)}{(\sinh\alpha d+\alpha d)^{1/2}}\qquad 0\leqslant x\leqslant d\tag{46}$$

We use the notation ψ_{nk} to denote the wave function for the nth band corresponding to a particular value of kd. When $k=\pi/d$ we have

$$\psi_{1\pi}(x)=\frac{2^{1/2}\alpha^{1/2}N^{-1/2}\sinh\alpha\left(\dfrac{d}{2}-x\right)}{(\sinh\alpha d-\alpha d)^{1/2}}\qquad 0\leqslant x\leqslant d\tag{47}$$

It is interesting to note that $\psi_{1\pi}(x)$ has a node at $x=d/2$, and corresponds to an energy which is higher than that corresponding to $\psi_{10}(x)$. This is a fairly general property of wave functions—the greater the number of nodes the higher the energy*. $\psi_{1\pi}$ and ψ_{10} are shown in Fig. 4.5 for a fairly large value of $\alpha_0 d$.

When $E>0$ we may similarly express $\psi(x)$ in the form

$$\psi(x)=C'[\sin k_0(d-x)+e^{ikd}\sin k_0 x]\qquad 0\leqslant x\leqslant d\tag{48}$$

where C' is a constant. Proceeding as before we obtain for the normalised wave function

$$\left.\begin{aligned}\psi_k &= \frac{k_0^{1/2}N^{-1/2}[\sin k_0(d-x)+e^{ikd}\sin k_0 x]}{\{(k_0 d-\sin k_0 d\cos k_0 d)+\cos kd(\sin k_0 d-k_0 d\cos k_0 d)\}^{1/2}}\\ &\qquad 0\leqslant x\leqslant d\end{aligned}\right\}\tag{49}$$

* This follows from a theorem on the order of the eigenvalues of a differential equation, known as Haupt's theorem (see, for example, A. H. Wilson, *The Theory of Metals*, Cambridge University Press (1954), p. 23). The condition is clear for the case of free electrons where the wave functions are simple linear combinations of $\cos kx$ and $\sin kx$. The higher values of k correspond to wave functions having the greater number of nodes, and the energy is proportional to k^2.

As we have seen, $k_0 \to k$ as $\alpha_0 \to 0$ or as $k_0 \to \infty$. We then have as $k \to k_0$,

$$\psi \to \frac{k^{1/2} N^{-1/2} e^{ikx} \sin kd}{\{k_0 d \sin^2 k_0 d\}^{1/2}}$$

Provided $k_0 d \neq r\pi$, where r is an integer, we have

$$\psi \to N^{-1/2} d^{-1/2} e^{ikx} \tag{50}$$

This is the normalised wave function for a perfectly free electron. The conditions $k_0 = r\pi/d$ and $k = 0$ and π/d are particularly interesting; they correspond to band edges and the limiting form of the wave function requires rather careful treatment.

Let us consider first the condition $k = 0$; we may either have $k_0 d \neq 2n\pi$, where n is an integer or $k_0 d = 2n\pi$, and these correspond, respectively, to the top edge of the second band and the bottom edge of the third band, when $n = 1$. For the former we may proceed at once to obtain the limiting form of the wave function by putting $k = 0$ in (49); we have then

$$\psi_{20} = \frac{2^{1/2} k_0^{1/2}}{N^{1/2}} \frac{\cos k_0 \left(\dfrac{d}{2} - x\right)}{(k_0 d + \sin k_0 d)^{1/2}} \qquad 0 \leqslant x \leqslant d \tag{51}$$

This wave function is shown in Fig. 4.10, and corresponds to a standing wave—a feature of all wave functions associated with a band edge. When $k \to 0$ and $k_0 \to 2\pi/d$ we must proceed more carefully. If $k_0 d = 2\pi + \epsilon$, when $kd = \eta$ equation (30) becomes approximately small values of ϵ

$$1 - \tfrac{1}{2}\eta^2 = 1 - \tfrac{1}{2}\epsilon^2 - \frac{\alpha_0 d}{2\pi}\epsilon$$

i.e. ϵ is of the order of η^2 and may be neglected to the first order in η. Thus

$$\psi_{30} \to C'' \sin \frac{2\pi x}{d} (e^{i\eta} - 1)$$

where C'' is a constant, which may be determined by a simple integration which gives

$$\psi_{30} = 2^{1/2} N^{-1/2} d^{-1/2} \sin(2\pi x/d) \tag{52}$$

ψ_{30} has nodes at $x = 0$, $d/2$, d and, as would be expected, corresponds to a higher energy than ψ_{20}; ψ_{30} is also a standing wave. The nodes at $x = 0$ and $x = d$ cause the potential wells to have no effect on the motion, and this is why $k = k_0$ in the expanded representation for this particular value of k_0; this wave function is also shown in Fig. 4.10.

When $k = \pi/d$ we may also determine the wave functions in the same way; if $k_0 \neq (2n+1)\pi/d$ we have

$$\psi_{3\pi} = \frac{2^{1/2} k_0^{1/2} N^{-1/2} \sin k_0 \left(\dfrac{d}{2} - x\right)}{(k_0 d - \sin k_0 d)^{1/2}} \qquad 0 \leqslant x \leqslant d \tag{53}$$

with $2\pi < k_0 < 3\pi$. In this case $\psi_{3\pi}$ has three nodes in the range $0 \leqslant x \leqslant d$, that at $x = d/2$, and two others. The function is shown in Fig. 4.10. When $k \to k_0 \to 2\pi/d$ we have

$$\psi_{2\pi} = 2^{1/2} N^{-1/2} d^{-1/2} \sin(\pi x/d) \tag{54}$$

This function has two nodes in the range $0 \leqslant x \leqslant d$, again at the potential wells, which explains why $k = k_0$.

4.5 Space harmonics and the propagation constant *k*

We must now examine rather more closely the significance of the propagation constant k. Apart from the fact that we may have negative as well as positive values of k we may restrict its value, as we have already seen, to the range $0 \leqslant k \leqslant \pi/d$ and use the reduced representation. The equally good propagation constants $k \pm 2n\pi/d$, where n is a positive integer, lead to no further information about the energy or wave functions, but it is frequently convenient to use the range $0 \leqslant k \leqslant \pi/d$ to represent the lowest band, $\pi/d < k \leqslant 2\pi/d$ for the next band and generally $(r-1)\pi/d \leqslant k \leqslant r\pi/d$ for the rth band. In this representation, which we have called in Chapter 3 the expanded representation, the energy is a single-valued function of k, whereas in the reduced representation it is a multiple-valued function of k (cf. Figs. 4.8 and 4.9). A similar relationship was seen to hold between the frequency of lattice vibrations and the propagation constant k. There is, however, one very important difference between the two types of problem; for the lattice vibrations, the wave function had physical significance *only at the periodic lattice points*, whereas for the present problems the electronic wave functions have meaning for *all values of the variable* x permitted by the boundary conditions. The consequence is that for the lattice vibrations the higher values of $|k|$ (greater than π/d) are *exactly* equivalent and have the same physical meaning as the corresponding values in the range $0 \leqslant |k| \leqslant \pi/d$. For the electron waves, however, the higher values of k can, as we shall see, be given a physical interpretation in terms of plane waves.

The solutions for the lattice vibrations consist of a finite set of plane waves. For the electron waves the solutions are in the form

$$\psi = A_k(x)\, e^{ikx} \tag{55}$$

or combinations of such functions, where $A_k(x)$ is a periodic function with the periodicity of the potential, so that the functional form of ψ in (55) does *not* represent, in general, a *plane* wave; it only does so when $A_k(x)$ is a constant. Although we have called k a propagation constant it is not strictly one as defined in Chapter 1, where we considered only plane waves or combinations of plane waves with slightly *different* values of k. The wave function in equation (55) may, however, be expanded as a series of plane waves; the function $A_k(x)$, being periodic with period d, may be expanded as a Fourier series of the form

$$A_k(x) = \sum_{n=-\infty}^{\infty} a_n \mathrm{e}^{2\pi inx/d} \tag{56}$$

Thus we have

$$\psi(x) = \sum_{n=-\infty}^{\infty} a_n \mathrm{e}^{ik_n x} \tag{57}$$

where $k_n = k+2\pi n/d$, so that all the propagation constants of the form $k+2\pi n/d$ are represented, and the quantity k_n may be regarded as the propagation constant of the nth plane wave making up the wave function. Now suppose we give k a particular value k_1, in the range $0 \leqslant k_1 \leqslant \pi/d$, in the wave function represented by equation (57). Then we have propagation constants k lying in the following ranges, $0 \leqslant k \leqslant \pi/d$, $2\pi/d \leqslant k \leqslant 3\pi/d, \ldots$ $-2\pi/d \leqslant k \leqslant -\pi/d$, $-4\pi/d \leqslant k \leqslant -3\pi/d \ldots$. (These are the ranges for which we have full-line curves in the lowest band of Fig. 4.11). We note that we have both positive and negative values of k_n although (55) represents, as we may suppose by comparison with free electrons, a stream of particles travelling to the *right*; we shall consider this point more carefully later. Again if we restrict k_1 to the range $-\pi/d \leqslant k_1 \leqslant 0$ we have propagation constants lying in the ranges $-\pi/d \leqslant k \leqslant 0$, $\pi/d \leqslant k \leqslant 3\pi/d$, etc., i.e. the ranges having broken curves in the lowest band of Fig. 4.11. The question arises, which of the quantities k_n should we call *the* propagation constant; the answer is that it really does not matter very much so long as we choose one. Frequently we choose k_1 and in this case the propagation constant is restricted to the range $-\pi/d \leqslant k \leqslant \pi/d$. (Note that we must include the range $-\pi/d \leqslant k \leqslant 0$ if we wish to have electrons travelling to the left; although we frequently use the range $0 \leqslant k \leqslant \pi/d$ to represent k—the reduced representation—the negative values lying in the range $-\pi/d \leqslant k \leqslant 0$ are always implied as being equally possible.)

Another alternative is to use the expanded representation and to use k_1 when $\psi(x)$ refers to the lowest allowed energy band, $k_1+2\pi/d$ for the next, and so on, and there is some logic as well as convenience in such a choice. In a series such as (57) there will generally be one value of $|a_n|$ which is greater than the others and the corresponding plane wave may be termed the dominant mode; the other plane waves are termed space harmonics. We may then choose as *the* propagation constant that corresponding to the dominant mode. If we look at Fig. 4.10 we shall see that the wave functions for the second band have two nodes, for the third three, etc. Since from the theory of Fourier series, it is well known that

$$a_n = \frac{1}{d}\int_0^d A_k(x)\, \mathrm{e}^{-2\pi inx/d}\, \mathrm{d}x \tag{58}$$

we see that a_1 is likely to be greatest when $A_k(x)$ has not more than one zero, a_2 will be greater when $A_k(x)$ has two zeros, and so on. Indeed for the special wave functions corresponding to the band edges, the values $n = \pm 1$, or $n = \pm 2$, etc., are the only values of n represented as will be seen from the form of the wave functions $\psi_{2\pi}$ and ψ_{30}. This will perhaps be clearer when we determine the constants a_n for the wave functions for the problem discussed in § 4.2. Such a representation is just that which we

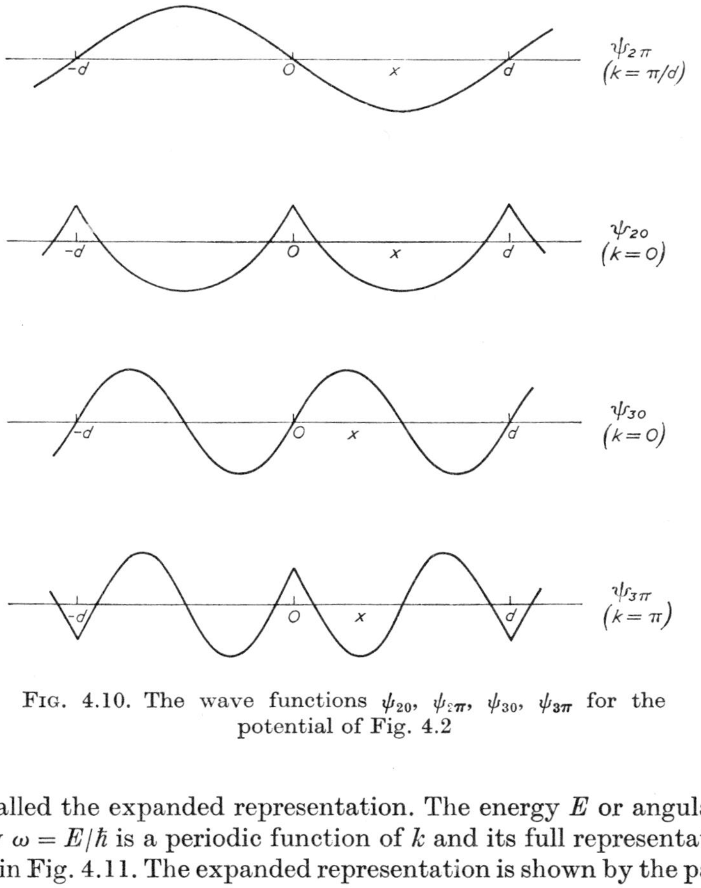

FIG. 4.10. The wave functions ψ_{20}, $\psi_{2\pi}$, ψ_{30}, $\psi_{3\pi}$ for the potential of Fig. 4.2

have called the expanded representation. The energy E or angular frequency $\omega = E/\hbar$ is a periodic function of k and its full representation is shown in Fig. 4.11. The expanded representation is shown by the parts of the curves in heavier type, each of the other parts corresponds to a space harmonic.

For the wave function given by equation (48) we have

$$a_n = \frac{D}{d}\int_0^d \sin k_0(d-x)\,e^{-i(k+2\pi n/d)x}\,dx + \frac{D\,e^{ikd}}{d}\int_0^d \sin k_0 x\,e^{-i(k+2\pi n/d)x}\,dx \tag{59}$$

The integrals may be readily evaluated and we obtain

$$\frac{a_n}{a_0} = \frac{k^2 - k_0^2}{(k + 2\pi n/d)^2 - k_0^2} \tag{60}$$

This shows that a_n is greatest when $(k + 2\pi n/d)$ is most nearly equal to k_0 and gives the justification of the expanded representation as shown in Fig. 4.8. In particular if we have $k = \pm k_0 = r\pi/d$ where r is an integer, we have $a_n = 0$ unless $n = \pm r$; also since $|a_n| = |a_{-n}|$ we have a standing wave.

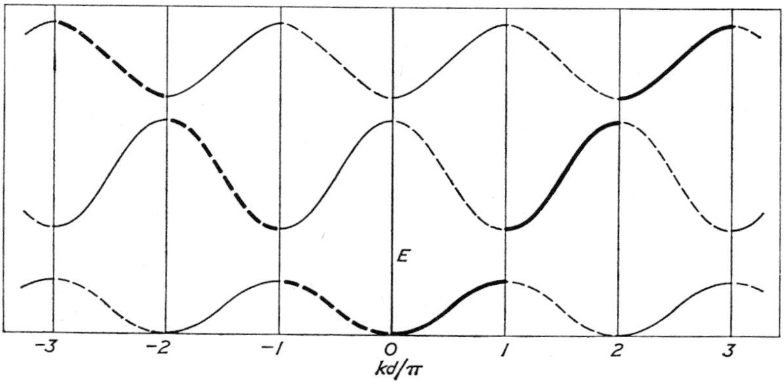

FIG. 4.11. Full ω/k curves for electron waves in a periodic potential

4.6 The average velocity of an electron

From the ω/k curves shown in Fig. 4.11 it will be seen that the waves corresponding to each space harmonic will, for a given value k_1 in the interval $0 \leqslant k_1 \leqslant \pi/d$, have a different phase velocity ω/k_n; indeed some of the harmonics will have positive phase velocities and some negative. As we have seen for free electrons, however, the phase velocity has no physical significance and it is the group velocity $U = d\omega/dk$ which has. We note that the group velocity is the same for each space harmonic for a given value of k_1 and is always positive if $0 < k_1 < \pi/d$. Similarly the group velocity U is always negative if $-\pi/d < k_1 < 0$. This is shown for a single band in Fig. 4.12. The phase velocity of the dominant mode, say at $k = k_1$, is ω_1/k_1 where ω_1 is the value of $E/\hbar$ for $k = k_1$ (and for $k = k_1 + 2\pi n/d$). The phase velocity of the first space harmonic with $k > k_1$ is $\omega_1/(k_1 + 2\pi/d)$ and that for the first space harmonic with $k < k_1$ is $-\omega_1/(2\pi/d - k_1)$; for each of these the group velocity is $(d\omega/dk)_1$, i.e. the value of $d\omega/dk$ at $k + k_1$. (It has the same value at $k = k_1 + 2\pi/d$ and at $k = k_1 - 2\pi/d$, etc.)

Now suppose we form a wave packet with solutions of the form (55) but having a small range of values of k having an average value $\bar{k}$. This may be regarded as the superposition of an infinite number of packets of plane waves with k values near $\bar{k}$, $\bar{k}+2\pi/d$, etc., and since all these packets move with the same group velocity U we see that the whole wave packet will move with velocity U. The wave packet will now be represented by an expression of the form

$$\psi(x) = \int_{-\infty}^{\infty} a(k)\, A_k(x)\, e^{ikx}\, dk \tag{61}$$

We shall suppose that $a(k)$ is effectively zero except over a range Δk of k such that $\Delta k \ll \pi/d$; from the considerations of § 1.3 it will be seen that this is equivalent to saying that the wave function $\psi(x)$ is appreciable

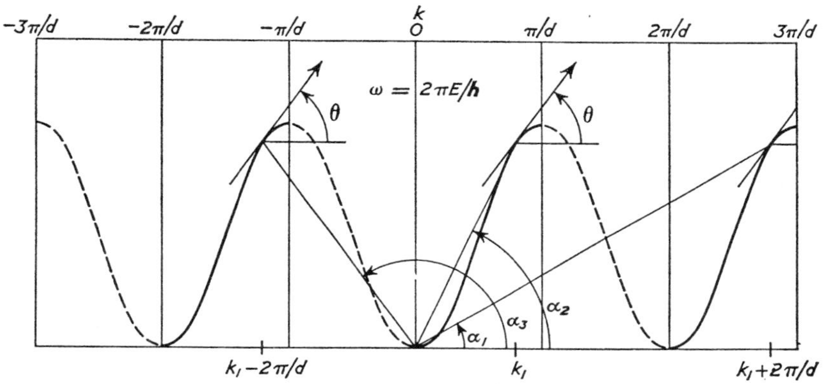

FIG. 4.12. Phase and group velocities of the space harmonic waves. $V_1 = \tan\alpha_1$, etc. $U = \tan(\theta)$

over a distance of the order D where $D \gg d$, i.e. over many lattice spacings. For this small range of k the variation of $A_k(x)$ with k will be small and we may write the wave function $\psi(x)$ in the approximate form

$$\psi(x) = A(x)_{\bar{k}} \int_{-\infty}^{\infty} a(k)\, e^{ikx} dk \tag{62}$$

where $\bar{k}$ is the mean value of k for the range Δk. Again, writing $k' = k-\bar{k}$, we have

$$\psi(x) = A(x)_{\bar{k}}\, e^{i\bar{k}x} \int_{-\infty}^{\infty} a(k'+\bar{k})\, e^{ik'x} dx$$

$(k'+\bar{k})$ being appreciable only for small values of k'. We may now apply to this wave packet the arguments used in § 1.9 to show that it moves with group velocity U given by $U = (d\omega/dk)_{k=\bar{k}}$. This velocity U may then be taken to represent the average velocity of the electron which the wave packet represents. For a given value of the energy E and hence of k it is a constant and is given by

$$U = \frac{d\omega}{dk} = \frac{1}{\hbar}\frac{dE}{dk} \tag{63}$$

For energy bands of the form shown in Fig. 4.8 we note that U is zero at the band edges; this is reasonable since the wave functions then represent standing waves. As we move from the bottom of a band to the top we have a positive value of U if $0 \leqslant \bar{k} \leqslant \pi/d$, passing through a maximum and decreasing to zero at the top edge, as illustrated in Fig. 4.13. This form is a feature of the one-dimensional problem; for real crystals we shall find that the energy is not always a monotonic increasing or decreasing function of k as we cross a band. The value of U at the band edges is, however, always zero, for a non-degenerate band. The actual velocity of the electron varies rapidly as it moves in the periodic potential and it is only its average value which will remain constant in the absence of an external field; this average value is the quantity that is observable on a macroscopic scale, and is given by the rate of motion of the wave

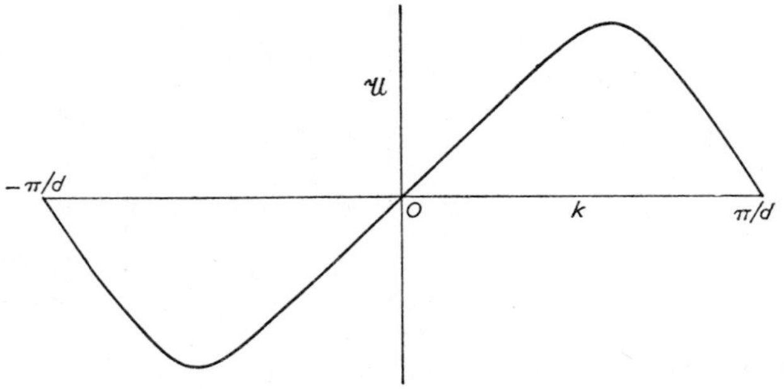

FIG. 4.13. Variation of average velocity U with k

packet, i.e. by U. We shall prove below that if we average the instantaneous velocity of the electron over a single period of the potential we also obtain the value given by equation (63). If the electron moves in a time t a distance B much greater than the periodic distance d (but which may be very small compared with the dimensions of the crystal giving rise to the periodic potential) then we have very nearly $U = B/t$.

The instantaneous value v of the velocity may be obtained from the well-known quantum mechanical expression

$$v = \frac{\hbar}{2mi}\left[\psi^{\star}\frac{\partial\psi}{\partial x} - \psi\frac{\partial\psi^{\star}}{\partial x}\right] \tag{64}$$

(see § 1.2.1 equations (43), (44)). Now let us average this over a periodic distance d; the average value $\bar{v}$ of v is given by

$$\bar{v} = \frac{\hbar}{2mi}\int_0^d \left[\psi_k^{\star}\frac{d\psi_k}{dx} - \psi_k\frac{d\psi_k^{\star}}{dx}\right] dx \Bigg/ \int_0^d \psi_k\psi_k^{\star}\, dx \tag{65}$$

We shall now show that $\bar{v} = \hbar^{-1}\mathrm{d}E/\mathrm{d}k$. Since ψ_k is a solution of Schrödinger's equation we have

$$\frac{\mathrm{d}^2\psi_k}{\mathrm{d}x^2} + \frac{2m}{\hbar^2}[E - V]\psi_k = 0 \tag{66}$$

Let us differentiate this equation with respect to k to obtain an expression for $\mathrm{d}E/\mathrm{d}k$; we obtain

$$\psi_k \frac{\mathrm{d}E}{\mathrm{d}k} = -\frac{\hbar^2}{2m}\cdot\frac{\mathrm{d}^2}{\mathrm{d}x^2}\frac{\partial\psi_k}{\partial k} - (E - V)\frac{\partial\psi_k}{\mathrm{d}k} \tag{67}$$

If we express ψ_k in the form of equation (55) we have

$$\frac{\partial\psi_k}{\partial k} = ix\psi_k + \mathrm{e}^{ikx}\frac{\partial A}{\partial k} \tag{68}$$

Differentiating equation (68) twice gives

$$\frac{\mathrm{d}^2}{\mathrm{d}x^2}\left(\frac{\partial\psi_k}{\partial k}\right) = ix\frac{\mathrm{d}^2\psi_k}{\mathrm{d}x^2} + 2i\frac{\mathrm{d}\psi_k}{\mathrm{d}x} + \frac{\mathrm{d}^2}{\mathrm{d}x^2}\left(\mathrm{e}^{ikx}\frac{\partial A}{\partial k}\right) \tag{69}$$

If we substitute this in the expression (67) for $\mathrm{d}E/\mathrm{d}k$ we obtain

$$\frac{2m}{\hbar^2}\psi_k\frac{\mathrm{d}E}{\mathrm{d}k} = -2i\frac{\mathrm{d}\psi_k}{\mathrm{d}x} + \left\{\frac{\mathrm{d}^2}{\mathrm{d}x^2} + \frac{2m}{\hbar^2}(E - V)\right\}\left(\mathrm{e}^{ikx}\frac{\partial A}{\partial k}\right) \tag{70}$$

Let us multiply this equation by $\psi^\star$ and integrate; we obtain from the left-hand side

$$\frac{2m}{\hbar^2}\frac{\mathrm{d}E}{\mathrm{d}k}\int_0^d \psi_k^\star\psi_k\,\mathrm{d}x \tag{71}$$

Now let us evaluate the integral

$$\int_0^d \psi^\star\frac{\mathrm{d}^2}{\mathrm{d}x^2}\left(\mathrm{e}^{ikx}\frac{\partial A}{\partial k}\right)\mathrm{d}x \tag{72}$$

Two integrations by parts give

$$\left[\psi^\star\frac{\mathrm{d}}{\mathrm{d}x}\left(\mathrm{e}^{ikx}\frac{\partial A}{\partial k}\right) - \mathrm{e}^{ikx}\frac{\partial A}{\partial k}\cdot\frac{\mathrm{d}\psi^\star}{\mathrm{d}x}\right]_0^d + \int_0^d \mathrm{e}^{ikx}\frac{\partial A}{\partial k}\cdot\frac{\mathrm{d}^2\psi^\star}{\mathrm{d}x^2}\mathrm{d}x \tag{73}$$

The first term will be seen to vanish because of the periodicity of the functions $A(x)$ and $\mathrm{d}A/\mathrm{d}x$, as will be readily seen on expressing the various quantities in terms of $A(x)$ and $A^\star(x)$. Replacing the integral (72) by the integral in (73) in equation (70) we have for the integral arising from the second term on the right of equation (70)

$$\int_0^d \mathrm{e}^{ikx}\frac{\partial A}{\partial k}\left[\frac{\mathrm{d}^2\psi^\star}{\mathrm{d}x^2} + \frac{2m}{\hbar^2}(E - V)\psi^\star\right]\mathrm{d}x \tag{74}$$

$\psi^\star$, however, satisfies Schrödinger's equation and so the integral vanishes. The first term on the right of equation (70) then gives the integral

$$-2i\int_0^d \psi_k^\star \frac{d\psi_k}{dx}\,dx \tag{75}$$

An integration by parts shows that this is also equal to

$$2i\int_0^d \psi_k \frac{d\psi_k^\star}{dx}\,dx \tag{76}$$

the integrated term vanishing because of the periodicity of $A(x)$. Combining equations (71), (75), (76) we obtain

$$\bar{v} = \frac{1}{\hbar}\frac{dE}{dk} = \frac{\hbar}{2mi}\int_0^d \left(\psi_k^\star \frac{d\psi_k}{dx} - \psi_k \frac{d\psi_k^\star}{dx}\right) dx \Big/ \int_0^d \psi_k \psi_k^\star\,dx \tag{77}$$

and we see that $\bar{v}$ and U, the group velocity of a wave packet, are the same.

The result which we have proved is a very remarkable one; it shows that for an electron moving in a perfect periodic potential under the influence of no *external* forces the average velocity is a constant, and as we shall see later, this conclusion also holds in three dimensions. In a perfect crystal such an electron would therefore move steadily through the crystal and would not be 'scattered' by the atoms of which it is composed. A perfect crystal would therefore offer no resistance to the electron's motion and we must look elsewhere than to collisions with the majority of the atoms for the electrical resistance of real crystals; we shall return to this later. We see again, however, why Sommerfeld's free-electron model is not a bad description of a metal, since electrons moving in a perfect periodic potential may be regarded as essentially free.

4.7 The crystal momentum *P*

Let us now introduce the quantity P given by

$$P = \hbar k \tag{78}$$

called the crystal momentum, in order to distinguish it from the real momentum of the electron mv. (We have already introduced a similar quantity, the phonon momentum in connection with lattice vibrations (see § 3.10).) In terms of P we have

$$\bar{v} = U = \frac{dE}{dP} \tag{79}$$

In classical mechanics the velocity v is given in terms of the momentum p by the relationship

$$v = \frac{dE}{dp} \tag{80}$$

when the energy E is expressed as a function of the momentum p. The similarity in the forms of equations (79) and (80) is one reason for calling P the crystal momentum. The true momentum p is given by the operator

$$p\psi = -i\hbar\frac{\partial\psi}{\partial x} \tag{81}$$

$$= P\psi - i\hbar\, \mathrm{e}^{ikx}\frac{\mathrm{d}A}{\mathrm{d}x} \tag{81a}$$

and is not a constant of the motion, whereas P is. Just as U is an average velocity so P is a kind of average momentum; it is not, however, the average of the true momentum p over a period. This is given by

$$\bar{p} = -i\hbar\int_0^d \psi^\star\frac{\partial\psi}{\partial x}\,\mathrm{d}x \bigg/ \int_0^d \psi\psi^\star\,\mathrm{d}x \tag{82}$$

and it will readily be seen on comparing (82) with (65) that

$$\bar{p} = \boldsymbol{m}\bar{v} \tag{82a}$$

This is equal to P only when $A(x)$ is a constant, that is for free electrons.

Near the bottom of an allowed energy band we may write

$$E = E_1 + Ak^2$$

where A is a constant. Let us write for A the quantity $\hbar^2/2m^\star$ defining another constant $m^\star$; then we have

$$E = E_1 + \hbar^2 k^2/2m^\star$$

Expressing $\bar{v}$ and $\bar{p}$ in terms of $m^\star$ we have

$$\bar{v} = \hbar k/m^\star = P/m^\star \tag{83}$$

$$\bar{p} = \boldsymbol{m}\bar{v} = \boldsymbol{m}P/m^\star \tag{84}$$

The quantity $m^\star$ is called the effective mass, since, in the crystal, an electron behaves in many respects like a free particle of mass $m^\star$.

Near the top of a band which occurs at $k = 0$ we may approximate to E by an expression of the form

$$E = E_2 - \hbar^2 k^2/2m^\star \tag{85}$$

(the value of $m^\star$ will, in general, be different now), and we have

$$\bar{v} = -\hbar k/m^\star = -P/m^\star \tag{86}$$

$$\bar{p} = -\boldsymbol{m}P/m^\star \tag{87}$$

The quantity P is not uniquely defined since we may add to it any integral multiple of $\boldsymbol{h}/d$; in this way we may make P positive when $\bar{v}$ is positive if we wish to, but, in general, we do not trouble to do this.

We shall now discuss another reason for calling P a form of momentum. When no external field acts on an electron other than that due to the periodic potential, P is constant and the electron moves with constant average velocity $\bar{v}$. Suppose we now apply an electric field which produces a force F on the electron. (We shall suppose this field to be small compared with the very large fields due to the periodic potential.) The quantity P will no longer be an eigenvalue of the complete Schrödinger equation and will begin to change its value at a rate which we may calculate. The rate at which the external force F does work, averaged over a time long compared with the time the electron takes to travel a distance d, when its average velocity is $\bar{v}$ will be equal to $F\bar{v}$. Thus we have

$$\frac{\mathrm{d}E}{\mathrm{d}t} = \frac{\mathrm{d}E}{\mathrm{d}P}\cdot\frac{\mathrm{d}P}{\mathrm{d}t} = F\bar{v}$$

and since $\bar{v} = \mathrm{d}E/\mathrm{d}P$, we have

$$\frac{\mathrm{d}P}{\mathrm{d}t} = F \tag{88}$$

Equation (88) is a fundamental one. The derivation which we have given here does not hold under all conditions in three dimensions, when we may include forces at right angles to the motion which do no work, and we shall later give a more general proof. Equation (88) is similar in form to Newton's second law of motion and again the quantity P behaves in this respect like a momentum.

When the energy is given by the form appropriate to the bottom of an allowed energy band we have

$$\frac{\mathrm{d}P}{\mathrm{d}t} = m^{\star}\frac{\mathrm{d}\bar{v}}{\mathrm{d}t} = F \tag{89}$$

Thus we see that an electron near the bottom of an allowed energy band is accelerated as though it had a mass $m^{\star}$. For very narrow bands $m^{\star}$ is, as we have seen, very large and electrons in such bands are very difficult to accelerate, quite apart from considerations of availability of states into which they can be accelerated. For electrons near the top of a band we have

$$-\frac{\mathrm{d}P}{\mathrm{d}t} = m^{\star}\frac{\mathrm{d}\bar{v}}{\mathrm{d}t} = -F \tag{90}$$

such electrons therefore behave as though they had a negative mass, i.e. they tend to be slowed down by a field which would speed up a free electron. We shall later discuss this in much more detail (see § 4.13.1).

Suppose we start with an electron near the bottom of a band and having energy E, velocity $\bar{v}$ equal to zero, and crystal momentum P equal to zero. As the field is applied P will increase and so will $\bar{v}$. At a certain value of P, $\bar{v}$ will reach a maximum value, and for further increase of P,

$\bar{v}$ will begin to decrease reaching zero when P has increased to $\boldsymbol{h}/2d$ (corresponding to $k = \pi/d$). A small further increase in P will bring it to $\boldsymbol{h}/2d+\epsilon$ (corresponding to $\boldsymbol{k} = \pi/d+\hbar^{-1}\,\epsilon$) and this is equivalent to $P = -\boldsymbol{h}/2d+\epsilon$. $\bar{v}$ will then have a small negative value which will increase in magnitude, then reduce to zero and again become positive, so that the electron will appear to oscillate about a mean position. The reversal in velocity is simply due to reflection of the electron when its velocity reaches the critical value such that the wavelength $\lambda = 2d$. In practice the electron will make collisions with impurity atoms or lattice defects before this cycle of events can be accomplished and, as we shall see later, its average velocity in a metal cannot be changed very much. In an insulator it may be appreciably changed and this leads to some interesting effects.

There is, however, another interesting possibility which we have neglected. We have assumed that the electron stays in a single band; under the influence of the external field, however, it may be excited into a higher band as was shown by C. Zener.* Unless the field is very large the probability of this happening is extremely small (see § 8.13).

Equation (89) may be generalised as follows: we have

$$\begin{aligned}\frac{\mathrm{d}\bar{v}}{\mathrm{d}t} &= \frac{\mathrm{d}}{\mathrm{d}t}\frac{\mathrm{d}E}{\mathrm{d}P}\\ &= \frac{\mathrm{d}^2 E}{\mathrm{d}P^2}\cdot\frac{\mathrm{d}P}{\mathrm{d}t} = F\frac{\mathrm{d}^2 E}{\mathrm{d}P^2}\end{aligned} \tag{91}$$

if we define an effective mass $m^\star$ by writing (91) in the form

$$m^\star\frac{\mathrm{d}\bar{v}}{\mathrm{d}t} = F \tag{92}$$

we have

$$\frac{1}{m^\star} = \frac{\mathrm{d}^2 E}{\mathrm{d}P^2} \tag{93}$$

which reduces to the same definition as before near the band edges.

4.8 A simple matrix notation

When we come to consider problems more complex than that discussed in § 4.2, we find that, although the principles are similar, the algebra involved in applying the boundary conditions becomes tedious unless this is systematised. A simple matrix notation helps greatly in this and enables us to deal with apparently complex problems in a simple manner.

We have already introduced in Chapter 3 a matrix notation for a linear transformation such as

$$\left.\begin{aligned} x' &= ax+by\\ y' &= cx+dy\end{aligned}\right\} \tag{94}$$

* C. Zener, *Proc. Roy. Soc.* **A145**, 521 (1934).

which we write in the form

$$\begin{pmatrix} x' \\ y' \end{pmatrix} = \mathbf{A}\begin{pmatrix} x \\ y \end{pmatrix} \tag{94a}$$

where **A** represents the matrix (or array)

$$\begin{pmatrix} a & b \\ c & d \end{pmatrix}$$

Now suppose we have two such transformations in sequence; for example suppose we have now

$$\left.\begin{aligned} x'' &= a'\,x' + b'\,y' \\ y'' &= c'\,x' + d'\,y' \end{aligned}\right\} \tag{95}$$

or

$$\begin{pmatrix} x'' \\ y'' \end{pmatrix} = \mathbf{A}'\begin{pmatrix} x' \\ y' \end{pmatrix} \tag{95a}$$

where **A′** is the matrix

$$\begin{pmatrix} a' & b' \\ c' & d' \end{pmatrix}$$

Then if we write

$$\begin{pmatrix} x'' \\ y'' \end{pmatrix} = \mathbf{A}''\begin{pmatrix} x \\ y \end{pmatrix} \tag{96}$$

it will readily be verified by substituting for x, y that **A″** is the matrix

$$\begin{pmatrix} a'\,a + b'c & a'\,b + b'\,d \\ c'\,a + d'\,c & c'\,b + d'\,d \end{pmatrix} \tag{96a}$$

The matrix **A″** is called the product of **A′** and **A** since we may write in an obvious notation

$$\begin{pmatrix} x'' \\ y'' \end{pmatrix} = \mathbf{A}'\begin{pmatrix} x' \\ y' \end{pmatrix} = \mathbf{A}'\,\mathbf{A}\begin{pmatrix} x \\ y \end{pmatrix} \tag{97}$$

so that $\mathbf{A}'' = \mathbf{A}'\mathbf{A}$. We note that the order of multiplication is important and that in general $\mathbf{A}\mathbf{A}' \neq \mathbf{A}'\mathbf{A}$. The row by column multiplication rule (similar to that for determinants except that order is important) will readily be seen.

There are two more elementary results which we require. If we reverse the transformation (94) and solve for x, y in terms of x' y', writing

$$\begin{pmatrix} x \\ y \end{pmatrix} = \mathbf{B}\begin{pmatrix} x' \\ y' \end{pmatrix} \tag{98}$$

the matrix **B** is given by

$$\mathbf{B} \equiv \begin{pmatrix} d & -b \\ -c & a \end{pmatrix} \div \Delta \tag{99}$$

where Δ is the determinant

$$\begin{vmatrix} a & b \\ c & d \end{vmatrix}$$

called the determinant of the matrix **A** and is written $|\mathbf{A}|$. It frequently happens that $|\mathbf{A}| = 1$ and in this case **B** is given by

$$\mathbf{B} \equiv \begin{pmatrix} d & -b \\ -c & a \end{pmatrix} \tag{100}$$

We note that $|\mathbf{B}| = |\mathbf{A}|$. Let us now form the product **BA**; we get simply the matrix

$$\begin{pmatrix} 1 & 0 \\ 0 & 1 \end{pmatrix}$$

This is called the unit matrix and we may write

$$\mathbf{BA} = 1 \tag{101}$$

It is then logical to write the matrix **B** as $\mathbf{A}^{-1}$, and it is called the reciprocal matrix of **A**. If the determinant $|\mathbf{A}|$ is equal to zero we see that the quantities x', y' are not independent but merely a constant multiple of each other, since then $a/c = b/d$.

4.8.1 Use of matrix notation for the application of boundary conditions

The general solution of the equation

$$\frac{\mathrm{d}^2 y}{\mathrm{d}x^2} + k^2 y = 0 \tag{102}$$

may be written as

$$y = A \cos kx + B \sin kx \tag{103}$$

If at $x = 0$ we have $y = y_0$, and also the derivative $y' = y_0'$, then clearly

$$\left.\begin{aligned} y &= y_0 \cos kx + k^{-1} y_0' \sin kx \\ y' &= -k y_0 \cos kx + y_0' \cos kx \end{aligned}\right\} \tag{104}$$

We may therefore use matrix notation to express y and its derivative y' for any value of x in terms of the values at $x = 0$. We have

$$\begin{pmatrix} y \\ y' \end{pmatrix} = \mathbf{A}_x \begin{pmatrix} y_0 \\ y_0' \end{pmatrix} \tag{105}$$

where $\mathbf{A}_x$ is the matrix

$$\begin{pmatrix} \cos kx & k^{-1} \sin kx \\ -k \sin kx & \cos kx \end{pmatrix} \tag{106}$$

and we note that $|\mathbf{A}_x| = 1$. This may clearly be extended to give the solution for any value of x in terms of the solution at a point $x = d$; in an obvious notation we have

$$\begin{pmatrix} y \\ y' \end{pmatrix} = \mathbf{B} \begin{pmatrix} y_d \\ y_d' \end{pmatrix} \tag{107}$$

where **B** is now the matrix

$$\begin{pmatrix} \cos k(x-d) & k^{-1}\sin k(x-d) \\ -k\sin k(x-d) & \cos k(x-d) \end{pmatrix} \tag{108}$$

Now suppose we have to apply boundary conditions at $x = d$. These may frequently be expressed in matrix notation in the form

$$\begin{pmatrix} y_{d+} \\ y'_{d+} \end{pmatrix} = \mathbf{P}\begin{pmatrix} y_{d-} \\ y'_{d-} \end{pmatrix} \tag{109}$$

where y_{d+} is the value of y to the right of $x = d$ and y_{d-} the value to the left of $x = d$, etc., and **P** is a matrix with constant elements. In particular, if both y and y' are continuous then **P** is the unit matrix. Again in the problem discussed in § 4.2, at each of the potential wells we have $\psi_{d+} = \psi_{d-}$, $\psi'_{d+} - \psi'_{d-} = -2\alpha_0\psi_{d+}$ so that

$$\mathbf{P} \equiv \begin{pmatrix} 1 & 0 \\ -2\alpha_0 & 1 \end{pmatrix} \tag{110}$$

For this problem we have ($E > 0$)

$$\begin{pmatrix} \psi_{d-} \\ \psi'_{d-} \end{pmatrix} = \mathbf{A}_d\begin{pmatrix} \psi_{0+} \\ \psi'_{0+} \end{pmatrix} \tag{111}$$

where $\mathbf{A}_d$ is the matrix (106) with $x = d$. Thus we have

$$\begin{pmatrix} \psi_{d+} \\ \psi'_{d+} \end{pmatrix} = \mathbf{PA}_d\begin{pmatrix} \psi_{0+} \\ \psi'_{0+} \end{pmatrix} \tag{112}$$

However, we know from equation (5) that

$$\left.\begin{aligned} \psi_{d+} &= e^{ikd}\psi_{0+} \\ \psi'_{d+} &= e^{ikd}\psi'_{0+} \end{aligned}\right\} \tag{113}$$

so that we have

$$\mathbf{PA}_d\begin{pmatrix} \psi_{0+} \\ \psi'_{0+} \end{pmatrix} = \begin{pmatrix} e^{ikd} & 0 \\ 0 & e^{ikd} \end{pmatrix}\begin{pmatrix} \psi_{0+} \\ \psi'_{0+} \end{pmatrix} \tag{114}$$

This is equivalent to two linear relationships between ψ_{0+} and ψ'_{0+} and has a non-zero solution only if the determinant of the coefficients vanishes. The matrix $\mathbf{PA}_d$ is given by

$$\begin{pmatrix} \cos k_0 d & k_0^{-1}\sin k_0 d \\ -2\alpha_0\cos k_0 d - \sin k_0 d & -2\alpha_0 k_0^{-1}\sin k_0 d + \cos k_0 d \end{pmatrix} \tag{115}$$

We have therefore

$$\begin{vmatrix} \cos k_0 d - e^{ikd} & k_0^{-1}\sin k_0 d \\ -2\alpha_0\cos k_0 d - \sin k_0 & -2\alpha_0 k_0^{-1}\sin k_0 d + \cos k_0 d - e^{ikd} \end{vmatrix} = 0 \tag{116}$$

Since $|\mathbf{P}| = 1$ and $|A_d| = 1$ we have $|PA_d| = 1$ and this greatly simplifies the algebra. The determinantal equation is thus seen to be equivalent to

$$\cos kd = \cos k_0 d - \alpha_0 k_0^{-1}\sin k_0 d \tag{117}$$

This is just the result we had previously (equation (30)). For this problem the gain in algebraic simplicity is not great, but in more complex problems it is quite considerable.

A similar notation may clearly be devised for the equation

$$\frac{d^2y}{dx^2} - \alpha^2 y = 0 \tag{118}$$

and indeed for any equation of the form

$$\frac{d^2y}{dx^2} + f(x)y = 0 \tag{119}$$

The general solution of equation (119) may be written in the form

$$y = Ag(x) + Bh(x) \tag{120}$$

where $g(x)$ and $h(x)$ are any two independent real solutions of equation (119). We now choose two particular real solutions $G(x)$, $H(x)$ such that $G(0) = 1$, $G'(0) = 0$, $H(0) = 0$, $H'(0) = 1$; this we may do by choosing the constants A and B appropriately. Suppose we have

$$\left.\begin{aligned} G(x) &= ag(x) + bh(x) \\ H(x) &= cg(x) + dh(x) \end{aligned}\right\} \tag{121}$$

Then we must have

$$\left.\begin{aligned} 1 &= ag(0) + bh(0) \\ 0 &= ag'(0) + bh'(0) \\ 0 &= cg(0) + dh(0) \\ 1 &= cg'(0) + dh'(0) \end{aligned}\right\} \tag{122}$$

and these four equations enable us to determine a, b, c, d, which are clearly real if $g(x)$ and $h(x)$ are. We now use $G(x)$, $H(x)$ instead of $g(x)$, $h(x)$ to express the general solution in the form

$$y = CG(x) + DH(x) \tag{123}$$

Now if $y = y_0$ at $x = 0$ and $y' = y_0'$ at $x = 0$ we clearly have $C = y_0$ and $D = y_0'$ so that

$$\left.\begin{aligned} y &= y_0 G(x) + y_0' H(x) \\ y' &= y_0 G'(x) + y_0' H'(x) \end{aligned}\right\} \tag{124}$$

or in matrix notation

$$\begin{pmatrix} y \\ y' \end{pmatrix} = \mathbf{A}_x \begin{pmatrix} y_0 \\ y_0' \end{pmatrix} \tag{124a}$$

where

$$\mathbf{A}_x \equiv \begin{pmatrix} G(x) & H(x) \\ G'(x) & H'(x) \end{pmatrix} \tag{125}$$

The appropriate functions $G(x)$, $H(x)$ for equation (103) are clearly $\cos kx$ and $k^{-1}\sin kx$. For equation (118) they are $\cosh \alpha x$ and $\alpha^{-1}\sinh \alpha x$ and we have

$$\mathbf{A}_x \equiv \begin{pmatrix} \cosh \alpha x & \alpha^{-1}\sinh \alpha x \\ \alpha \sinh \alpha x & \cosh \alpha x \end{pmatrix} \tag{126}$$

We may readily show that once again we have $|A_x| = 1$ for all values of x. We have

$$\frac{d^2 G}{dx^2} + f(x)G = 0 \tag{127}$$

and

$$\frac{d^2 H}{dx^2} + f(x)H = 0 \tag{128}$$

Let us multiply equation (127) by $H(x)$ and (128) by $G(x)$ and subtract; we get at once

$$(HG'' - GH'') = \frac{d}{dx}|\mathbf{A}_x| = 0 \tag{129}$$

Hence $|\mathbf{A}_x| = 1$ for all values of x, since clearly $|\mathbf{A}_x| = 1$ when $x = 0$. In § 4.10 we shall apply this notation to the problem of the motion of electrons in a periodic array of potential wells of finite width and depth.

4.9 Floquet's theorem

We now give a formal proof of Floquet's theorem which we previously deduced as a limiting case*, in § 3.8. Let $G(x)$ and $H(x)$ be two real solutions of equation (3) chosen to satisfy the conditions $G(0) = 1$, $G'(0) = 0$, $H(0) = 0$, $H'(0) = 1$ as shown in § 4.8.1, and corresponding to a particular value of the energy E. Now since $f(x)$ is periodic, with period d, $G(x+d)$ and $H(x+d)$ are clearly also solutions for the same value of E and may therefore be expressed as linear functions of $G(x)$ and $H(x)$. Thus we have

$$\begin{pmatrix} G(x+d) \\ H(x+d) \end{pmatrix} = \begin{pmatrix} \alpha_1 & \alpha_2 \\ \beta_1 & \beta_2 \end{pmatrix} \begin{pmatrix} G(x) \\ H(x) \end{pmatrix} \tag{130}$$

By putting $x = 0$ in equation (130) and its derivative we see that

$$\begin{pmatrix} \alpha_1 & \alpha_2 \\ \beta_1 & \beta_2 \end{pmatrix} \equiv \begin{pmatrix} G(d) & G'(d) \\ H(d) & H'(d) \end{pmatrix} \tag{131}$$

Now the general solution of equation (3) may be written as

$$\psi(x) = AG(x) + BH(x) \tag{132}$$

where A and B may be complex, and we may also write, using (131),

$$\psi(x+d) = [AG(d) + BH(d)]\,G(x) + [AG'(d) + BH'(d)]\,H(x) \tag{133}$$

We may now write $\psi(x+d)$ in the form

$$\psi(x+d) = K\psi(x) \tag{134}$$

provided we can choose K so as to make both

$$AG(d) + BH(d) = KA \tag{135}$$

and

$$AG'(d) + BH'(d) = KB \tag{136}$$

* The proof given here follows closely that given by A. H. Wilson, *The Theory of Metals*. Cambridge (1954), p. 21, but is not essentially different from that given in E. T. Whittaker and G. N. Watson, *Modern Analysis* (3rd Ed.), p. 412.

This we can do (apart from the trivial case $A = B = 0$) if, and only if

$$\begin{vmatrix} G(d)-K & H(d) \\ G'(d) & H'(d)-K \end{vmatrix} = 0 \tag{137}$$

Since

$$\begin{vmatrix} G(d) & H(d) \\ G'(d) & H'(d) \end{vmatrix} = 1 \tag{138}$$

(cf. equation (129)) this reduces to

$$K^2-[G(d)+H'(d)]K+1 = 0 \tag{139}$$

If we write K as $e^{\mu d}$ this becomes

$$2\cosh \mu d = G(d)+H'(d) \tag{140}$$

Equation (139) has two roots for K and these correspond to $e^{\mu d}$ and $e^{-\mu d}$ where μ is given by (140); since the right-hand side of (140) is real then μ is *either* wholly real or wholly imaginary*.

We may then write

$$\psi(x+d) = e^{\mu d}\psi(x) \tag{141}$$

Let $U(x) = \psi(x)e^{-\mu x}$; then we have

$$U(x+d) = \psi(x+d)e^{-\mu(x+d)} = \psi(x)e^{-\mu x} = U(x) \tag{142}$$

through equation (141), so that $U(x)$ is a periodic function of period d. We may thus write the solutions of equation (2) in the forms

$$\psi(x) = e^{\pm \mu x}U(x) \tag{143}$$

or

$$\psi(x) = e^{\pm ikx}U(x) \tag{143a}$$

where μ and k are real constants and $U(x)$ is a periodic function of period d.

When we have to deal with values of x extending from $+\infty$ to $-\infty$ the solutions of the form (143) are inapplicable and we have to use solutions of the form (143a). When we have breaks in the periodicity of the potential or are concerned with semi-infinite or finite lattices, solutions of the form (143) may have to be taken into account. For the moment we shall be concerned only with solutions of the form (143a), and, in this case, equation (140) may be written in the form

$$\cos kd = \tfrac{1}{2}[G(d)+H'(d)] \tag{144}$$

which is the equation for the propagation constant k.

Exercise For the problem discussed in § 4.2 show that, when $E > 0$,

$$G(d) = \cos k_0 d$$
$$H'(d) = \cos k_0 d - 2\alpha_0 k_0^{-1} \sin k_0 d$$

(cf. equation (115)), and hence that

$$\cos kd = \cos k_0 d - \alpha_0 k_0^{-1} \sin k_0 d$$

* In fact μ may have a real part and an imaginary part of the form $2ni/d$ but this leads to no new form of solution as will readily be seen by inserting it in equation (140).

4.10 Motion of electrons in an infinite linear array of potential wells of finite width and depth (the Krönig–Penney problem)

The motion of electrons in an infinite linear array of potential wells was first fully discussed by R. de L. Krönig and W. J. Penney* as an application of Bloch's theory of metals (§ 4.3); they also discussed a limiting form similar to that treated in § 4.2. We shall vary the treatment somewhat and concentrate on wells with negative potential energy instead of potential humps which were emphasised in the original treatment.

The potential which we shall consider is that shown in Fig. 4.1; we shall first deal with negative values of the energy E, which we shall measure from the *top* of the wells. We shall, in order to make use of the symmetry of the potential function, take the origin of the coordinate x at the centre of one of the wells. The differential equations which we have to solve are therefore

$$\frac{d^2\psi}{dx^2}-\alpha^2\psi = 0 \quad \frac{a}{2} \leqslant x \leqslant d-\frac{a}{2} \tag{145}$$

$$\frac{d^2\psi}{dx^2}+\beta^2\psi = 0 \quad -\frac{a}{2} < x < \frac{a}{2} \tag{146}$$

and similarly for other values of x, where

$$\alpha^2 = -2mE/\hbar^2 \tag{147}$$

$$\beta^2 = 2m(E+W)/\hbar^2 \tag{148}$$

We shall also write

$$\gamma^2 = 2mW/\hbar^2 \tag{149}$$

so that we have

$$\alpha^2+\beta^2 = \gamma^2 \tag{150}$$

The functions $G(x)$ and $H(x)$ as defined in §§ 4.9, 4.8.1) are simply $\cos\beta x$ and $\beta^{-1}\sin\beta x$ for $-a/2 \leqslant x \leqslant a/2$; in terms of them we may write the general solution in the form

$$\psi(x) = \psi_0 G(x)+\psi_0' H(x) \tag{151}$$

or

$$\begin{pmatrix}\psi(x)\\ \psi'(x)\end{pmatrix} = \mathbf{A}_x\begin{pmatrix}\psi_0\\ \psi_0'\end{pmatrix} \tag{151a}$$

where

$$\mathbf{A}_x \equiv \begin{pmatrix}\cos\beta x & \dfrac{1}{\beta}\sin\beta x\\ -\beta\sin\beta x & \cos\beta x\end{pmatrix} \tag{152}$$

ψ_0 and ψ_0' are the values of ψ and ψ' at $x = 0$.

The values of ψ and ψ' at $x = a/2$ are given by

$$\begin{pmatrix}\psi_{a/2}\\ \psi_{a/2}'\end{pmatrix} = \mathbf{A}_{a/2}\begin{pmatrix}\psi_0\\ \psi_0'\end{pmatrix} \tag{153}$$

* R. de L. Krönig and W. J. Penney, *Proc. Roy. Soc.* **A130**, 499 (1930).

where

$$A_{a/2} = \begin{pmatrix} \cos\dfrac{\beta a}{2} & \dfrac{1}{\beta}\sin\dfrac{\beta a}{2} \\ -\beta\sin\dfrac{\beta a}{2} & \cos\dfrac{\beta a}{2} \end{pmatrix} \tag{154}$$

Similarly we may write for $a/2 < x < d-a/2$, since now both $\psi(x)$ and $\psi'(x)$ are continuous at $x = a/2$

$$\begin{pmatrix} \psi(x) \\ \psi'(x) \end{pmatrix} = B_x \begin{pmatrix} \psi_{a/2} \\ \psi'_{a/2} \end{pmatrix} \tag{155}$$

where

$$B_x \equiv \begin{pmatrix} \cosh\alpha\left(x-\dfrac{a}{2}\right) & \dfrac{1}{\alpha}\sinh\alpha\left(x-\dfrac{a}{2}\right) \\ \alpha\sinh\alpha\left(x-\dfrac{a}{2}\right) & \cosh\alpha\left(x-\dfrac{a}{2}\right) \end{pmatrix} \tag{156}$$

This may be seen by transferring the origin to $x = a/2$ and using for $G(x)$ and $H(x)$, $\cosh\alpha(x-a/2)$ and $\alpha^{-1}\sinh\alpha(x-a/2)$, the functions appropriate to this range of x. In particular, we have

$$\begin{pmatrix} \psi_{d/2} \\ \psi'_{d/2} \end{pmatrix} = \mathbf{B}_{d/2} \begin{pmatrix} \psi_{a/2} \\ \psi'_{a/2} \end{pmatrix} \tag{157}$$

where $\mathbf{B}_{d/2}$ is the value of B_x when $x = d/2$

$$\mathbf{B}_{d/2} = \begin{pmatrix} \cosh\dfrac{\alpha b}{2} & \dfrac{1}{\alpha}\sinh\dfrac{\alpha b}{2} \\ \alpha\sinh\dfrac{\alpha b}{2} & \cosh\dfrac{\alpha b}{2} \end{pmatrix} \tag{158}$$

where $b = d-a$. Combining (153) and (157) we have

$$\begin{pmatrix} \psi_{d/2} \\ \psi'_{d/2} \end{pmatrix} = \mathbf{B}_{d/2}\,\mathbf{A}_{a/2} \begin{pmatrix} \psi_0 \\ \psi'_0 \end{pmatrix} \tag{159}$$

On performing the matrix multiplication we have

$$\mathbf{B}_{d/2}\,\mathbf{A}_{a/2} \equiv \begin{pmatrix} p & q \\ r & s \end{pmatrix} \tag{160}$$

where

$$p = \cosh\frac{\alpha b}{2}\cos\frac{\beta a}{2} - \frac{\beta}{\alpha}\sinh\frac{\alpha b}{2}\sin\frac{\beta a}{2} \tag{161}$$

$$q = \frac{1}{\beta}\cosh\frac{\alpha b}{2}\sin\frac{\beta a}{2} + \frac{1}{\alpha}\sinh\frac{\alpha b}{2}\cos\frac{\beta a}{2} \tag{162}$$

$$r = \alpha\sinh\frac{\alpha b}{2}\cos\frac{\beta a}{2} - \beta\cosh\frac{\alpha b}{2}\sin\frac{\beta a}{2} \tag{163}$$

$$s = \frac{\alpha}{\beta}\sinh\frac{\alpha b}{2}\sin\frac{\beta a}{2} + \cosh\frac{\alpha b}{2}\cos\frac{\beta a}{2} \tag{164}$$

We now make use of the symmetry of the potential function $V(x)$; suppose that we reverse the direction of x, i.e. write $-x$ for x then clearly the relationship between ψ_d, $-\psi'_d$ and $\psi_{d/2}$, $-\psi'_{d/2}$ is the same as between ψ_0, ψ'_0 and $\psi_{d/2}$, $\psi'_{d/2}$, i.e.

$$\begin{pmatrix}\psi_{d/2}\\ -\psi'_{d/2}\end{pmatrix} = \begin{pmatrix}p & q\\ r & s\end{pmatrix}\begin{pmatrix}\psi_d\\ -\psi'_d\end{pmatrix} \tag{165}$$

or

$$\begin{pmatrix}\psi_{d/2}\\ \psi'_{d/2}\end{pmatrix} = \begin{pmatrix}p & -q\\ -r & s\end{pmatrix}\begin{matrix}\psi_d\\ \psi'_d\end{matrix} \tag{166}$$

Now let us reverse the transformation (166) using the result of equation (100) for the reciprocal matrix. The determinant of the matrix (160) is 1, since we have

$$|\mathbf{B}_{d/2}| = |\mathbf{A}_{d/2}| = 1$$

We then have

$$\begin{pmatrix}\psi_d\\ \psi'_d\end{pmatrix} = \begin{pmatrix}s & q\\ r & p\end{pmatrix}\begin{matrix}\psi_{d/2}\\ \psi'_{d/2}\end{matrix} \tag{167}$$

We may now combine (167) and (160) to express ψ_d, ψ'_d in terms of ψ_0, ψ'_0; we thus obtain the relationship

$$\begin{pmatrix}\psi_d\\ \psi'_d\end{pmatrix} = \begin{pmatrix}ps+qr & 2qs\\ 2pr & ps+qr\end{pmatrix}\begin{pmatrix}\psi_0\\ \psi'_0\end{pmatrix} \tag{168}$$

When $\psi'_0 = 0$, $\psi_0 = 1$ we have $\psi = G(x)$ and when $\psi_0 = 0$, $\psi'_0 = 1$, we have $\psi = H(x)$ (equation (151)) so that

$$H'(d) = G(d) = ps+qr \tag{169}$$

The equation for the propagation constant k is therefore

$$\cos kd = ps+qr \tag{170}$$

Since the determinant of the matrix in (165) is equal to unity, we have

$$ps-qr = 1 \tag{171}$$

so that equation (170) may also be written in the two equivalent forms

$$\cos kd = 1+2qr \tag{170a}$$

$$= -1+2ps \tag{170b}$$

On substituting for q and r from (162), (163) these may be reduced to the form

$$\cos kd = f(\alpha) = \cosh \alpha b \cos \beta a - \frac{\beta^2-\alpha^2}{2\alpha\beta}\sinh \alpha b \sin \beta a \tag{172}$$

From equation (172) we may determine α and hence E in terms of k, provided $|f(\alpha)| \leqslant 1$. The function $f(\alpha)$ is plotted in Fig. 4.14. It will be seen that again there is a series of bands for which $|f(\alpha)| > 1$; these represent forbidden energy bands, separating regions of allowed energy values. We shall discuss the limits of these bands later.

To complete the determination of the wave function we have to eliminate one of the unknown constants ψ_0, ψ_0'; the other is then simply given by the normalising condition. We now substitute for ψ_d in (168) $\psi_0 e^{ikd}$ and obtain

$$[(ps+qr)-e^{ikd}]\psi_0 = -2qs\psi_0' \tag{173}$$

or, using (170),

$$i \sin kd\, \psi_0 = 2qs\psi_0' \tag{174}$$

Thus we have

$$\psi = \psi_0\left[\cos\beta x+\frac{i\sin kd}{2\beta qs}\sin\beta x\right], \quad -\frac{a}{2} \leqslant x \leqslant \frac{a}{2} \tag{175}$$

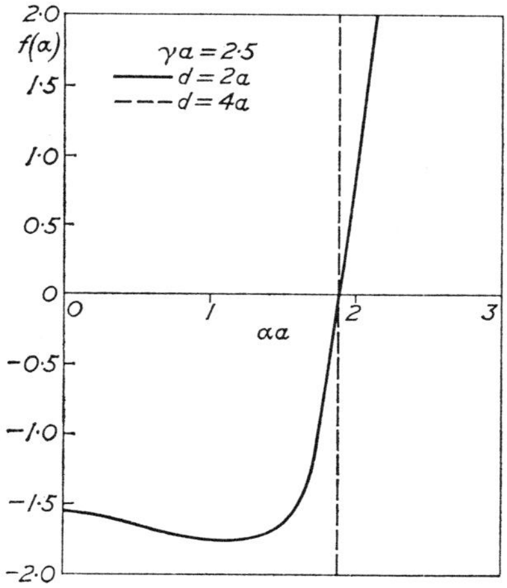

Fig. 4.14. The functions $f(\alpha)$, for $\gamma a = 2{\cdot}5$, with $d = 2a$ and $d = 4a$

Similarly we have

$$\psi_{d/2} = p\psi_0+q\psi_0'$$

$$= \psi_0\left[p+\frac{i\sin kd}{2s}\right] \tag{176}$$

$$\psi_{d/2}' = \psi_0\left[r+\frac{i\sin kd}{2q}\right] \tag{177}$$

Combining (176) and (177) it may be readily shown that

$$\psi_{d/2}' = \frac{i\sin kd}{2pq}\psi_{d/2} \tag{178}$$

Hence we have, using (156), (176) and (178)

$$\left.\begin{aligned}\psi(x) = p\psi_0\left[1+\frac{i\sin kd}{2ps}\right]\left[\cosh\alpha\left(x-\frac{d}{2}\right)+\frac{i\sin kd}{2\alpha pq}\sinh\alpha\left(x-\frac{d}{2}\right)\right] \\ a/2 \leqslant x \leqslant d-\frac{a}{2}\end{aligned}\right\} \tag{179}$$

Thus the wave function is now completely determined apart from the multiplying constant ψ_0, which is obtained from the normalising condition. The expression for ψ_0 is somewhat complex and we shall not give it, since it is not required in the subsequent discussion.

For positive values of E, the differential equation for the wave function ψ in the range $a/2 \leqslant x \leqslant d-a/2$ is

$$\frac{d^2\psi}{dx^2}+k_0^2\psi = 0 \tag{180}$$

where now we have

$$E = \hbar^2 k_0^2/2m \tag{181}$$

From the above it will be seen that the equation for $\cos kd$ and the expressions for the wave functions may be obtained at once by replacing α by ik_0. Thus we have

$$\cos kd = g(k_0) = \cos k_0 b \cos \beta a - \frac{\beta^2+k_0^2}{2k_0\beta}\sin k_0 b \sin \beta a \tag{182}$$

where

$$\beta^2 = \gamma^2+k_0^2 \tag{183}$$

We must now examine more closely the behaviour of the functions $f(\alpha)$, $g(k_0)$ giving the value of $\cos kd$. This value depends on the ratio d/a as well as on the quantity γa, which as we have seen in § 2.9, determines the behaviour of the energy levels in an isolated potential well. In particular, the value of γa determines the number of bound states in the isolated well; if $n\pi \leqslant \gamma a \leqslant (n+1)\pi$ we have $(n+1)$ bound states (see § 2.9). Before we determine the allowed energy bands for particular values of γa and d/a it is instructive to consider certain limiting conditions. We shall suppose first of all that a is fixed and we vary W, the depth of the potential well; if $W \to 0$ then $\gamma a \to 0$ and there is no solution of equation (172) other than $\alpha = 0$. From equation (183), however, we see that when $E > 0$ $\beta \to k_0$ and equation (182) becomes simply

$$\cos kd = \cos k_0 b \cos k_0 a - \sin k_0 b \sin k_0 a = \cos k_0 d$$

Thus we have $k \to k_0$ and the electrons behave as though they were free. The relationship between E and k is then parabolic and is shown in Fig. 4.16 in the reduced representation under $\gamma a = 0$. If we let $W \to \infty$ and $a \to 0$ in such a way that $\gamma a \to 0$ but Wa tends to a finite value given by $\hbar^2\alpha_0/m$ then equation (172) and (182) simply reduce to equations (18) and (30) giving the dispersion equations for an array of deep narrow potential wells.

If we keep a and W fixed and vary d we see that as d becomes large $\sinh \alpha d \to \cosh \alpha d \to \frac{1}{2}\exp \alpha d$. It is then clear from equation (172) that $|\cos kd| \leqslant 1$ only if

$$\tan \beta a \simeq \frac{2\alpha\beta}{\beta^2-\alpha^2} \tag{184}$$

This is simply the equation which determines the energy levels in an isolated well (cf. equation (107) of § 2.9). Let α_n be the solution of equation (184) giving the nth bound state, then we may write equation (172) approximately in the form

$$2\,e^{-\alpha_n b}\cos kd = \cos\beta a - \frac{\beta^2-\alpha^2}{2\alpha\beta}\sin\beta a = F(\alpha) \tag{185}$$

since $2\sinh\alpha b \simeq 2\cosh\alpha b \simeq e^{\alpha b}$, and since the term on the left-hand side is small we may write α_n for α in it. Let us now expand the function $F(\alpha)$ neglecting powers of $\alpha-\alpha_n$ greater than the first. We have then since $F(\alpha_n) = 0$

$$(\alpha-\alpha_n)\,F'(\alpha_n) = 2\,e^{-\alpha_n b}\cos kd \tag{186}$$

giving

$$\alpha = \alpha_n - 2\,e^{-\alpha_n b}\cos kd/F'(\alpha_n) \tag{187}$$

From this we obtain for the energy E the approximate expression

$$E = E_n - 4E_n\,e^{-\alpha_n b}\cos kd/\alpha_n F'(\alpha_n) \tag{188}$$

The level at $E = E_n$ is thus broadened into a band of width W_n given by

$$W_n/E_n = 8\,e^{-\alpha_n b}/\alpha_n F'(\alpha_n) \tag{189}$$

the lower edge corresponding to $k = 0$ if $F'(\alpha_n) > 0$ and to $k = \pi$ if $F'(\alpha_n) < 0$; if $\alpha_n b$ becomes very large the band becomes very narrow.

Let us now examine in detail the behaviour of the allowed energy bands as γa is varied; we shall first consider the condition when $d = 2a$. A plot of the functions $f(\alpha)$ and $g(k_0)$ which occur in the dispersion equations (172) and (183) are shown for $\gamma a = 2{\cdot}5$ in Figs. 4.14 and 4.15; similar figures may be drawn for other values of γ and the corresponding bands obtained. This is done more readily if we consider the condition for which an allowed band crosses the line corresponding to $E = 0$. If we put $\alpha = 0$ in equation (172) or $k_0 = 0$ in equation (183) we obtain the equation

$$\cos kd = \cos\gamma a - \frac{\gamma b}{2}\sin\gamma a \tag{190}$$

We have a real value of k given by equation (190) provided

$$\left|\cos\gamma a - \frac{\gamma b}{2}\sin\gamma a\right| \leqslant 1 \tag{191}$$

Limiting conditions occur when $k = 0$ and $kd = \pi$. A band just touches the line $E = 0$ at $k = 0$ provided

$$1-\cos\gamma a = -\frac{\gamma b}{2}\sin\gamma a \tag{192}$$

This is equivalent to the two equations

$$\sin\frac{\gamma a}{2} = 0 \tag{193}$$

$$\tan\frac{\gamma a}{2} = -\frac{\gamma b}{2} \tag{194}$$

Equation (193) is satisfied for $\gamma a = 0, 2\pi, \ldots$, etc.; the values of γa satisfying (194) depend on the ratio b/a; when $b = a$ the first root is $\gamma a = 4{\cdot}06$. Similarly when $kd = \pi$ we have either

$$\cos\frac{\gamma a}{2} = 0 \tag{195}$$

or

$$\cot\frac{\gamma a}{2} = \frac{\gamma b}{2} \tag{196}$$

Thus we have always a band touching the line $E = 0$ at $kd = \pi$ if $\gamma a = \pi$, $3\pi, \ldots$, etc.; the first root of equation (196) when $b = a$ is $\gamma a = 1{\cdot}72$.

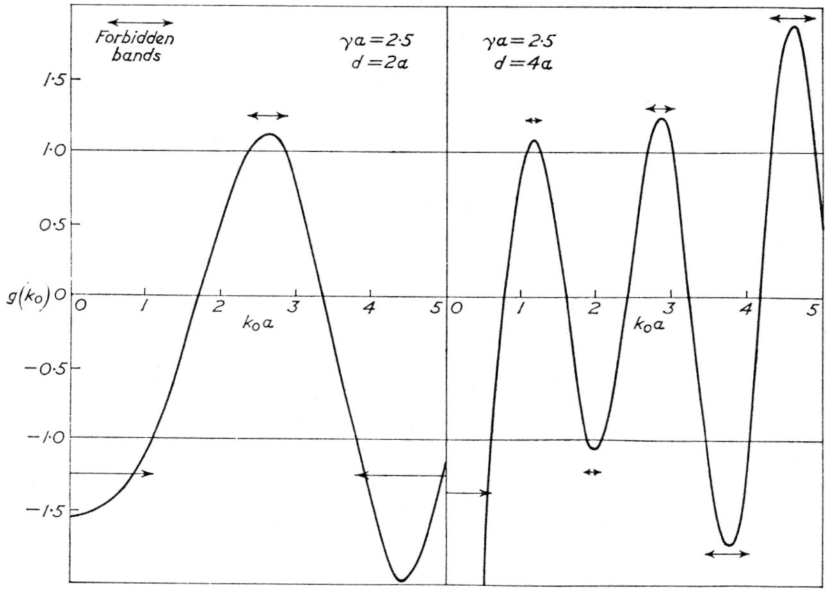

FIG. 4.15. The function $g(k_0)$ for $\gamma a = 2{\cdot}5$, with $d = 2a$ and $d = 4a$

For a small value of γa equation (190) is clearly satisfied by a small real value of k given by

$$k^2 = \gamma^2 a/d \tag{197}$$

Moreover equation (172) is satisfied by a small real value of α; we therefore have an allowed band of energies having its lower edge below the line $E = 0$ and crossing it for a small value of k. This is illustrated in Fig. 4.16 in which we show the allowed energy bands for various values of γa; that for $\gamma a = \pi/2$ is typical of the condition in which the lowest allowed band lies partially above and partially below the line $E = 0$. For higher values of E it will be seen that we have various allowed energy bands separated by regions of forbidden energy. As γa is increased, the lowest band goes more and more into the region corresponding to negative values of E, till when $\gamma a = 1{\cdot}72$ it lies wholly below the line $E = 0$, touching it at the point corresponding to $k = \pi$; the value $\gamma a = 1{\cdot}72$ is smallest solution of equations (193)–(196) other than $\gamma a = 0$. (It is a solution of equation (196).) As γa is increased still further the whole of the lowest band falls below the line $E = 0$ and becomes narrower; the other bands also fall to lower values of E, the curves for $\gamma a = 2{\cdot}5$ in Fig. 4.16 being typical of this condition. When $\gamma a = 2\pi$, the next solution of equations (193)–(196), the next band reaches the line $E = 0$ and touches it also where $k = \pi$ and for still larger values of γa this second band begins to

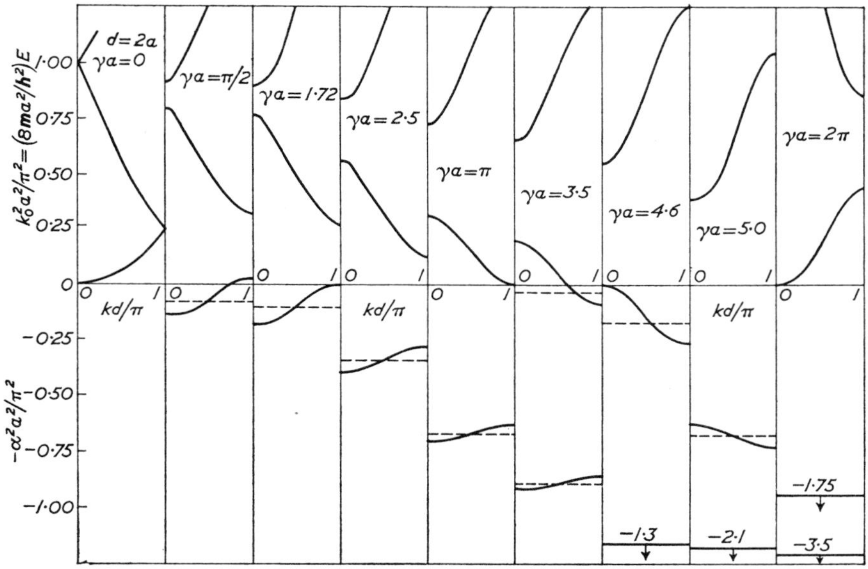

FIG. 4.16. Energy bands for various values of γa, with $d = 2a$

cross the line $E = 0$; the curves for $\gamma a = 3{\cdot}5$ in Fig. 4.16 are typical of this condition. When $\gamma a = 4{\cdot}06$ (a solution of equation (194)) the second band is all below the line $E = 0$ and touches it at $k = 0$; the lowest band has now become very narrow and is at the position of the corresponding level for an isolated well. The levels for the latter are shown in broken lines in Fig. 4.16, and should be compared with the corresponding energy bands for the array of wells. As γa is still further increased we have two allowed energy bands with negative energies, as for $\gamma a = 5{\cdot}0$, the lower being very narrow, and when $\gamma a = 2\pi$ the third band just begins to cross into the negative-energy region.

Similar curves are shown for various values of γa and for $d = 4a$ in Fig. 4.17; it will be seen that in this case the bands are narrower. The values $\gamma a = 1{\cdot}72, 4{\cdot}06$, etc., for the final crossing of bands completely into the negative energy region no longer hold and bands tend to cross completely for smaller values of γa; the values of γa for which bands first *begin* to cross into the negative energy region are, however, always the same, namely $\gamma a = \pi$, 2π, 3π, etc. Thus we have the very interesting result that when we have n bound levels in the isolated well we have at least $n-1$ complete bands with $E < 0$ and may have one more complete band or one partially above and partially below the line $E = 0$. It is interesting to note from Figs. 4.16, 4.17 how each of the bands develops as γa is increased, and how the lower bands decrease in width as they fall

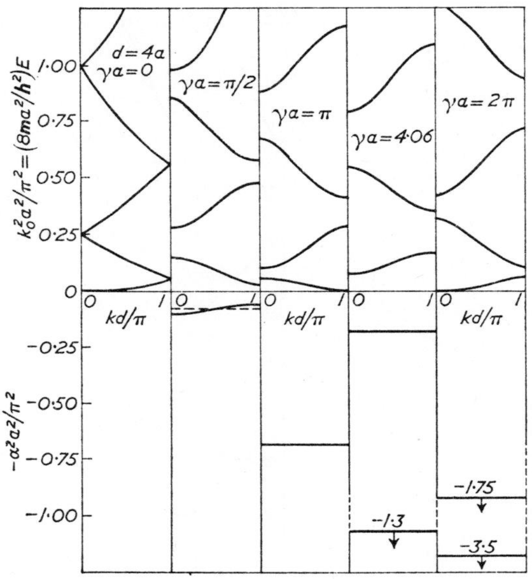

FIG. 4.17. Energy bands for various values of γa, with $d = 4a$

below the line $E = 0$. Thus, as for the simpler problem discussed in § 4.2, we see that the lower levels tend to those for an isolated well, and it is only for the levels a little below $E = 0$ that broadening is appreciable. With this model we may have as many bound levels as we please so that it gives a much better approximation to the atoms in a crystal. We shall later discuss its application to describe the motion of electrons in a crystalline solid.

Let us now consider the form of the wave functions for the electrons; we shall restrict ourselves to one or two interesting limiting conditions. Let us discuss first the bottom edge of the lowest band; here we have $k = 0$ so that from equation (170a) we see that either $q = 0$ or $r = 0$, q and r being the quantities defined in equations (162) and (163). By considering the condition for small values of α or by inserting the calculated values of α for the bottom edge of a particular band it will be seen that it is r that is equal to zero in this case. From equation (174) we see that $\psi_0' = 0$ and

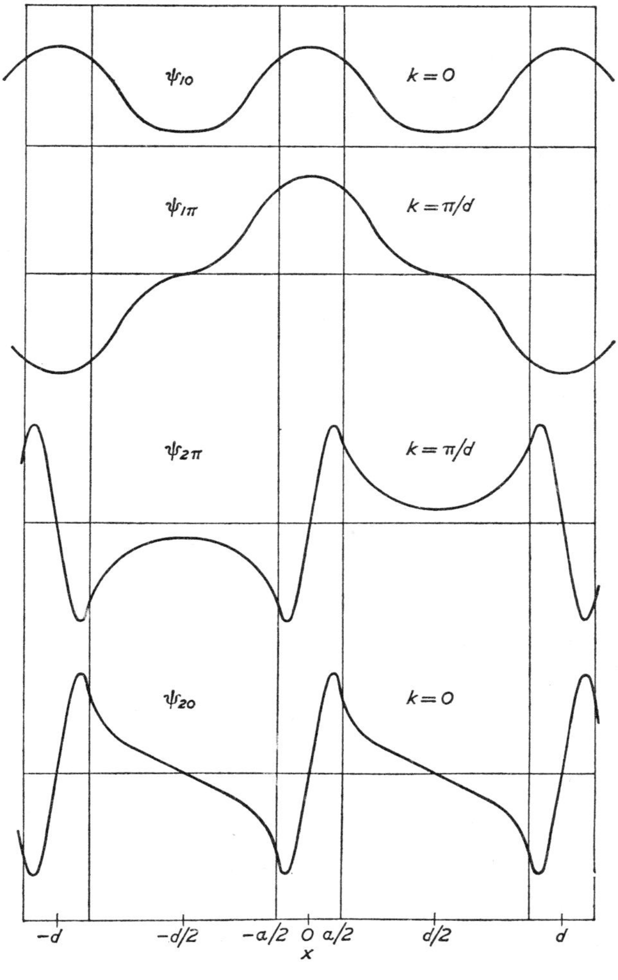

FIG. 4.18. The wave functions ψ_{10}, $\psi_{1\pi}$, ψ_{20}, $\psi_{2\pi}$ for the potential of Fig. 4.1

from equation (177) that $\psi_{d/2}' = 0$; also we have from equation (176), $\psi_{d/2} = p\psi_0$, and, if αd is large, then $\psi_{d/2}/\psi_0$ is very small. The form of the wave function ψ_{10} corresponding to these conditions is shown in Fig. 4.18; for large values of αd it approximates to the symmetrical wave function for an isolated well near to each well.

The top edge of the bottom band corresponds to $kd = \pi$, and from equation (170b) we see that either $p = 0$ or $s = 0$; from the condition $\gamma a = 1{\cdot}72$ and small value of α we see that we actually have $p = 0$. Thus again we have from equation (174) $\psi_0' = 0$; also from equation (176) we have $\psi_{d/2} = 0$. The form of the corresponding wave function $\psi_{1\pi}$ is shown in Fig. 4.18; when αd is large, near each well $\psi_{1\pi}$ is approximately equal to the symmetrical wave function for an isolated well, but for each alternate well is multiplied by -1, arising from the factor $\exp(ikd)$. When the wave function between the wells is small this has little effect on the value of $|\psi|^2$. We also note that the wave function has a node at $x = d/2$ and corresponds to a higher energy than does ψ_{10}.

For the bottom of the second band we also have $k = \pi/d$ and it will readily be seen that in this case we have $s = 0$. Since for small values of $\sin kd$ we have $\sin kd \simeq 2(ps)^{1/2}$ and we see from equation (174) that $\psi_0 = 0$ and that $\psi_{d/2} = q\psi_0'$. (We have now one constant, ψ_0'.) The wave function $\psi_{2\pi}$ corresponding to this condition is shown in Fig. 4.18; it corresponds for large values of αd to the antisymmetrical solution giving the second lowest level for an isolated well, the function near each alternate well being multiplied by -1.

For the top of the second band we similarly have $q = 0$ giving $\psi_0 = 0$ and $\psi_{d/2} = 0$. The corresponding wave function ψ_{20} is shown in Fig. 4.18; it has an extra node (at $x = d/2$) and corresponds to a higher energy than $\psi_{2\pi}$ does. This sequence is continued for the higher bands.

4.11 Approximate methods of solution

The number of forms of the periodic potential $V(x)$ for which exact solutions may be obtained as in §§ 4.2, 4.10 is quite small. An exact solution may also be obtained for the potential function

$$V(x) = A \sin 2\pi x/d$$

The solution involves Mathieu functions and consequently leads to much greater mathematical difficulties than were encountered with the potential functions treated in §§ 4.2, 4.10; this function leads to an energy band structure very similar to that shown in Figs. 4.16 and 4.17. It has been applied to discuss the motion of electrons in crystalline solids by P. M. Morse* and a detailed discussion of the band structure of the allowed energy levels has been given by L. Brillouin†. The band structure for this function confirms the general conclusions obtained from our treatment of the array of 'square' potential wells but leads to no essentially new conclusions.

These few functions then represent all those for which exact solutions may readily be found and we must therefore look for approximate methods of treating any other potential function $V(x)$. There are two

* P. M. Morse, *Phys. Rev.*, **35**, 1310 (1930).

† L. Brillouin, *Wave Propagation in Periodic Structures*. Dover (1953), p. 172.

conditions under which approximate solutions may readily be obtained; the first is when $V(x)$ is small compared with the average kinetic energy and corresponds to large values of $k_0 a$ in the problems of §§ 4.2, 4.10; the second is when the overlap of the wave functions from neighbouring potential wells, regarded as isolated, is small, and in this case the complete wave function is small except near the wells and corresponds to large values of αd in the problems of §§ 4.2, 4.10. The energy levels in this case lie well below the zero level of kinetic energy and the states correspond to electrons which are tightly bound in the wells; this approximation is accordingly often referred to as the tight-binding approximation.

4.11.1 Solution when $V(x)$ is small*

We shall now consider the condition in which $V(x)$ is small compared with the total energy E. If, as a zero-order approximation, we neglect $V(x)$ entirely we have

$$\psi = A_0 \mathrm{e}^{ikx} \tag{198}$$

and the energy E equal to $\hbar^2 k^2/2\boldsymbol{m}$. If we write $-2\boldsymbol{m}V(x)/\hbar^2 = \epsilon g(x)$ where $g(x)$ is a periodic function of period d and ϵ is a small constant, we may write the wave equation in the form

$$\frac{\mathrm{d}^2\psi}{\mathrm{d}x^2} + [k_0^2 + \epsilon g(x)]\psi = 0 \tag{199}$$

the energy E being given by $\hbar^2 k_0^2/2\boldsymbol{m}$. Since $g(x)$ is a periodic function it may be expanded as a Fourier series in the form

$$g(x) = \sum_{-\infty}^{\infty} c_n \mathrm{e}^{2\pi inx/d} \tag{200}$$

where

$$c_n = \frac{1}{d}\int_0^d g(x)\,\mathrm{e}^{-2\pi inx/d}\,\mathrm{d}x \tag{201}$$

For simplicity we assume that $c_0 = 0$, the constant term being absorbed in k_0^2. The solution $\psi(x)$, may be expanded in its space harmonics as in § 4.5, and we have

$$\psi(x) = A(x)\,\mathrm{e}^{ikx}$$

where

$$A(x) = \sum_{-\infty}^{\infty} A_n \mathrm{e}^{2\pi inx/d} \tag{202}$$

If $\epsilon = 0$ we have $A(x) = A_0$ and $k = k_0$, and we therefore expect to have $A_n \to 0$ as $\epsilon \to 0$, $n \neq 0$. We therefore write $\psi(x)$ in the form

$$\psi(x) = A_0 \mathrm{e}^{ikx} + \epsilon \sum{}' a_n \mathrm{e}^{i(k+2\pi n/d)x} \tag{203}$$

* This treatment is practically the same as that given by Lord Rayleigh in discussing small periodic variations of refractive index. (*Theory of Sound*, Macmillan, 1937.)

where $\sum'$ means a sum over all values of n except $n = 0$. If we substitute the solution (203) and the expression (200) for $g(x)$ in the wave equation (199), neglecting terms in ϵ^2 we obtain

$$A_0(k_0^2 - k^2)\,\mathrm{e}^{ikx} + \epsilon \sum_{-\infty}^{\infty}{}' \left[(k_0^2 - k_n^2)\,a_n + A_0\,c_n\right]\mathrm{e}^{ik_n x} = 0 \tag{204}$$

where as before

$$k_n = k + 2\pi n/d \tag{205}$$

If we multiply this equation by $\mathrm{e}^{-ik_n x}$ and integrate from 0 to d, we obtain zero for all terms except one; this gives for $n = 0$

$$k = k_0 \tag{206}$$

and for other values of n

$$a_n = \frac{A_0\,c_n}{k_n^2 - k_0^2} \tag{207}$$

To this approximation the perturbation has no effect on the propagation vector k_0 but merely introduces some space harmonics.

If, however, we include the term in ϵ^2 neglected in equation (204) we now obtain on multiplying by e^{-ikx} and integrating from 0 to d

$$A_0\,d(k_0^2 - k^2) + \epsilon^2 \sum_n \sum_{n'} a_{n'}\,c_n \int_0^d \mathrm{e}^{2\pi i(n+n')x/d}\,\mathrm{d}x = 0$$

The integrals under the sum vanish unless $n = -n'$, and we have

$$A_0(k_0^2 - k^2) + \epsilon^2 \sum{}' a_n\,c_{-n} = 0$$

If we now substitute for a_n from (207) we have, since if $g(x)$ is real, $c_n = c^\star_{-n}$

$$k_0^2 = k^2 + \epsilon^2 \sum{}' \frac{|c_n|^2}{k^2 - k_n^2} \tag{208}$$

or

$$E = E_0 + \frac{\epsilon^2 \hbar^4}{4m^2} \sum{}' \frac{|c_n^2|}{E_0 - E_n} \tag{209}$$

where $E_n = \hbar^2 k_n^2/2m$; this gives the change in energy to the order of ϵ^2.

The constants c_n may be expressed in terms of the Fourier coefficients V_n of the potential function $V(x)$. We have from equation (200)

$$-\epsilon c_n = 2mV_n/\hbar^2 \tag{210}$$

so that

$$E = E_0 + \sum{}' \frac{|V_n|^2}{E_0 - E_n} \tag{209a}$$

We note that if $k^2 \to k_n^2$, $a_n \to \infty$ and $E \to -\infty$. This indicates that for this condition the approximation we have made is invalid. When $k^2 \simeq k_n^2$ we have either

$$k_n \simeq k \tag{211}$$

or

$$k_n \simeq -k \tag{212}$$

Condition (211) can hold only if $n = 0$ and this is ruled out. (The sum does not include $n = 0$.) The condition (212) gives

$$k \simeq -n\pi/d \tag{213}$$

(n going from $-\infty$ to $+\infty$). These are just the values of k which we have associated with band edges.

From previous experience we should expect a standing wave at a band edge so that (198) is not a good approximation. Equation (207) shows that if we have $k \simeq -n\pi/d$ we have two large coefficients in the series for ψ—that of e^{ikx} and that of $e^{ik_n x}$; but for $k = -n\pi/d$ $k_n = -k$, so the second large coefficient is associated with e^{-ikx}. A better zero-order approximation when $k \simeq n\pi/d$ is therefore

$$\psi = A_0 e^{ikx} + A_1 e^{i(k-2n\pi/d)x} + \epsilon \sum{}'' a_s e^{ik_s x} \tag{214}$$

where $\sum''$ means the sum omitting $s = 0$ and $s = -n$, and we no longer assume A_1 to be small. Substituting as before and neglecting ϵ^2 we have, on multiplying by e^{-ikx} and integrating

$$\left.\begin{aligned} A_0(k_0^2 - k^2) + \epsilon A_1 c_n &= 0 \\ A_0 c_{-n}\epsilon + A_1(k_0^2 - k_{-n}^2) &= 0 \end{aligned}\right\} \tag{215}$$

The determinant of the coefficients of A_0 and A_1 must vanish and we have

$$(k_0^2 - k^2)(k_0^2 - k_{-n}^2) = \epsilon^2 |c_n|^2 \tag{216}$$

Suppose we write $k = n\pi/d + k'$, k' being small, and $k_0 = n\pi/d + \eta$, then we may write equation (216), neglecting powers higher than the order of $k'\eta$, η^2, etc., in the form

$$k'^2 = \eta^2 - \frac{d^2 \epsilon^2 |c_n|^2}{4\pi^2 n^2} \tag{217}$$

There is no solution with k real for $|\eta| < \eta_0$ where

$$\eta_0 = \frac{d\epsilon\,|c_n|}{2\pi n} \tag{218}$$

i.e. there is a forbidden band of energies corresponding to values of k_0 lying between

$$k_0 = \frac{n\pi}{d} - \frac{d\epsilon\,|c_n|}{2\pi n}$$

and

$$k_0 = \frac{n\pi}{d} + \frac{d\epsilon\,|c_n|}{2\pi n}$$

i.e. to values of the energy E in the range given by (neglecting ϵ^2)

$$n^2 - \frac{\epsilon d^2 |c_n|}{\pi^2} < \frac{8md^2 E}{h^2} < n^2 + \frac{\epsilon d^2 |c_n|}{\pi^2} \tag{219}$$

or

$$E_n - \tfrac{1}{2}\Delta E_n < E < E_n + \tfrac{1}{2}\Delta E_n \tag{219a}$$

The energies E'_n corresponding to $8md^2E/h^2 = n^2$ for integral values of n are just the 'free electron' energies for $k = \pm n\pi/d$; the forbidden energy gaps occur at these energies, the gaps being of width ΔE_n given by

$$\Delta E_n = \hbar^2 \epsilon|c_n|/m = 2|V_n| \tag{220}$$

The gap for each value of n is therefore proportional to V_n the Fourier coefficient of the potential $V(x)$, so that, for a given form of $V(x)$, it is proportional to the magnitude of $V(x)$, at least when this is small. This shows us that the existence of the energy gaps does not depend on the special form of $V(x)$ but is a universal property of periodic potentials;

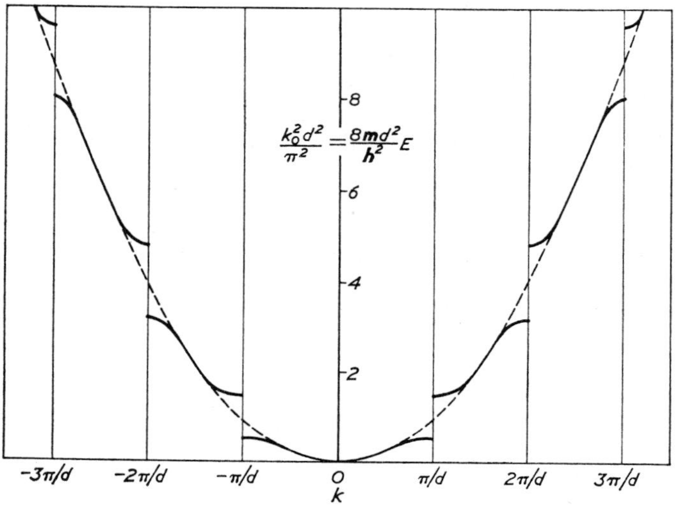

FIG. 4.19. Form of the energy bands for a small periodic potential in the expanded representation. (Broken curve corresponds to free electrons)

certainly we have only proved this when $V(x)$ is small but we shall also demonstrate that we have the same behaviour when $V(x)$ is large.

From the form of equation (217) which may be written in the approximate form

$$\eta = \pm\left[\frac{d\epsilon|c_n|}{2\pi n} + \frac{\pi n k'^2}{d\epsilon|c_n|}\right] \tag{221}$$

it is also clear that the band edges near $k = n\pi/d$ are parallel to the k-axis. The form of the bands is shown in Fig. 4.19. Here we have no restricted range for the propagation constant k and the energy is a single-valued function of k, except at $k = n\pi/d$, where it is two-valued and we have a discontinuity; this then corresponds to what we have called an expanded representation. Alternatively we could plot the energy as a function of k' for values of k from $k = n\pi/d$ to $k = (n+1)\pi/d$, where we have another band beginning. We need then only use values of $|k'|$ in the range 0 to π/d, labelling each band by the integer n. The energy is now a multi-valued function of k' and this would correspond to the reduced representation with $k' = k - n\pi/d$. The main advantage of the expanded representation is that it tends to the familiar parabolic form for free electrons as $\epsilon \to 0$. Taking first the positive sign in equation (221) we see that the energy E when k is nearly equal to $n\pi/d$ is given approximately by

$$E = \frac{\hbar^2}{2m}\left[\frac{n^2\pi^2}{d^2} + \frac{2n\pi}{d}\eta\right] \tag{222}$$

This may be written in the form

$$E = E_n + \tfrac{1}{2}\Delta E_n + \frac{4E_n}{\Delta E_n}\left(\frac{\hbar^2 k'^2}{2m}\right) \tag{223}$$

For small values of k' this corresponds to the bottom of an allowed energy band which occurs at $E = E_n + \frac{1}{2}\Delta E_n$, the effective mass $m^\star$ near the bottom of the band being given by

$$m^\star = m\Delta E_n/4E_n \tag{224}$$

Thus we see that, when the band separation ΔE_n is small, the effective mass is small near the bottom of the band. If we take the negative sign in equation (221) we have

$$E = E_n - \tfrac{1}{2}\Delta E_n - \frac{4E_n}{\Delta E_n}\left(\frac{\hbar^2 k'^2}{2m}\right) \tag{225}$$

This corresponds to the top of the band just below the former and here we have again a small (negative) value of $m^\star$. We must note, however, that the range of values of E over which the effective mass is small is also small, in fact of the order ΔE_n, since the approximate form (221) holds only when $\hbar^2 k'^2/2m$ is considerably less than $\Delta E_n^2/E_n$.

This approximation is valid when the forbidden energy gap is small compared with the width of the allowed bands. Let us now consider the other extreme in which the widths of the allowed bands are small compared with the spacing between the bands.

4.11.2 The tight-binding approximation

From the treatment of the two potential functions for which we have given exact solutions in §§ 4.2 and 4.10 we have seen that the allowed energy bands become very narrow when the electrons are tightly bound in the individual potential wells centred at each lattice point of the array. This, as we shall see, is also a quite general result, as might be expected from the fact that when the overlap of the wave functions associated with each individual potential well is small it will have only a small influence on the total energy.

Suppose now that we have an infinite number of attractive centres of potential arranged on a one-dimensional lattice with spacing d between them; for each isolated centre we shall suppose that the potential is given by $U_0(x)$, x being the distance from the centre. For finite values of d, but sufficiently large to ensure that there is no appreciable interaction between the centres we may write for the array

$$V(x) = \sum_{s=-\infty}^{\infty} U_0(x-sd) \tag{226}$$

$V(x)$ is then a periodic function of period d.

When d is such that there is a small amount of mutual interaction $V(x)$ will differ from (226) but will still be a periodic function of period d. The function $V(x)$ is shown in Fig. 4.20, and also, for comparison, the function $U_0(x)$ (broken curve). Under these conditions $U_0(x)$ and $V(x)$ will not differ by very much, when $|x|$ is less than $d/2$, and if we write

$$W(x) = V(x) - U_0(x)$$

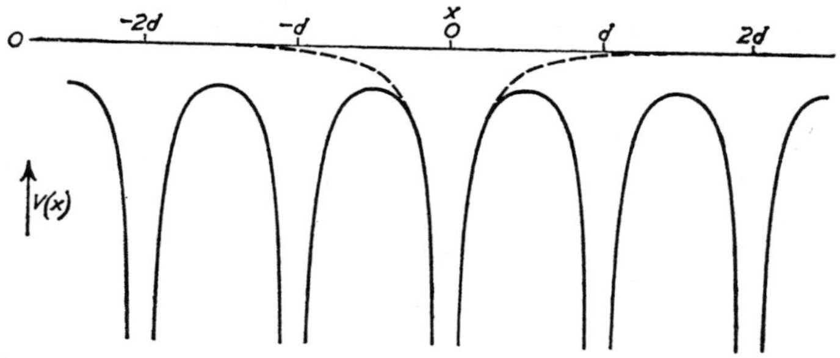

FIG. 4.20. The potential functions $V(x)$ and $U_0(x)$ (broken curve)

we see that $W(x)$ is small for the range $-d/2 \leqslant x \leqslant d/2$. The wave function near each centre will have approximately the same form, apart from a (complex) constant, as that for an isolated centre (cf. solutions given in §§ 4.2, 4.10 when αd is large). Thus an approximate wave function may be written in the form

$$\psi(x) = \sum_{s} a_s \psi_0(x-sd) \tag{227}$$

where $\psi_0(x)$ is the wave function for an isolated centre at $x = 0$. This wave function does not satisfy the wave equation exactly; moreover unless we choose the constants a_s correctly it is not in the form of a Bloch wave function such that

$$\psi(x+rd) = e^{ikrd}\psi(x)$$

as it should be since $V(x)$ is a periodic function of period d. A better approximation may therefore be obtained by choosing the constants a_s equal to $A\exp(iksd)$. We then have

$$\psi_k(x) = A\sum_{s} e^{iksd}\psi_0(x-sd) \tag{228}$$

This has now the form of a Bloch wave function, and is the best approximation we can make with the functions $\psi_0(x)$.* Let us substitute the wave function in the wave equation; we then have

$$E\psi_k(x) = -\frac{\hbar^2}{2m}\nabla^2\psi_k(x) + V(x)\psi_k(x) \tag{229}$$

Using the equation

$$E_0\psi_0(x) = -\frac{\hbar^2}{2m}\nabla^2\psi_0(x) + U_0(x)\psi_0(x) \tag{230}$$

we may write equation (229) in the form

$$(E-E_0)\psi_k(x) = A\sum_s e^{iksd}\,W(x-sd)\,\psi_0(x-sd) \tag{231}$$

We now multiply by $\psi_k^\star(x)$ and integrate over the whole array, obtaining

$$E = E_0 + A\sum e^{iksd}\int \psi_k^\star\,W(x-sd)\,\psi_0(x-sd)\,dx \Big/ \int \psi_k^\star\psi_k\,dx \tag{232}$$

The second term in equation (232) is assumed to be small so that we may neglect the overlap integrals in the denominator and write

$$\int \psi_k^\star\psi_k\,dx = A^2 N \tag{233}$$

where N is the number of centres in the array. Thus we have

$$E = E_0 + \frac{1}{N}\sum_t\sum_s e^{ik(s-t)d}\int \psi_0^\star(x-td)\,W(x-sd)\,\psi_0(x-sd)\,dx \tag{234}$$

All the integrals in the sum over s are the same and there are N of them. We may therefore write, putting $r = t-s$,

$$E = E_0 + \sum_r e^{-ikrd}\int_{-\infty}^{\infty} \psi_0^\star(x-rd)\,W(x)\,\psi_0(x)\,dx \tag{235}$$

The limits of integration may be taken as from $-\infty$ to $+\infty$ without appreciable error because of the rapid decrease of $\psi_0(x)$ for large values of $|x|$. We now neglect all contributions except from neighbouring centres, i.e. taking only $r = 0, 1, -1$. Let us write

$$-\alpha = \int_{-\infty}^{\infty} \psi_0^\star(x)\,W(x)\,\psi_0(x)\,dx \tag{236}$$

$$-\beta = \int_{-\infty}^{\infty} \psi_0^\star(x-d)\,W(x)\,\psi_0(x)\,dx \tag{237}$$

$$\beta' = \int_{-\infty}^{\infty} \psi_0^\star(x+d)\,W(x)\,\psi_0(x)\,dx \tag{238}$$

* See also exercise at the end of this section.

Let us now assume that $\psi_0(x)$ is an even function of x so that

$$\psi(x) = \psi(-x)$$

i.e. that it is a symmetrical wave function as discussed in § 2.2. Clearly $W(x)$ is negative (see Fig. 4.20) so that in this case both α and β are positive, and $\beta' = -\beta$, and we have

$$E = E_0 - \alpha - 2\beta \cos kd \tag{239}$$

We therefore once again obtain the familiar form for an allowed energy band. We note that when $\psi_0(x)$ is an even function of x the energy increases with k $(0 \leqslant k \leqslant \pi/d)$, the bottom of the band being at $k = 0$.

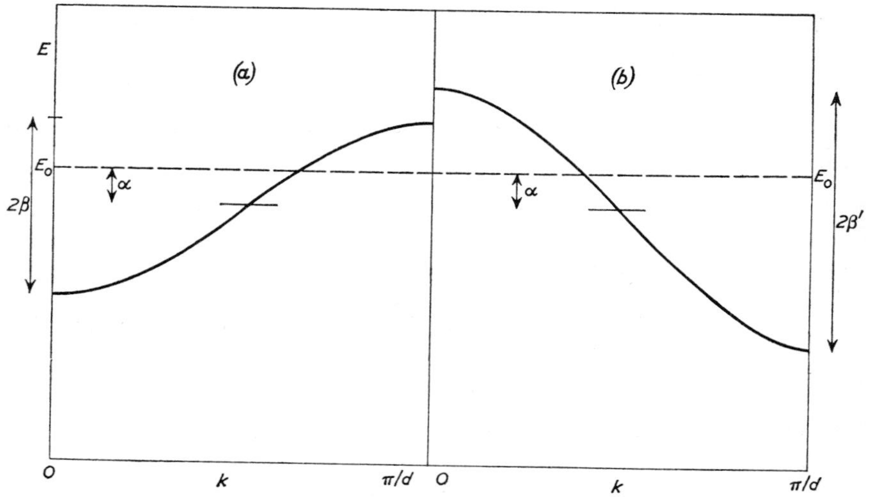

FIG. 4.21. Energy bands arising from states having even and odd wave functions. (a) from even wave functions, (b) from odd wave functions

Now let us suppose that $\psi_0(x)$ is an odd function of x so that

$$\psi_0(x) = -\psi_0(-x)$$

i.e. that it is an antisymmetrical function of the form discussed in § 2.2; as we saw for the particular problem treated in § 2.2, the energy levels associated with even and odd wave functions alternate. In this case $\beta' > 0$ and $\beta = -\beta'$; also we have $\alpha > 0$ and may express the energy E in the form

$$E = E_0 - \alpha + 2\beta' \cos kd \tag{240}$$

The form of the energy bands corresponding to these two types of wave function are shown in Fig. 4.21.

Exercises (1) Show that if we take for the wave function for the array of potential wells the form (227), with the constants a_s as yet undetermined, to the same approximation as we have used, the energy E is given by

$$(E-E_0)\sum_s a_s a_s^\star = \sum_{st} a_s^\star a_t \mathbf{W}(s-t)$$

where
$$\mathbf{W}(r) = \int_{-\infty}^{\infty} \psi_0^\star(x-rd)\, W(x)\, \psi_0(x)\, dx$$

(2) Regarding the quantities a_s, $a_s^\star$ as independent parameters show that E has a stationary value if

$$(E-E_0)a_s = \sum_t a_{s-t}\mathbf{W}(t)$$

and similar equations for $a_s^\star$.

Show that this set of equations is satisfied by taking $a_s = e^{i\alpha s}$ for all values of s, and that in this case

$$(E-E_0) = \sum_s e^{-i\alpha s}\mathbf{W}(s)$$

On writing $k = \alpha/d$ we get the same result as before. This shows that the constants a_s chosen to give a Bloch wave function are also those determined by the 'variation principle' to make E stationary.

(3) Show that if we neglect $e^{-\alpha d}$ compared with unity, the wave function for an array of N deep narrow potential wells may be written in the form

$$\psi = A\sum_{r=0}^{N-1} \psi(x-rd)\, e^{ikrd}$$

where
$$\psi(x) = \alpha_0^{1/2} e^{-\alpha_0|x|}$$

Hence show that, apart from a phase factor which cannot be determined, the normalising factor A is given by $N^{-1/2}$.

(4) Show that the wave functions (228) are accurately orthogonal for different values of k and that they may be normalised by taking

$$|A|^{-2} = N\left\{1+\tfrac{1}{2}\sum_{s=1}^{N} \alpha(s)\, e^{-iksd}\right\}$$

where s is an integer and

$$\alpha(s) = \int_{-\infty}^{\infty} \psi^\star(x)\, \psi(x-sd)\, dx$$

When the quantities $\alpha(s)$ are small show that approximately we have

$$|A| = N^{-1/2}\left\{1-\tfrac{1}{4}\sum_{s=1}^{N} \alpha(s)\, e^{-iksd}\right\}$$

4.11.3 Crossing of energy bands

An interesting situation may arise when we have two energy levels of the isolated potential well lying fairly closely together. Let us suppose that one has an odd wave function ψ_1 corresponding to an energy E_1 and that the other has an even wave function ψ_2 corresponding to an energy E_2. If we regard the levels as independent, the energy in two allowed bands for the array of potential wells arising from these two levels would be given by equations (239), (240) with constants α_1 β_1', α_2, β_2. Let us write $E_A = E_1 - \alpha_1 + 2\beta_1'$ and $E_S = E_2 - \alpha_2 - 2\beta_2$. Then the A-band will extend from E_A to $E_A - 4\beta_1'$ and the S-band from E_S to $E_S + 4\beta_2$ (see Fig. 4.22, broken curves). We shall suppose that $E_A > E_S$; then two situations may arise. If $E_A - 4\beta_1' > E_S + 4\beta_2$ the two bands do not cross (Fig. 4.22(a)); if they come close together some interaction may take

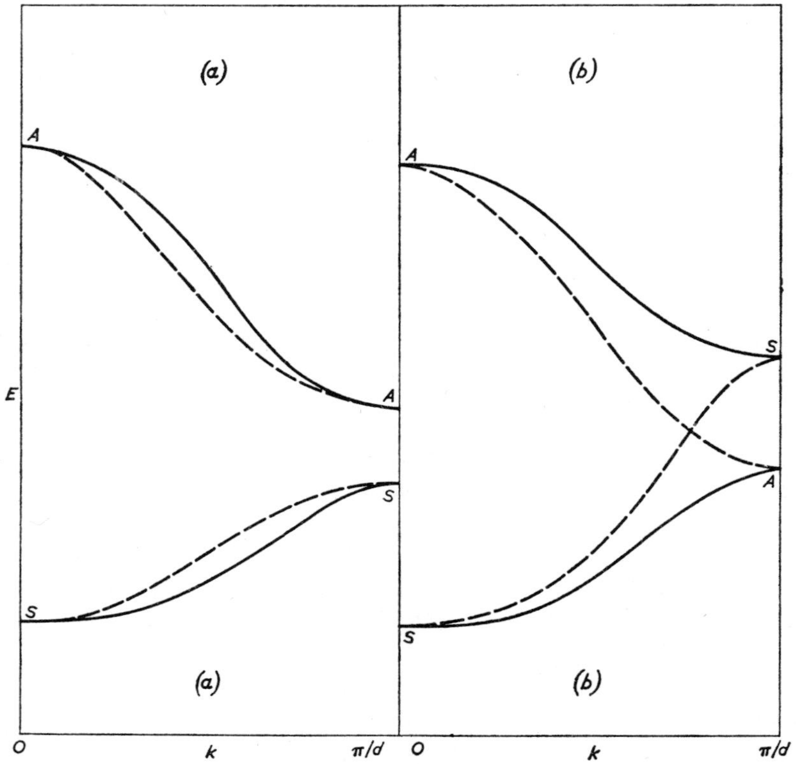

FIG. 4.22. Interaction of allowed energy bands. (Broken curve shows energy bands without interaction)

place and the form of the bands will be slightly changed, they will, however, retain their essential character as A- and S-bands. If, however, we have $E_A - 4\beta_1' < E_S + 4\beta_2$ the bands will cross and will become degenerate at the crossing point (Fig. 4.22(b)). The approximation of treating the bands separately is no longer valid and we have to form a wave function for the array made up of both the wave function formed from ψ_1 and that formed from ψ_2. Let us write

$$\psi_{k1} = \sum e^{ikrd}\,\psi_1(x - rd) \tag{241}$$

$$\psi_{k2} = \sum e^{ikrd}\,\psi_2(x - rd) \tag{242}$$

and take for the complete wave function

$$\Psi = a_1 \psi_{k1} + a_2 \psi_{k2} \tag{243}$$

To obtain the energy E we proceed as from equation (229) to equation (232), but using the wave function Ψ for substitution in the wave equation, multiplying in turn by $\psi_{k1}^{\star}$ and $\psi_{k2}^{\star}$ and integrating; we thus obtain the two equations

$$a_1[E - E_1(k)] + a_2 E_{12} = 0 \tag{244}$$

$$a_1 E_{21} + a_2[E - E_2(k)] = 0 \tag{245}$$

where $E_1(k)$ and $E_2(k)$ are given by equation (235) with the appropriate wave functions, and hence by equations (239)–(240) with the appropriate constants. (They are the E/k relations for the two bands without interaction.) The quantities E_{12}, E_{21}, will readily be seen, on comparing with equation (235), to be given by

$$E_{12} = \sum_r \mathrm{e}^{-ikrd} \int_{-\infty}^{\infty} \psi_1^{\star}(x - rd)\, W(x)\, \psi_2(x)\, \mathrm{d}x \tag{246}$$

$$E_{21} = \sum_r \mathrm{e}^{-ikrd} \int_{-\infty}^{\infty} \psi_2^{\star}(x - rd)\, W(x)\, \psi_1(x)\, \mathrm{d}x \tag{247}$$

Again we need consider only the values $-1, 0, 1$ for r. When $r = 0$ we have

$$\int_{-\infty}^{\infty} \psi_1^{\star}(x)\, W(x)\, \psi_2(x)\, \mathrm{d}x = 0 \tag{248}$$

because of the opposite symmetry of ψ_1 and ψ_2. Let us write

$$\left.\begin{aligned} \gamma &= \int_{-\infty}^{\infty} \psi_1^{\star}(x - d)\, W(x)\, \psi_2(x)\, \mathrm{d}x \\ &= -\int_{-\infty}^{\infty} \psi_1^{\star}(x + d)\, W(x)\, \psi_2(x)\, \mathrm{d}x \end{aligned}\right\} \tag{249}$$

the equality of the two integrals being clear from the symmetry properties of $\psi_1(x)$, $\psi_2(x)$, $W(x)$ being an even function. Similarly we may write

$$\left.\begin{aligned} \gamma' &= -\int_{-\infty}^{\infty} \psi_2^{\star}(x - d)\, W(x)\, \psi_1(x)\, \mathrm{d}x \\ &= \int_{-\infty}^{\infty} \psi_2^{\star}(x + d)\, W(x)\, \psi_1(x)\, \mathrm{d}x \end{aligned}\right\} \tag{250}$$

γ and γ' will be seen to have the same sign on examination of the form of the integrands. Equations (244), (245) then become

$$a_1[E - E_1(k)] - 2i\gamma \sin kd\, a_2 = 0 \tag{251}$$

$$2i\gamma' \sin kd\, a_1 + a_2[E - E_2(k)] = 0 \tag{252}$$

We have then unless $a_1 = a_2 = 0$

$$[E - E_1(k)][E - E_2(k)] - 4\gamma\gamma' \sin^2 kd = 0 \tag{253}$$

There are two solutions of equation (253) which we shall call $E^+(k)$ and $E^-(k)$. They are given by the equations

$$E^+(k) = \tfrac{1}{2}[E_1(k) + E_2(k)] + \tfrac{1}{2}\{[E_1(k) - E_2(k)]^2 + 4\gamma\gamma' \sin^2 kd\}^{1/2} \tag{254}$$

$$E^-(k) = \tfrac{1}{2}[E_1(k) + E_2(k)] - \tfrac{1}{2}\{[E_1(k) - E_2(k)]^2 + 4\gamma\gamma' \sin^2 kd\}^{1/2} \tag{255}$$

We note that, as k approaches the values 0 and π/d, $\sin kd$ tends to zero. By expanding the square root we see that, provided $E_1(k) > E_2(k)$ for all values of k,

$$\left.\begin{aligned} E^+(0) &\to E_1(0) \\ E^+(\pi/d) &\to E_1(\pi/d) \\ E^-(0) &\to E_2(0) \\ E^-(\pi/d) &\to E_2(\pi/d) \end{aligned}\right\} \tag{256}$$

The curves for $E^+(k)$ and $E^-(k)$ are shown in Fig. 4.21(a).

If $E_A - E_S < 4(\beta_2 + \beta_1')$ then for some value of k, say $k = k_1$, $E_1(k_1) = E_2(k_1)$ and we have a different situation. We may write $E^+(k)$, $E^-(k)$ in the form

$$\left.\begin{aligned} E^+(k) &= E_1(k_1) + (\gamma\gamma')^{1/2} \sin k_1 d \\ E^-(k) &= E_1(k_1) - (\gamma\gamma')^{1/2} \sin k_1 d \end{aligned}\right\} \tag{257}$$

when $k = k_1$; this shows that we must always take the positive value of the square root in E^+ and the negative value in E^-. Thus as $k \to 0$ we have

$$\left.\begin{aligned} E^+(0) &= E_1(0) \\ E^-(0) &= E_2(0) \end{aligned}\right\} \tag{258}$$

as before, but as $k \to \pi/d$ we now have

$$E^+(\pi/d) \to \tfrac{1}{2}[E_1(\pi/d) + E_2(\pi/d)] + \tfrac{1}{2}|E_1(\pi/d) - E_2(\pi/d)| = E_2(\pi/d) \tag{259}$$

and, similarly,

$$E^-(\pi/d) \to E_1(\pi/d) \tag{260}$$

Thus the curves do not actually cross but are forced away from one another as shown in Fig. 4.22(b) (full curves).

When the curves for $E_1(k)$, $E_2(k)$ do not cross we clearly have from equation (251) $a_2 = 0$ when $E = E_1(k)$ and similarly from equation (252) $a_1 = 0$ when $E = E_2(k)$. The complete wave function for the state corresponding to E^+ is therefore made up entirely from the odd type of wave functions when $k = 0$ and $k = \pi/d$ and that for E^- of even type of wave functions (called A- and S-states in Fig. 4.21).

When the curves for $E_1(k)$ and $E_2(k)$ cross, however, it will readily be seen that, for E^+, $a_1 \to 0$ as $k \to \pi/d$ and, for E^-, $a_2 \to 0$ as $k \to \pi/d$. The state corresponding to E^+ therefore changes over from being entirely A-type at $k = 0$ to being entirely S-type at $k = \pi/d$. Similarly, the state corresponding to E^- changes from S-type at $k = 0$ to A-type at $k = \pi/d$. For intermediate values of k the states are mixed, the ratio $|a_1/a_2|^2$ as given by equations (251) and (252) giving a measure of the extent to which each type is represented. The energy curves therefore follow the usual pattern in spite of the tendency to cross but there is a switch over in the character of the wave functions.

4.12 Some general properties of the E/k relationships

We may now summarise some general properties of the relationship between the energy E and the propagation constant k, which is related to the crystal momentum P by the equation $P = \hbar k$. First of all, it is easy to see that E is an even function of k, i.e. if a particular value k_1 of k corresponds to energy E_1, then the value $-k_1$ also corresponds to energy E_1. This follows at once from the observation that if we have a solution ψ_1 of the form $\psi_1 = e^{ik_1x}A_1(x)$, $A_1(x)$ being a periodic function of period d, then $\psi_1^\star$ is also a solution since $V(x)$ is real. Thus we have

$$\psi_1^\star = e^{-ik_1x}A_1^\star(x)$$

a solution of the correct form with $k = -k_1$, of the wave equation with energy $E = E_1$.

We may also readily show that a solution with $k = k_1 + 2n\pi/d$ where n is an integer also corresponds to $E = E_1$.

Let us write

$$\psi_n(x) = e^{i(k_1+2\pi n/d)x}\,U(x) \tag{261}$$

where $U(x)$ is a periodic function of period d. This is of the usual form with $k = k_1 + 2\pi n/d$. If we take

$$U(x) = A_1(x)\,e^{-2\pi inx/d}$$

this being a periodic function of period d we have

$$\psi_n(x) = \psi_1(x)$$

which is a solution corresponding to an energy $E = E_1$. Thus we see that there is a wave function of the correct form with $k = k_1 + 2\pi n/d$ which corresponds to $E = E_1$. E is therefore an even periodic function of k of period $2\pi/d$. Thus we may restrict, if we wish to do so, values of k to the interval $-\pi/d \leqslant k \leqslant \pi/d$ to express E as a function of k; in this case E is a many-valued function of k, and we have called this the reduced representation. Alternatively we may use all values of k from $-\infty$ to ∞ allotting each range $n\pi \leqslant k \leqslant (n+1)\pi$, and $-(n+1)\pi \leqslant k \leqslant -n\pi$, to one energy band; this we have called the expanded representation. In this representation E is, in general, a discontinuous function of k the discontinuities occurring at $k = n\pi/d$.

We may further show that, when k is real, E may be expressed as a function of $\cos kd$; this follows from equation (144), obtained in the proof of Floquet's theorem. From this also follows both the above properties and also the property that in general when $k \simeq n\pi/d$, $\mathrm{d}E/\mathrm{d}k = 0$, and this corresponds to the edge of a forbidden energy zone.

If we write equation (144) in the form

$$\cos kd = \phi(E) \tag{262}$$

and differentiate with respect to k we have

$$-d \sin kd = \phi'(E)\frac{\mathrm{d}E}{\mathrm{d}k} \tag{263}$$

Hence unless $\phi'(E) = 0$ when $\phi(E) = \pm 1$ we have $\mathrm{d}E/\mathrm{d}k = 0$ when $k = n\pi/d$. The condition $\phi'(E) = 0$ means that the curve giving $\phi(E)$ as a function of E touches one of the lines $\phi(E) = \pm 1$, i.e. we have two continuous allowed energy bands and there is no forbidden zone. In this case we have degeneracy at the band edge and *need not have* $\mathrm{d}E/\mathrm{d}k = 0$ (see for example the curves for $\gamma a = 0$ in Fig. 4.16). We shall now apply these considerations to a further discussion of the motion of electrons in crystalline solids.

4.13 Further discussion of electrons in crystalline solids

We have already seen in § 4.3 how the simple model consisting of an array of very deep narrow potential wells may be used to discuss the behaviour of the electrons in deep-lying atomic levels in a crystalline solid. The problem treated in § 4.10 gives a more realistic model, in that we may have a number of levels, some deep and some having only a small binding energy. The deep-lying levels form narrow bands, a property which holds for any periodic potential as shown in § 4.11.2;* thus, to a good approximation, we may regard the deep-lying electrons in a crystalline solid as having the same energies and approximately the same wave functions near each atom as for an isolated atom. In the space between the atoms the wave functions are small in amplitude, and although the electrons are free in the sense that they may be represented by travelling waves which extend throughout the crystal, they take no part in conduction processes and do not contribute to the specific heat. The reason for this is not only that they have large effective masses (narrow energy bands) but because the energy levels forming each band are completely filled. In all the complete inner electron shells in atoms electron spins are paired and we have two electrons per level when we do not take account of fine structure. We have seen that each band arising from a group of N atoms contains N levels and we can just fill it with $2N$ electrons (one with each value of the spin on each level). For the doublet atomic levels referred to we therefore just fill all the corresponding levels in the band.

* We shall assume for the moment that these considerations may be extended to three dimensions; this we shall show in later Chapters.

4.13.1 Metals

For the outer or valence electrons, which we may consider as occupying levels not far below the line $E = 0$ in Fig. 4.16, we have considerable broadening. Let us consider for example the alkali metals, taking sodium as typical; the electrons in the $1s$-shell form a very narrow full band lying at a depth of $1{\cdot}65 \times 10^3$ eV. The second ionisation potential of sodium is 47 eV and the lattice spacing we may take as 4×10^{-8} cm giving $\alpha d = 15$; the $2s$- and $2p$-levels having eight electrons in all therefore form quite narrow bands and again these are *full* bands.

This requires a little more careful consideration since the three $2p$-states are degenerate in the absence of a magnetic field. First of all let us suppose that the $2s$- and $2p$-states are well separated so that the $2s$-state, which is a singlet, forms a separate completely filled band. If we were to regard each of the p-states as independent we should obtain overlapping bands having $3N$ states in all. Each level in the p-band would (apart from spin) be triply degenerate. When we take spin into account each level would be sixfold degenerate and there would thus be $6N$ levels in all in the band arising from the $2p$-states. As there are six $2p$-electrons per sodium atom this band just contains all the $6N$ electrons from the $2p$-levels of the atoms of the crystal; we have therefore again a completely filled band. When we have two nearby bands which try to cross each other more care has to be taken in allocating the bands to particular energy levels (see § 4.11.3).

We are then left with the single $3s$-valence electron whose energy is well separated from those of the $2s$- and $2p$-electrons. The effective potential well for this electron may be taken as having a radius of about 10^{-8} cm and $E_0 = 5$ eV; for a crude estimate we may take the barrier width $b = 2 \times 10^{-8}$ cm, and thus we have $\alpha_0 b = 2{\cdot}4$. The approximate width of the allowed band corresponding to this level would be about $8E_0 \exp(-\alpha_0 b)$ or about 4 eV; the only point we wish to make from this crude estimate is that the atomic level is broadened into quite a wide band. Moreover this band is just half filled since we have only *one* electron for each $3s$-level in sodium; this is typical of the alkali metals. A similar situation exists for the noble metals, Au, Ag, Cu, but we may no longer entirely neglect the next lower electron shell.

We are now in a position to compare Bloch's theory of a metal (see § 2.13) in which all the electrons are regarded as free, but in which account is taken of the periodic potential of the crystal, with that of Sommerfeld (see § 2.3) in which the potential is taken as constant. In Sommerfeld's theory it was simply *assumed* that we could ignore the electrons in the inner shells, while Bloch's theory shows us why this can be done, and moreover tells us which electrons must be taken into account. In the alkali metals we need take account of only the single valence electron for each atom so it is not now so surprising that Sommerfeld's theory gives a good description of the alkali metals. For the transition metals, with incomplete shells below the outer shell, several electrons from different levels have to be taken into account and Sommerfeld's theory is too crude to do this so as to give anything like quantitative agreement with observation. Sommerfeld's theory assumes a constant potential in which electrons are free to move; in Bloch's theory the electrons are also free to move throughout the crystal, being unimpeded by the atoms forming a perfect crystal lattice, but now the relationship between energy and momentum is not the simple one for a constant potential which gives a parabolic law. However, if we identify the bottom of the potential well representing the metal in Sommerfeld's model with the bottom of the half-filled band due to a single valence electron in Bloch's theory, we see that near the bottom of the band the energy relationships are similar, except that in Bloch's theory we must use an effective mass m^* instead of the free-electron mass $\boldsymbol{m}$. A comparison is made in Fig. 4.23 of a typical band shape as given by Bloch's theory with that given by Sommerfeld's theory *corrected to have the appropriate effective mass at the* bottom of the band. It will be seen that for a fairly broad band, up to the point where it is half filled there is not a great deal of difference; thus we see again why, by using an effective mass, the simple Sommerfeld theory gives quite a good description of the alkali metals.

We must here point out, however, a source of confusion over the term effective mass. The mass m^* referred to in § 2.7 and derived from the electronic specific heat is that which gives the parabolic band equivalent to the real band; it corresponds most closely to the effective mass near the *bottom* of the band. The actual electrons which are thermally excited, however, have energies, as we have seen in § 2.7, in a range $\pm \boldsymbol{k}T$ on either side of the level at which the band is just half filled. Here the *curvature* of the band may be much less than for the equivalent parabolic band and may even have changed sign if the band is filled above its point of inflexion (see Fig. 4.23), and in this case the effective mass as defined in § 4.7 may be negative for energy levels near the top of the filled part of the band. It is just these electrons near the top of the filled section of the band that take part in conduction phenomena and they may have a very different effective mass from that obtained from the specific heat data.

Here we must pause to consider the cause of electrical and thermal resistance in a crystalline solid. Although Sommerfeld considered the electrons in the solid as free in order to calculate their energy levels, he assumed that the atoms of the solid could not altogether be ignored, and that the resistance of a metal is due to electrons making collisions with them and having their propagation constants k changed as a result. In Bloch's theory, however, the atoms of the crystal are taken into account by means of the periodic potential and do *not* cause scattering of electrons so long as they form a *perfect* lattice. For electrons in an ideal crystalline lattice there *is no electrical or thermal resistance* according to Bloch's theory. It is fairly easy to see, however, how electrical resistance arises; first of all, the atoms will execute random thermal oscillations about their positions of equilibrium, and we should expect the resistance of a metal to increase as the temperature is increased, which it certainly does. Again we should expect that crystalline imperfections such as are caused by strains or the presence of foreign atoms would cause departures from the ideal lattice and would cause transitions between states with different values of k, leading to resistance; this effect is also observed. We shall defer further discussion of these aspects for the moment; we may observe, however, that a qualitative explanation of both thermal and impurity effects can be given in terms of Bloch's model.

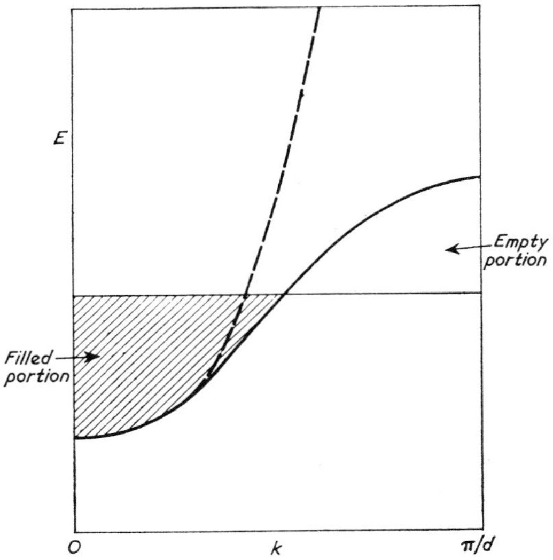

FIG. 4.23. Comparison of a half-filled band with a parabolic band (shown by broken curve)

As we might expect, an electron may travel a distance many times the spacing between the atoms before being scattered. Suppose we assume the electron to be quite free to be accelerated in an applied electric field $\mathscr{E}$ for a time τ and then to suffer a collision which changes its velocity in a random fashion. It will acquire in the field an average velocity of the order of $e\mathscr{E}\tau/2m$. If μ is the velocity acquired in unit field, known as the mobility, then we have

$$\mu = e\tau/2m \tag{264}$$

If we have n free electrons per unit volume then the current density I will be given by

$$I = -ne\mu\mathscr{E} \tag{265}$$

and the conductivity σ by

$$\sigma = ne\mu \tag{266}$$

If we assume for n the value obtained by taking one electron per atom, we may calculate μ from the measured conductivity and hence obtain τ. If $\bar{v}$ is the average velocity of an electron, which we may calculate from the energy distribution we may estimate the average distance $l = \bar{v}\tau$ travelled by the electron between collisions; it turns out to be of the order of several hundred lattice spacings. This is quite inexplicable if we assume that the electron collides classically with the atoms of the crystal but is just the sort of value we might expect on the basis of Bloch's theory. We therefore see that although Sommerfeld's theory does give a simple picture of some of the properties of metals, particularly the alkali metals, we have to use Bloch's theory to discuss such topics as electrical and thermal conductivity. We shall see that it also accounts for various other phenomena such as the Hall effect.

When a transverse magnetic field is applied to a thin strip of metal carrying an electric current, a transverse voltage is produced. The sign of this voltage depends on the sign of the current carriers; this effect is known as the Hall effect and we shall discuss it in much greater detail later. Most metals show a normal Hall effect corresponding to negatively charged current carriers; this is taken to indicate that the current is carried by electrons. Some metals, such as zinc, have, however, a Hall coefficient indicating positive charge carriers; this also is explained in terms of Bloch's theory. If we replace the electronic mass $\boldsymbol{m}$ in equation (264) by $m^{\star}$, the effective mass as defined in § 4.7, then when $m^{\star}$ is negative the sign of the current relative to the field E is reversed. Such electrons therefore behave as though they had a *positive* charge. We shall later consider yet another way of looking at this situation—one which is particularly useful for the case of a band which is nearly filled (see § 8.11).

Let us now consider a metal in which the half-filled band is filled to a level which lies below the point of inflexion (see Fig. 4.23). For such a metal the charged carriers in the highest filled levels will have $\mathrm{d}^2E/\mathrm{d}P^2 > 0$ and so will behave normally. If, however, the band is filled *above* the inflexion point the charged carriers in the highest levels, which are the ones taking part in conduction, will have $\mathrm{d}^2E/\mathrm{d}P^2 < 0$ and so will have a negative effective mass; this would account for a change of sign of the Hall effect. In particular, for a nearly filled band the sign of the charge carriers will apparently be positive (see § 8.11).

Let us now consider the alkaline earth metals Be, Mg, etc. They each have two electrons in their outer shell and these we should expect to fill completely a band of allowed energy levels. Now suppose this were so, and that a forbidden energy gap of appreciable width lies above this band; then a large amount of energy would be required to excite electrons to the higher band and no rearrangement of the k-values of the electrons in the full band would be possible without such excitation. This is just the condition we should expect to hold in an insulator, and clearly it does not hold for the alkaline earth metals. The reason is that, while most of the properties we have derived for energy bands in one dimension hold equally well in three dimensions, there is one very important difference. In one dimension neighbouring bands cannot overlap but are separated by a finite range of forbidden energies (see § 4.11.3); in three dimensions, however, as we shall see, energy bands *may* overlap. This is what happens for the alkali earth metals; the band arising from the nearby p-states, which correspond to excited states of the atom, dips down to overlap the widened band due to the s-states. Overlapping of bands also occurs for the transition metals. We have then two partially filled energy bands which overlap, and in which conduction may take place. The overlapping bands lead to a certain complexity in all metals except the alkali metals and we shall not discuss the matter further at the moment.

4.13.2 Insulators

There are a good many substances, both elements and compounds, for which the highest band containing an appreciable number of electrons is completely filled. For this to happen we clearly need to have an even number of valence electrons; for a compound, we must consider the valence electrons of *all* the atoms making up the compound crystal. In diamond, for example, we have four valence electrons and these form a filled band. (In graphite we have a curious quasi two-dimensional structure.) In the alkali halides we have the equivalent of a closed group of eight electrons for each pair of metal and halogen atoms, and the top filled band is again completely filled. Such substances therefore behave as insulators, and we see that Bloch's theory not only explains metallic conduction but also shows why we have insulators. The latter have plenty of effectively free electrons which can travel through the crystal as electron waves yet cannot contribute to conduction because they form completely filled bands.

When the energy gap separating a filled band from the empty band above it is not very great, a few electrons from the filled band may be thermally excited into the nearby empty band; we may then have some conductivity. Such substances form a very important class called semiconductors, and we shall discuss their properties in some detail later.

5

Crystal Lattices

5.1 Real crystals

In Chapters 3 and 4 we have used the concept of a linear crystal in which the atoms have their equilibrium positions at the points of a simple linear lattice or at the points of a lattice with a basis (see § 3.7), the whole crystal being made up of unit cells which are repeated periodically. Ideal three-dimensional crystals are similarly made up by periodic repetition of a basic three-dimensional unit cell. In practice, real crystals usually have this perfect periodic repetition only over quite small volumes and contain many imperfections, which may be either due to inclusion of impurities, which upset the perfect periodicity, or to mechanical defects which distort the structure. For the moment, however, we shall restrict our discussion to perfect periodic structures and shall later discuss the effect of deviations from perfection. In real crystals the atoms at any given time do not occupy the points of a perfect periodic lattice, due to thermal vibrations, but over regions of the crystal small enough to be free from imperfections their equilibrium positions are at the points of such a lattice.

5.2 Two-dimensional lattices

In one dimension, the simplest linear crystals consist of identical atoms whose equilibrium positions have coordinates x given by $x = ld$ where l is an integer; similarly in two and three dimensions the simplest crystals consist of identical atoms having equilibrium positions whose vector coordinates $\mathbf{r}$ are given, respectively, by

$$\mathbf{r} = l\mathbf{d}_1 + m\mathbf{d}_2 \tag{1}$$

and

$$\mathbf{r} = l\mathbf{d}_1 + m\mathbf{d}_2 + n\mathbf{d}_3 \tag{2}$$

l, m, n being integers, and $\mathbf{d}_1$, $\mathbf{d}_2$, $\mathbf{d}_3$ vectors.

More complex crystals are formed, as we shall see, by introducing the two- and three-dimensional forms of the lattice with a basis (see § 3.7). The assembly of points given by equations (1) or (2) constitute what is known as a simple lattice or Bravais lattice. There is considerably greater variety of such lattices in two and three dimensions since the vectors $\mathbf{d}_1$, $\mathbf{d}_2$, $\mathbf{d}_3$ need not have the same magnitude, and the angles between them may also vary.

For the reader studying the properties of such lattices for the first time, a consideration first of all of two-dimensional lattices will be found to be worthwhile. These have some physical interest, since the arrangement of atoms on the surface of perfect crystals may form a two-dimensional lattice, but what is more important at this stage, drawings of the configurations of atoms making up a two-dimensional crystal may easily be made. Many of the fundamental properties of three-dimensional crystals are not very different from those of two-dimensional crystals, but may differ significantly from those of one-dimensional crystals; this difference arises from the extra freedom in choice of basic vectors. Apart from scaling there is in fact only one simple one-dimensional lattice, while there is a

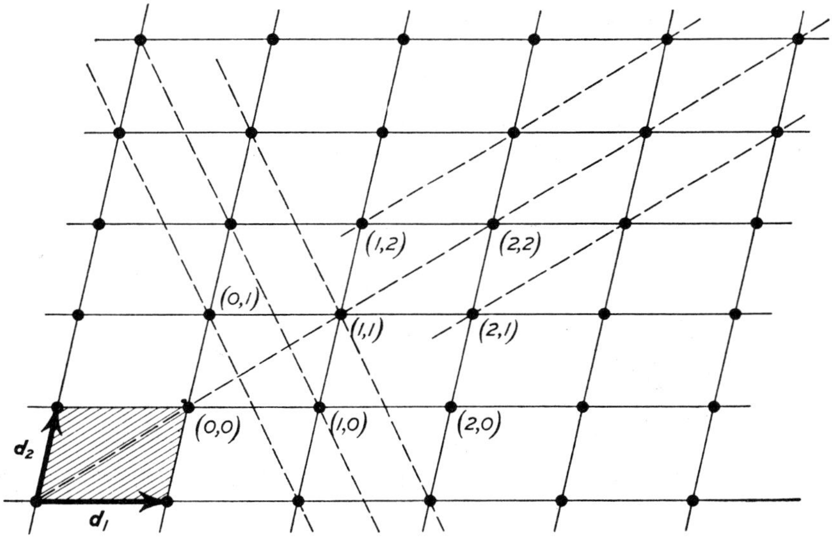

FIG. 5.1. Two-dimensional simple lattice

double infinity of two-dimensional simple lattices obtained by varying the angle between the vectors $\mathbf{d}_1$ and $\mathbf{d}_2$ and the ratio d_1/d_2. We shall therefore begin with a brief discussion of two-dimensional lattices. The reader beginning to study these is urged to make drawings for himself of a number of such lattices and in this way to study their properties.

The general form of a two-dimensional simple lattice is shown in Fig. 5.1. Each point of the lattice may be labelled in terms of the integers l, m in the form (l, m); the point $(0, 0)$ is called the origin of the lattice and may be chosen arbitrarily. It will be clear that when the two basic vectors $\mathbf{d}_1$ and $\mathbf{d}_2$ have been chosen the lattice is uniquely defined. The converse is, however, not true: a given lattice may be defined in terms of more than one set of basic vectors. Clearly, any of the pairs of vectors $\pm\mathbf{d}_1$, $\pm\mathbf{d}_2$ will define the same lattice, but now the numbering of the points of the lattice will be different. These are not all the basic vectors which may be used; for example the lattice may be defined in terms of the vectors $\mathbf{d}_1$, $\mathbf{d}_2'$ where $\mathbf{d}_2' = \mathbf{d}_1 + \mathbf{d}_2$ or $\mathbf{d}_2$, $\mathbf{d}_2'$ and also with various permutations of sign. This is illustrated in Fig. 5.2 in which the vectors $\mathbf{d}_1$, $\mathbf{d}_2'$ are used as basic vectors, the points being labelled accordingly.

Only vectors of the lattice, i.e. of the form $\mathbf{d} = l\mathbf{d}_1 + m\mathbf{d}_2$ may clearly be used as basic vectors, since the point $(1, 1)$ in the new representation must be a point of the lattice. However, not all vectors of this form will give *all* the points of the lattice if used as basic vectors; for example if vectors $\mathbf{d}_1$, $2\mathbf{d}_2$ were used as basic vectors, points such as $(0, 1)$ in Fig. 5.1 would not be included and we should have a sub-lattice. These omitted points may, as we shall see, be re-introduced as a lattice with a basis (see § 3.7) but the simple representation is generally preferred.

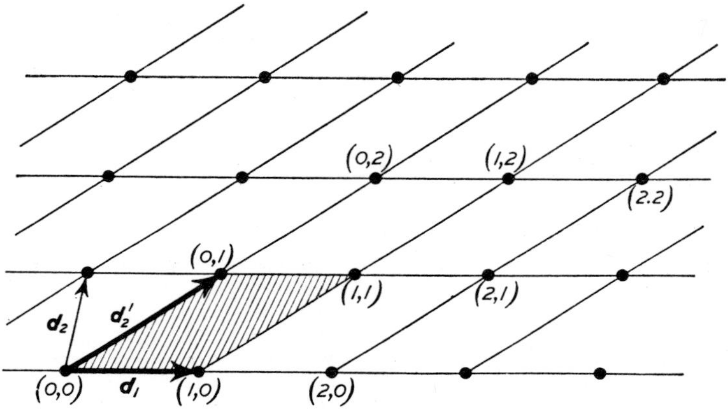

FIG. 5.2. Alternative representation of lattice shown in Fig 5.1

The figure formed by completing the parallelogram two of whose sides are formed by the basic vectors $\mathbf{d}_1$, $\mathbf{d}_2$ is called the unit cell. It has area A_d equal to $|\mathbf{d}_1 \times \mathbf{d}_2|$ or $d_1 d_2 \sin\alpha$, d_1 and d_2 being the lengths of the vectors $\mathbf{d}_1$, $\mathbf{d}_2$ and α the angle between them; the unit cells are shown shaded in Figs. 5.1 and 5.2. For a simple lattice each unit cell accounts for *one* lattice point; the area of all unit cells constructed from pairs of possible basic vectors which may be used to define the lattice (*without* use of basis vectors) must therefore be equal. This gives us a simple criterion for deciding whether two vectors $\mathbf{d}_1' = l\mathbf{d}_1 + m\mathbf{d}_2$, $\mathbf{d}_2' = p\mathbf{d}_1 + q\mathbf{d}_2$ may be used as basic vectors, without a basis. We have

$$\mathbf{d}_1' \times \mathbf{d}_2' = \begin{vmatrix} l & m \\ p & q \end{vmatrix} (\mathbf{d}_1 \times \mathbf{d}_2) \tag{3}$$

so that the required condition is

$$\begin{vmatrix} l & m \\ p & q \end{vmatrix} = 1 \tag{4}$$

When this condition holds $\mathbf{d}_1'$ and $\mathbf{d}_2'$ are called *primitive* vectors. From a study of Figs. 5.1 and 5.2 it will be seen how all space may be filled by *parallel* translation of the unit cells. In certain rather special cases a two-dimensional space may be filled by *parallel* translation of figures other than parallelograms, e.g. by regular hexagons, the lattice points lying at the corners of the hexagons, but in this case a representation in terms of parallelograms may also be made (see Fig. 5.7).

5.2.1 Square lattices

We shall now consider some special cases. The simplest two-dimensional lattice is the square lattice which we obtain when the magnitudes

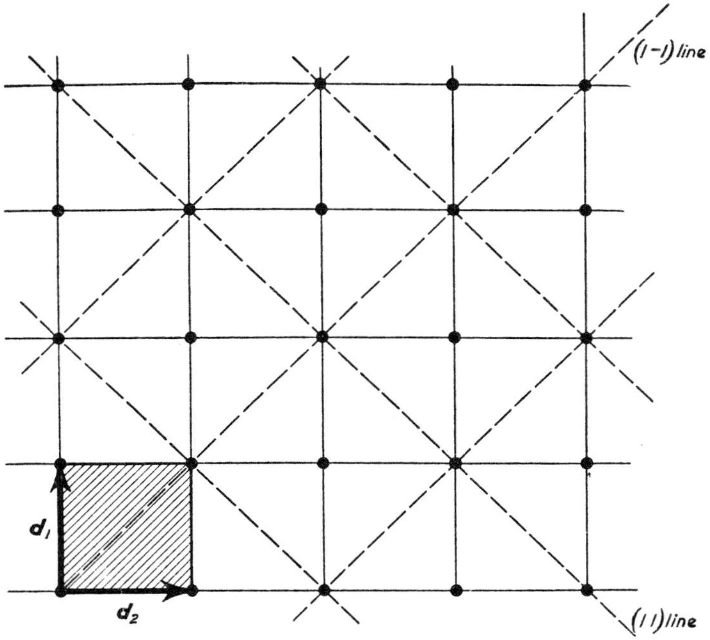

Fig. 5.3. The simple square lattice

d_1, d_2, of the basic vectors are equal and the vectors are at right angles; this is illustrated in Fig. 5.3. In this case the Cartesian coordinates of the points of the lattice are (ld, md) where d^2 is the area of the unit cell. The simplest two-dimensional crystals are built up by having the same type of atom at each point of a simple lattice. To obtain more complex structures, extra points to be repeated with the fundamental translations are introduced so as to form a lattice with a basis (see § 3.7); at these extra points either similar or different atoms may be introduced. The simplest symmetrical extension of the square lattice is made by introducing an extra point at the centre of the unit square, i.e. with the basis vector

$\frac{1}{2}(\mathbf{d}_1+\mathbf{d}_2)$. If this point, and those obtained from it by the lattice translations, are occupied by atoms of the same type as the corners of the squares, this does not lead to a new kind of crystal since it will readily be seen that we again have a simple square lattice whose unit square has side $d/\sqrt{2}$ (cf. broken lines in Fig. 5.3). If, however, the atoms at the corners of the original unit square are of type A and those at the centres are of type B we get a new form of crystal; since there are clearly equal numbers of type A and type B, the chemical composition of such a crystal will be AB. This is illustrated in Fig. 5.4 in which the atoms of type A are marked ○ and those of type B are marked ×.

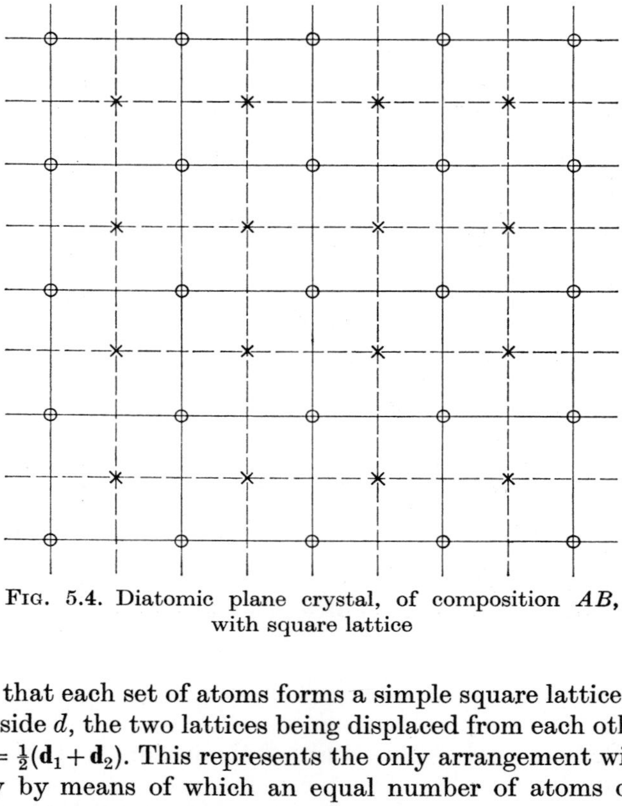

FIG. 5.4. Diatomic plane crystal, of composition AB, with square lattice

We see that each set of atoms forms a simple square lattice with unit square of side d, the two lattices being displaced from each other by the vector $\mathbf{a} = \frac{1}{2}(\mathbf{d}_1+\mathbf{d}_2)$. This represents the only arrangement with square symmetry by means of which an equal number of atoms of type B may be introduced into the lattice; it occurs commonly in nature, the atoms in certain cleavage faces of many diatomic crystals having this arrangement.

Another simple symmetrical arrangement would be to introduce atoms of type B midway between each pair of atoms of type A. This would give twice as many atoms of type B as of type A and would give a crystal of composition AB_2, as shown in Fig. 5.5; the extra basis vectors are in this case $\mathbf{a}_1 = \frac{1}{2}\mathbf{d}_1$ and $\mathbf{a}_2 = \frac{1}{2}\mathbf{d}_2$. If the types of atom A and B are identical we

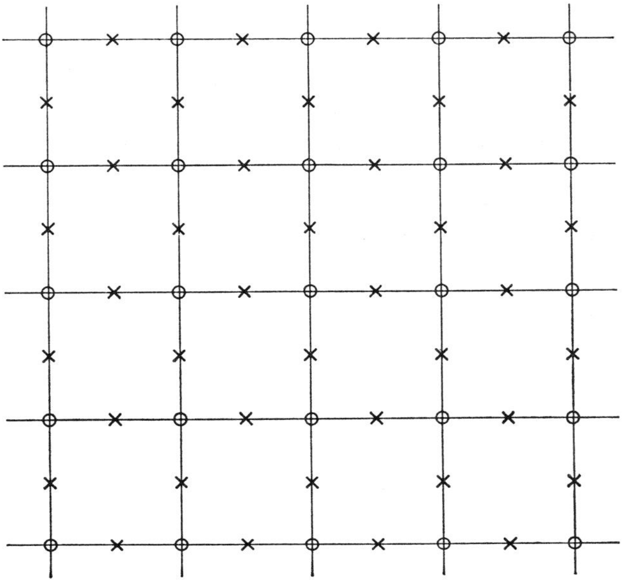

FIG. 5.5. Diatomic plane crystal of composition AB_2 with square lattice

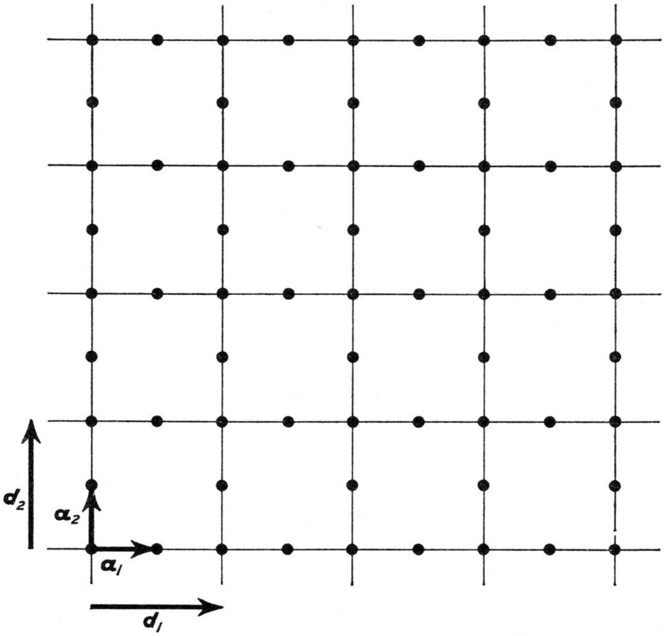

FIG. 5.6. Square lattice with a basis

then have a square lattice with atoms missing from certain symmetrical positions (see Fig. 5.6). In order to have a crystal of composition AB we could now introduce extra atoms of type A in Fig. 5.5 at the centres of the squares. This however, simply gives us back the arrangement of Fig. 5.4 (turned through 45° and with a unit square of side $d/\sqrt{2}$).

5.2.2 Hexagonal lattices

Another interesting special case of considerable practical importance occurs when the magnitudes of the basic vectors $\mathbf{d}_1$ and $\mathbf{d}_2$ are equal and the angle between them is 60°; the resulting lattice is shown in Fig. 5.7,

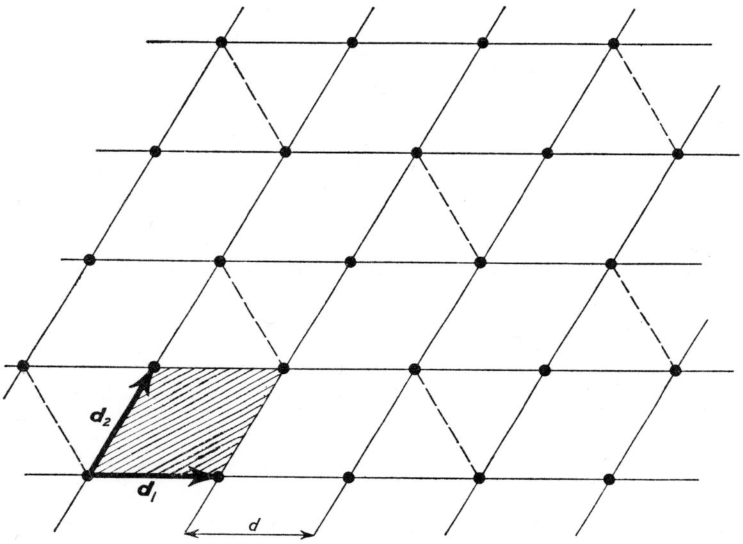

FIG. 5.7. The simple hexagonal lattice

and it will be seen that the lattice points lie at the corners of a series of regular hexagons, which fill the plane, and also at the centre of each hexagon. The unit cell is a rhombus of area $\sqrt{3}d^2/2$, where $d_1 = d_2 = d$. The physical interest in this lattice lies in the fact that it represents the arrangement of closest packing of equal spheres on a plane, as will be seen from Fig. 5.8. (It will be readily verified that the greatest number of equal circles that can be made to touch a given circle is six, and that the centres of these lie at the corners of a regular hexagon with the centre of the given circle at the centre of the hexagon.) If the radius of each sphere is r the magnitude of the basic translation vectors is equal to $2r$. Each sphere therefore accounts for an area equal to that of the unit cell, namely $2\sqrt{3}r^2$. It is interesting to compare this arrangement with the corresponding one derived from the simple square lattice as shown in Fig. 5.9.

It will be seen that this represents a less tightly packed array than that derived from the hexagonal lattice, since each sphere now accounts for an area equal to $4r^2$. In this case it has four nearest neighbours in the plane as distinct from six in the hexagonal arrangement. As we shall see

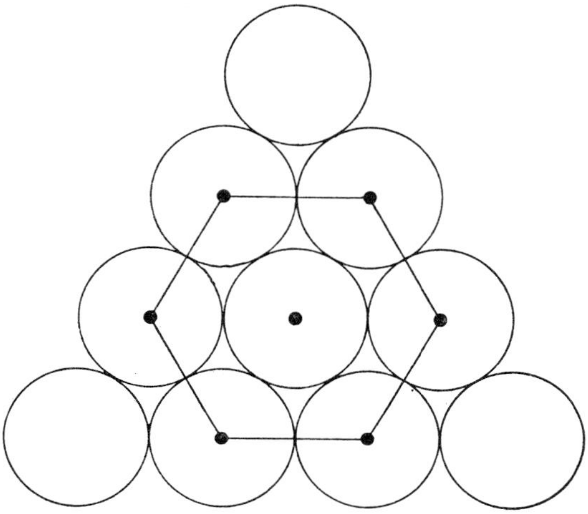

FIG. 5.8. Closest packing of equal spheres in a plane

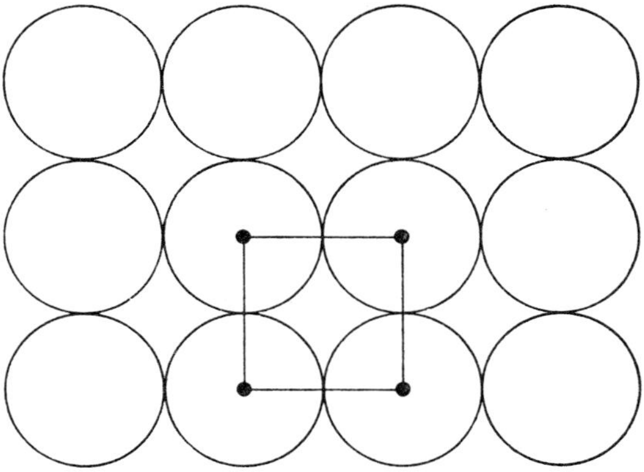

FIG. 5.9. Packing of equal spheres in a simple square lattice

(§ 5.3.3) the hexagonal arrangement represents the close-packed planes in many metal crystals, the square arrangement occurring in some of the less tightly packed planes.

If, in the simple hexagonal lattice, we omit the lattice point which occurs at the centre of each unit hexagon, we obtain a lattice which may be regarded as being derived from the translation of a unit cell consisting of a regular hexagon, the lattice points being at the corners of the basic cell and those derived from it by translation; this is shown in Fig. 5.10 and is the lattice already mentioned in § 5.2.1. It is sometimes known as the hexagonal 'net' lattice, and occurs in certain compounds with a 'ring' type of structure, each atom having three nearest neighbours. At first sight this lattice might be considered to be simpler than the simple hexagonal lattice, but this is not so; it may be derived from the simple hexagonal lattice by introducing a basis vector $\mathbf{a}_1$ given by $\mathbf{a}_1 = \frac{1}{3}(\mathbf{d}_1 + \mathbf{d}_2)$ (see Fig. 5.10).

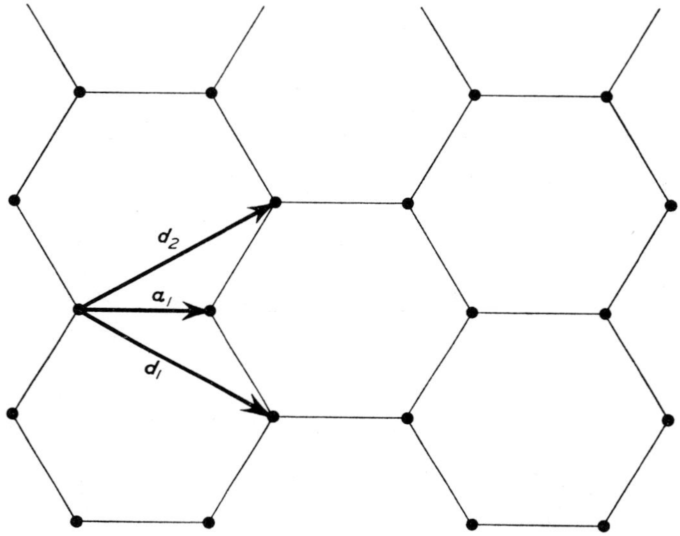

FIG. 5.10. Hexagonal 'net' lattice

5.2.3 Lattice lines

It will be seen from Fig. 5.1 that, in addition to the rows of lattice points on lines parallel to the basic vectors, there are many other well-defined linear arrays of atoms which may be picked out; two such sets of lines are shown (broken lines) in Fig. 5.1. There are indeed an infinite number of such sets of lines, each passing through an infinite number of points of the lattice as may be seen by considering all lines parallel to the line joining the lattice points $(l,0)$ $(0,m)$, where l and m are integers. When l and m are small integers, having no common factor, the linear density of lattice points is seen to be high; some examples are shown in Fig. 5.11. The integers l, m could be used to define the lattice lines; it is, however, found to be more convenient to use their reciprocals, as follows. Any particular lattice line may not meet the axes (i.e. the lines through the origin parallel to the basic vectors) at points of the lattice but it is clear from the way we have defined the lattice lines that some will meet both of the axes in lattice points which we shall suppose to be $(l,0)$ and $(0,m)$. If we now form the quantities $l' = s/l$ and $m' = s/m$ where s is the smallest integer which makes *both* l' and m' integers, then we may label the lattice lines in the form $(l'm')$; l and m and hence l' and m' may be either positive or negative and changing sign will give another set of lines, but these may have equivalent properties, particularly in a lattice with a high degree of symmetry. For example, in the simple square lattice, the lattice lines parallel to the basic vectors are (10) lines and (01) lines and are equivalent as are the lines (11) and $(1-1)$ (shown as broken lines in Fig. 5.3). When we do not wish to distinguish between equivalent sets of

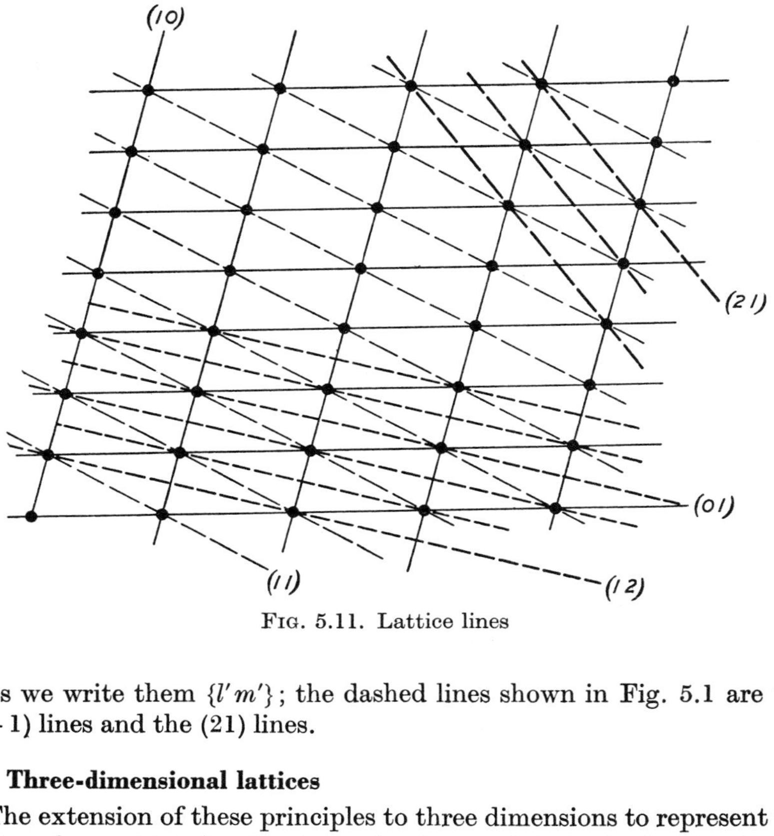

FIG. 5.11. Lattice lines

lines we write them $\{l'm'\}$; the dashed lines shown in Fig. 5.1 are the $(1-1)$ lines and the (21) lines.

5.3 Three-dimensional lattices

The extension of these principles to three dimensions to represent the lattices from which the structure of real crystals is derived is simple and straightforward; the coordinates of the lattice points of the three-dimensional simple or Bravais lattice are given by equation (2). The unit cell in three dimensions consists of the parallelopiped whose corners are at the vector positions 0, $\mathbf{d}_1$, $\mathbf{d}_2$, $\mathbf{d}_3$, $\mathbf{d}_1+\mathbf{d}_2$, $\mathbf{d}_2+\mathbf{d}_3$, $\mathbf{d}_3+\mathbf{d}_1$, $\mathbf{d}_1+\mathbf{d}_2+\mathbf{d}_3$; as is well known, the volume V_d of this parallelopiped is given by

$$V_d = (\mathbf{d}_1 . \mathbf{d}_2 \times \mathbf{d}_3) \tag{5}$$

In terms of the lengths d_1, d_2, d_3, and angles α, β, γ between the basic vectors this may be expressed in the form

$$V_d = d_1 d_2 d_3(1-\cos^2\alpha-\cos^2\beta-\cos^2\gamma+2\cos\alpha\cos\beta\cos\gamma) \quad (6)$$

$$= d_1 d_2 d_3 \phi(\alpha,\beta,\gamma) \quad (6a)$$

In order that three lattice vectors $\mathbf{d}_1'$, $\mathbf{d}_2'$, $\mathbf{d}_3'$ may be used as alternative basic vectors, without the need of extra basis vectors, we must have, if

$$\mathbf{d}_r' = \sum_s a_{rs}\mathbf{d}_s, \qquad r,s = 1,2,3 \quad (7)$$

$$|a_{rs}| = 1 \quad (8)$$

where $|a_{rs}|$ represents the determinant of the coefficients a_{rs}. This follows exactly as for the two-dimensional case (see § 5.2). The whole of space

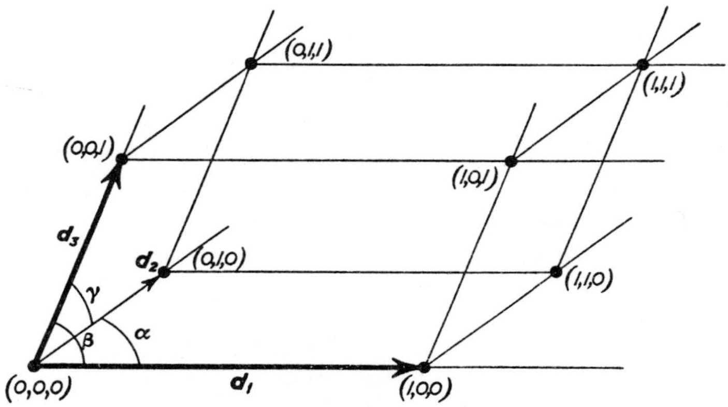

FIG. 5.12. Simple three-dimensional lattice

is filled by translation of the unit parallelopiped, and the various crystal forms come from having particular relationships between the lengths of the basic vectors $\mathbf{d}_1$, $\mathbf{d}_2$, $\mathbf{d}_3$, and the angles between them. More complex lattices may be built up by introducing additional basis vectors $\mathbf{a}_1, \mathbf{a}_2, \ldots$. Clearly we may vary the two ratios d_1/d_2, d_1/d_3 and the three angles α, β, γ included between the vectors $\mathbf{d}_1$, $\mathbf{d}_2$, and $\mathbf{d}_3$ so that there is a quintuple infinity of simple three-dimensional lattices. The general crystal form in which $d_1 \neq d_2 \neq d_3$ and $\alpha \neq \beta \neq \gamma$ is called *triclinic* (Fig. 5.12); if $\alpha = \beta = \pi/2 \neq \gamma$ the structure is called *monoclinic*. If the three basic vectors are mutually at right angles we have the *orthorhombic* form, of which the *tetragonal* form is a special case when $\mathbf{d}_1 = \mathbf{d}_2$; when also $d_1 = d_2 = d_3$ we have the *cubic* form. When $d_1 = d_2 = d_3$ and

$$\alpha = \beta = \gamma \neq \pi/2$$

the form is known as *rhombohedral*.

5.3.1 Cubic lattices

The simple cubic lattice is illustrated in Fig. 5.13; this structure does not occur in natural monatomic crystals, for there are structures derived from it which represent considerably closer packing of the atoms. In three dimensions, when we introduce a single additional lattice point at the centre of the unit cell we get a new structure and *not* the same structure, as occurs in one and two dimensions. This structure is known as the body-centred cubic structure and is illustrated in Fig. 5.14. The basis vector in this case is $\frac{1}{2}(\mathbf{d}_1+\mathbf{d}_2+\mathbf{d}_3)$; it has length $\sqrt{3}d/2$, and makes equal angles $\cos^{-1}(1/\sqrt{3})$ with each of the edges of the unit cube. The body-centred cubic lattice may be regarded as being made up of two identical

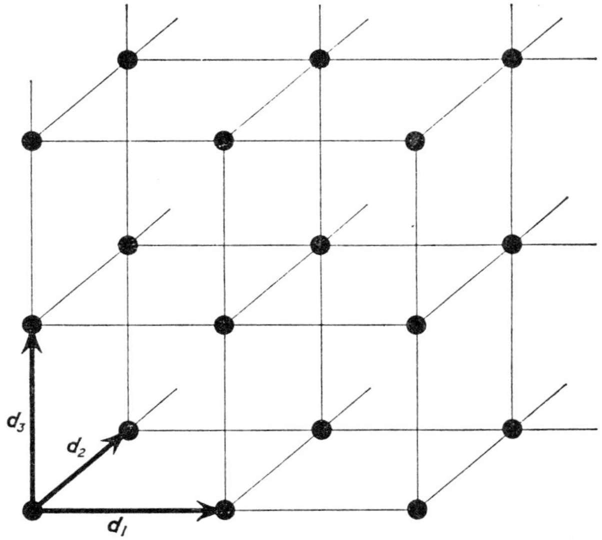

FIG. 5.13. The simple cubic lattice

simple cubic lattices displaced from each other by a distance equal to half the length of a body diagonal of the unit cube in a direction along the diagonal. There are clearly two lattice points per unit cell, and each has eight equidistant nearest neighbours. (It will readily be seen that each of the lattice points at the corners of the original unit cube also lies at the centre of an exactly similar cube with eight lattice points at the corners.)

This is actually not the most fundamental way of looking at the body-centred cubic lattice which is really a simple lattice. If we take as basic vectors three half body diagonals, i.e. vectors with components, in terms of d, $(\frac{1}{2},\frac{1}{2},\frac{1}{2})$, $(\frac{1}{2},\frac{1}{2},-\frac{1}{2})$, $(\frac{1}{2},-\frac{1}{2},\frac{1}{2})$, it will readily be seen that all the other lattice points, for example $(1,1,1)$, $(0,0,1)$, $(1,1,0)$, $(1,0,1)$, $(0,1,1)$, may be obtained by addition of lattice vectors. For some purposes it is preferable to use this representation but for others it is better to use the former representation which retains the symmetry of the cube. In the fundamental representation we have

$$d_1 = d_2 = d_3 = \sqrt{3}d/2, \alpha = \beta = \cos^{-1}(-\tfrac{1}{3}) = \pi - \gamma$$

It will readily be verified from equation (6) that the volume of the true unit cell is $d^3/2$, as it should be, since we have now only one lattice point per unit cell. Monatomic crystals of this type occur in nature, particularly among the noble metals (see § 5.3.3). If the atoms at the corners of the unit cube are of type A and those at the centre of type B, we have a

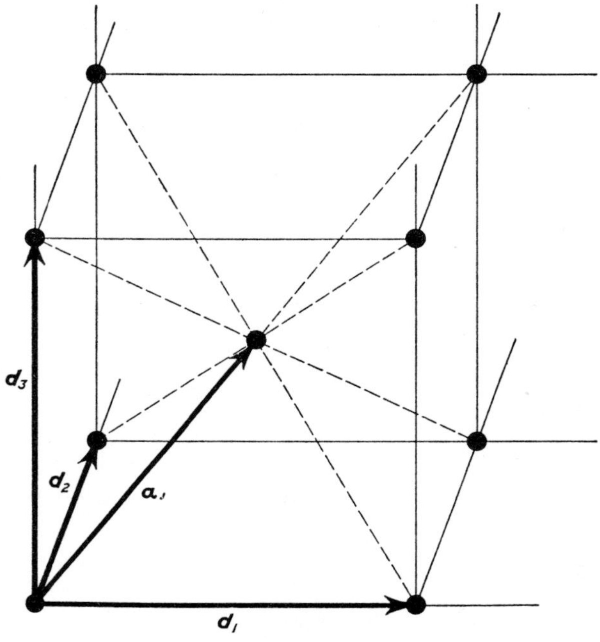

FIG. 5.14. The body-centred cubic lattice

body-centred cubic diatomic crystal, the atoms of each type forming a simple cubic lattice; this structure occurs in certain ionic crystals, for example in CsCl, and is frequently called the *caesium chloride* structure (see Fig. 5.15).

There is yet another way in which a new symmetrical structure may be built from the simple cubic lattice; if a new lattice point is introduced at the centre of each *face* of the unit cube we get the face-centred cubic structure. We may note that we need to introduce three basis vectors in this case namely, $\frac{1}{2}(\mathbf{d}_1+\mathbf{d}_2)$, $\frac{1}{2}(\mathbf{d}_1+\mathbf{d}_3)$, $\frac{1}{2}(\mathbf{d}_2+\mathbf{d}_3)$ so that there are *four* lattice points per unit cell. (Note that there is no lattice point at the centre of the unit cube; if such a point were introduced additionally we should revert to a simple cubic structure. This again is not the most fundamental representation of the face-centred cubic lattice. If we take as basic vectors half the face diagonals of the three faces meeting at the origin we may obtain all the other lattice points by addition; the face-centred cubic lattice is therefore also a *simple* lattice. For many purposes, however, it is better to retain the representation having cubic symmetry. It will readily be verified that now we have $d_1 = d_2 = d_3 = d/\sqrt{2}$ and that $\alpha = \beta = \gamma = 60°$, also from equation (6) that the volume of the true unit cell is $d^3/4$, as it should be, since there is now only one lattice point per

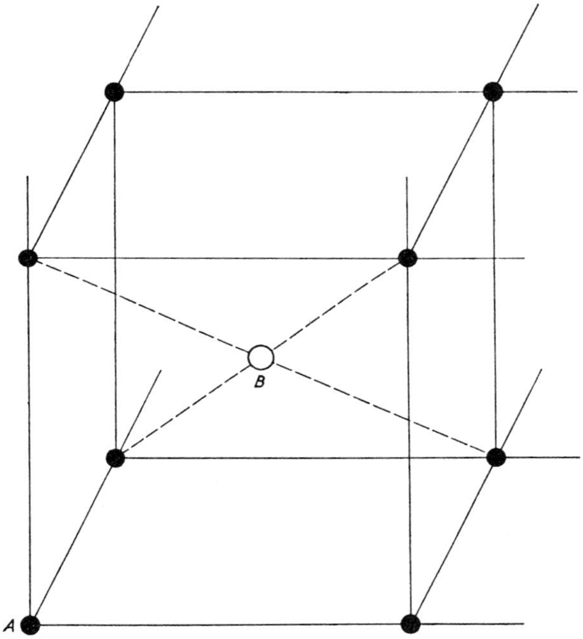

FIG. 5.15. Diatomic crystal with body-centred cubic structure (*caesium chloride* structure)

unit cell. The face-centred cubic structure occurs commonly in monatomic crystals, particularly among the noble metals; it is illustrated in Fig. 5.16.

If, in a simple cubic lattice, we place atoms of types A and B at alternate lattice points, we get a structure which consists of two face-centred lattices displaced from each other by half of one of the basic vectors of either. This structure is shown in Fig. 5.17; it will be seen atoms of type B lie at the centres of the edges of the unit cube of the face-centred lattice made up of the atoms of type A, and that there is additionally an atom of type B at the centre of the unit cube, i.e. thirteen in all. The one at the centre is not shared with neighbouring cells, while the others are each shared with four neighbouring cells, giving four atoms

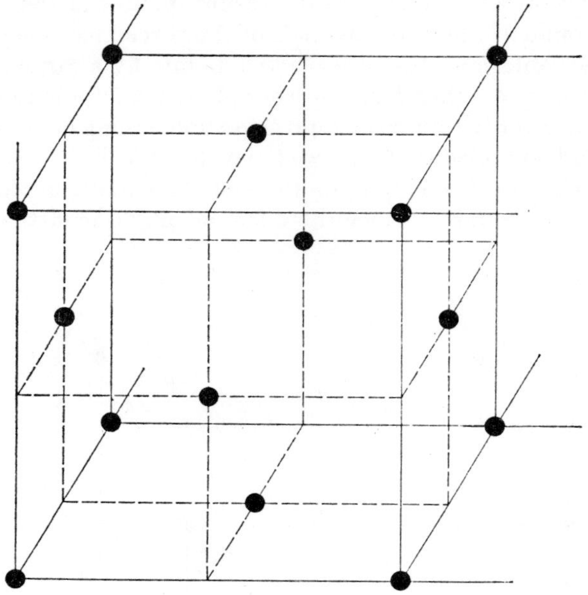

FIG. 5.16. The face-centred cubic lattice

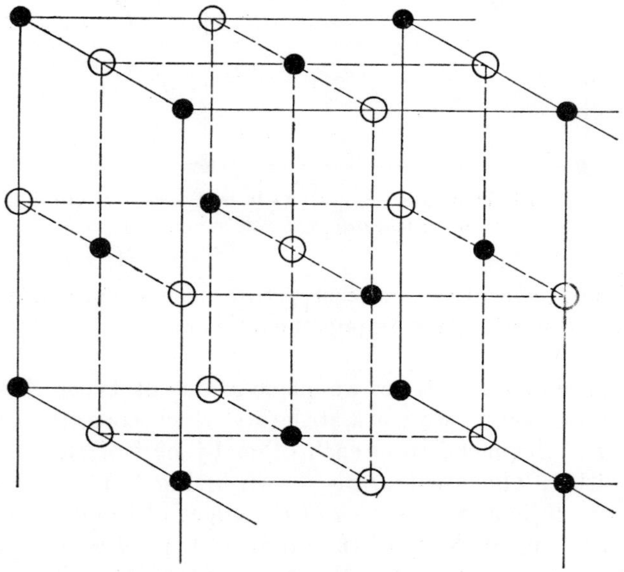

FIG. 5.17. Diatomic crystal with face-centred cubic structure (*sodium chloride* structure)

per unit cube as for type A; the crystal has clearly a chemical structure of type AB. Many ionic crystals, for example NaCl, have this structure, which is frequently known as the *sodium chloride* structure.

If we take two face-centred cubic lattices and displace them as indicated above, we obtain a simple cubic lattice whose unit cube has sides half those of the original unit cubes; also, if we displace one of the face-centred lattices in a direction of a body diagonal an amount equal to half the length of the diagonal we again obtain a simple cubic lattice; if, however, the displacement is only one-quarter the length of the body diagonal, we obtain a new and very interesting lattice. The form of this lattice may best be seen by dividing the unit cube of one of the original face-centred lattices into eight cubes of side $d/2$ and examining the positions of the lattice points, which all occur at corners of these smaller cubes. This is illustrated in Fig. 5.18, which shows four smaller cubes which fit together to form a face of the unit cube. It will be seen that in each case the lattice points occur at the ends of opposite-face diagonals, each smaller cube having four such points; these points lie at the corners of regular tetrahedra whose centres coincide with the centres of the smaller cubes. It will readily be seen that when we introduce the displaced face-centred lattice the new lattice points will be at the centres of the smaller cubes, but that *only alternate small cubes* will have a lattice point at their centre (see Fig. 5.18). It will thus be seen that each new lattice point is surrounded by four equidistant former lattice points, and it is not very difficult to see that each former lattice point is also surrounded by *four* equidistant new lattice points. In each case the four equidistant points are symmetrically disposed, i.e. they lie at the corners of a regular tetrahedron with the particular lattice point at its centre. The distance between nearest neighbours is $\sqrt{3}d/4$, and when no distinction between the types of atom in the two cubic face-centred lattices is made, we obtain the *diamond* structure shown in Fig. 5.19. This is an extremely important structure which, in addition to diamond, occurs for the semiconductors silicon, germanium and gray tin. The valence electrons of these substances make exactly *four* bonds with their neighbours and so fit naturally into this structure*, the bond length being $\sqrt{3}d/4$. An alternative approach to this structure is to start with a single point, add four further symmetrically disposed points, and so on for each new point. (This may be readily done by making a model with table-tennis balls, and students are urged to construct such models for themselves.) In this way the complete diamond lattice may be built up; it is not so easy to see *a priori* in this way that this is a cubic lattice. When we use orthogonal basic vectors, the basis vectors for the diamond structure are $\frac{1}{2}(\mathbf{d}_1+\mathbf{d}_2)$, $\frac{1}{2}(\mathbf{d}_1+\mathbf{d}_3)$, $\frac{1}{2}(\mathbf{d}_2+\mathbf{d}_3)$, $\frac{1}{4}(\mathbf{d}_1+\mathbf{d}_2+\mathbf{d}_3)$, $\frac{1}{4}(3\mathbf{d}_1+3\mathbf{d}_2+\mathbf{d}_3)$, $\frac{1}{4}(3\mathbf{d}_1+\mathbf{d}_2+3\mathbf{d}_3)$, $\frac{1}{4}(\mathbf{d}_1+3\mathbf{d}_2+3\mathbf{d}_3)$, the first three giving the positions of the face-centre

* See for example, R. A. Smith, *Semiconductors*. Cambridge University Press (1959), Chapter 3.

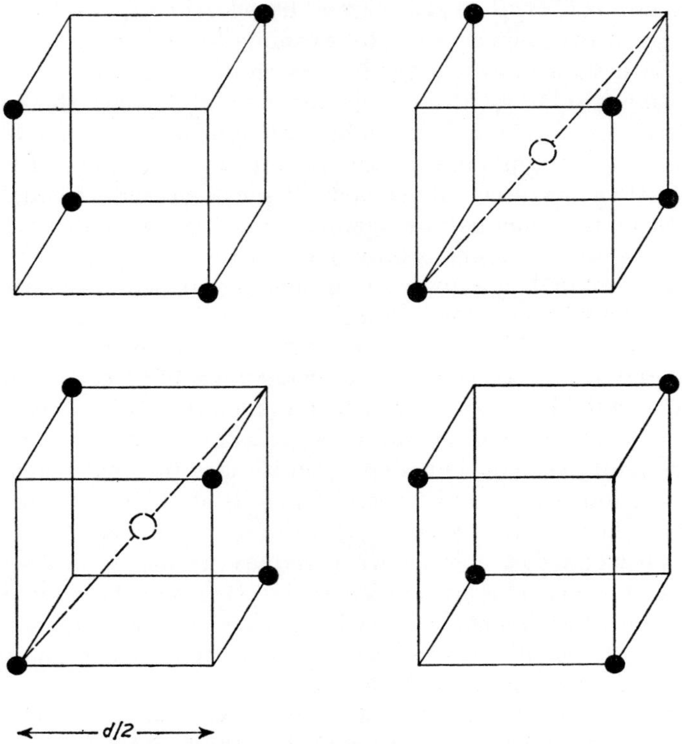

FIG. 5.18. Breakdown of unit cube of face-centred cubic lattice into smaller cubes

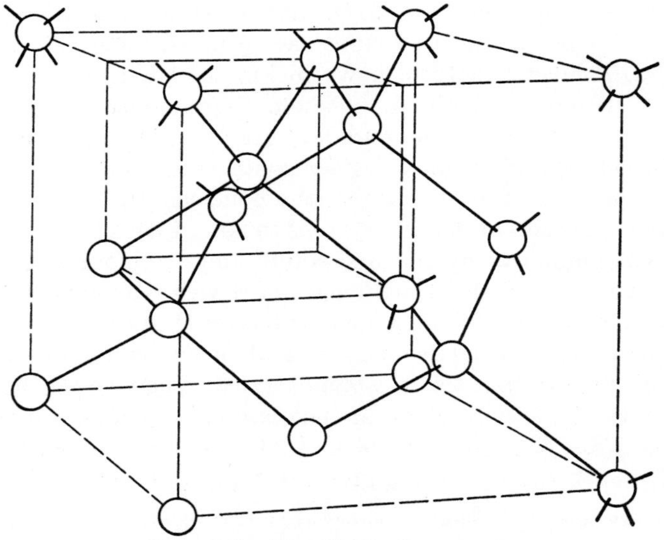

FIG. 5.19. The *diamond* structure

points of the indisplaced face-centred cubic lattice and the remainder the positions of the points of the displaced face-centred cubic lattice; there are thus eight lattice points per unit cubic cell. The *internal* points have coordinates, in terms of d, $(\frac{1}{4}, \frac{1}{4}, \frac{1}{4})$, $(\frac{3}{4}, \frac{3}{4}, \frac{1}{4})$, $(\frac{3}{4}, \frac{1}{4}, \frac{3}{4})$, $(\frac{1}{4}, \frac{3}{4}, \frac{3}{4})$, with respect to the origin of the cubic unit cell.

At first sight it might appear that the diamond lattice could be reduced to a simple lattice since each atom appears to be in an equivalent position; this, however, is not so, the relative positions of the four nearest neighbours having different directions for atoms of the original and displaced lattices. The diamond structure, unlike the face-centred cubic structure, is a true lattice with a basis and cannot be reduced to a simple lattice.

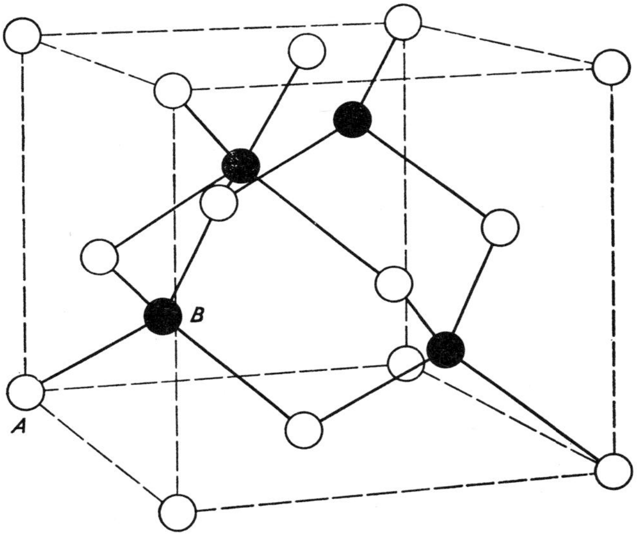

FIG. 5.20. The *zincblende* structure

This may readily be checked by means of the following test: if, for a simple lattice, the vector $\mathbf{r}$ represents a lattice point then clearly the vector $-\mathbf{r}$ also represents a lattice point (l, m, n are replaced by $-l$, $-m$, $-n$), while for a true lattice with a basis this is not necessarily so. For the diamond lattice, if the point $d(\frac{1}{4}, \frac{1}{4}, \frac{1}{4})$ is a lattice point then $-d(\frac{1}{4}, \frac{1}{4}, \frac{1}{4})$ is not; hence the lattice cannot be simple but must have an irreducible basis. The simplest representation is obtained by using the three primitive vectors of the face-centred cubic lattice and a single basis vector giving *two* lattice points per unit cell; for many purposes, however, the orthogonal representation given above is preferable, and in this case we have eight lattice points per unit cell.

If the atoms of the original face-centred lattice are of type A and those of the displaced lattice of type B we obtain the *zincblende* structure.

Many compounds of type AB crystallize in this form including one form of ZnS; in particular, many of the inter-metallic semiconductors such as InSb have this crystal structure, which is shown in Fig. 5.20.

A third face-centred lattice may also be introduced so as to fill the centres of the empty small cubes, and we obtain yet another structure. This does not give monatomic crystals but a number of diatomic crystals of the chemical form AB_2 have this structure. The A-type atoms form a face-centred lattice and the B-type, occupying the sites of the *two* displaced lattices, form a simple cubic lattice. Each B-type atom has four equidistant A-type neighbours but each A-type atom has eight B-type neighbours. This structure is found for a number of ionic crystals such as CaF_2 and is known as the *fluorite* structure. Other more complex variants of the cubic lattice will be found in books on crystallography.

5.3.2 Hexagonal lattices

A three-dimensional lattice may be derived from the plane simple hexagonal lattice discussed in § 5.2.2, and is obtained by periodic displacement with a basic vector $\mathbf{d}_3$ perpendicular to the vectors $\mathbf{d}_1$, $\mathbf{d}_2$ defining the plane hexagonal lattice*; thus we have $d_1 = d_2 = a$, $d_3 = c$, $\alpha = 60°$, $\beta = \gamma = 90°$; this lattice is known as the simple hexagonal lattice, and is illustrated in Fig. 5.21. The volume of the unit cell is equal to $\sqrt{3}a^2c/2$. Crystals formed from this lattice have a single axis of symmetry, generally known as the c-axis.

Another lattice, which is of great practical importance may be derived from the simple hexagonal lattice by inserting a second identical simple hexagonal lattice, having the same direction for the c-axis, and having its planes containing the plane hexagonal lattices half-way between those of the original lattice. The points of the second lattice are also displaced, so that, when projected along the c-axis they come at the centres of an equivalent set of equilateral triangles formed by points of the original lattice; this is illustrated in Fig. 5.22, the points of the original lattice being marked ● and those of the displaced lattice marked ×. It will be noted that the centres of another set of equilateral triangles, marked ○, are unoccupied by projected points and give rise to 'channels' through the lattice parallel to the c-axis. When $c = 2\sqrt{2}a/\sqrt{3}$ this lattice is known as the close-packed hexagonal lattice for reasons which we shall discuss in the next section; it occurs in many of the noble metals and is illustrated in Fig. 5.23. The close-packed hexagonal lattice may be derived from the simple hexagonal lattice by means of a single additional basis vector $\mathbf{a}_1 = \frac{1}{3}(\mathbf{d}_1 + \mathbf{d}_2) + \frac{1}{2}\mathbf{d}_3$. If two close-packed hexagonal lattices with the same c-axis are interleaved one having atoms of type A and one of type B, we obtain the *wurtzite* structure shown in Fig. 5.24. The displacement of the A-type lattice from the B-type is $3c/8$, along the c-axis. This structure is in many ways similar to the zinc blende structure, each

* See, however, § 5.4.5 for conventional notation using four indices.

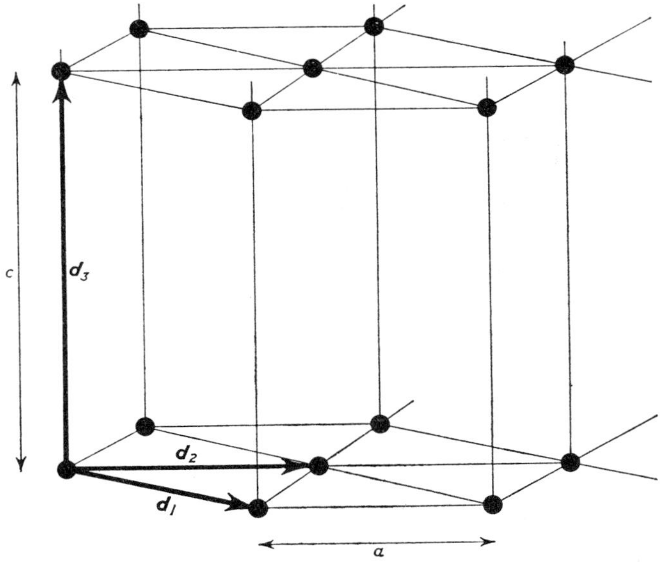

FIG. 5.21. The simple hexagonal lattice in three dimensions

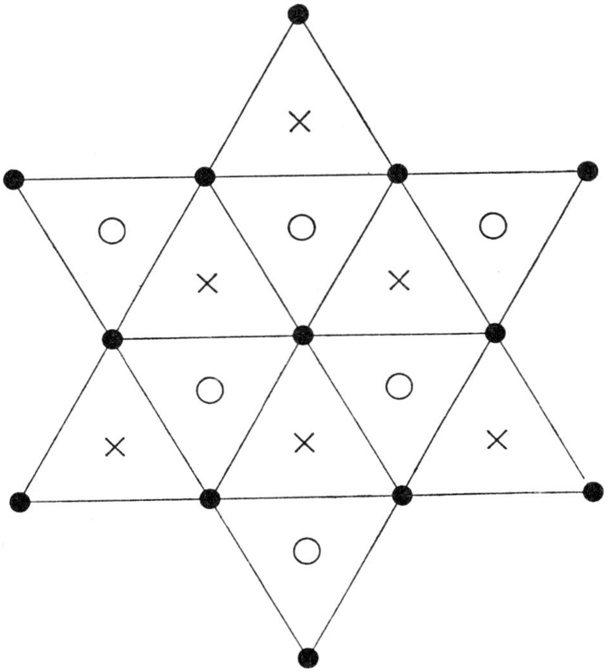

FIG. 5.22. Displaced hexagonal lattices

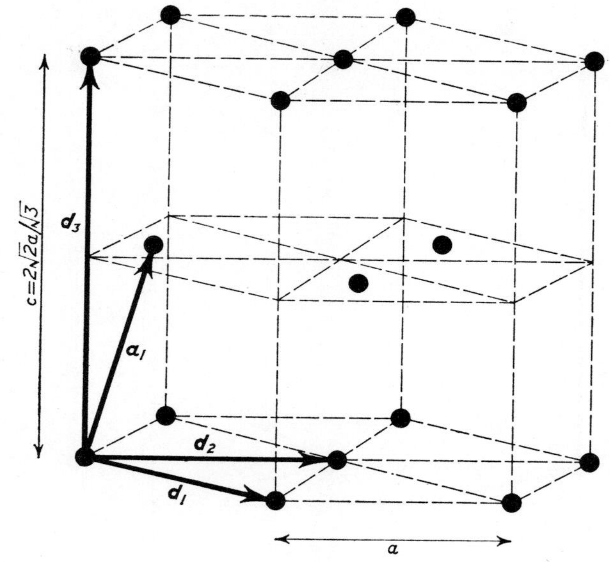

FIG. 5.23. The close-packed hexagonal lattice

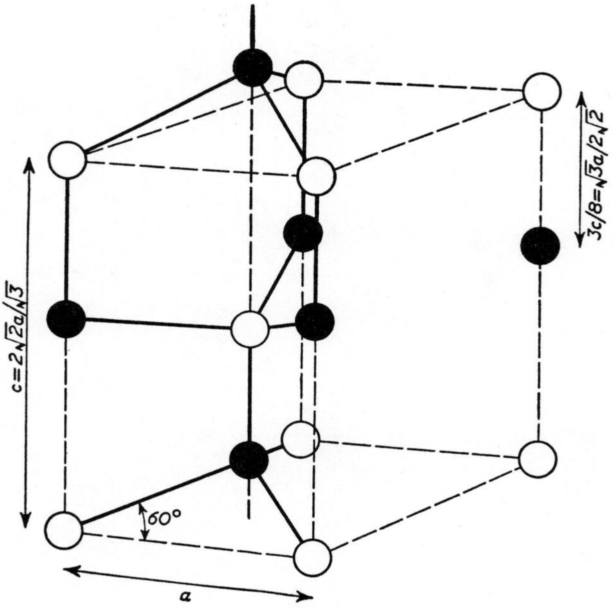

FIG. 5.24. The *wurtzite* structure

atom having four nearest neighbours of the other type, at the corners of a regular tetrahedron of which it occupies the centre. The bond length in this case is $3c/8 = \sqrt{3}a/2\sqrt{2}$.

5.3.3 Close packing of spherical atoms

It is beyond the scope of this book to discuss in any detail the crystal structures of the great variety of solids which occur in nature, or the reasons why a particular crystal form is adopted. This is a very large subject and has been dealt with fully in many textbooks on crystallography*. It is of some interest, however, to consider in an elementary manner the structures obtained by packing together spherical atoms. If we start with a plane of atoms arranged horizontally in a simple square lattice, each having four neighbours, and place above it a series of other planes with the spheres touching at their highest and lowest points, this would give a simple cubic lattice, each sphere occupying a volume $8r^3$, where r is the radius of the sphere. Clearly this is far from being the most closely packed arrangement; for example, we could place the second layer so that each sphere fits into the gaps between the spheres of the first layer as closely as possible. A consideration of the geometry shows that the distance between the planes containing the centres of the two layers of spheres is equal to $\sqrt{2}r$. Another layer placed so as to fill the gaps in the second layer would come exactly over the first layer but would not touch it; the arrangement would thus not have cubic symmetry. If, however, the spheres in each layer were separated a little we could have the spacing between the first and third layers equal to the spacing between the centres of neighbouring spheres in the first layer and this would give a cubic arrangement, each sphere touching eight neighbours symmetrically disposed with respect to its centre; this is clearly the body-centred cubic form, three spheres being in contact along each body diagonal of the elementary cube. The body diagonal has thus length $4r$ and the sides of the elementary cube are equal to $4r/\sqrt{3}$; the spacing between the spheres in the original plane should thus be $(4/\sqrt{3}-2)r$. Since there are two spheres per unit cell (see § 5.3.1) each sphere accounts for a volume of space equal to $32r^3/3\sqrt{3}$ ($= 6{\cdot}16r^3$); this is considerably smaller than for the simple cubic arrangement. As we shall see below, there are two methods of packing the spheres still closer; nevertheless a number of metals (including the alkali metals) form monatomic crystals of this type.

We should clearly expect to obtain a more closely packed structure by starting with the most closely packed arrangement of spheres in a plane. We have seen in § 5.2.2 that this is the hexagonal arrangement, each sphere being in contact with six other spheres in the plane. A second plane of spheres similarly packed may be placed over the first so as to

* See for example, W. A. Wooster, *Crystal Physics*. Cambridge University Press (1938); R. C. Evans, *Crystal Chemistry*. Cambridge University Press (1946).

fit as tightly as possible. If the centres of the spheres of the first layer are represented by the points marked ● in Fig. 5.22 those of the second layer may be represented by the points marked ×. Each point × will be at the apex of a regular tetrahedron of which three points marked ● form the base. The spacing between the layers will thus be equal to the height of the tetrahedron, $2\sqrt{2}r/\sqrt{3}$, the length of each side being $2r$. We now have two choices open; we may either place a next layer in contact with the second so that the centres of the spheres come exactly above those of the first layer or we may place them so as to occupy the points marked ○ in Fig. 5.22. In the first case we get the close-packed hexagonal form, the value of a being $2r$. The value of c, being equal to the spacing between layers one and three, is $4\sqrt{2}r/\sqrt{3}$; the ratio $c:a$ is thus $2\sqrt{2}/\sqrt{3}$ (1·632). In the second case it turns out that we get the face-centred cubic arrangement, three spheres being in contact along each *face* diagonal of the unit cube; this may readily be seen from a model. The closeness of packing is clearly the same for each of the two forms. The length of each face diagonal of the unit cube is $4r$ so that its volume is $16\sqrt{2}r^3$, and there are four spheres per unit cube, so that each accounts for a volume $4\sqrt{2}r^3$ $(= 5{\cdot}66r^3)$; the packing is thus somewhat closer than for the body-centred cubic arrangement. Both arrangements are commonly found among the metals; for example Cu, Ag and Au have the face-centred cubic form, while Mg, Be, Ca have the hexagonal form.

These considerations may also be applied to the discussion of certain types of ionic crystals; these are made up from positive and negative ions such as Na^+ and Cl^-. The negative ions, having more valence electrons than the positive ions, are usually somewhat larger; they frequently form a face-centred cubic lattice, the small positive ions fitting in the spaces between to form another face-centred cubic lattice so as to give the *sodium chloride* structure. If the positive ions are too large to fit in these spaces another structure is usually found; also, when valence binding predominates the crystal structure is usually determined by the number of bonds available as for the *diamond* structure (see § 5.3.1).

5.3.4 Lattice planes and directions

In the two-dimensional lattices we have seen (§ 5.2.3) that there are well-defined lines of lattice points; similarly we have planes of lattice points in three dimensions, and the same type of notation may be used to define the planes. It is through a study of such planes by means of X-rays (see below) that most analysis of crystal structure is carried out. If a plane meets the axes of the unit cell in the points whose lattice coordinates (see § 5.2) are $(l, 0, 0)$, $(0, m, 0)$, $(0, 0, n)$ and we form the quantities l', m', n' where $l' = s/l$, $m' = s/m$, $n' = s/n$, s being the smallest integer required to reduce all the ratios s/l, s/m, s/n to integers, then the l', m', n' are used to define the plane, and are known as the Miller indices of the plane. Instead of l, m, n as defined above, we may use the intercepts

α, β, γ, with the axes of any plane parallel to the required plane, and write $l' = s'a/\alpha$, $m' = s'b/\beta$, $n' = s'c/\gamma$, a, b, c being the lengths of the basic vectors; s' is again the smallest number which reduces l', m', n' to integers. Such a number can be found since α/a, β/b, γ/c are in an integral ratio there being parallel lattice planes which meet all three axes in lattice points; from the intercepts we clearly obtain the same Miller indices. For a cubic crystal referred to three axes along the edges of the unit cube we have a particularly simple interpretation of the Miller indices; they are proportional to the direction cosines of the normal to the plane, as may readily be seen from a consideration of the equation of the plane in Cartesian coordinates (x,y,z) taken along the axes. If the intercepts are α, β, γ this equation is

$$\frac{x}{\alpha}+\frac{y}{\beta}+\frac{z}{\gamma} = 1$$

or

$$l'x+m'y+n'z = k$$

In this form we see that the direction cosines are proportional to l', m', n'. When a particular plane is referred to, it is denoted by the symbol $(l'm'n')$, any one of a set of equivalent planes being denoted by the symbol $\{l'm'n'\}$.

In a three-dimensional crystal we also have lines of lattice points; corresponding to any such line there will clearly be a parallel lattice line passing through the origin. Let the lattice coordinates of any lattice point on this line be (l'',m'',n'') then the direction in the crystal parallel to this line is called the $[lmn]$ direction, l, m, n being the smallest integers having the same ratio as l'', m'', n''; when any one of an equivalent set of directions is referred to it is denoted by $\langle lmn\rangle$. For a cubic lattice l, m, n are proportional to the direction cosines defining the particular direction. For the cubic lattice referred to rectangular axes the planes normal to the $[lmn]$ direction are clearly the (lmn) planes, since the equation of the plane through the lattice point (l,m,n) with normal in the $[lmn]$ direction is

$$l(x-la)+m(y-ma)+n(x-na) = 0$$

the intercepts on the axes are thus in the ratio $l^{-1}:m^{-1}:n^{-1}$. For other lattices the relationship between directions and planes is more complex.

5.3.5 Bragg X-ray reflections from crystal planes

As mentioned above, the structure of crystals is now mainly determined by means of a study of the reflections of X-rays from the lattice planes; details of the methods of analysis will not be given here since they are discussed in all modern books on crystallography. We shall merely note that the direction of the normal to a set of planes is readily determined since it makes equal angles with the direction of incidence of a narrow beam of X-rays and the direction at which a strong reflection is observed.

This is the same relationship as exists between the directions of incidence and reflection of a light beam in a plane mirror and is merely an expression of the condition that all the atoms lying in a particular plane scatter the X-radiation in phase; this condition applies to each plane of a set. The fact that we have a large number of parallel planes, each contributing only a small amount to the total scattered intensity, must also be taken into account. If ϕ is the angle of incidence (and also of reflection), as shown in Fig. 5.25, there must exist a relationship between this angle,

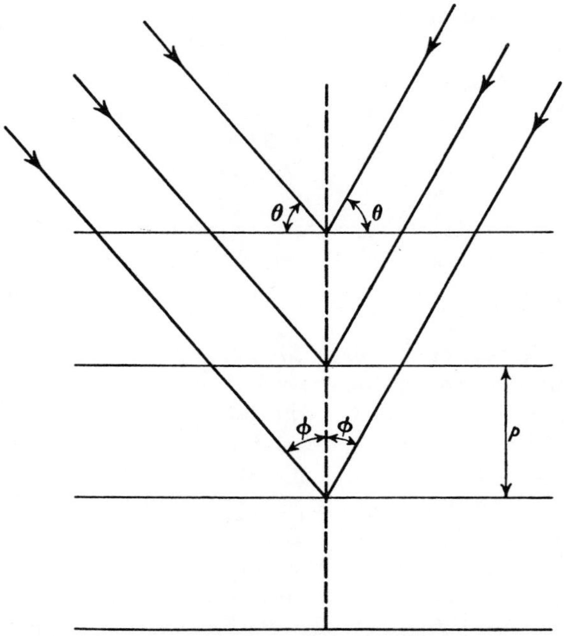

FIG. 5.25. Bragg reflection of X-rays from crystal planes

the spacing p between the planes, and the wavelength λ of the X-rays in order that *all the planes* of a set may scatter in phase. The condition is that there should be an integral number of wavelengths in the extra path length corresponding to reflection from successive planes; this is the well-known Bragg relationship and is given by

$$2p\cos\phi = s\lambda \tag{9}$$

where s is an integer, as will be seen from Fig. 5.25. It is more commonly expressed in terms of θ the glancing angle, where $\theta = \pi/2-\phi$, and is then

$$2p\sin\theta = s\lambda \tag{9a}$$

It will thus be seen that only for certain values of θ will strong reflection be found, and, by observing these values of θ, the directions of the crystal planes and the spacing between them may be found. From the variation of intensity of reflections from the various planes information may be obtained about the density of the various atoms which occupy them, and in this way the crystal structure may be determined.

5.4 The reciprocal lattice

Associated with any simple lattice there is another related lattice, called the reciprocal lattice, which has important uses in the theory of propagation of waves. Since it is easier to visualise the geometrical relationship between the two lattices in two dimensions we shall first of all consider plane lattices.

5.4.1 The reciprocal lattice in two dimensions

Let $\mathbf{d}_1$ and $\mathbf{d}_2$ be basic vectors defining a lattice; this lattice we shall call the *direct* lattice to distinguish it from the reciprocal lattice. Let us

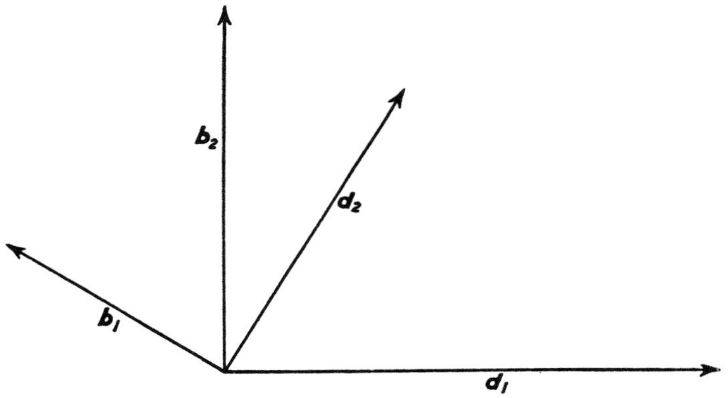

FIG. 5.26. Reciprocal lattice, basic vectors

now define two vectors $\mathbf{b}_1$ and $\mathbf{b}_2$ such that $\mathbf{b}_1$ is perpendicular to $\mathbf{d}_2$ and $\mathbf{b}_2$ is perpendicular to $\mathbf{d}_1$ (see Fig. 5.26). The magnitudes of $\mathbf{b}_1$ and $\mathbf{b}_2$ are chosen so that the two scalar products $(\mathbf{b}_1 . \mathbf{d}_1)$, $(\mathbf{b}_2 . \mathbf{d}_2)$ are both equal to 1. These conditions may be expressed by means of the four equations

$$(\mathbf{b}_i . \mathbf{d}_j) = \delta_{ij} \tag{10}$$

i and j taking the values 1 and 2, and δ_{ij} being, as usual, equal to 0 when $i \neq j$ and to 1 when $i = j$. The conditions are clearly sufficient to define the vectors $\mathbf{b}_1, \mathbf{b}_2$, which may now be used to define a simple lattice, called the reciprocal lattice. We may note that the relationship between the direct and reciprocal lattices is indeed a 'reciprocal' one in the sense that if we form the reciprocal of the lattice defined by $\mathbf{b}_1$, $\mathbf{b}_2$ we get the same lattice as that formed from $\mathbf{d}_1$, $\mathbf{d}_2$; there is, however, another sense in which the lattices are reciprocal. If α is the angle between the vectors $\mathbf{d}_1$

and $\mathbf{d}_2$, whose lengths are d_1 and d_2, then the area A_d of the unit cell of the direct lattice is equal to $d_1 d_2 \sin\alpha$; moreover the angle between the vectors $\mathbf{b}_1$, $\mathbf{b}_2$ is also equal to α and the area A_b of the unit cell of the reciprocal lattice is equal to $b_1 b_2 \sin\alpha$. Also, we have from equations (10), since the angle between $\mathbf{b}_2$ and $\mathbf{d}_2$ is $\pi/2-\alpha$, and between $\mathbf{b}_1$ and $\mathbf{d}_1$ is $\pi/2+\alpha$,

$$\left.\begin{aligned} b_1 &= (d_1 \sin\alpha)^{-1} \\ b_2 &= (d_2 \sin\alpha)^{-1} \end{aligned}\right\} \tag{11}$$

so that

$$A_d A_b = (b_1 b_2 \sin\alpha)(d_1 d_2 \sin\alpha) = 1 \tag{12}$$

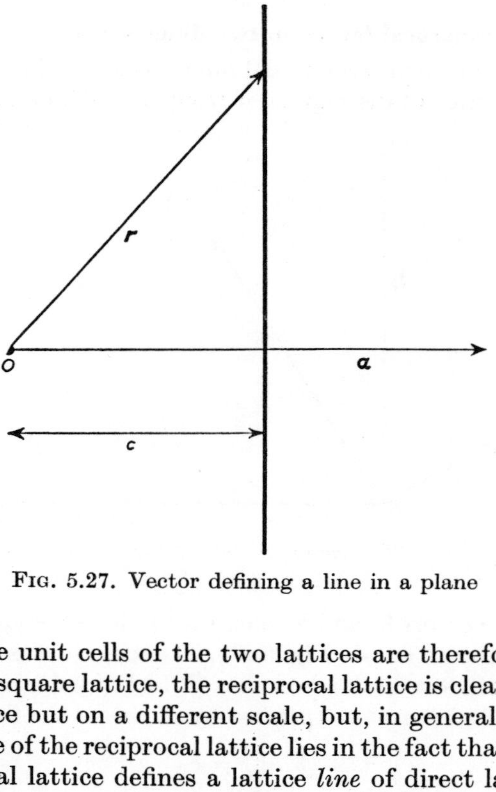

FIG. 5.27. Vector defining a line in a plane

the areas of the unit cells of the two lattices are therefore reciprocals. For the simple square lattice, the reciprocal lattice is clearly the same as the direct lattice but on a different scale, but, in general, this is not so. The importance of the reciprocal lattice lies in the fact that a lattice *point* of the reciprocal lattice defines a lattice *line* of direct lattice and vice versa. To define a line in a plane we require a constant vector $\mathbf{a}$; the line is then the locus of points such that the projection of their position vector $\mathbf{r}$ on a line through the origin parallel to $\mathbf{a}$ is constant (see Fig. 5.27); the equation of the line may therefore be written in the form

$$\mathbf{r}.\mathbf{a} = c \tag{13}$$

and the vector $\mathbf{a}$ may be said to define a set of parallel lines given by different values of c. Now suppose the vector $\mathbf{a}$ is a vector of the reciprocal lattice, for example the vector defined by the lattice point (l', m'), so that we have $\mathbf{a} = l'\mathbf{b}_1 + m'\mathbf{b}_2$, l' and m' being integers with no common factor. The equation of the corresponding line in the plane of the direct lattice is then

$$l'(\mathbf{r}.\mathbf{b}_1) + m'(\mathbf{r}.\mathbf{b}_2) = c \tag{14}$$

The position vector $\mathbf{r}$ of any point in the plane of the direct lattice may be expressed in terms of the lattice vectors $\mathbf{d}_1$, $\mathbf{d}_2$ in the form $\mathbf{r} = \xi\mathbf{d}_1 + \eta\mathbf{d}_2$; for lattice points, $\xi = l$ and $\eta = m$, l and m being integers. The intercepts $\xi_0 d_1$ and $\eta_0 d_2$ on the axes of the line given by equation (14) are obtained by taking $\xi = 0$ and $\eta = 0$; inserting the corresponding values of r and using equations (10) we find that

$$\left.\begin{aligned} \xi_0 &= c/l' \\ \eta_0 &= c/m' \end{aligned}\right\} \tag{15}$$

and these are in the ratio $(1/l'):(1/m')$. Since l' and m' are integers with no common factor, this line is therefore parallel to the lattice lines $(l'm')$ (see § 5.2.3); moreover the parallel line through the origin is obtained when we have $c = 0$. The condition that this line should pass through a lattice point (l, m) is obtained by substituting $\mathbf{r} = l\mathbf{d}_1 + m\mathbf{d}_2$ in equation (14) with $c = 0$. We have

$$ll' + mm' = 0 \tag{16}$$

and we see that an infinite number of integers l, m satisfy this equation; for other lattice lines the condition is

$$ll' + mm' = c \tag{17}$$

and we see that c is restricted to integral values, the different values defining a set of lattice lines. We now enquire what is the smallest possible value of c, for given values of l', m'. Since l' and m' are assumed to have no common factor the vector $\mathbf{a}$ is the *smallest* vector of the reciprocal lattice having components in the ratio $l':m'$ and, since the value of c is integral, its smallest value must be unity; the next parallel lattice line to that passing through the origin is therefore at a distance equal to $1/|\mathbf{a}|$ or $|l'\mathbf{b}_1 + m'\mathbf{b}_2|^{-1}$. Summarising, we see that the *point* (l', m') of the reciprocal lattice defines a set of lattice lines $(l'm')$ in the direct lattice, the spacing p between them being equal to $|l'\mathbf{b}_1 + m'\mathbf{b}_2|^{-1}$; similarly a point (l, m) in the direct lattice defines a set of lines (lm) in the reciprocal lattice the spacing p' between them being $|l\mathbf{d}_1 + m\mathbf{d}_2|^{-1}$. In each case the integers l', m' and l, m are assumed to have no common factor, i.e. the corresponding vectors are the *smallest* lattice vectors having these directions. We may readily express p in terms of d_1, d_2 and α; using equation (11) we have

$$p = \sin\alpha\{(l'/d_1)^2 + (m'/d_2)^2 + 2(l'm'/d_1 d_2)\cos\alpha\}^{-1/2} \tag{18}$$

We shall now consider some simple examples corresponding to the direct lattices discussed in §§ 5.2.1, 5.2.2. As we have seen, the reciprocal lattice of the simple square lattice is also a simple square lattice; if the unit cell of the direct lattice has side d, that of the reciprocal lattice has side $1/d$. When we have a plane crystal consisting of atoms of type A at the points of a simple square lattice, with unit cell of side d, and atoms of type B at the centres of the lattice squares, as in Fig. 5.4, the reciprocal lattice *is still a simple square lattice* whose unit cell has side $1/d$, since the basic direct lattice is a simple square lattice, the points at the centres of the squares not being points of the simple lattice but forming a true basis. The reciprocal of the simple hexagonal lattice is also a simple hexagonal lattice, the sides of the unit cell (rhombus) being $1/d$. (The lattice is rotated through an angle 30° from the direct lattice.) The reciprocal of a rectangular lattice the sides of whose unit cell are a, b is also a rectangular lattice the sides of the unit cell being $1/b$, $1/a$; it is therefore a similar lattice.

5.4.2 Plane waves and Fourier expansions

We have seen (§ 1.1.1) that a plane travelling wave may be expressed in the form

$$\psi = A\,e^{2\pi i[(\mathbf{r}.\mathbf{a})-\nu t]} \tag{19}$$

Omitting the time variation and expressing the position vector in terms of Cartesian coordinates, we have

$$\psi = A\,e^{2\pi i(\alpha x+\beta y)} \tag{20}$$

If this is to be periodic with periods d_1, d_2, in the x, y, directions then we must have $\alpha = l/d_1$, $\beta = m/d_2$, where l and m are integers. This form of expression is useful for lattices with the basic vectors at right angles, d_1 and d_2 being the lengths of the basic vectors. The expression for ψ is then periodic, with the periodicity of the lattice, and any function $f(x,y)$ with the lattice periodicity may be expanded as a Fourier series in the form

$$f(x,y) = \sum_{l,m} A_{lm}\,e^{2\pi i(lx/d_1+my/d_2)} \tag{21}$$

We must now find a generalisation of this form of expansion which is valid for any plane lattice; the required form is suggested by the fact that the quantities d_1 and d_2 occur in equation (21) as reciprocals. Let us consider the plane wave obtained by taking the vector $\mathbf{a}$ in equation (19) as a vector of the reciprocal lattice; we have

$$\psi(\mathbf{r}) = A\,e^{2\pi i(\mathbf{b}_{l'm'}.\mathbf{r})} \tag{22}$$

where $\mathbf{b}_{l'm'} = l'\mathbf{b}_1 + m'\mathbf{b}_2$, l' and m' being integers, so that

$$\psi(\mathbf{r}) = A\,e^{2\pi i[l'(\mathbf{b}_1\cdot\mathbf{r})+m'(\mathbf{b}_2\cdot\mathbf{r})]} \tag{23}$$

If we replace the position vector $\mathbf{r}$ by the vector $\mathbf{r}' = \mathbf{r}+l\mathbf{d}_1+m\mathbf{d}_2$, l and m being integers, we have

$$\psi(\mathbf{r}') = \psi(\mathbf{r})\,e^{2\pi i[ll'(\mathbf{b}_1.\mathbf{d}_1)+ml'(\mathbf{b}_1.\mathbf{d}_2)+lm'(\mathbf{b}_2.\mathbf{d}_1)+mm'(\mathbf{b}_2.\mathbf{d}_2)]} \tag{24}$$

Using the relationships (6) between $\mathbf{b}_1$, $\mathbf{b}_2$, $\mathbf{d}_1$, $\mathbf{d}_2$ we see that

$$\psi(\mathbf{r}') = \psi(\mathbf{r}+ll'\,\mathbf{d}_1+mm'\,\mathbf{d}_2) = \psi(\mathbf{r}) \tag{25}$$

so that $\psi(\mathbf{r})$ is a periodic function with the periodicity of the lattice. All plane waves of the form (22) are such periodic functions and a general periodic function $f(\mathbf{r})$ may be expanded in the form

$$f(\mathbf{r}) = \sum_{l'm'} A_{l'm'}\,e^{2\pi i(\mathbf{b}_{l'm'}.\mathbf{r})} \tag{26}$$

where $\mathbf{b}_{l'm'}$ is a vector of the reciprocal lattice; this is a fairly obvious generalisation of the double Fourier expansion, and may readily be derived from such an expansion by writing $\mathbf{r} = \xi\mathbf{d}_1+\eta\mathbf{d}_2$, and using ξ and η as variables. We thus have a second very important application of the reciprocal lattice.

The coefficients $A_{l'm'}$ may be determined by multiplying both sides of (26) by $\exp[-2\pi i(\mathbf{b}_{lm}.\mathbf{r})]$ and integrating over a unit cell; integrals of the type which occur on the right side of the equation vanish unless $l = l'$ and $m = m'$. (This may be readily shown using ξ and η as variables.) When $l' = l$ and $m' = m$ we have

$$A_{lm} = A_d^{-1}\int f(\mathbf{r})\,e^{-2\pi i(\mathbf{b}_{lm}.\mathbf{r})}\,d\mathbf{r} \tag{27}$$

A_d denoting, as before, the area of the unit cell.

5.4.3 The reciprocal lattice in three dimensions

These ideas may be generalised at once to three dimensions; here, since the vector $\mathbf{d}_1\times\mathbf{d}_2$ is now a vector of the space under discussion, whereas in two dimensions it is not, the geometry is slightly different. We define the reciprocal lattice by means of three basic vectors $\mathbf{b}_1$, $\mathbf{b}_2$, $\mathbf{b}_3$, which satisfy the equations

$$\mathbf{b}_i.\mathbf{d}_j = \delta ij \tag{28}$$

i, j having the values 1, 2, 3, and $\mathbf{d}_1$, $\mathbf{d}_2$, $\mathbf{d}_3$ being three basic vectors of the direct lattice. The conditions

$$\mathbf{b}_1.\mathbf{d}_2 = 0$$

$$\mathbf{b}_1.\mathbf{d}_3 = 0$$

imply that the vector $\mathbf{b}_1$ is perpendicular to both the vectors $\mathbf{d}_2$ and $\mathbf{d}_3$ and so is parallel to the vector $\mathbf{d}_2\times\mathbf{d}_3$; similar relationships hold for $\mathbf{b}_2$ and $\mathbf{b}_3$ and we may write

$$\mathbf{b}_1 = k_1(\mathbf{d}_2\times\mathbf{d}_3),\text{ etc.} \tag{29}$$

where k_1 is a constant. Thus we have

$$1 = (\mathbf{d}_1.\mathbf{b}_1) = k_1(\mathbf{d}_1.\mathbf{d}_2\times\mathbf{d}_3)$$

Now $(\mathbf{d}_1 . \mathbf{d}_2 \times \mathbf{d}_3)$ is just equal to the volume V_d of the unit cell of the direct lattice (cf. equations (5) and (6)) so that $k_1 = V_d^{-1}$ and we have

$$\mathbf{b}_1 = V_d^{-1}(\mathbf{d}_2 \times \mathbf{d}_3) \tag{30}$$

Similarly, we have

$$\left.\begin{aligned} \mathbf{b}_2 &= V_d^{-1}(\mathbf{d}_3 \times \mathbf{d}_1) \\ \mathbf{b}_3 &= V_d^{-1}(\mathbf{d}_1 \times \mathbf{d}_2) \end{aligned}\right\} \tag{30a}$$

so that

$$\left.\begin{aligned} b_1 &= d_1^{-1} \sin \alpha / \phi(\alpha, \beta, \gamma) \\ b_2 &= d_2^{-1} \sin \beta / \phi(\alpha, \beta, \gamma) \\ b_3 &= d_3^{-1} \sin \gamma / \phi(\alpha, \beta, \gamma) \end{aligned}\right\} \tag{31}$$

where $\phi(\alpha,\beta,\gamma)$ is the expression defined by equation (6a). The reciprocal relationship between the vectors $\mathbf{b}_1$, $\mathbf{b}_2$, $\mathbf{b}_3$ and $\mathbf{d}_1$, $\mathbf{d}_2$, $\mathbf{d}_3$ may readily be seen since clearly $\mathbf{d}_1$ is perpendicular both to $\mathbf{b}_2$ and $\mathbf{b}_3$, etc., and also $\mathbf{b}_1 . \mathbf{d}_1 = 1$; similar relationships hold for $\mathbf{d}_2$ and $\mathbf{d}_3$. We may therefore write

$$\left.\begin{aligned} \mathbf{d}_1 &= V_b^{-1}(\mathbf{b}_2 \times \mathbf{b}_3) \\ \mathbf{d}_2 &= V_b^{-1}(\mathbf{b}_3 \times \mathbf{b}_2) \\ \mathbf{d}_3 &= V_b^{-1}(\mathbf{b}_1 \times \mathbf{b}_2) \end{aligned}\right\} \tag{32}$$

where $V_b = (\mathbf{b}_1 . \mathbf{b}_2 \times \mathbf{b}_3)$ is the volume of the unit cell of the reciprocal lattice. We may also express V_b in the form

$$V_b = V_d^{-1}(\mathbf{d}_1 \times \mathbf{d}_2 . \mathbf{b}_1 \times \mathbf{b}_2)$$

Using the relationship, well known in vector analysis,

$$(\mathbf{d}_1 \times \mathbf{d}_2 . \mathbf{b}_1 \times \mathbf{b}_2) = (\mathbf{d}_1 . \mathbf{b}_1)(\mathbf{d}_2 . \mathbf{b}_2) - (\mathbf{d}_1 . \mathbf{b}_2)(\mathbf{b}_1 . \mathbf{d}_2) \tag{33}$$

we see that

$$(\mathbf{d}_1 \times \mathbf{d}_2 . \mathbf{b}_1 \times \mathbf{b}_2) = 1$$

and hence that

$$V_b = V_d^{-1} \tag{34}$$

thus establishing completely the reciprocal relationship between the vectors $\mathbf{d}_1$, $\mathbf{d}_2$, $\mathbf{d}_3$ and $\mathbf{b}_1$, $\mathbf{b}_2$, $\mathbf{b}_3$.

The rest of the analysis follows exactly as for two dimensions; the point (l', m', n') in the reciprocal lattice defines a set of lattice planes $(l'm'n')$ in the direct lattice, the distance p between adjacent planes being given by

$$p = |l'\mathbf{b}_1 + m'\mathbf{b}_2 + n'\mathbf{b}_3|^{-1} \tag{35}$$

l', m', n' in this case being the smallest integers in the same ratio. A periodic function with the periodicity of the lattice may be expanded as a Fourier series of the form

$$f(\mathbf{r}) = \sum_{l'm'n'} A_{l'm'n'} e^{2\pi i(\mathbf{b}_{l'm'n'} . \mathbf{r})} \tag{36}$$

where $\mathbf{b}_{l'm'n'}$ is the vector $l'\mathbf{b}_1+m'\mathbf{b}_2+n'\mathbf{b}_3$ of the reciprocal lattice, the constants $A_{l'm'n'}$ being given by

$$A_{l'm'n'} = V_d^{-1} \int_{\text{unit cell}} f(\mathbf{r})\, e^{-2\pi i(\mathbf{b}_{l'm'n'}.\mathbf{r})}\, d\mathbf{r} \tag{37}$$

Exercises (1) Show that the angles α', β', γ' between the reciprocal vectors $(\mathbf{b}_2, \mathbf{b}_3)$, $(\mathbf{b}_3, \mathbf{b}_1)$, $(\mathbf{b}_1, \mathbf{b}_2)$ taken in pairs are given by

$$\cos\alpha' = (\cos\beta\cos\gamma-\cos\alpha)/(\sin\beta\sin\gamma), \text{ etc.}$$

$\cos\beta'$ and $\cos\gamma'$ being obtained by permutation.

(2) Show from the general formula, and verify by geometrical consideration, that the distances between the $\{110\}$ and $\{111\}$ planes for a simple cubic lattice are, respectively, $d/\sqrt{2}$ and $d/\sqrt{3}$.

(3) Show more generally that, for a simple cubic lattice the distance between adjacent $\{lmn\}$ planes is $d(l^2+m^2+n^2)^{-1/2}$.

(4) Show that for a simple cubic lattice the lattice points in each $\{100\}$ plane form a simple square lattice, the points in each $\{110\}$ plane a rectangular lattice and the points in each $\{111\}$ plane a simple hexagonal lattice.

5.4.4 Reciprocals of the cubic lattices

As we have seen, the reciprocal of a simple cubic lattice is also a simple cubic lattice, and we must now consider the reciprocals of the two other cubic Bravais lattices, the body-centred cubic lattice and the face-centred cubic lattice. If d is the side of the unit cube and we use rectangular axes along three edges, basic vectors of the representation of the body-centred cubic lattice as a *simple* lattice are

$$\mathbf{d}_1 \equiv \left(\frac{d}{2},\frac{d}{2},\frac{d}{2}\right),\ \mathbf{d}_2 \equiv \left(\frac{d}{2},-\frac{d}{2},\frac{d}{2}\right) \quad\text{and}\quad \mathbf{d}_3 \equiv \left(\frac{d}{2},\frac{d}{2},-\frac{d}{2}\right)$$

(see § 5.3.1). Forming vector products of these in pairs and remembering that the volume of the unit cell is $d^3/2$ we have for the basic vectors $\mathbf{b}_1$, $\mathbf{b}_2$, $\mathbf{b}_3$ of the reciprocal lattice from equations (30), (30a),

$$\mathbf{b}_1 = d^{-1}\,(0,1,1)$$
$$\mathbf{b}_2 = d^{-1}\,(1,\ -1,0)$$
$$\mathbf{b}_3 = d^{-1}\,(1,0,-1)$$

By addition it will be seen that other points in the reciprocal lattice are $d^{-1}(2,0,0)$, $d^{-1}(0,2,0)$, $d^{-1}(0,0,2)$, etc. The reciprocal lattice is therefore a face-centred lattice, the unit cube having its edges $2d^{-1}$ in length and parallel to those of the unit cube of the direct lattice. The unit cube of the face-centred lattice contains four unit cells (see § 5.3.1), which have each, therefore, a volume $2/d^3$, as they should have, since this is reciprocal to the volume of the unit cell of the direct lattice. This is illustrated in Fig. 5.28, in which (*a*) shows the direct lattice and (*b*) the reciprocal lattice. It follows at once from the reciprocal properties of these lattices that the reciprocal lattice of a face-centred cubic lattice is a body-centred cubic lattice. If d' is the side of the unit cube of the direct lattice the unit cube in the reciprocal lattice will have side $2/d'$, the unit cells in the direct and reciprocal lattices having, respectively, volumes $d'^3/4$ and $4/d'^3$. This may be also illustrated in Fig. 5.28 if we take (*b*) as the direct lattice and (*a*) as the reciprocal lattice.

It is interesting to compare the reciprocal lattices of the body-centred and face-centred cubic lattices with the reciprocal lattices of the simple cubic lattice obtained by omitting the body-centre or face-centre lattice points. This clearly consists of a simple cubic lattice made up by introducing *extra* lattice points to make up the reciprocal lattices into simple cubic lattices. All points of the reciprocal lattices of the face-centred and

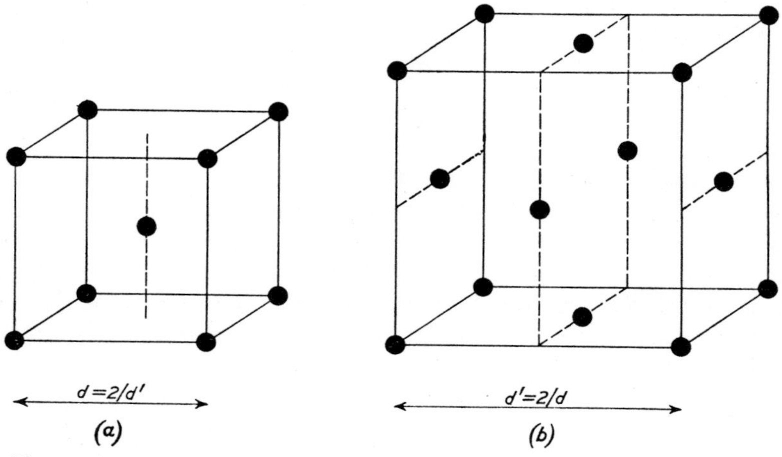

FIG. 5.28. Reciprocal lattices for the face-centred cubic and body-centred cubic lattices; (a) is the reciprocal of (b), and (b) of (a)

body-centred lattices are points of the reciprocal lattice of the simple cubic lattice but the converse is not true. This is a general result which may readily be proved (see exercise below), that if we have a sub-lattice of a given direct lattice the reciprocal lattice contains all the points of the given lattice plus extra points. This indicates that great care must be taken when we represent a *simple* lattice as a lattice with a basis, for example, in using the vectors $d(1,0,0)$, $d(0,1,0)$, $d(0,0,1)$ plus the basis $d(\frac{1}{2},\frac{1}{2},\frac{1}{2})$ to represent the body-centred cubic lattice. The three basic vectors in this case define a unit cell greater than the fundamental unit cell and the reciprocal lattice of that defined by these three vectors without the basis contains points which do not belong to the true reciprocal lattice. This only applies when the basis vector gives a point of the simple lattice. When this is not so, as for the diamond lattice, the reciprocal lattice is that of the basic vectors alone; the reciprocal lattice of the diamond lattice is the same as for a face-centred cubic lattice, a body-centred cubic lattice of side $2/d$, where d is the side of the unit cube. Again, the reciprocal lattice for the sodium chloride structure is a body-centred cubic lattice, the unit cube having side $2/d$ where $d/2$ is the distance between nearest atoms of the two different types (see § 5.3.1). For the caesium chloride structure (see § 5.3.1) the reciprocal lattice is a simple cubic lattice, the unit cube having side $1/d$ where d is the shortest distance between atoms of the *same* kind. It will be seen that, when using an orthogonal set of basic vectors, in order to retain cubic symmetry, great care must be taken to determine whether any basis vectors correspond to lattice points of a true simple lattice.

Exercise If $\mathbf{d}_1$, $\mathbf{d}_2$, $\mathbf{d}_3$, are three basic vectors defining a simple lattice L and three vectors $\mathbf{d}_1'$, $\mathbf{d}_2'$, $\mathbf{d}_3'$ are constructed from them by means of the relationships $\mathbf{d}_r' = \sum a_{rs}\, \mathbf{d}_s\, (r, s = 1, 2, 3)$, the quantities a_{rs} being integers, show that if the value of the determinant $|a_{rs}|$ (which must be integral) is an integer n greater than unity, then the lattice L' defined by the vectors $\mathbf{d}_1'$, $\mathbf{d}_2'$, $\mathbf{d}_3'$ does not contain all the points of the lattice L but that the reciprocal lattice of L' contains all the points of the reciprocal lattice of L. Show also that if the unit cell of the lattice L has volume V that of the lattice L' has volume nV; show also that the unit cell of the reciprocal lattice of L' has volume $1/nV$, and so is smaller than the unit cell of the reciprocal lattice of L which has volume $1/V$.

In using orthogonal basic vectors together with a basis to represent the face-centred and body-centred cubic lattices care must be taken with regard to notation for the lattice planes. While the lattice point (l', m', n') in the reciprocal lattice in the fundamental representation corresponds to the plane $(l'\, m'\, n')$ in the direct lattice, this is not obviously so in terms of the orthogonal representation. We may, however, readily obtain the correspondence by geometrical means since any vector $\mathbf{b}$ of the reciprocal lattice still gives the direction of the normal to a set of lattice planes in the direct lattice; l', m', n' are not now necessarily integral, e.g. points such as $(\frac{1}{2}, \frac{1}{2}, \frac{1}{2})$ occur in the face-centred cubic reciprocal lattice. The simplicity introduced by the orthogonal axes is that intercepts on the axes of a plane whose normal has a direction $l' : m' : n'$ are in the ratio $(1/l') : (1/m') : (1/n')$. The Miller indices in the orthogonal representation are therefore the smallest integers of the form $(s/l', s/m', s/n')$. The planes corresponding to the lattice point (l', m', n') are therefore the $(l''\, m''\, n'')$ planes where $l' : m' : n' = l'' : m'' : n''$, l'', m'', n'' being reduced to the smallest integers having this ratio; that it is possible to reduce the ratio to integral values follows from the fact that the point (l', m', n') belongs to the reciprocal lattice. If l', m', n' are the smallest values giving a lattice point having this ratio, the distance between the corresponding lattice planes is $|\mathbf{b}_{l'm'n'}|^{-1}$, $\mathbf{b}_{l'm'n'}$ being the *smallest* vector of the reciprocal lattice having coordinates in the ratio $l' : m' : n'$. For example, for the face-centred cubic lattice, the (100) planes correspond to the point $d^{-1}\,(2, 0, 0)$ in the reciprocal lattice and have spacing $d/2$ while the (110) planes correspond to the point $d^{-1}(2,2,0)$ and have spacing $d/2\sqrt{2}$. The (111) planes on the other hand correspond to the point $d^{-1}(1,1,1)$ and have the largest spacing of all the sets of planes, namely $d/\sqrt{3}$, and consequently crystals with the face-centred cubic structure tend to yield under stress by slipping along the {111} planes.

The lattice points in the {100} planes form a simple square lattice, the side of the unit square being $d/\sqrt{2}$, those on the {110} planes form a simple rectangular lattice, while those on the {111} planes form a simple hexagonal lattice. The latter are the closest packed planes (since they are most widely spaced) and correspond to the planes from which the close-packed cubic structure (see § 5.3.3) is built.

For the body-centred cubic lattice the (100) planes correspond to the point $d^{-1}(2,0,0)$ in the reciprocal lattice, the spacing between them being $d/2$; the (110) planes correspond to the point $d^{-1}(1,1,0)$, the spacing being $d/\sqrt{2}$, and the (111) planes to the point $d^{-1}(2,2,2)$, the spacing being $d/2\sqrt{3}$. In this case the {110} planes are the most widely spaced. The points of the {100} planes form a simple square lattice, those of the {110} planes a rectangular lattice, and those of the {111} planes a simple hexagonal lattice, the spacing between lattice points in this case being twice that for the face-centred cubic lattice.

Exercises (1) Show that for the face-centred cubic lattice the spacing between the (lmn) planes (orthogonal representation) is $d(l^2+m^2+n^2)^{-1/2}$ if l, m, n are all odd but $\frac{1}{2}d(l^2+m^2+n^2)^{-1/2}$ if l, m, n are not all odd.

(2) Show that for the body-centred cubic lattice the spacing between the (lmn) planes (orthogonal representation) is $\frac{1}{2}d(l^2+m^2+n^2)^{-1/2}$ if l, m, n are all odd or if one of them is odd and $d(l^2+m^2+n^2)^{-1/2}$ if two of them are odd.

When we have a crystal with more than one type of atom we obtain sets of lattice planes for each type of atom; some of these may coincide and we shall then have planes containing both types of atom. For example, in the caesium chloride structure the {100} planes corresponding to atoms of type A are interleaved midway between the {100} planes consisting of atoms of type B; on the other hand the {110} planes contain equal numbers of atoms of type A and type B each arranged in a rectangular plane lattice. It will readily be seen that for each basis vector $\mathbf{a}_s$ there will be corresponding to the $(l'm'n')$ planes of the simple lattice an extra set of planes parallel to these and displaced from them by an amount equal to the projection of the basis vector $\mathbf{a}_s$ on the normal to the planes; this is so because each lattice point in any one of the $(l'm'n')$ planes has a corresponding point displaced from it by the vector $\mathbf{a}_s$. The spacing between these planes is equal to $p(\mathbf{a}_s . l'\mathbf{b}_1 + m'\mathbf{b}_2 + n'\mathbf{b}_3)$ (cf. equation (35)) and if this is equal to p the displaced plane coincides with a lattice plane.

A particularly interesting situation arises in the case of the diamond lattice in which the atoms at the points corresponding to the basis vector are the *same* as those at the points of the basic lattice. Here we have extra planes of atoms of the *same* kind inserted between the planes of the basic lattice; only when the projections of the basis vectors along the normal are p or $\frac{1}{2}p$ will an equi-spaced set of planes result. For the diamond lattice in its simplest representation with three primitive vectors $d(\frac{1}{2}, \frac{1}{2}, 0)$, $d(\frac{1}{2}, 0, \frac{1}{2})$, $d(0, \frac{1}{2}, \frac{1}{2})$, a single basis vector $d(\frac{1}{4}, \frac{1}{4}, \frac{1}{4})$ will give the lattice. It will readily be seen that for the $\{100\}$ and $\{110\}$ planes this results in an extra set of planes midway between the planes of the basic face-centred cubic lattice so that the spacings are now $d/4$ and $d/2\sqrt{2}$, respectively.

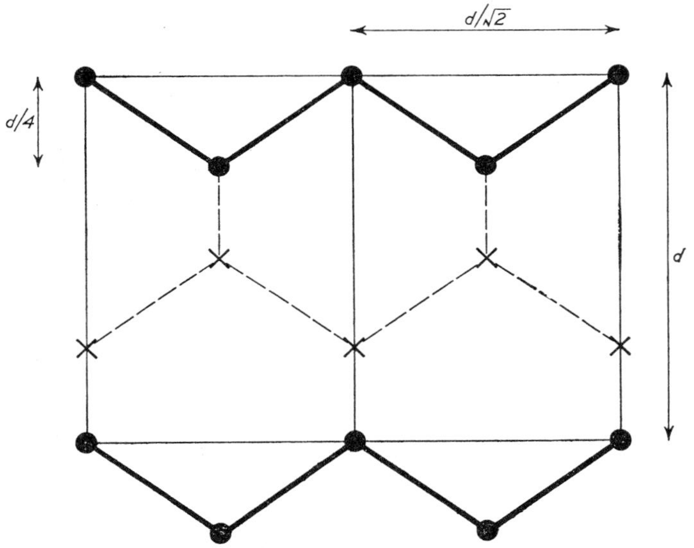

FIG. 5.29. Arrangement of atoms in the $\{110\}$ planes of the diamond lattice

The arrangement of atoms on a $\{100\}$ plane is in the form of a simple square lattice as for the face-centred cubic lattice. The arrangement in the $\{110\}$ planes is very interesting and is shown in Fig. 5.29. It consists of two interleaved rectangular lattices, the angles between the lines joining nearest neighbours being equal to the bond angle for the symmetrical tetrahedral arrangement, namely $\cos^{-1}(-\frac{1}{3}) = 109{\cdot}5°$. The atoms in one plane are marked ● and those in the succeeding plane are marked $\times$; this arrangement gives the diamond lattice its curious 'hexagonal' appearance when viewed along a $\langle 110 \rangle$ direction. For the $\{111\}$ planes the displacement due to the basis vector is $\sqrt{3}d/4$ so we have planes at perpendicular distances $(d/4\sqrt{3})$ (3, 4, 7, 8, 11, 12 ..., etc.) from the origin, i.e. we have pairs of planes $d/4\sqrt{3}$ apart separated by a spacing of $3d/4\sqrt{3}$ as shown in Fig. 5.30. The atoms in each plane form a hexagonal lattice, those in two neighbouring planes spaced $\sqrt{3}d/4$ apart being opposite each other and joined by a bond normal to the planes. Those in the more closely spaced planes are displaced with respect to each other as the points marked ●, ×, ○ in Fig. 5.22. The corresponding planes in Fig. 5.30 are marked in this way. Well defined lines of atoms occur in the $\langle 110\rangle$ directions in each $\{111\}$ plane, and these determine the direction of slip in the plane.

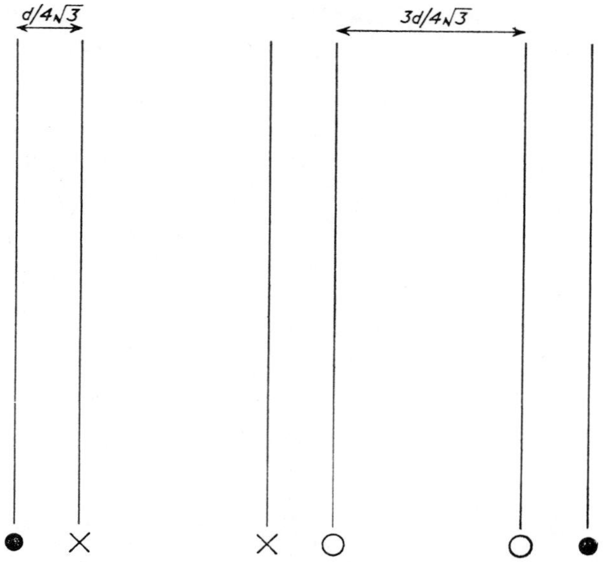

FIG. 5.30. The $\{111\}$ planes of the diamond lattice

5.4.5 Reciprocals of hexagonal lattices

For convenience it has become customary to use *four* indices to describe the planes of three-dimensional hexagonal lattices (see § 5.3.2). *Three* axes are chosen in a plane parallel to the hexagons, the angles between them being each equal to 120°, the fourth axis being the c-axis perpendicular to this plane (see Fig. 5.31).

One of the axes in the plane of the hexagons is, of course, redundant but is retained for convenience and symmetry. If α, β, γ, δ are the intercepts of a plane on the four axes and l', m', n', q' are the smallest integers in the ratio $\alpha^{-1}:\beta^{-1}:\gamma^{-1}:\delta^{-1}$ the plane is called the $(l'm'n'q')$ plane. From elementary geometrical considerations it will be clear that $-\gamma^{-1} = \alpha^{-1}+\beta^{-1}$ so that $l'+m'+n' = 0$. For example, planes normal to the c-axis, and each containing a simple plane hexagonal lattice, are the $(0,0,0,1)$ planes, and the planes parallel to the hexagon sides are

(1, 0, −1, 0), etc. When the first three indices may be permuted to give equivalent planes they are written as $\{l'm'n'q'\}$.

It will readily be seen that the reciprocal lattice of a simple hexagonal lattice is another simple hexagonal lattice rotated 30° about the c-axis. If the basic lattice translations for the direct lattice in the planes perpendicular to the c-axis and along the c-axis, are respectively a and c then for the reciprocal lattice they are a', c' where $a' = 2/\sqrt{3}a$ and $c' = 1/c$. The close-packed hexagonal lattice, being a true lattice with a basis (see § 5.3.2) has the same reciprocal lattice as the simple hexagonal lattice.

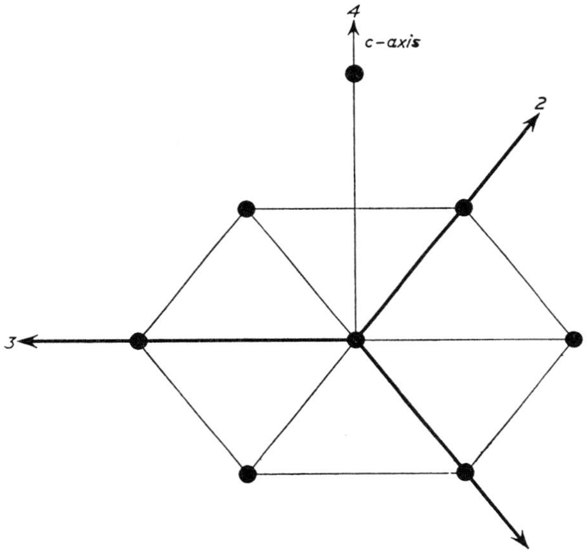

FIG. 5.31. Use of four axes with the hexagonal lattice

5.5 X-ray scattering and the reciprocal lattice

As we have seen in § 5.3.4 scattering of X-rays is closely connected with the existence of lattice planes; these in turn are closely related to the reciprocal lattice vectors and so it is not surprising that we have a simple relationship between the directions of marked X-ray scattering and the vectors of the reciprocal lattice. Suppose we have a monochromatic beam of X-rays, incident on a crystal, the direction of incidence and wavelength λ being defined by means of a propagation vector $\mathbf{k}$, so that $|\mathbf{k}| = k = 2\pi/\lambda$. Now suppose that we have a strongly reflected beam arising from a set of planes associated with the vector $\mathbf{b}$ of the reciprocal lattice, the direction of reflection being defined by means of the propagation vector $\mathbf{k}'$; we also have $|\mathbf{k}'| = k' = 2\pi/\lambda$ since there is no change of wavelength on reflection. Since the vector $\mathbf{b}$ points in the direction of the normal to the reflecting planes we have clearly (see Fig. 5.32),

$$\frac{\mathbf{b}}{|\mathbf{b}|} = \frac{\mathbf{k}'-\mathbf{k}}{|\mathbf{k}'-\mathbf{k}|} \tag{38}$$

Now we also have (Fig. 5.32), if θ is the 'glancing' angle,

$$|\mathbf{k}'-\mathbf{k}| = 2k\sin\theta = 4\pi\sin\theta/\lambda \tag{39}$$

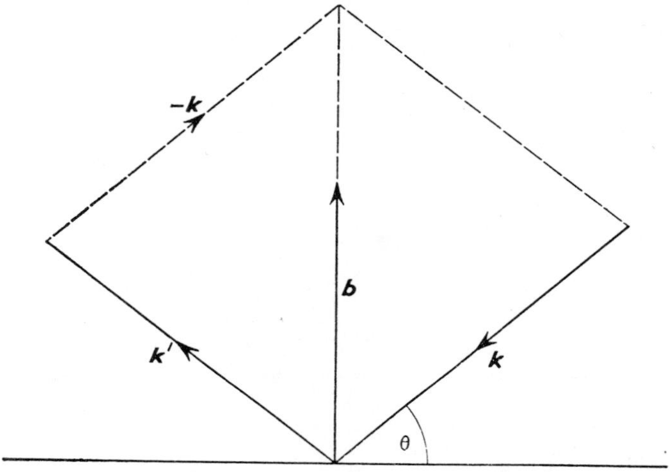

FIG. 5.32. X-ray reflection and the reciprocal lattice

and, from equation (35), $|\mathbf{b}|^{-1} = p$, the separation between successive planes; this is given in terms of λ by equation (9a). Combining these we obtain the relationship

$$\mathbf{k}'-\mathbf{k} = 2\pi s\mathbf{b} \tag{40}$$

where s is an integer. Thus, by observation of the directions of strong X-ray reflections we may obtain the vectors of the reciprocal lattice, the values of the 'order' integers s being obtained by observation of a few of the simplest lattice planes. The incident X-ray beam may, fairly readily be arranged to lie along one of the principal crystal directions, thus simplifying the analysis. The reader is referred to the many excellent books on X-ray crystal analysis for further details.

An alternative form of equation (40) which is sometimes useful may be obtained as follows. Since $|\mathbf{k}| = |\mathbf{k}'|$ we have

$$|\mathbf{k}'|^2 = |\mathbf{k}|^2 = |\mathbf{k}|^2+4\pi s(\mathbf{b}.\mathbf{k})+4\pi^2 s^2|\mathbf{b}|^2$$

or

$$(\mathbf{b}.\mathbf{k})+\pi s|\mathbf{b}|^2 = 0 \tag{41}$$

Since, **b** is a lattice vector of the reciprocal vector so also is $s\mathbf{b}$ and we may write equation (41) in the form

$$(\mathbf{b}.\mathbf{k}/2\pi)+\tfrac{1}{2}|\mathbf{b}|^2 = 0 \qquad (42)$$

where **b** is a vector of the reciprocal lattice. If we map the vectors $\mathbf{k}/2\pi$ in the coordinate system of the reciprocal lattice we see that values giving strong X-ray reflections are restricted to certain planes given by equation (42). Each plane has its normal in the direction of the lattice vector **b** and bisects the line joining the origin to the lattice point defined by the vector $-\mathbf{b}$ (see Fig. 5.33); there is clearly a similar plane bisecting

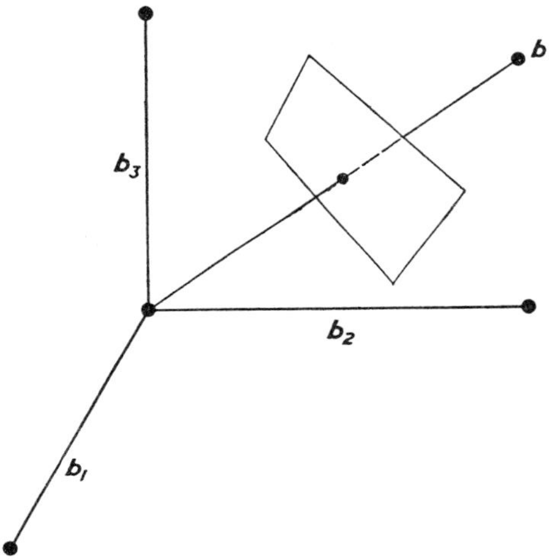

FIG. 5.33. Construction to determine allowed values of the vector **k** for strong X-ray reflections

the line joining the origin to the point defined by the vector **b**. Thus we see that all the values of $\mathbf{k}/2\pi$ giving strong reflections lie on a series of planes which are the perpendicular bisectors of the lines joining the points of the reciprocal lattice to the origin.

For a fixed monochromatic beam of X-rays we may regard the vector **k** as fixed. If now we rotate the crystal about an axis, the end point of the vector **k** will meet in turn the various planes which are the perpendicular bisectors of the lines joining the points of the reciprocal lattice (which, of course, is also rotated) to the origin. Thus we get a series of 'spots' showing strong reflections as in the well-known Laue photographs. When powder samples are used, containing crystals of all orientations, these spots are spread out into rings as observed in Debye–Scherrer photographs.

6

Lattice Vibrations in Three Dimensions

6.1 Interatomic forces

In order to be able to discuss vibrations of the atoms in real crystals about their positions of equilibrium we must now extend to three dimensions the analysis, given in Chapter 3, for linear lattices. We may state at the outset that this is a very difficult problem, and even approximate numerical solutions have been obtained only for the simplest lattices by making very drastic assumptions regarding the forces between the atoms. A full discussion of interatomic forces is beyond the scope of this book*. These forces may, however, be divided into two categories, central forces and angular forces; the former depend only on the distance between two atoms and act along the line joining them, whereas the latter depend on the position of the line joining the atoms. Central forces are simpler to treat mathematically and all but a few of the calculations which have been made are for this type of force. For small displacements, the forces may be assumed to obey a form of Hooke's law, the restoring force being proportional to the displacement. The atoms, in this case, may be regarded as being connected by light springs which are unstretched when the interatomic distances have their equilibrium values. The corresponding problem for a linear lattice is that treated in § 3.3.3 to determine the longitudinal modes of vibration. For atoms in a crystal, the strength of the equivalent springs will be the same for all equidistant nearest neighbours and will vary according to the distance of other neighbours, becoming very weak for the more distant atoms. There are, however, many more of the latter, and problems of convergence arise; although all numerical calculations made so far have assumed that forces between all but a few neighbours may be neglected, there is now a good deal of doubt as to the validity of this assumption. Fortunately a formal solution of the problem of determining the vibration spectrum of two- and three-dimensional lattices may be given and will be discussed later; the difficulties arise in the practical application of the solution. In order to obtain some insight into these difficulties we shall first of all consider a few of the simplest cases.

* For an elementary discussion of interatomic forces see N. F. Mott and R. W. Gurney, *Electronic Processes in Ionic Crystals*, Oxford University Press (1940), Chapter 1; a more advanced discussion is given by M. Born and K. Huang, *Dynamical Theory of Crystal Lattices*. Oxford University Press (1954).

6.2 Vibration spectrum of the simple cubic lattice

The three-dimensional analogue of the linear crystal with one kind of equi-spaced atom is the simple cubic lattice having identical atoms at each lattice point. We shall assume that we have central forces between the atoms and that the bonds between the atoms are not in tension in equilibrium so that the only forces called into play are due to *changes* in the interatomic distances. Let us consider the effect of atomic displacements on the force between an atom (n) and an atom (m). Let the displacements of the atoms from their equilibrium positions be represented by the vectors $\mathbf{u}_n$, $\mathbf{u}_m$. Then to the first order in the displacements we have the change in bond length δr equal to $(\mathbf{i}\,.\,\mathbf{u}_m - \mathbf{u}_n)$ where $\mathbf{i}$ is a unit

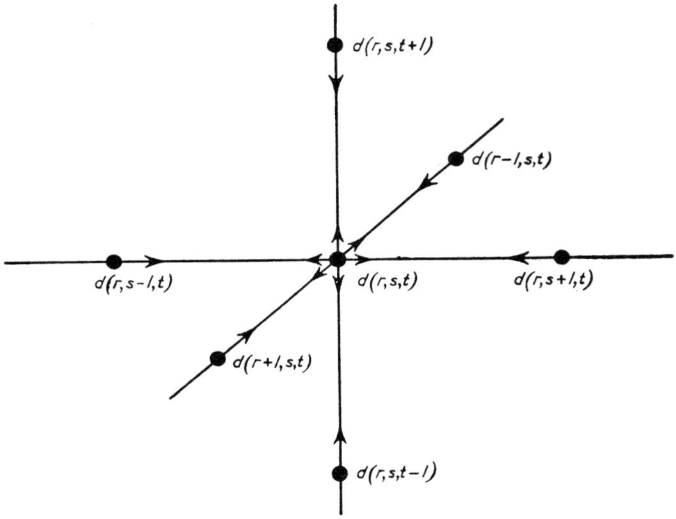

FIG. 6.1. Nearest neighbour forces in a simple cubic lattice

vector along the line joining the atoms (n) and (m) in equilibrium. The magnitude F of the restoring force is then assumed to be given by

$$F = \alpha\delta r = \alpha(\mathbf{i}\,.\,\mathbf{u}_m - \mathbf{u}_n) \tag{1}$$

Vectorially the force may therefore be written as

$$\mathbf{F} = \alpha\mathbf{i}(\mathbf{i}\,.\,\mathbf{u}_m - \mathbf{u}_n) \tag{2}$$

Let us now consider the simple cubic lattice, and to begin with we shall assume that all forces are zero except those between nearest neighbours (see Fig. 6.1). We shall now use equation (2) to derive the components of the forces, along the cube axes, acting on an atom at the lattice point $d(r,s,t)$. The only other atoms involved are those at

$$d(r\pm 1,s,t),\;\; d(r,s\pm 1,t),\;\; d(r,s,t\pm 1)$$

six in all. The components (X, Y, Z) of the resulting forces acting on the atom at $d(r,s,t)$ are thus

$$\left.\begin{aligned} X &= (\alpha u_{r+1,s,t} - u_{r,s,t}) + \alpha(u_{r-1,s,t} - u_{r,s,t}) \\ Y &= \alpha(v_{r,s+1,t} - v_{r,s,t}) + \alpha(v_{r,s-1,t} - v_{r,s,t}) \\ Z &= \alpha(w_{r,s,t+1} - w_{r,s,t}) + \alpha(w_{r,s,t-1} - w_{r,s,t}) \end{aligned}\right\} \quad (3)$$

$(u_{rst}, v_{rst}, w_{rst})$ being the components of the displacement vector $\mathbf{u}$ of the atom at the lattice point $d(r,s,t)$. The equations of motion of the atoms of the lattice are therefore

$$\left.\begin{aligned} \ddot{u}_{r,s,t} + \alpha'(2u_{r,s,t} - u_{r+1,s,t} - u_{r-1,s,t}) &= 0 \\ \ddot{v}_{r,s,t} + \alpha'(2v_{r,s,t} - v_{r,s+1,t} - v_{r,s-1,t}) &= 0 \\ \ddot{w}_{r,s,t} + \alpha'(2w_{r,s,t} - w_{r,s,t+1} - w_{r,s,t-1}) &= 0 \end{aligned}\right\} \quad (4)$$

where $\alpha' = \alpha/M$, M being the mass of each atom. To this approximation we see that the three components of the motion are independent, and equations (4) may be solved by using 'descriptive waves' as in § 3.3. Thus if we write

$$\left.\begin{aligned} u &= A\, e^{i(\mathbf{k}.\mathbf{r}) - i\omega t} \\ v &= B\, e^{i(\mathbf{k}'.\mathbf{r}) - i\omega t} \\ w &= C\, e^{i(\mathbf{k}''.\mathbf{r}) - i\omega t} \end{aligned}\right\} \quad (5)$$

the value of u_{rst} being given by the value of u when the vector $\mathbf{r}$ has components $d(r,s,t)$ and similarly for v and w. On substituting in equations (4) we obtain the following three equations for ω^2,

$$\left.\begin{aligned} \omega^2 &= 2\alpha'(1 - \cos k_1' d) \\ \omega^2 &= 2\alpha'(1 - \cos k_2' d) \\ \omega^2 &= 2\alpha'(1 - \cos k_3'' d) \end{aligned}\right\} \quad (6)$$

where (k_1, k_2, k_3) are the components of a propagation vector $\mathbf{k}$, and similarly for $\mathbf{k}'$ and $\mathbf{k}''$. The three solutions correspond to three directions of polarisation of the lattice waves, and to this approximation all three have the same frequency spectrum referred to the appropriate direction in the lattice. The frequency corresponding to polarisation in the x-direction depends only on the component of the propagation vector in the x-direction. It will be noted that ω^2 is an even function of k_1, k_2', k_3'' and is periodic with period π/d in each of these. When we take negative values of k_1, etc., into account we see that we may restrict the values of the propagation vectors $\mathbf{k}$, $\mathbf{k}'$, $\mathbf{k}''$ to the region $-\pi/d \leqslant k_1, k_2, k_3 \leqslant \pi/d$. If we write $\boldsymbol{\varkappa} = \mathbf{k}/2\pi$ and plot $\boldsymbol{\varkappa}$ in the reciprocal lattice space we see that the value of $\boldsymbol{\varkappa}$ may be restricted to a zone of the space which is a cube whose centre is at the origin and whose edges are equal to the distance between neighbouring points of the reciprocal lattice. This is the three-dimensional equivalent of the zone used for the reduced representation as discussed in § 3.3 (see Fig. 6.2).

The approximation to the forces which we have used is clearly a very poor one; when $k_1 = 0$, $\omega = 0$ even though k_2 and k_3 are not zero, this would mean that transverse waves do not propagate in the crystal. The reason for this is that we have neglected the effect of *second-nearest neighbours* which, in the simple cubic lattice, are not very much further apart than the nearest neighbours. Forces arising from the second-nearest neighbours may easily be added, and solutions for both the two-dimensional and three-dimensional problem have been given to this approximation by M. Blackman*. The forces from the twelve second-nearest

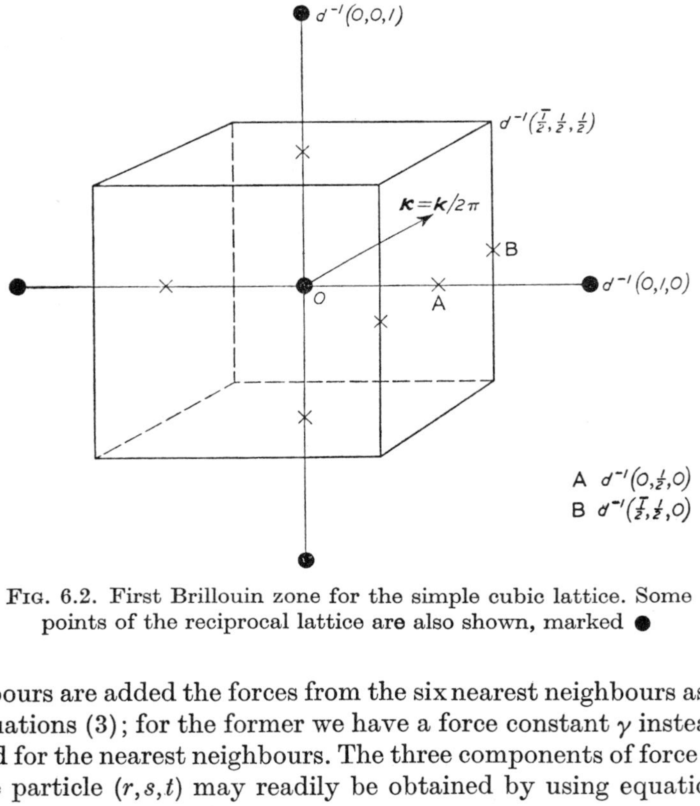

FIG. 6.2. First Brillouin zone for the simple cubic lattice. Some points of the reciprocal lattice are also shown, marked ●

neighbours are added the forces from the six nearest neighbours as given by equations (3); for the former we have a force constant γ instead of α as used for the nearest neighbours. The three components of force acting on the particle (r,s,t) may readily be obtained by using equation (2); we have

$$\begin{aligned}
-X = {} & \alpha(2u_{r,s,t} - u_{r+1,s,t} - u_{r-1,s,t}) \\
& + \gamma(8u_{r,s,t} - u_{r+1,s+1,t} - u_{r-1,s+1,t} - u_{r+1,s-1,t} - u_{r-1,s-1,t} \\
& - u_{r+1,s,t+1} - u_{r-1,s,t+1} - u_{r+1,s,t-1} - u_{r-1,s,t-1}) \\
& - \gamma(v_{r+1,s+1,t} - v_{r-1,s+1,t} - v_{r+1,s-1,t} + v_{r-1,s-1,t}) \\
& - \gamma(w_{r+1,s,t+1} - w_{r-1,s,t+1} - w_{r+1,s,t-1} + w_{r-1,s,t-1})
\end{aligned} \tag{7}$$

* M. Blackman, *Proc. Roy. Soc.*, **A148**, 384 (1935); *Ibid.*, **159**, 416 (1937).

and similar expressions for Y and Z. Using these, the equations of motion for the particle (r,s,t) become

$$\begin{aligned} M\ddot{u}_{r,s,t} = & -\alpha(2u_{r,s,t}-u_{r+1,s,t}-u_{r-1,s,t}) \\ & -\gamma(8u_{r,s,t}-u_{r+1,s+1,t}-u_{r-1,s+1,t}-u_{r+1,s-1,t}-u_{r-1,s-1,t} \\ & -u_{r+1,s,t+1}-u_{r+1,s,t-1}-u_{r-1,s,t+1}-u_{r-1,s,t-1}) \\ & +\gamma(v_{r+1,s+1,t}-v_{r-1,s+1,t}-v_{r+1,s-1,t}+v_{r-1,s-1,t}) \\ & +\gamma(w_{r+1,s,t+1}-w_{r-1,s,t+1}-w_{r+1,s,t-1}+w_{r-1,s,t-1}) \end{aligned} \quad (8)$$

$$\begin{aligned} M\ddot{v}_{r,s,t} = & \gamma(u_{r+1,s+1,t}-u_{r-1,s+1,t}-u_{r+1,s-1,t}+u_{r-1,s-1,t}) \\ & -\alpha(2v_{r,s,t}-v_{r,s+1,t}-v_{r,s-1,t}) \\ & +\gamma(8v_{r,s,t}-v_{r+1,s+1,t}-v_{r-1,s+1,t}-v_{r+1,s-1,t}-v_{r-1,s-1,t} \\ & -v_{r,s+1,t+1}-v_{r,s-1,t+1}-v_{r,s+1,t-1}-v_{r,s-1,t-1}) \\ & +\gamma(w_{r,s+1,t+1}-w_{r,s-1,t+1}-w_{r,s+1,t-1}-w_{r,s-1,t-1}) \end{aligned} \quad (8a)$$

$$\begin{aligned} M\ddot{w}_{r,s,t} = & \gamma(u_{r+1,s,t+1}-u_{r-1,s,t+1}-u_{r+1,s,t-1}-u_{r-1,s,t-1}) \\ & +\gamma(v_{r,s+1,t+1}-v_{r,s-1,t+1}-v_{r,s+1,t-1}+v_{r,s-1,t-1}) \\ & -\alpha(2w_{r,s,t}-w_{r,s,t+1}-w_{r,s,t-1}) \\ & +\gamma(8w_{r,s,t}-w_{r+1,s,t+1}-w_{r-1,s,t+1}-w_{r+1,s,t-1}-w_{r-1,s,t-1} \\ & -w_{r,s+1,t+1}-w_{r,s-1,t+1}-w_{r,s+1,t-1}-w_{r,s-1,t-1}) \end{aligned} \quad (8b)$$

We may now solve these equations by the method of 'descriptive' waves using the wave functions

$$\left.\begin{matrix} u \\ v \\ w \end{matrix}\right\} = \left.\begin{matrix} A \\ B \\ C \end{matrix}\right\} e^{i(\mathbf{k}.\mathbf{r})-i\omega t} \quad (9)$$

On substituting in equations (8), (8a), (8b) using the values of $\mathbf{r}$ corresponding to the coordinates of the various points, we obtain the equations

$$\begin{aligned} & [2\alpha(1-\cos k_1 d)+4\gamma(2-\cos k_1 d\cos k_2 d-\cos k_1 d\cos k_3 d)-M\omega^2]A \\ & \qquad +4\gamma(\sin k_1 d\sin k_2 d)B+4\gamma(\sin k_1 d\sin k_3 d)C = 0 \end{aligned} \quad (10)$$

$$\begin{aligned} & 4\gamma(\sin k_1 d\sin k_2 d)A+[2\alpha(1-\cos k_2 d)+4\gamma(2-\cos k_1 d\cos k_2 d \\ & \qquad -\cos k_2 d\cos k_3 d-M\omega^2]B+4\gamma(\sin k_2 d\sin k_3 d)C = 0 \end{aligned} \quad (10a)$$

$$\begin{aligned} & 4\gamma(\sin k_1 d\sin k_3 d)A+4\gamma(\sin k_2 d\sin k_3 d)B+[2\alpha(1-\cos k_3 d) \\ & \qquad +4\gamma(2-\cos k_2 d\cos k_3 d-\cos k_1 d\cos k_3 d)-M\omega^2]C = 0 \end{aligned} \quad (10b)$$

k_1, k_2, k_3 being, as before, the components of the vector $\mathbf{k}$.

These are three linear, homogeneous equations in A, B, C and have a non-zero solution only provided the determinant of the coefficients is equal to zero. This gives a cubic equation for ω^2 of the form

$$\begin{vmatrix} M\omega^2 - f(k_1, k_2, k_3) & g(k_1, k_2) & g(k_1, k_3) \\ g(k_1, k_2) & M\omega^2 - f(k_2, k_1, k_3) & g(k_2, k_3) \\ g(k_1, k_3) & g(k_2, k_3) & M\omega^2 - f(k_3, k_2, k_1) \end{vmatrix} = 0 \quad (11)$$

where

$$\begin{aligned} f(x, y, z) &= 2\alpha(1 - \cos xd) + 4\gamma(2 - \cos xd \cos yd - \cos xd \cos zd) \\ g(x, y) &= -4\gamma \sin xd \sin yd \end{aligned}$$

The three solutions of this equation for ω^2 correspond to three states of polarisation and, in general, lead to different ratios of the quantities $A:B:C$ which give the direction of the displacement vector. The three solutions do not generally correspond to transverse and longitudinal waves; this will be so only in certain directions having a high degree of symmetry. For any direction of propagation, given by the ratios $k_1:k_2:k_3$, we obtain, in general, three curves giving ω as a function of the propagation constant k.

It is interesting to examine some of the directions having a high degree of symmetry. For the [100] direction $k_1 = k$, $k_2 = k_3 = 0$. In this case $g(k_1, k_2) = g(k_1, k_3) = g(k_2, k_3) = 0$ and the equations for ω^2 reduce to $\omega = \omega_1$ or $\omega = \omega_2$ or $\omega = \omega_3$ where

$$\left.\begin{aligned} M\omega_1^2 &= (2\alpha + 8\gamma)(1 - \cos kd) \\ M\omega_2^2 &= 4\gamma(1 - \cos kd) \\ M\omega_3^2 &= 4\gamma(1 - \cos kd) \end{aligned}\right\} \quad (12)$$

When $\omega = \omega_1$, $B = C = 0$, and we have longitudinal waves, for which we may write ω_l for ω_1. As $k \to 0$ we have

$$\omega_l = \left(\frac{\alpha + 4\gamma}{M}\right)^{1/2} kd \quad (13)$$

and this corresponds to a longitudinal acoustic wave.

If c_l is the velocity of longitudinal sound waves we have

$$c_l = \omega_l/k = d(\alpha + 4\gamma)^{1/2}/M^{1/2} \quad (14)$$

If ρ is the density we have $\rho = M/d^3$ so that

$$c_l = (\alpha + 4\gamma)^{1/2}/(\rho d)^{1/2} \quad (14a)$$

Again when $\omega = \omega_2$ or $\omega = \omega_3$ we have $A = 0$. In this case we have degeneracy since the ratio $B:C$ is undetermined, the waves corresponding to ω_2 and ω_3 having the same frequency for a given value of k; these waves are therefore two degenerate components of the transverse waves.

Now when $k \to 0$, we have, writing $\omega_2 = \omega_3 = \omega_t$, we have

$$\omega_t = \left(\frac{2\gamma}{M}\right)^{1/2} kd \quad (15)$$

and c_t, the velocity of transverse sound waves, is given by

$$c_t = (2\gamma)^{1/2}d/M^{1/2} = (2\gamma)^{1/2}/(\rho d)^{1/2} \tag{16}$$

The ratio of the velocities of longitudinal to transverse waves in the [100] direction is therefore given by

$$c_l/c_t = (\alpha+4\gamma)^{1/2}/(2\gamma)^{1/2} \tag{17}$$

Thus c_l is greater than c_t and, if the second-neighbour forces are small compared with those due to nearest neighbours, so that $\gamma \ll \alpha$, then c_t is considerably smaller than c_l. Thus if $\gamma/\alpha = 1/20$ (the value taken by Blackman, *loc. cit.*) we have $c_l/c_t = 3{\cdot}5$. If the values of c_l and c_t are known, the values of the force constants α and γ may be obtained from equations (14a) and (16).

Exercises (1) Discuss as above the propagation of lattice vibrations in the $\langle 111\rangle$ directions of a simple cubic crystal showing that one solution corresponds to longitudinal waves and the other two to degenerate transverse waves.

(2) Obtain the vibrational spectrum for a simple square lattice, assuming forces between nearest and second nearest neighbours. Show that the velocities of propagation of longitudinal and transverse acoustic waves in the $\langle 10\rangle$ direction are given by $c_l = (\alpha+\gamma)^{1/2}/\sigma^{1/2}$, $c_t = \gamma^{1/2}/\sigma^{1/2}$ where σ is the surface density and α and γ are, respectively, the force constants for nearest and second-nearest neighbours.

We may note at this stage one or two features of the equations giving the frequency ω as a function of the wave vector $\mathbf{k}$. If we replace $\mathbf{k}$ by the vector $\mathbf{k}'$ given by $\mathbf{k}' = \mathbf{k}+2\pi\mathbf{b}$ where $\mathbf{b}$ is any vector of the reciprocal lattice, we see that the equations are unchanged. This follows at once from the fact that if

$$\mathbf{b} = l\mathbf{b}_1+m\mathbf{b}_2+n\mathbf{b}_3$$

l, m, n being integers, then (see § 5.4.3)

$$k_1' = k_1+2\pi l/d$$

$$k_2' = k_2+2\pi m/d$$

$$k_3' = k_3+2\pi n/d$$

on replacing k_1, k_2, k_3, by k_1', k_2', k_3' we see that the equations for ω^2 are unchanged. Thus the functions for ω^2 expressed in terms of the vector $\boldsymbol{\varkappa} = \mathbf{k}/2\pi$ have the periodicity of the reciprocal lattice and we may restrict the values of $\boldsymbol{\varkappa}$ to the zone bounded by the planes $\kappa_1 = \pm 1/2d$, $\kappa_2 = \pm 1/2d$, $\kappa_3 = \pm 1/2d$ as already discussed. Also we see that ω^2 is an even function of $\mathbf{k}$, and from continuity of the periodic functions at the edges of the zone it follows that $\partial\omega/\partial k_n$ is zero, k_n representing the value of $\mathbf{k}$ in the direction of the normal to the zone boundary. This zone is called a Brillouin zone and we shall later discuss it in greater detail in § 6.5 (see Fig. 6.2).

6.2.1 Boundary conditions and normal modes for a simple cubic lattice

So far, we have not obtained the actual frequencies of oscillation of the atoms of the crystal. The precise values of the frequencies depend on the conditions at the boundary, but their distribution over the range of k-values is hardly influenced by these since, as we shall see, for a large crystal, the allowed values of k are very closely spaced and spread throughout the zone to which we have chosen to limit them.

We may see at once that for a *free* crystal containing N_a atoms we have $3N_a - 6$ normal modes of oscillation, since N_a particles require $3N_a$ coordinates to describe their position and bodily displacement and rotation account for six coordinates which will not lead to oscillations in a free crystal; since N_a is assumed to be very large it is quite a good approximation to say that we have $3N_a$ normal modes. As a crude approximation we may consider each atom of the crystal as moving in a field of force independent of the other atoms of the crystal. It will then have three normal modes of oscillation of which two or even all three may be degenerate, depending on the symmetry of the force field acting in the crystal. When the three modes are non-degenerate we may then regard the whole crystal as having three normal frequencies only, the modes corresponding to each being N_a-fold degenerate. This point of view has been taken by some investigators* who have, in this way, obtained quite good agreement between this simplified mode spectrum and specific heat data in certain cases and have also been able to explain some features of Raman infra-red spectra by this means. For crystals having more than one atom per unit cell, more normal frequencies will occur. That this is a crude approximation will be clear from the analysis of § 6.2 and the more general treatment of § 6.3; that it gives results in some cases in reasonable agreement with experiment is due to the fact that the density of allowed frequencies is high around certain values of the frequency (see § 7.4).

To determine the allowed frequencies we have first to find the allowed values of the wave vector **k**. The conditions at the surface of a crystal are difficult to formulate and since, for large crystals, they have little effect on the distribution of normal frequencies, we use periodic boundary conditions already discussed for one-dimensional problems in § 3.3.2. We assume that we have a crystal whose dimensions in the x, y, z directions are Pd, Qd, Rd, the quantities P, Q, R being very large integers. The volume of the crystal is therefore equal to $PQRd^3$ and the number of atoms N_a in the crystal (assumed to be perfect) is equal to PQR. We then consider an infinite crystal made up of units of this type and assume that

* See, for example, C. V. Raman, *Memoirs of the Raman Research Institute*, No. 85 (1956).

the solution in each unit is the same, i.e. that all quantities are periodic in the coordinates with periods Pd, Qd, Rd. From the form of the wave function (9) we see that for this to be so we must have

$$\left.\begin{aligned} k_1 &= 2\pi l/Pd \\ k_2 &= 2\pi m/Qd \\ k_3 &= 2\pi n/Rd \end{aligned}\right\} \tag{18}$$

where l, m, n, are integers; these equations give the values of the wave vectors associated with the normal modes; the corresponding normal frequencies are obtained by substituting for k_1, k_2, k_3 in equation (11). It will be seen that the allowed values are spread throughout the zone, the spacing between neighbouring values being very small, since P, Q, R are large integers. The value of l, in order that $-\pi/d < k_1 \leqslant \pi/d$, ranges from $-P/2 < l \leq P/2$; i.e. there are P allowed values (cf. § 3.3.2) and similarly $-Q/2 < m \leqslant Q/2$, $-R/2 < n \leqslant R/2$. There are therefore PQR or N_a normal modes for each type of solution, or $3N_a$ in all. We shall later give an expression for the density of modes per unit volume of **k**-space, and per unit frequency range (see § 6.7).

6.3 Vibration spectra of other lattices

The above discussion of the simple cubic lattice indicates that, even for this lattice, the calculation of the vibration spectrum is quite tedious. Even so, we have not yet taken account of forces from atoms beyond the second nearest neighbours and when this is done the calculation rapidly increases in complexity as forces from more distant atoms are introduced. Treatments similar to the above have been given for only a few other simple lattices. For example, the body-centred cubic lattice has been discussed by F. C. Fine* and the face-centred lattice by R. B. Leighton†; as an example of a lattice with a basis, the diamond lattice has been treated by Helen M. J. Smith‡.

It is only recently that direct experimental verification of the form of the vibration spectrum has been possible (see § 6.10) although through the theoretical work of M. Blackman§ and others some correlation of calculated spectra with specific heat measurements has been possible. Recent comparisons of observed spectra for Ge with those calculated have thrown some doubts on the usefulness of such numerical calculations, the main difficulty being to know how many neighbours should be included in the calculation in order to obtain a significant result.

It will now be clear that detailed calculations of vibration spectra are very difficult but fortunately it turns out that a fairly general formal

* F. C. Fine, *Phys. Rev.*, **56**, 355 (1939).
† R. B. Leighton, *Rev. Mod. Phys.*, **20**, 165 (1948).
‡ Helen M. J. Smith, *Philos. Trans.*, A**241**, 105 (1948).
§ M. Blackman, *Handb. Phys.*, **7**, (1), 325 (1955).

solution of the problem is possible from which we may draw a number of very important conclusions; this we now proceed to discuss.

6.4 Vibration spectrum—general treatment for a simple lattice

We shall now consider the vibrations of a three-dimensional crystal having a simple lattice defined by means of the three basic vectors $\mathbf{d}_1$, $\mathbf{d}_2$, $\mathbf{d}_3$; we shall suppose that atoms of the crystal occupy the points of the lattice, in equilibrium, and we shall denote the lattice point $l\mathbf{d}_1+m\mathbf{d}_2+n\mathbf{d}_3$ by the symbol $\mathbf{d}_{lmn}$, or alternatively by $\mathbf{d}_\mathbf{n}$, the symbol $\mathbf{n}$ representing the three integers l, m, n. The displacement of an atom of the crystal from its equilibrium position at the lattice point $\mathbf{d}_\mathbf{n}$ will be denoted by $\mathbf{u}_{lmn}$ or more briefly by $\mathbf{u}_\mathbf{n}$; moreover we shall assume that such displacements are very small compared with the lattice spacing, i.e. that $|\mathbf{u}_\mathbf{n}| \ll |\mathbf{d}_\mathbf{n}-\mathbf{d}_{\mathbf{n}'}|$ for all values of $\mathbf{n}$ and $\mathbf{n}'$ other than $\mathbf{n} = \mathbf{n}'$ (see Fig. 6.3). The potential

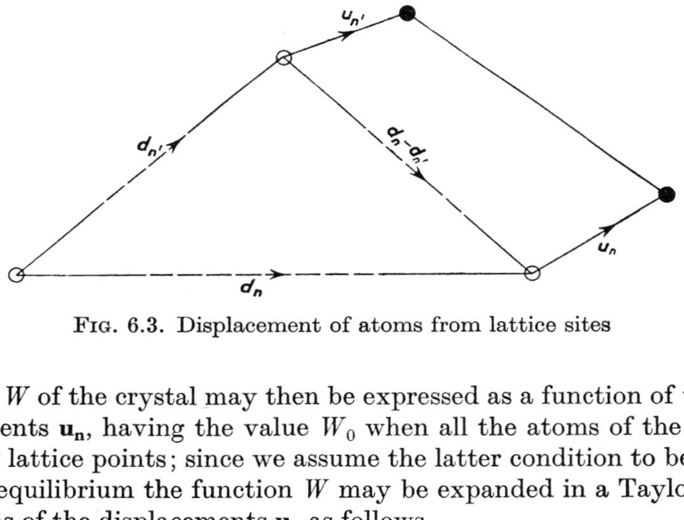

FIG. 6.3. Displacement of atoms from lattice sites

energy W of the crystal may then be expressed as a function of the displacements $\mathbf{u}_\mathbf{n}$, having the value W_0 when all the atoms of the crystal occupy lattice points; since we assume the latter condition to be one of stable equilibrium the function W may be expanded in a Taylor series in terms of the displacements $\mathbf{u}_\mathbf{n}$ as follows

$$W = W_0+\tfrac{1}{2}\sum_{\mathbf{nn}'} \mathbf{W}_{\mathbf{nn}'}\,\mathbf{u}_\mathbf{n}\,\mathbf{u}_{\mathbf{n}'}+\text{higher terms} \tag{19}$$

where each of the quantities $\mathbf{W}_{\mathbf{nn}'}$ is a symmetrical tensor of rank two and $\mathbf{W}_{\mathbf{nn}'}\,\mathbf{u}_{\mathbf{n}'}\,\mathbf{u}_\mathbf{n}$ represents a scalar product of the tensor and the two vectors $\mathbf{u}_\mathbf{n}\,\mathbf{u}_{\mathbf{n}'}$. The tensor $\mathbf{W}_{\mathbf{nn}'}$ has six components $W^{rs}_{\mathbf{nn}'}$ $(r,s = 1,2,3)$ since, being symmetrical $W^{rs}_{\mathbf{nn}'} = W^{sr}_{\mathbf{nn}'}$. When written in terms of the components of the tensor and the components of the vectors $\mathbf{u}_\mathbf{n}\,\mathbf{u}_{\mathbf{n}'}$ the scalar sum has the form

$$\begin{aligned}\mathbf{W}_{\mathbf{nn}'}\,\mathbf{u}_\mathbf{n}\,\mathbf{u}_{\mathbf{n}'} &= W^{11}_{\mathbf{nn}'}\,u_\mathbf{n}\,u_{\mathbf{n}'}+W^{12}_{\mathbf{nn}'}\,u_\mathbf{n}\,v_{\mathbf{n}'}+\ldots \\ &\quad +\ldots+W^{23}_{\mathbf{nn}'}\,v_\mathbf{n}\,w_{\mathbf{n}'}+W^{33}_{\mathbf{nn}'}\,w_\mathbf{n}\,w_{\mathbf{n}'}\end{aligned} \tag{20}$$

It will be seen that there are no linear terms in the expansion since W_0 represents an equilibrium condition, and moreover the quadratic terms are positive definite; also since the quantities $\mathbf{u_n}$ are small we shall neglect the higher terms. The quantities $W^{rs}_{\mathbf{nn'}}$ have to be determined from the forces between the atoms and herein lies the main difficulty; without having explicit expressions for these we may, however, deduce some general properties which they possess. We note that the quantities $\mathbf{W_{nn'}}$ depend only on the *separation* of the lattice points, i.e. we have $\mathbf{W_{nn'}} = \mathbf{W_{pp'}}$ provided $\mathbf{d_n}-\mathbf{d_{n'}} = \pm(\mathbf{d_p}-\mathbf{d_{p'}})$. We may therefore write $W_{\mathbf{nn'}}$ in the form

$$\mathbf{W_{nn'}} = \mathbf{W}(\mathbf{n}-\mathbf{n'}) = \mathbf{W}(\mathbf{n'}-\mathbf{n}) \tag{21}$$

It is now very easy to write down the equations of motion in the form

$$\left.\begin{aligned} M\ddot{u}_{\mathbf{n}} &= -\frac{\partial W}{\partial u_{\mathbf{n}}} \\ M\ddot{v}_{\mathbf{n}} &= -\frac{\partial W}{\partial v_{\mathbf{n}}} \\ M\ddot{w}_{\mathbf{n}} &= -\frac{\partial W}{\partial w_{\mathbf{n}}} \end{aligned}\right\} \tag{22}$$

where M is the mass of each atom. Using the expansion (19) for W these equations may be written in the form

$$M\ddot{u}_{\mathbf{n}} = -\sum_{\mathbf{n'}} W^{11}(\mathbf{n}-\mathbf{n'})\,u_{\mathbf{n'}} - \sum_{\mathbf{n'}} W^{12}(\mathbf{n}-\mathbf{n'})\,v_{\mathbf{n'}} - \sum_{\mathbf{n'}} W^{13}(\mathbf{n}-\mathbf{n'})\,w_{\mathbf{n'}} \tag{23}$$

and similar equations for $\ddot{v}_{\mathbf{n}}$ and $\ddot{w}_{\mathbf{n}}$. To solve these we assume that we may express $u_{\mathbf{n}}, v_{\mathbf{n}}, w_{\mathbf{n}}$ in terms of descriptive waves as in § 3.3. In this case we have wave motion in three dimensions so that we require a propagation *vector* $\mathbf{k}$ for their description. The form of solution is then given in terms of the wave functions

$$\left.\begin{aligned} u &= A\,\mathrm{e}^{i\mathbf{k}.\mathbf{r}-i\omega t} \\ v &= B\,\mathrm{e}^{i\mathbf{k}.\mathbf{r}-i\omega t} \\ w &= C\,\mathrm{e}^{i\mathbf{k}.r-i\omega t} \end{aligned}\right\} \tag{24}$$

A, B, C being constants, and $u_{\mathbf{n}}, v_{\mathbf{n}}, w_{\mathbf{n}}$ are obtained by giving $\mathbf{r}$ the value $\mathbf{d_n}$ in u, v, w, respectively. To see that this form of solution is possible we substitute the values of $u_{\mathbf{n}}, v_{\mathbf{n}}, w_{\mathbf{n}}$, given by the wave functions in equations (23) and obtain the equation

$$\begin{aligned} M\omega^2 A = A\sum_{\mathbf{n'}} W^{11}(\mathbf{n}-\mathbf{n'})\,\mathrm{e}^{i\mathbf{k}.(\mathbf{d_{n'}}-\mathbf{d_n})} - B\sum_{\mathbf{n'}} W^{12}(\mathbf{n}-\mathbf{n'})\,\mathrm{e}^{ik.(\mathbf{d_{n'}}-\dot{\mathbf{d}}_{\mathbf{n}})} \\ -C\sum_{\mathbf{n'}} W^{13}(\mathbf{n}-\mathbf{n'})\,\mathrm{e}^{i\mathbf{k}.(\mathbf{d_{n'}}-\mathbf{d_n})} \end{aligned} \tag{25}$$

and two similar equations for $M\omega^2 B$ and $M\omega^2 C$. The summations in the right-hand side of these equations may clearly be rearranged so that they may be written in the form

$$\left.\begin{aligned} M\omega^2 A &= A\sum_{\mathbf{n}} W^{11}(\mathbf{n})\,\mathrm{e}^{i\mathbf{k}.\mathbf{d_n}} + B\sum_{\mathbf{n}} W^{12}(\mathbf{n})\,\mathrm{e}^{i\mathbf{k}.\mathbf{d_n}} + C\sum_{\mathbf{n}} W^{13}(\mathbf{n})\,\mathrm{e}^{i\mathbf{k}.\mathbf{d_n}} \\ M\omega^2 B &= A\sum_{\mathbf{n}} W^{21}(\mathbf{n})\,\mathrm{e}^{i\mathbf{k}.\mathbf{d_n}} + B\sum_{\mathbf{n}} W^{22}(\mathbf{n})\,\mathrm{e}^{i\mathbf{k}.\mathbf{d_n}} + C\sum_{\mathbf{n}} W^{23}(\mathbf{n})\,\mathrm{e}^{i\mathbf{k}.\mathbf{d_n}} \\ M\omega^2 C &= A\sum_{\mathbf{n}} W^{31}(\mathbf{n})\,\mathrm{e}^{i\mathbf{k}.\mathbf{d_n}} + B\sum_{\mathbf{n}} W^{32}(\mathbf{n})\,\mathrm{e}^{i\mathbf{k}.\mathbf{d_n}} + C\sum_{\mathbf{n}} W^{33}(\mathbf{n})\,\mathrm{e}^{i\mathbf{k}.\mathbf{d_n}} \end{aligned}\right\} \tag{26}$$

The fact that we get the *same* equations from a group of any three equations like (23) for *any* value of $\mathbf{n}$ justifies our choice of wave functions. The equations (26) are a set of three homogeneous linear equations in the three quantities A, B, C and have a non-zero solution only if the determinant of the coefficients is zero. This gives a cubic equation for ω^2 which may be written in the form

$$\left| \sum_{\mathbf{n}} W^{rs}(\mathbf{n})\, e^{i\mathbf{k}.\mathbf{d_n}} - M\omega^2 \delta_{rs} \right| = 0 \tag{27}$$

For each of the solutions ω_1, ω_2, ω_3 we obtain a ratio $A:B:C$ corresponding to three states of polarisation of the vibrations propagated through the crystal. Except in certain directions having a high degree of symmetry (cf. § 6.2) these directions of polarisation will not be orthogonal.

We must now see how one of these solutions corresponds to the long-wave acoustic vibrations when the value of k is small. When $k \to 0$ we have a solution in which also $\omega \to 0$; this may be seen as follows. If we have an arbitrary displacement which is the *same* for all values of $\mathbf{n}$ there is no change in the potential energy W, since this corresponds to a simple translation of the crystal; thus we must have

$$\sum_{\mathbf{n}} W^{rs}(\mathbf{n}) = 0 \tag{28}$$

It follows from equation (27) that $\omega \to 0$ as $k \to 0$. Moreover, we may readily show that ω^2 is an even function of $\mathbf{k}$; equation (27) is unchanged by replacing $\mathbf{k}$ by $-\mathbf{k}$ and $\mathbf{d_n}$ by $-\mathbf{d_n}$ since $W^{rs}(\mathbf{n}) = W^{rs}(-\mathbf{n})$ (cf. equation 21)). For small values of k, ω is therefore a linear function of $\mathbf{k}$. We have therefore a constant velocity $\mathbf{v} = \nabla_k \omega$ which corresponds to the velocity of acoustic waves. The three solutions ω_1, ω_2, ω_3 give three velocities c_1, c_2, c_3 which only in exceptional circumstances correspond to longitudinal and transverse acoustic waves.

We may readily deduce another very important property of the function giving the frequency ω as a function of the wave vector $\mathbf{k}$, namely that ω is a periodic function of $\boldsymbol{\varkappa} = \mathbf{k}/2\pi$, having the periodicity of the reciprocal lattice. Suppose we replace $\mathbf{k}$ in equation (27) by a quantity $\mathbf{k}'$ given by the equation

$$\boldsymbol{\varkappa}' = \mathbf{k}'/2\pi = \mathbf{k}/2\pi + \mathbf{b_p} \tag{29}$$

where $\mathbf{b_p}$ is a vector of the reciprocal lattice, i.e.

$$\mathbf{b_p} = p\mathbf{b}_1 + q\mathbf{b}_2 + r\mathbf{b}_3$$

p, q, r being integers. We see at once that equation (27) is unchanged by this substitution since

$$\mathbf{k}'.\mathbf{d_n} = \mathbf{k}.\mathbf{d_n} + 2\pi(pl + qm + rn) \tag{30}$$

and $\exp[(2\pi i)(pl+qm+rn)] = 1$ since $pl+qm+rn$ is an integer. Moreover, the values of the components of $\mathbf{u_n}$ are unchanged for the same reason when $\mathbf{k}$ is replaced by $\mathbf{k}'$ in the wave functions (24); $\mathbf{k}'$ is therefore exactly equivalent to $\mathbf{k}$ and ω is a periodic function of $\boldsymbol{\varkappa} = \mathbf{k}/2\pi$ with the periodicity of the reciprocal lattice. This corresponds exactly to the situation discussed in § 3.3 for the one-dimensional lattice.

6.5 Brillouin zones

It is clear from the above discussion that we may restrict the values of $\boldsymbol{\varkappa} = \mathbf{k}/2\pi$ to a unit cell of the reciprocal lattice since any other value, obtained by a lattice translation, is exactly equivalent. What we have now to decide is to which part of the space mapped out by the reciprocal lattice we may most conveniently restrict the values of $\boldsymbol{\varkappa}$. The basic unit cell would not be suitable since this would restrict $\mathbf{k}$ to values having one sign only and this is clearly inconvenient when dealing with travelling waves; in one dimension we found it more convenient to use the interval $-\pi/d < k \leqslant \pi/d$ rather than $0 < k \leqslant 2\pi/d$ for the same reason. We wish therefore to find a region of $\boldsymbol{\varkappa}$-space which is centred on the origin, whose volume is equal to that of the unit cell and which contains no equivalent points and all non-equivalent points. The region should clearly intersect the axes defined by the basic vectors of the reciprocal lattice in the points $(\pm\frac{1}{2}|b_1|, 0, 0)$, $(0, \pm|b_2|, 0)$, $(0, 0, \pm|b_3|)$ and should intersect them at right angles. The chosen region will therefore be bounded by planes bisecting at right angles the lines joining the origin to the lattice points $(\pm 1, 0, 0)$, $(\pm 0, 1, 0)$, $(0, 0, \pm 1)$. The space enclosed by these planes may in some cases give the required region of $\boldsymbol{\varkappa}$-space, as for the simple cubic lattice discussed in § 6.2 (see Fig. 6.2). For other lattices, however, this region may be intersected by planes bisecting at right angles lines joining the origin to other nearby lattice points; if so, the more restricted region should clearly be used if the smallest values of k which give all the non-equivalent points are to be preferred. Such a zone is generally known as the first Brillouin zone after L. Brillouin* who showed how such zones may be constructed. Further equivalent zones may be obtained from planes bisecting at right angles *all* the lines joining the origin to points of the reciprocal lattice (the origin itself being a lattice point). The *first* Brillouin zone is given by the smallest enclosed polyhedron whose centre is at the origin. If, as is sometimes done, the Brillouin zones are represented in $\mathbf{k}$-space instead of in the space of the reciprocal lattice, their linear dimensions are increased by multiplying by a factor 2π. Since any point of reciprocal space may be obtained by means of a lattice translation from a point in the first Brillouin zone it is clear that the whole of reciprocal lattice space may be filled with figures identical to the first zone and displaced by lattice translations.

Similar zones exist in two dimensions and the first zone is a closed polygon. The second and third zones giving equivalent areas of $\boldsymbol{\varkappa}$-space

* See, for example, L. Brillouin, *Wave Propagation in Periodic Structures*, Dover (1953).

are made up of separate parts. As an illuminating exercise the reader unfamiliar with such zones should construct a number of zones for the simple square lattice; the first two zones are shown in Fig. 6.4. In three dimensions, the first Brillouin zone is most commonly used. We shall give more details of zones for some of the simpler lattices in § 6.8.

In the meantime we shall consider some further properties of the relationship between ω and $\mathbf{k}$ connected with the zone structure. The planes bounding a zone correspond to values of the wavelength

$$\lambda = 2\pi/|\mathbf{k}| = 1/|\boldsymbol{\varkappa}|$$

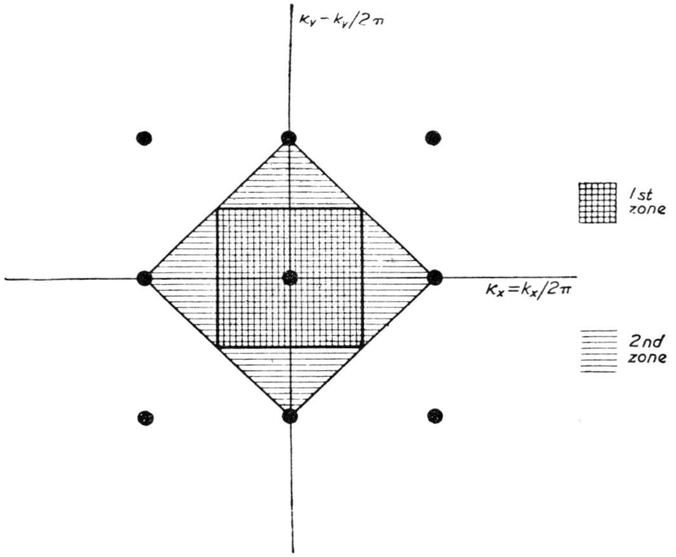

FIG. 6.4. First two Brillouin zones for the simple square lattice

giving strong X-ray reflections as discussed in § 5.3.4. For a wave vector $\mathbf{k}$ such that the end of the vector $\boldsymbol{\varkappa} = \mathbf{k}/2\pi$ lies in such a plane the vector $-\boldsymbol{\varkappa}$ lies in the opposite face of the zone and is equivalent. The wave corresponding to such a wave vector is therefore a standing wave; this follows exactly as for the discussion in § 3.3. It also follows that at the boundary of a zone $\partial\omega/\partial k_n = 0$, provided there is no degeneracy, where $\partial/\partial k_n$ represents differentiation along the normal to the zone. This is also a consequence of continuity of the derivative between equivalent points which are separated by a little more than a zone diameter.

It we plot ω as a function of $\mathbf{k}$ along a particular direction in terms of the magnitude k of the vector $\mathbf{k}$ we shall obtain three curves corresponding to the three solutions ω_1, ω_2, ω_3 as shown in Fig. 6.5. The values k_m correspond to the intersection of a line through the origin in this direction with the face of the first Brillouin zone; Fig. 6.5 represents a direction which meets a face at right angles. It will be seen that each of the curves is similar in form to those found for a one-dimensional crystal. For a direction having a high degree of symmetry two of the curves may meet at the zone boundary and then $\partial\omega/\partial n$ is not necessarily equal to zero.

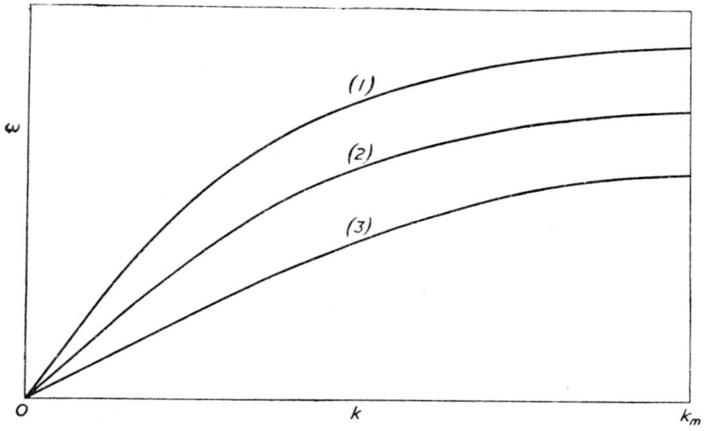

FIG. 6.5. ω/k curves for a particular direction in a three-dimensional crystal with a simple lattice

6.6 Vibration spectrum for crystals having a lattice with a basis and for polyatomic crystals

The analysis of § 6.4 may readily be generalised to deal with a crystal having more than one atom per unit cell of the basic lattice. Suppose we have q such atoms their masses being $M_1, M_2 \ldots M_p \ldots M_q$ situated at the points $\mathbf{a}_1, \mathbf{a}_2 \ldots \mathbf{a}_p \ldots \mathbf{a}_q$ in the unit cell relative to the origin; in the crystal we shall then have atoms of mass M_p at the points $\mathbf{d_n} + \mathbf{a}_p$. If all the atoms are identical we have the case of a crystal having a lattice with a basis provided none of the vectors $\mathbf{a}_p$ are lattice vectors; the diamond lattice represents such a structure. If, on the other hand, we have different types of atom at the sites corresponding to $\mathbf{a}_1 \ldots \mathbf{a}_q$ we have a polyatomic crystal. These two types do not differ greatly as regards their vibrational spectra.

Equations (23) may be generalised to include the different types of atoms; the analysis is straightforward and leads to equations of motion of the form

$$M_p \ddot{u}_{\mathbf{n}p} = -\sum_{\mathbf{n}'p'} W^{11}_{pp'}(\mathbf{n}-\mathbf{n}')\, u_{\mathbf{n}'p'} - \sum_{\mathbf{n}'p'} W^{12}_{pp'}(\mathbf{n}-\mathbf{n}')\, v_{\mathbf{n}'p'}$$
$$- \sum_{\mathbf{n}'p'} W^{13}_{pp'}(\mathbf{n}-\mathbf{n}')\, w_{\mathbf{n}'p'} \qquad (31)$$

and similar equations for $\ddot{v}_{\mathbf{n}p}$ and $\ddot{w}_{\mathbf{n}p}$ where $u_{\mathbf{n}p}$, $v_{\mathbf{n}p}$, $w_{\mathbf{n}p}$ are the three components of the vector displacement $\mathbf{u}_{\mathbf{n}p}$ of the atom at the site denoted by $\mathbf{d_n}+\mathbf{a}_p$. The quantities $W^{12}_{pp'}(\mathbf{n}-\mathbf{n}')$, etc., are the coefficients in the expansion of the potential energy corresponding to interaction between atoms at the sites denoted by $\mathbf{d_n}+\mathbf{a}_p$ and $\mathbf{d_{n'}}+\mathbf{a}_{p'}$. Proceeding as in the one-dimensional case discussed in § 3.7 we solve these equations by means of $3q$ wave functions of the form

$$\left.\begin{aligned} u_p &= A_p \mathrm{e}^{-i\mathbf{k}.\mathbf{d}p}\,\mathrm{e}^{i\mathbf{k}.\mathbf{r}.-i\omega t} \\ v_p &= B_p \mathrm{e}^{-i\mathbf{k}.\mathbf{d}p}\,\mathrm{e}^{i\mathbf{k}.\mathbf{r}-i\omega t} \\ w_p &= C_p \mathrm{e}^{-i\mathbf{k}.\mathbf{d}p}\,\mathrm{e}^{i\mathbf{k}.\mathbf{r}-i\omega t} \end{aligned}\right\} p = 1, 2 \ldots q \tag{32}$$

We note that the wave functions u_p, v_p, w_p give the components of the displacement of the pth atom in the cell corresponding to $\mathbf{d_n}$ on substituting the value $\mathbf{d_n}+\mathbf{a}_p$ for $\mathbf{r}$. On substituting in equations (31) we find that we obtain a set of linear homogeneous equations for the $3q$ quantities A_p, B_p, C_p $(p = 1\ldots q)$ similar to equations (26); they may be written in the form

$$\begin{aligned} M_p A_p \omega^2 = & \sum_{\mathbf{n}p'} A_{p'} W^{11}_{pp'}(\mathbf{n})\,\mathrm{e}^{i\mathbf{k}.\mathbf{d_n}} + \sum_{\mathbf{n}p'} B_{p'} W^{12}_{pp'}(\mathbf{n})\,\mathrm{e}^{i\mathbf{k}.\mathbf{d_n}} \\ & + \sum_{\mathbf{n}p'} C_{p'} W^{13}_{pp'}(\mathbf{n})\,\mathrm{e}^{i\mathbf{k}.\mathbf{d_n}} \quad (p = 1, \ldots q) \end{aligned} \tag{33}$$

and two similar sets having each q equations. We have in all $3q$ equations and it will be seen that there are $3q$ terms in the quantities A_p, B_p, C_p on the right-hand side of each equation. The relationship between ω and $\mathbf{k}$ is obtained by equating to zero the determinant of the coefficients; in this way we obtain an equation of order $3q$ in ω^2 which may be written in the form

$$\left| \sum_{\mathbf{n}} W^{rs}_{pp'}(\mathbf{n})\,\mathrm{e}^{i\mathbf{k}.\mathbf{d_n}} - M_p \omega^2 \delta_{rs}\delta_{pp'} \right| = 0 \tag{34}$$

$r, s = 1, 2, 3$, $p = 1, 2 \ldots q$; thus we have, in general, $3q$ solutions. Three of these are of special interest when $k \simeq 0$. For these the quantities A_p, have the same value for all values of p, as do the quantities B_p, C_p, i.e. all the atoms in each cell move in the same direction. These correspond to the long-wave acoustic vibrations. It may be shown as before that

$$\sum_{p\mathbf{n}} W^{rs}_{pp'}(\mathbf{n}) = 0 \tag{35}$$

and it follows from equations (33) that $\omega \to 0$ as $k \to 0$. For the other $3q-3$ solutions ω does not tend to zero as $k \to 0$ and we have optical modes similar to those discussed for the one-dimensional case in § 3.6.

Of particular interest is the case when $q = 2$; here we have six solutions, three giving acoustical modes and three optical modes. The general form of the ω/k curves for a particular direction in the crystal when $q = 2$ is shown in Fig. 6.6. For a lattice with a high degree of symmetry, each group of three curves may coincide entirely for some directions or may meet at points having a specially high order of symmetry; for example, the curves for the diamond lattice in the ⟨100⟩ directions are shown in Fig. 6.7. The curves labelled 1, 2, 3, 4 correspond, respectively, to the

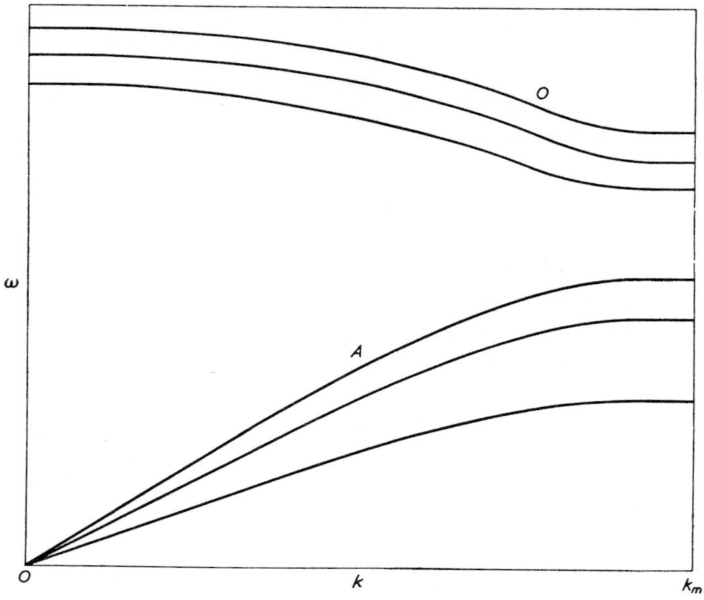

FIG. 6.6. ω/k curves for a crystal with two atoms per unit cell

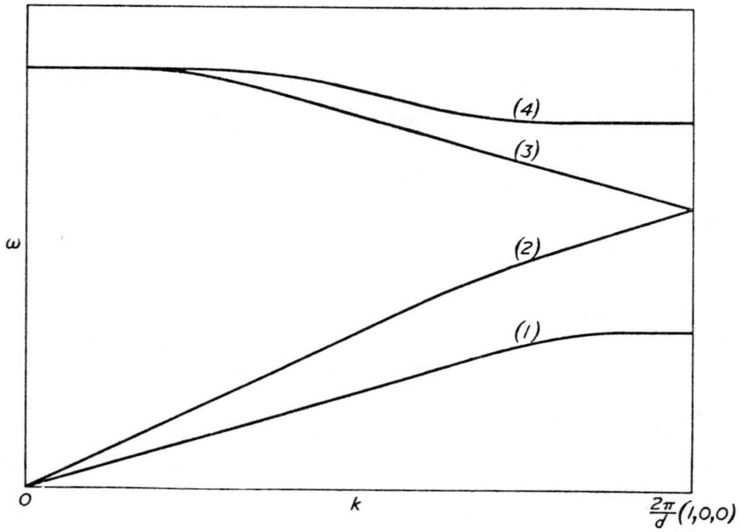

FIG. 6.7. ω/k curves for the ⟨100⟩ directions of the diamond lattice (after Helen M. J. Smith, *Philos. Trans.* **A241**, 105 (1948))

transverse acoustic, longitudinal acoustic, longitudinal optical and transverse optical modes. Fig. 6.7 illustrates another interesting point concerning the general form of the curves. When we showed that $\partial\omega/\partial k_n = 0$ at a zone edge we implicitly assumed that ω was a single-valued function of $\mathbf{k}$; when this is not so it is no longer necessary for $\partial\omega/\partial k_n$ to be zero at the zone edge, as shown also for one dimension in § 4.12, and this case occurs for the curves 2 and 3 at the zone edge.

6.7 Boundary conditions and normal modes

We should perhaps point out again that the reciprocal lattice corresponding to a crystal having a lattice with a basis is just the reciprocal of the simple lattice from which the crystal is formed, without the basis; the Brillouin zones are therefore just the same as those for the simple lattice. This corresponds to the condition we had in one dimension in which the periodicity of the ω/k curves is $2\pi/d$ where d is the distance corresponding to the repetition of a *group* of atoms. For example, the Brillouin zone for the caesium chloride structure is a cube of side $1/d$ centred on the origin of $\boldsymbol{\varkappa}$-space, as shown in Fig. 6.2, where d is the shortest distance between *like* atoms in the structure; if we represent the first Brillouin zone in $\mathbf{k}$-space it is a cube of side $2\pi/d$.

The normal modes of vibration are in general determined by the application of boundary conditions as already discussed for the simple cubic lattice in § 6.2.1. Periodic boundary conditions, in the general case, are applied by assuming that we have a crystal in the form of a parallelepiped whose edges are parallel to the basic lattice vectors $\mathbf{d}_1, \mathbf{d}_2, \mathbf{d}_3$ and are of lengths Ld_1, Md_2, Nd_3, L, M, N being large integers. For the solutions of the type (24) or (32) to be periodic with periods corresponding to a repetition of this parallelepiped we must have

$$e^{i\mathbf{k}.\mathbf{r}} = e^{i\mathbf{k}.(\mathbf{r}+L\mathbf{d}_1)} = e^{i\mathbf{k}.(\mathbf{r}+M\mathbf{d}_2)} = e^{i\mathbf{k}.(\mathbf{r}+N\mathbf{d}_3)}$$

and this will be so if

$$\left.\begin{aligned} \mathbf{k}.\mathbf{d}_1 &= 2\pi l/L \\ \mathbf{k}.\mathbf{d}_2 &= 2\pi m/M \\ \mathbf{k}.\mathbf{d}_3 &= 2\pi n/N \end{aligned}\right\} \tag{36}$$

where l, m, n are integers. If we write $\mathbf{k}$ and $\boldsymbol{\varkappa}$ in terms of the vectors of the reciprocal lattice in the form

$$\begin{aligned} \mathbf{k} &= \xi\mathbf{b}_1 + \eta\mathbf{b}_2 + \zeta\mathbf{b}_3 \\ \boldsymbol{\varkappa} &= \xi'\,\mathbf{b}_1 + \eta'\,\mathbf{b}_2 + \zeta'\,\mathbf{b}_3 \end{aligned}$$

we must have,

$$\left.\begin{aligned} \xi' &= l/L = \xi/2\pi \\ \eta' &= m/M = \eta/2\pi \\ \zeta' &= n/N = \xi/2\pi \end{aligned}\right\} \tag{37}$$

Inside the first Brillouin zone there are therefore LMN normal modes for *each* solution, i.e. a number equal to the number of unit cells in the crystal; these represent the normal modes. If we have q atoms per unit cell we have seen that we have $3q$ solutions and therefore $3LMNq = 3N_a$ normal modes, where N_a is the total number of atoms in the crystal.

It is interesting to calculate the density of allowed values of $\mathbf{k}$. If we increase l, m and n each by one unit we account for a volume V_b/LMN of $\boldsymbol{\varkappa}$-space, i.e. $8\pi^3 V_b/LMN$ of $\mathbf{k}$-space, where V_b is the volume of the unit cell of the reciprocal lattice. The density is therefore equal to

$$V_b^{-1} LMN/8\pi^3 = V_d LMN/8\pi^3$$

where V_d is the volume of the unit cell of the direct lattice, since $V_d = V_b^{-1}$. The density is therefore equal to $V/8\pi^3$ where V is the volume of the crystal, the allowed values being distributed uniformly in $\mathbf{k}$-space over the first Brillouin zone; this is a very important result.

6.8 Brillouin zones for the cubic and hexagonal lattices

We have already discussed in § 6.2 the first Brillouin zone for the simple cubic lattice (Fig. 6.2) and have given in § 6.5 the method of finding the first Brillouin zone for any lattice; we shall now find the first Brillouin zones for some of the lattices discussed in § 5.3. Let us consider first the face-centred cubic lattice, the reciprocal lattice of which is (see § 5.4.4) a body-centred cubic lattice the edge of the unit cube having a length $2/d$ where d is the length of the edge of the unit cube of the direct lattice. Taking one of the points of the reciprocal lattice as the origin O and centre of the zone, the eight nearest lattice points are $(1/d)(1,1,1)$ $(1/d)(-1,1,1)$ etc. To obtain the first Brillouin zone we form the closed figure obtained from the planes which bisect the lines joining the origin to these eight nearest lattice points, i.e. the $\{111\}$ planes at distance $\sqrt{3}/2d$ from the origin; these eight planes enclose a figure in the form of a regular octahedron. This, however, is not the first Brillouin zone since the planes arising from the next nearest neighbours intersect it. The plane faces of the octahedron meet on the axes at points distant $3/2d$ from the origin. The next nearest neighbours are at six points such as $(1/d)(2,0,0)$ and the bisecting planes are $\{100\}$ planes distant $1/d$ from the origin; they therefore intersect the regular octahedron in six square faces (see Fig. 6.8). The resulting figure is the first Brillouin zone; it is known as a truncated octahedron and has six square faces and eight faces which are regular hexagons. It is shown in Fig. 6.8 together with some of the points of the reciprocal lattice; the zone is drawn in terms of $\boldsymbol{\varkappa}$ and to obtain it in terms of the wave vector $\mathbf{k}$ its linear dimensions must be increased by a factor 2π. Points on the surface of the zone having a high degree of symmetry are A, the point at the centre of one of the square faces, corresponding to a wave vector $\mathbf{k} = (2\pi/d)(1,0,0)$, B the point at the centre of one of the hexagonal faces corresponding to a wave vector $\mathbf{k} = (\pi/d)(1,1,1)$

and C the mid-point of an edge common to two hexagonal faces and corresponding to a wave vector $\mathbf{k} = (\pi/d)(\frac{3}{2}, \frac{3}{2}, 0)$. The volume of the first zone may readily be shown to be equal to half the volume of the inscribing cube formed by the points $(1/d)(\pm 1, \pm 1, \pm 1)$ (see Fig. 6.8); it has therefore the value $4/d^3$. Now the volume of the unit cell of the direct lattice is $\frac{1}{4}d^3$ (four points per unit cube) so we check that the zone is indeed equivalent to a unit cell of the reciprocal lattice, its volume being equal to the reciprocal of that of the direct lattice. To obtain higher-order

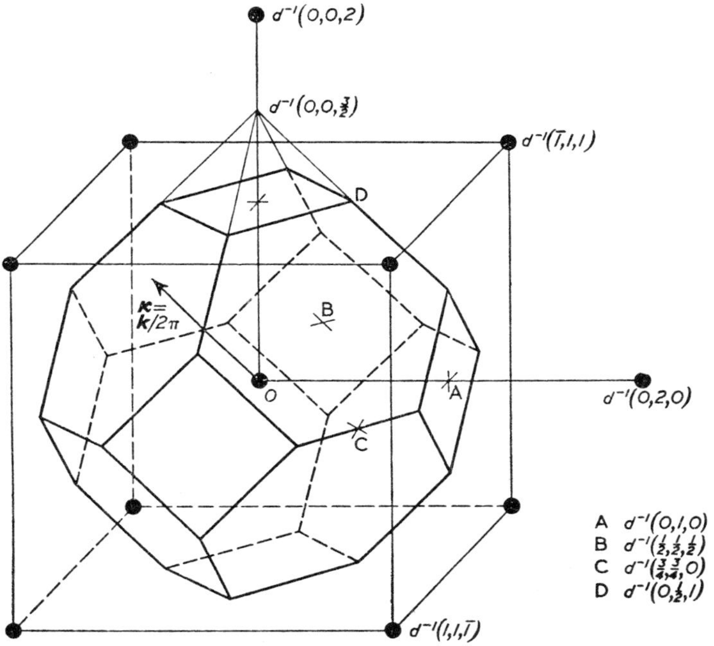

FIG. 6.8. First Brillouin zone for the face-centred cubic lattice. Some points of the reciprocal lattice are also shown, marked ●

Brillouin zones we must introduce bisecting planes for more distant neighbours. The figures obtained are rather complicated but are not frequently required, since it is usual to restrict $\boldsymbol{\varkappa}$ (and hence $\mathbf{k}$) to the first zone. It is interesting to note how a series of truncated octahedra may be stacked together to fill up the whole of reciprocal lattice space. (This can best be seen by use of models.)

The body-centred cubic lattice has for its reciprocal lattice (see § 5.4.4) a face-centred cubic lattice, the side of the unit cube being $2/d$ where d is the length of the edge of the unit cube of the direct lattice. To form the first Brillouin zone we take one of the points of the reciprocal lattice as origin and consider first the twelve nearest neighbours at the points

$(1/d)(1,1,0)$, etc. The bisecting planes are the twelve $\{110\}$ planes at distance $1/\sqrt{2}d$ from the origin. These planes make a closed figure in the form of a regular dodecahedron having twelve equal faces each in the form of a rhombus. By writing down the equations of the planes in three groups of four each, it may readily be shown that they intersect in fours on the axes at points distant $1/d$ from the origin and in threes at points in the $\langle 111\rangle$ directions distant $\sqrt{3}/2d$ from the origin. The next nearest points of the lattice are at points such as $(1/d)(2,0,0)$ so that the bisecting planes from these do not intersect the figure, which therefore gives the surface of the first Brillouin zone. It is shown in Fig. 6.9 together with some points of the reciprocal lattice; again it is drawn in terms of $\boldsymbol{\varkappa}$ and to

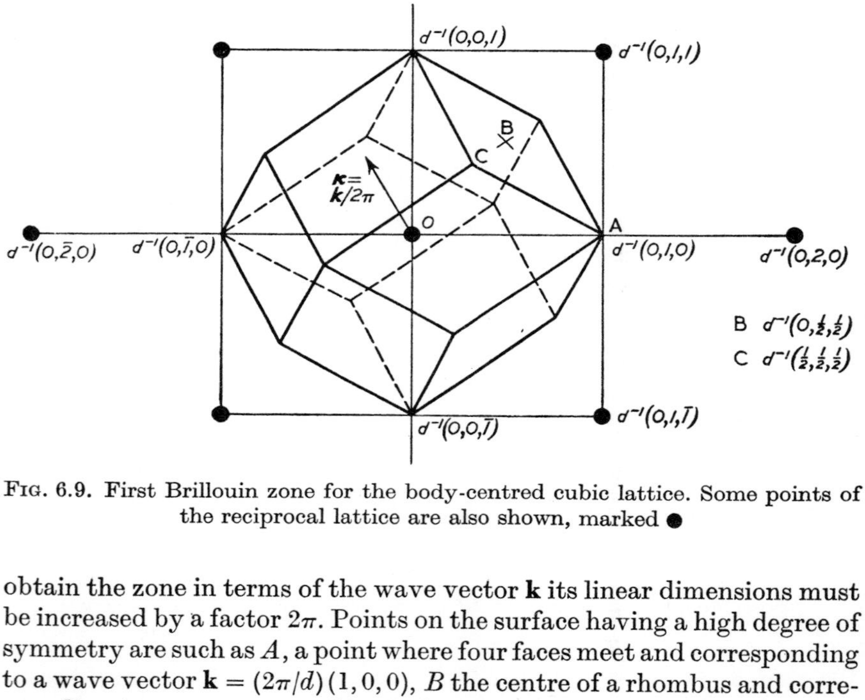

FIG. 6.9. First Brillouin zone for the body-centred cubic lattice. Some points of the reciprocal lattice are also shown, marked ●

obtain the zone in terms of the wave vector **k** its linear dimensions must be increased by a factor 2π. Points on the surface having a high degree of symmetry are such as A, a point where four faces meet and corresponding to a wave vector $\mathbf{k} = (2\pi/d)(1,0,0)$, B the centre of a rhombus and corresponding to a wave vector $\mathbf{k} = (\pi/d)(0,\frac{1}{2},\frac{1}{2})$, and C a point where three faces meet corresponding to a wave vector $\mathbf{k} = (\pi/d)(\frac{1}{2},\frac{1}{2},\frac{1}{2})$. The volume of the first zone may be shown to be only one-quarter of that of the inscribing cube of side $2/d$ and so is equal to $2/d^3$. This is equal to the reciprocal of the volume of the unit cell of the direct lattice since there are, in this case, two points per unit cube. It may readily be shown that the acute angle of the rhombus forming each face (at the points where four faces meet) is $2\tan^{-1}(1/\sqrt{2})$. To obtain higher-order zones again more distant points must be considered. The second zone in this case has for its outer surface the truncated octahedron shown in Fig. 6.8 whose volume we have seen to be equal to $4/d^3$. The space between the two figures makes up the second zone and has volume $2/d^3$ as required.

The diamond structure, being a lattice with a basis derived from the face-centred cubic lattice, (see §§ 5.3.1, 5.4.4) has the same reciprocal lattice and the same Brillouin zones as the latter. In this case we have two atoms per unit cell and so that if Ω is the atomic volume, i.e. the volume per atom of the crystal, the unit cell has volume 2Ω and the first Brillouin zone volume $1/2\Omega$.

The close-packed hexagonal lattice is also a lattice with a basis (see § 5.3.2) and has the same reciprocal lattice and Brillouin zones as a simple

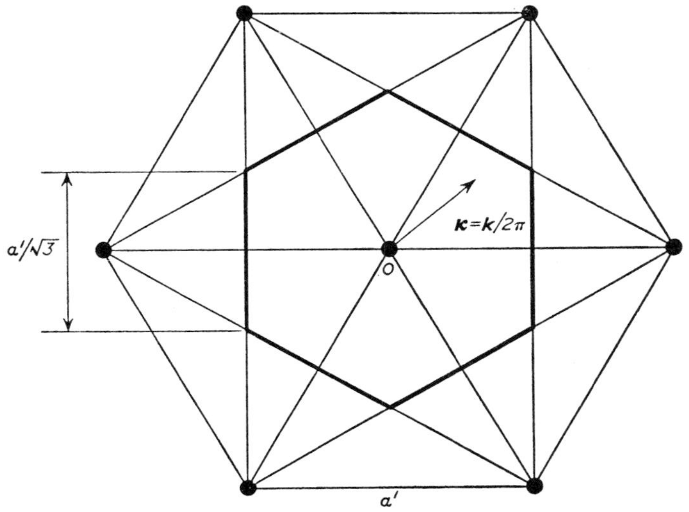

FIG. 6.10. Construction of the first Brillouin zone for the plane simple hexagonal lattice. Points marked ● are points of the reciprocal lattice

hexagonal lattice (see § 5.4.5). If c is the spacing between hexagonal planes and a the side of each unit hexagon, the reciprocal lattice consists of a simple hexagonal lattice with constants c' and a' where $c' = 1/c$ and $a' = 2/\sqrt{3}a$. The hexagons of the reciprocal lattice are turned through 90° with respect to those of the direct lattice. In Fig. 6.10 seven points of the reciprocal lattice of a plane simple hexagonal lattice are shown. When the central point O is taken as origin the other six lie at the corners of a regular hexagon of side $a' = 2/\sqrt{3}a$, and are the six nearest neighbours. The first Brillouin zone is formed by drawing the bisectors of the lines joining O to the other six points. The zone will be seen to be a hexagon whose sides have length $a'/\sqrt{3} = 2/3a$ and hence with area equal to that of the unit cell of the reciprocal lattice, namely $\sqrt{3}a'^2/2$; this is also equal to $2/\sqrt{3}a^2$ and so is equal to the reciprocal of the area of the unit cell of the direct lattice. It should be noted that the hexagon forming the first Brillouin zone has the *same* orientation as the hexagons of the direct lattice. The whole of reciprocal space may be filled with such hexagons, each of which may be regarded as a type of unit cell. (We have already discussed this type of cell in § 5.2.2.)

The first Brillouin zone of the three-dimensional simple hexagonal lattice is clearly obtained by introducing planes parallel to the c-axis through the sides of the hexagon giving the first zone for the plane lattice

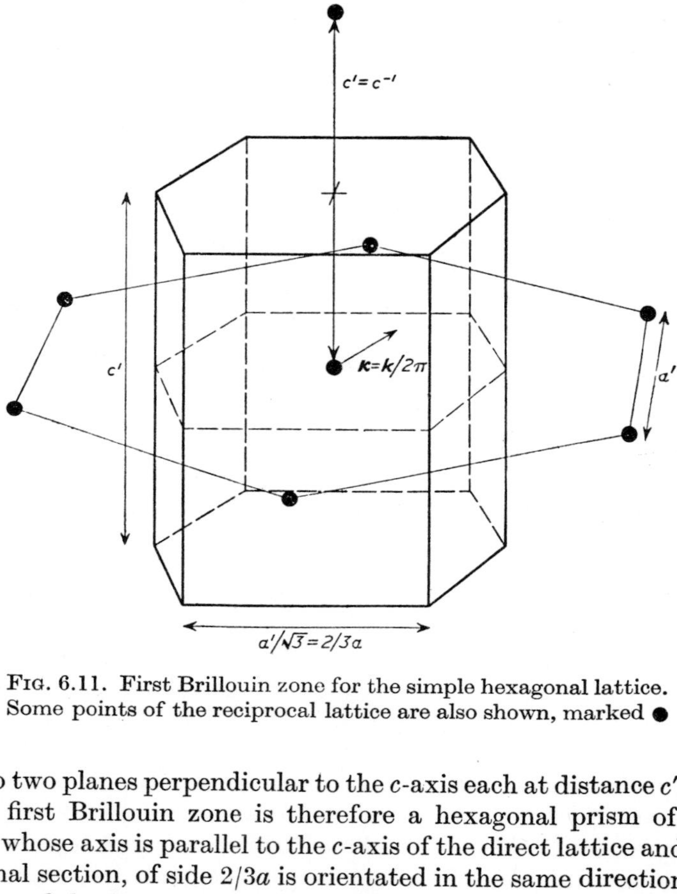

FIG. 6.11. First Brillouin zone for the simple hexagonal lattice. Some points of the reciprocal lattice are also shown, marked ●

and also two planes perpendicular to the c-axis each at distance $c'/2$ from O. The first Brillouin zone is therefore a hexagonal prism of height $c' = 1/c$ whose axis is parallel to the c-axis of the direct lattice and whose hexagonal section, of side $2/3a$ is orientated in the same direction as the hexagons of the direct lattice. Its volume is $2/\sqrt{3}a^2c$ and so is equal to the reciprocal of the volume of the unit cell of the direct lattice. The first zone is shown in Fig. 6.11, together with some points of the reciprocal lattice. For the close-packed hexagonal lattice we have $c = 2\sqrt{2}a/\sqrt{3}$ so that the height of the prism forming the first Brillouin zone is $\sqrt{3}/2\sqrt{2}a$ and the volume of the zone is $1/\sqrt{2}a^3$. For a monatomic crystal having the close-packed hexagonal lattice there are two atoms per unit cell so that the volume of the first Brillouin zone is equal to half the reciprocal of the atomic volume.

It is interesting to compare the face-centred cubic lattice and close-packed hexagonal lattice. They have the same atomic volume if we take $a = d/\sqrt{2}$ where d is the length of the side of the unit cube for the former. The volume of the unit cell of the hexagonal lattice is then $\sqrt{2}a^3 = d^3/2$ whereas for the cubic lattice it is $d^3/4$. The unit cell of the hexagonal lattice contains two atoms so the volume per atom is the same. The volume of the first Brillouin zone of the cubic lattice is $4/d^3$, whereas, that for the hexagonal lattice is $2/d^3$, i.e. only half that of the cubic lattice.

6.9 Quantization of the lattice vibrations

We have given in §§ 6.4, 6.6 a general classical treatment of the vibrations of the atoms of a crystal about their positions of equilibrium. For real crystals this is inadequate for many purposes and must be supplemented by a treatment according to quantum mechanics. Fortunately the quantum theory may be derived from the classical treatment by elementary methods in exactly the same way as we have shown in § 3.4 for a linear crystal. We have effectively found the normal modes of oscillation by means of the classical treatment and in terms of these the whole system may be resolved into a set of independent harmonic oscillators, each of which may then be treated by means of quantum mechanics as a simple harmonic oscillator.

The normal frequencies ω_s of the system are obtained from the $\omega/\mathbf{k}$ relationships when we know the allowed values $\mathbf{k}_s$ of the wave vector $\mathbf{k}$; these are determined only by the boundary conditions as shown in § 6.2.1. For a crystal with a simple lattice containing N unit cells the normal frequencies will be given by the $3N$ values ω_{sj}, $s = 1 \ldots N$, $j = 1, 2, 3$, corresponding to the values of ω on the three branches of the ω/k curves corresponding to the N allowed values of $\mathbf{k}$ (see § 6.7) in the first Brillouin zone. For periodic boundary conditions these will be degenerate in pairs. For more complex crystals j will take a greater number of values corresponding to the greater number of branches of the ω/k curves; for a crystal with q atoms per unit cell, for example, j will take the values from 1 to $3q$, corresponding to the three acoustic branches and $3q-3$ optical branches of the ω/k curves.

As before, we see that the energy E_{sj} of the normal mode corresponding to the frequency ω_{sj} is given by

$$E_{sj} = \hbar\omega_{sj}(n_{sj}+\tfrac{1}{2}) \tag{38}$$

where n_{sj} is an integer. The total energy $E(n_{sj})$ of the system can therefore take the values

$$E[(n_{sj})] = \hbar \sum_{sj} \omega_{sj}(n_{sj}+\tfrac{1}{2}) \tag{39}$$

where (n_{sj}) represents a set of $3qN$ integers.

As for the linear crystal, we define E_0, the zero-point energy by means of the equation

$$E_0 = (\hbar/2) \sum_{sj} \omega_{sj} \tag{40}$$

We may also express the displacements $u_{\mathbf{n}p}$, $v_{\mathbf{n}p}$, $w_{\mathbf{n}p}$ (see § 6.6) in terms of normal coordinates q_{sj} representing travelling waves and corresponding to the normal frequencies ω_{sj} in exactly the same way as for the linear crystal (see § 3.4.1). The expressions for the normal coordinates in terms of the displacements are somewhat complex and will not be given but are exactly analogous to the equation (82) of § 3.4.1, the summation being taken over all values of the three components of the displacements. In terms of the normal coordinates the total energy H may be expressed in the form

$$H = \tfrac{1}{2} \sum_{sj} M_j \dot{q}_{sj}^2 + \tfrac{1}{2} \sum_{sj} M_j \omega_{sj}^2 q_{sj}^2 \tag{41}$$

and q_s satisfies the equations of motion

$$\ddot{q}_{sj} + \omega_{sj}^2 q_{sj} = 0 \tag{42}$$

M_j being the mass of the jth particle in the unit cell. We shall not give the detailed derivation of these equations since it follows exactly as in § 3.4.

6.9.1 Mean energy as a function of temperature

We may now express the mean energy as a function of the absolute temperature T; we simply have to find the average value $\bar{E}$ of $E[(n_{sj})]$; we proceed exactly as in § 3.5 for a linear crystal. The probability $P[(n_{sj})]$ that the system is in the state having energy $E[(n_{sj})]$ is proportional to $\exp\{-E[(n_{sj})]/\boldsymbol{k}T\}$. Introducing a normalising factor as before and writing

$$\bar{E} = \sum_{(n_{sj})} P[(n_{sj})]E[(n_{sj})] \tag{43}$$

we obtain

$$\bar{E} = E_0 + \sum_{sj} \frac{\hbar\omega_{sj}}{e^{\hbar\omega_{sj}/\boldsymbol{k}T} - 1} \tag{44}$$

where E_0 is the zero-point energy, given by equation (40). In the classical limit when $\boldsymbol{k}T \gg \hbar\omega_{sj}$ for all values of ω_{sj} we have, expanding the term under the summation sign,

$$\bar{E} = E_0 + \sum_{sj} \hbar\omega_{sj}\left[\frac{\boldsymbol{k}T}{\hbar\omega_{sj}} - \tfrac{1}{2} + \ldots\right] \tag{45}$$

To this order of approximation we have therefore

$$\bar{E} = \boldsymbol{k}T \sum_{sj} 1 = 3qN\boldsymbol{k}T = 3N_a \boldsymbol{k}T \tag{46}$$

where N_a is the total number of atoms in the crystal. It is interesting to note that in this approximation the constant term giving the zero-point energy is cancelled out; this is just the classical result, the energy per atom of the crystal being $3kT$. We shall discuss in the next chapter the evaluation of the sum in equation (44) when this approximation no longer holds.

6.9.2 Phonons

The quantized travelling waves corresponding to the various normal modes have many of the properties of particles which obey Einstein–Bose statistics and are known as phonons; the discussion of these given in § 3.10 for a linear crystal holds equally well for a three-dimensional crystal. We must now distinguish, however, between acoustic phonons and optical phonons, these being associated, respectively, with the acoustic and optical branches of the ω/k curves. The energy E_p associated with a phonon is, as before, $\hbar\omega$, where ω is the frequency; the momentum $\mathbf{p}_s$ is now a vector and is given by $\mathbf{p}_s = \hbar\mathbf{K}_s$ where $\mathbf{K}_s$ is the wave vector associated with the phonon, which may be regarded as a particle having this momentum and travelling in the direction of the vector $\mathbf{K}_s$. We shall later see that this momentum is conserved during interactions with electrons. For certain directions in the crystal we shall be able to divide the phonons into transverse and longitudinal phonons corresponding to transverse and longitudinal modes of oscillation; we have seen, however, that this is so only for these special directions.

6.10 Scattering of neutrons by lattice vibrations

The inelastic scattering of neutrons from single crystals has recently given the first direct observation of the relationship connecting the angular frequency ω of the lattice vibrations and the wave vector $\mathbf{K}_s$. Such a relationship may also, in principle, be deduced from the scattering of X-rays but the great disparity between the energy of a quantum of X-radiation and of the phonons makes the method very difficult to apply. For thermal neutrons on the other hand the energies are comparable and the wavelengths associated with the wave-mechanical description of the neutrons are of the order of the lattice spacing of the crystal.

Owing to the weak interaction between the neutrons and the crystal they may be regarded as free particles, and their wave functions ψ will be of the form

$$\psi = A\,e^{i\mathbf{k}.\mathbf{r}} \tag{47}$$

The wavelength λ associated with a neutron will be given by $\lambda = 2\pi/k$, the wave vector $\mathbf{k}$ being related to the velocity vector $\mathbf{v}$ by the equation

$$\hbar\mathbf{k} = M_n\mathbf{v} \tag{48}$$

where M_n is the mass of the neutron (cf. § 1.5.1 where these relationships are given for free electrons). Each side of equation (48) is an expression for the momentum of the neutron.

If a monoenergetic beam of such neutrons is incident on a single crystal it will be strongly reflected at the Bragg angles just as for X-rays—and if each neutron is elastically scattered the wavelength associated with the reflected beam of neutrons will be the same as for the incident beam. If $\mathbf{k}'$ is the wave vector for the reflected beam we shall have $k = k'$ ($k = |\mathbf{k}|$, etc.). Moreover, the condition for Bragg reflection is

$$\mathbf{k}' - \mathbf{k} = 2\pi\mathbf{b} \tag{49}$$

where $\mathbf{b}$ is a vector of the reciprocal lattice. (This equation is derived for X-ray scattering in § 5.4.6.) As for X-rays, this enables one to use a crystal as a spectrometer or monochromator, i.e. either to analyse a complex group of neutrons or to provide a monoenergetic beam.

In addition to the coherent elastic scattering, neutrons may be scattered after creating or absorbing one or more phonons, the single-phonon process being by far the most important. For slow neutrons, only the absorption process may be possible if the neutrons have insufficient energy to create a phonon of the appropriate wavelength. When the scattering is coherent we must expect a relationship similar to (49) to hold, but before discussing this let us consider equation (49) in terms of the change of momentum on scattering. If we multiply this equation by $\hbar$, the right-hand side gives the change in momentum on scattering of a single neutron, and this is equal to $h\mathbf{b}$. If this is added to the momentum $\hbar\mathbf{K}_s$ of any phonon it merely moves the vector $\mathbf{K}_s$ to an equivalent point in $\mathbf{K}$-space and so does not change its energy. Thus for coherent elastic scattering, momentum is transferred to the crystal in amounts $h\mathbf{b}$. If, however, a phonon with wave vector $\mathbf{K}_s$ is absorbed, the momentum transfer would be reduced by an amount $\hbar\mathbf{K}_s$, so that we should have

$$\mathbf{k}' - \mathbf{k} = 2\pi\mathbf{b} + \mathbf{K}_s \tag{50}$$

as we shall see in §§ 10.9.2, 13.5.2, a relationship of this form always holds when phonon absorption or creation takes place. In addition to equation (48) we also have from conservation of energy

$$\frac{\hbar^2 k^2}{2M_n} = \frac{\hbar^2 k'^2}{2M_n} - \hbar\omega \tag{51}$$

where ω is the angular frequency of the absorbed phonon.

If a monoenergetic beam of neutrons falls on a single crystal it is found that various weak groups of neutrons may be observed in addition to the main reflected beam and may be identified as the inelastically scattered neutrons. If these are analysed by means of a spectrometer, their energy (and hence k') may be determined and from this, and from their direction of flight, the vector k' may be found. The value of k' immediately gives the frequency ω of the phonon. If $k' > k$ a phonon has been absorbed, but if $k' < k$ a phonon has been emitted and we must replace $-\mathbf{K}_s$ by $\mathbf{K}_s$ in equation (50) and $\hbar\omega$ by $-\hbar\omega$ in equation (51). From the value of $\mathbf{k}'$ the wave vector $\mathbf{K}_s$ may be determined from equation (50) apart from the uncertainty as to which vector $\mathbf{b}$ of the reciprocal lattice is to be taken. By using different groups of incident neutrons and by trial of a small number of possible values of **b** required to give a consistent interpretation, the actual value of **b** for each group may be found. This method has been used successfully by B. N. Brockhouse and A. T. Stewart* to study metal crystals such as Al and also by B. N. Brockhouse and P. K. Iyengar† to obtain a set of $\omega/\mathbf{K}_s$ curves for Ge for some of the high-symmetry directions in the crystal. This is clearly a very powerful method of investigating lattice vibrations and is likely to provide a great deal of much more detailed information on their spectra and dispersion laws than has been available in the past.

6.11 Phenomena due to lattice vibrations

One of the main applications of a knowledge of the properties of lattice vibrations is to a study of the specific heats of crystalline solids. The lattice specific heat is deduced from the mean energy content of the lattice due to the vibrations, as discussed in § 6.9.1; we shall defer a discussion of specific heat to the next chapter. Other quantities which may be deduced from a study of lattice vibrations are the thermal expansion coefficient and thermal conductivity; these arise from second-order effects and are rather more difficult to treat than the specific heat. If we expand the potential energy as in equation (19) of § 6.4 only as far as the second-order term and use this to calculate the mean value of the displacements of the atoms from their positions in static equilibrium we clearly obtain the average value zero. (This average is obtained by means of a Boltzmann factor $\exp[-W/kT]$ as weighting factor.) If, however, we expand W as far as third-order terms (the so-called anharmonic terms) this is no longer so and we find an average displacement proportional to the absolute temperature T. The numerical value of the resulting constant expansion coefficient is difficult to calculate and we shall not give details of such calculations.

Another effect of the so-called anharmonic terms in the expansion of the energy in terms of the atomic displacements is a contribution to the thermal conductivity. The lattice vibrations, as we have treated them so far, would travel without attenuation through the crystal. The anharmonic terms, however, introduced as a perturbation, cause transitions between the various modes and so lead to scattering; scattering is also caused by impurities and lattice defects. One of the main difficulties with treatments of thermal conductivity arises from the fact that long-wave phonons make a large contribution and these are strongly affected by the surface of the crystal. The result is that thermal conductivity is a very complex phenomenon and a full theoretical study is beyond the scope of this book. The effect of the lattice vibrations on thermal conductivity has recently been discussed in detail in a review article by P. G. Klemens‡.

* B. N. Brockhouse and A. T. Stewart, *Rev. Mod. Phys.*, **30**, 236 (1958).

† B. N. Brockhouse and P. K. Iyengar, *Phys. Rev.* **111**, 747 (1958).

‡ P. G. Klemens, *Solid State Phys.*, **7**, 1 (1958).

The broadening of X-ray lines due to lattice vibrations has been extensively studied, and the effect of high-energy optical phonons on the Raman scattering of light by crystals is well known. Recently, the absorption and emission of phonons during the process of absorption of infra-red radiation by electrons in crystalline solids in the form of semiconductors has become of great importance; this we shall discuss in § 13.5.2. The absorption of infra-red radiation directly by the lattice vibrations themselves gives the strong *reststrahl* absorption band for polar crystals. The complex lattice vibration spectra for non-polar crystals involving creation of two or more phonons have also been studied and on analysis give values of the more important phonon energies at the zone boundary. Values derived in this way agree very well with those derived from neutron scattering experiments*.

* F. A. Johnson, *Proc. Phys. Soc.*, **73**, 265 (1959).

7

Specific Heat of Crystalline Solids due to Lattice Vibrations

7.1 Lattice specific heat

FROM our discussion in § 6.9.1 of the mean energy of the lattice vibrations we may proceed to calculate the contribution to the specific heat of a crystalline solid due to the lattice vibrations. We shall later show that this is by far the largest contribution, that due to free electrons being, in general, quite small. In the classical limit, for a monatomic crystal, the atomic heat at constant volume C_v is obtained from equation (46) of § 6.9.1 and is given by

$$C_v = \frac{\partial \bar{E}}{\partial T} = 3Nk = 3R \tag{1}$$

if we take $N_a = N$ the number of atoms per mole (Avogadro's number); R is the gas constant equal to Nk. This is the well-known law of Dulong and Petit and holds quite well at high temperatures for monatomic crystalline solids. For crystals containing different types of atoms we have to modify this result; we have then a contribution $3k$ for each atom present; the molecular heat is thus equal to the sum of the atomic heats of the constituents. When we have molecules at the lattice sites the situation is more complex, since the vibrational and rotational energy of the molecules has to be taken into account. We shall only discuss crystals with atoms at the lattice sites.

It is well known that at low temperatures the specific heat is much smaller than the value given by equation (1) and that it tends quite rapidly to zero as T tends to zero. In order to explain this it is necessary to use the quantum theory, and indeed this was one of the first outstanding successes of this theory as applied by A. Einstein* in 1907 and 1911.

7.2 Einstein's approximation

Einstein used a very much simplified model to describe the lattice vibrations of a crystal; he assumed that each atom moves in the same field of force and vibrates about its mean position with a single frequency ν. Each atom has three degrees of freedom so the crystal is then equivalent

* A. Einstein, *Ann. Phys.*, **22**, 180 (1907); *Ibid.*, **34**, 170 (1911).

to $3N_a$ harmonic oscillators. The energy of each oscillator given by early quantum theory is $nh\nu$ where n is an integer and the mean energy $\bar{E}_a$ by

$$\bar{E}_a = \frac{h\nu}{e^{h\nu/kT} - 1} \tag{2}$$

(this is the same as for quantum mechanics apart from the term giving the zero-point energy). For the atomic heat C_v we have therefore

$$\left.\begin{aligned} C_v &= 3N\frac{\partial \bar{E}_a}{\partial T} \\ &= 3N\frac{h^2\nu^2}{kT^2}\frac{e^{h\nu/kT}}{(e^{h\nu/kT}-1)^2} \end{aligned}\right\} \tag{3}$$

This certainly tends to zero as T tends to zero but does so as

$$T^{-2}\exp(-h\nu/kT)$$

this is too rapid a decrease to agree with experiment which shows that, for sufficiently small values of T, C_v tends to zero as T^3. To obtain this result we must consider the lattice vibrations to have a frequency distribution in terms of the wave number* **K**. The first attempt to do this was made by P. Debye† in 1912.

7.3 Debye's approximation for crystals with one atom per unit cell

Debye was not able at that time to use a calculated relationship between ω and **K**, and, like Einstein's, his treatment was based on the old quantum theory. Apart from the absence of the zero-point energy, which does not contribute to the specific heat, the latter gives the same result for the energy as quantum mechanics. Debye used first of all the approximation that the relationship between ω and **K** is given, for *all* values of **K**, by the form for long acoustic waves and moreover that there is only a single velocity of propagation c_s, i.e. that $\omega = Kc_s$. Moreover, he took account only of the acoustic modes but adjusted the *total* number of modes to give the right value. The calculation which we shall give below based on this assumption is carried out more in accordance with modern methods but the approximate method gives essentially the same result as obtained originally by Debye. In order to evaluate the expression given in equation (44) of § 6.9.1 for the mean energy, we must sum over all the values of ω_{sj} corresponding the allowed values of the wave vector **K**. We shall assume that these are obtained from the periodic boundary conditions as discussed in § 6.7. There we saw that the allowed values are uniformly distributed over the first Brillouin zone and have density equal to $V/8\pi^3$, where V is the volume of the crystal. Thus, the number of allowed values of **K** in a small volume $d\mathbf{K}$ is equal to $V\,d\mathbf{K}/8\pi^3$. Since

* We use **K** here to avoid confusion with Boltzmann's constant k.
† P. Debye, *Ann. Phys.*, **29**, 789 (1912).

these lie very closely together we may replace the sum in equation (44) of § 6.9.1 by an integral and write

$$\bar{E} = E_0 + \sum_j \frac{V}{8\pi^3} \int \frac{\hbar\omega_j \, d\mathbf{K}}{e^{\hbar\omega_j/kT} - 1} \tag{4}$$

where ω_j is given as a function of $\mathbf{K}$ for the jth branch of the $\omega/\mathbf{K}$ curves. When we have a spherically symmetrical distribution we may replace $d\mathbf{K}$ by $4\pi K^2 dK$ and write equation (4) in the form

$$\bar{E} = E_0 + \sum_j \frac{V}{2\pi^2} \int \frac{\hbar\omega_j \, K^2 dK}{e^{\hbar\omega_j/kT} - 1} \tag{5}$$

Let us consider first a crystal with one atom per unit cell, so that we have only acoustic modes. For the $\omega/\mathbf{K}$ relationship used by Debye the $\omega/\mathbf{K}$ curves for the three values of j are identical. We have thus, writing $\omega_j = 2\pi\nu = Kc_s$, for all three values of j,

$$\bar{E} = E_0 + 12\pi c_s^{-3} V \int_0^{\nu_m} \frac{h\nu^3 \, d\nu}{e^{h\nu/kT} - 1} \tag{6}$$

The integral in (5) is to be taken over the first Brillouin zone. In Debye's approximation, however, this is not known and the integral in (6) is taken from $\nu = 0$ to $\nu = \nu_m$, a maximum value of ν still to be determined. Debye determined ν_m from the condition that the total number of allowed frequencies should be equal to $3N_a$, where N_a is the total number of atoms in the crystal. This condition may be expressed in the form

$$\frac{V}{2\pi^2} \int K^2 dK = N_a \tag{7}$$

and for the assumed dispersion law this reduces to

$$4\pi V c_s^{-3} \int_0^{\nu_m} \nu^2 d\nu = N_a$$

or

$$\frac{4\pi}{3}\left(\frac{\nu_m}{c_s}\right)^3 = \frac{N_a}{V} \tag{8}$$

If we write $x = h\nu/kT$ and $x_m = h\nu_m/kT$ we may express equation (6) in the form

$$\bar{E} = E_0 + \frac{12\pi V k^4 T^4}{c_s^3 h^3} \int_0^{x_m} \frac{x^3 \, dx}{e^x - 1} \tag{9}$$

The specific heat at constant volume C_v is obtained by differentiating equation (6) with respect to T. (This is easier than using (9) since the limit x_m is a function of T.) Equation (9) refers to N_a atoms and if we consider one mole we have

$$C_v = \left(\frac{\partial \bar{E}}{\partial T}\right)_V = \frac{12\pi \boldsymbol{N}}{c_s^3}\left(\frac{V}{\boldsymbol{N}}\right)\int_0^{\nu_m} \frac{\boldsymbol{h}^2 \nu^4 e^{\boldsymbol{h}\nu/\boldsymbol{k}T}\, d\nu}{\boldsymbol{k}T^2(e^{\boldsymbol{h}\nu/\boldsymbol{k}T}-1)^2} \tag{10}$$

If we introduce once more the quantity x_m and substitute from equation (8) for $\boldsymbol{N}/V$ we obtain

$$C_v = \frac{9\boldsymbol{N}\boldsymbol{k}^4 T^3}{\nu_m^3 \boldsymbol{h}^3}\int_0^{x_m} \frac{x^4 e^x\, dx}{(e^x-1)^2} \tag{11}$$

If we now define a temperature Θ_D (known as the Debye temperature) by means of the equation

$$\Theta_D = Tx_m = \boldsymbol{h}\nu_m/\boldsymbol{k} \tag{12}$$

we may write equation (9) in the form

$$\bar{E} = E_0 + 3\boldsymbol{R}TF\left(\frac{\Theta_D}{T}\right) \tag{13}$$

where the function $F(y)$ is defined as

$$F(y) = \frac{3}{y^3}\int_0^{y} \frac{x^3\, dx}{e^x-1} \tag{14}$$

Also we may write equation (11) in the form

$$C_v = 3\boldsymbol{R}\phi\left(\frac{\Theta_D}{T}\right) \tag{15}$$

where

$$\phi(y) = \frac{3}{y^3}\int_0^{y} \frac{x^4 e^x\, dx}{(e^x-1)^2} \tag{16}$$

For large values of T, $x_m \to 0$ and we clearly have

$$\int_0^{x_m} \frac{e^x x^4\, dx}{(e^x-1)^2} \to \int_0^{x_m} x^2\, dx = \tfrac{1}{3}x_m^3$$

so that $C_v \to 3\boldsymbol{R}$ as before.

On the other hand, for small values of T we have $x_m \to \infty$ and the integral in equation (11) may be evaluated; it has the value $4\pi^4/15$ so that we have

$$C_v \to \frac{12\pi^4 \boldsymbol{R}}{5}\left(\frac{T}{\Theta_D}\right)^3 \tag{17}$$

This equation may also be written in the form

$$C_v \to \frac{16\pi^5 \boldsymbol{k}}{5}\left(\frac{\boldsymbol{k}T}{\boldsymbol{h}c_s}\right)^3 V \tag{17a}$$

where V is the volume per mole; this may be derived directly from equation (9) by letting $x_m \to \infty$; it does not therefore depend on the particular process used to determine ν_m. Its validity depends on the assumption that the temperature is so low that only the long-wave acoustic modes are excited. Equations (17) or (17a) give the observed variation of C_v as T^3 for small values of T; it is found, however, that it holds only for values of T considerably less than Θ_D.

It will be seen that, according to equation (15), the specific heat of any crystalline solid of the type under consideration may be expressed in terms of a single parameter Θ_D. This turns out to be remarkably true in practice except at fairly low temperatures where, if this method is used to represent the specific heat curves, Θ_D is found to vary somewhat with temperature; this is hardly surprising in view of the crudity of the $\omega/\mathbf{K}$ relationship used. These variations have in some cases been accounted for by using a better approximation to the $\omega/\mathbf{K}$ curves (see below). The variation of specific heat of crystalline solids with temperature is discussed in considerable detail in most textbooks on heat, in terms of Debye's model, and will not be further discussed here*.

By combining equations (8) and (12), the Debye temperature may be expressed in terms of the velocity of sound, c_s in the crystal. Clearly a better approximation may be obtained by introducing two velocities c_l, c_t corresponding to longitudinal and transverse acoustic waves, the latter corresponding to two degenerate modes as for an isotropic solid. When we sum over the three acoustic branches of the $\omega/\mathbf{K}$ curves we clearly obtain instead of equation (8)

$$\frac{4\pi}{3}\left(\frac{1}{c_l^3}+\frac{2}{c_t^3}\right)\nu_m^3 = \frac{3N_a}{V} \tag{18}$$

This equation together with equation (12) is frequently used to calculate Θ_D for use in equations (13) and (15). This is, however, not quite correct since these equations should also be modified; the limits of $\mathbf{K}$ should be the same for *both* branches and this leads to *different* limits for ν_m, i.e. there should be exactly N_a modes for *each* branch. If we introduce two limiting frequencies ν_{lm} and ν_{tm} corresponding to the longitudinal and transverse branches we should have instead of equation (8) the two equations

$$\left.\begin{aligned}\frac{4\pi V}{3}\frac{\nu_{lm}^3}{c_l^3} &= N_a\\ \frac{4\pi V}{3}\frac{\nu_{tm}^3}{c_t^3} &= N_a\end{aligned}\right\} \tag{19}$$

* See, for example, J. K. Roberts and A. M. Miller, *Heat and Thermodynamics* (4th Ed.). Blackie, London (1951); D. Bijl, *Progress in Low Temperature Physics* (Edited by C. J. Gorter). North Holland Pub. Co. (1957), Vol. 2, p. 395.

On integrating over the two branches with these upper limits we obtain

$$\bar{E} = E_0 + \boldsymbol{R}T\left[F\left(\frac{\Theta_l}{T}\right) + 2F\left(\frac{\Theta_t}{T}\right)\right] \tag{20}$$

where $F(x)$ is the function defined by equation (14).

On differentiating equation (20) with respect to T we obtain an expression for C_v; this may be written down at once, on comparison with equation (15), in the form

$$C_v = \boldsymbol{R}\left[\phi\left(\frac{\Theta_l}{T}\right) + 2\phi\left(\frac{\Theta_t}{T}\right)\right] \tag{21}$$

where $\Theta_l = \boldsymbol{h}\nu_l/\boldsymbol{k}$, $\Theta_t = \boldsymbol{h}\nu_t/\boldsymbol{k}$, and $\phi(x)$ is the function defined by equation (16). This gives a better approximation, and it will be seen that two parameters are now required to specify the variation of specific heat. Use of an equation of this form removes some, but not all of the observed variations of Θ_D. For small values of T equation (21) reduces to

$$C_v = \frac{4\pi^4 \boldsymbol{R}}{5}\left[\left(\frac{T}{\Theta_l}\right)^3 + 2\left(\frac{T}{\Theta_t}\right)^3\right] \tag{22}$$

This is the same result as we had before provided we use equation (18) to calculate Θ_D, since we note from equations (18) and (19) that

$$\frac{3}{\Theta_D^3} = \frac{1}{\Theta_l^3} + \frac{2}{\Theta_t^3} \tag{23}$$

Equation (22) may also be written in the form (cf. equation (17a))

$$C_v = \frac{16\pi^5 \boldsymbol{k}}{15}\left(\frac{\boldsymbol{k}T}{\boldsymbol{h}}\right)^3\left(\frac{1}{c_l^3} + \frac{2}{c_t^3}\right) V \tag{22a}$$

7.4 Calculation from the dispersion law

The need for using correct relationships between the angular frequency ω and the wave vector $\mathbf{K}$ in order to calculate the specific heat was pointed out as long ago as 1912 by M. Born and T. von Kármán*. The difficulty lies in finding the relationship between ω and $\mathbf{K}$ for all points in the first Brillouin zone for real crystals. The need for using such relationships was later stressed again by M. Blackman† who showed by application of the $\omega/\mathbf{K}$ curves for linear, square and cubic lattices (see § 6.2) that deviations from the simple Debye theory such as are frequently observed are just what one should expect.

We should point out that, in using equation (4) to obtain the mean energy, no extra parameters such as ν_m are required, since, if the $\omega/\mathbf{K}$ curves are known, the limits of integration over $\mathbf{K}$ are set by the first Brillouin zone and the maximum frequencies are given by the maximum values in the zone, usually those corresponding and the zone surface,

* M. Born and T. von Kármán, *Phys. Z.*, **13**, 297 (1912).

† M. Blackman, *Proc. Roy. Soc.*, **A148**, 365, 385 (1935); *Ibid.*, **149**, 117, 126 (1935).

for the acoustical modes. In general, the $\omega/\mathbf{K}$ curves are only known for certain special directions but these may be used to extrapolate to other parts of the zone.

It is instructive to consider the specific heat of a linear crystal using Debye's approximation and to compare it with the correct value obtained in § 3.5. In one dimension, the frequency distribution $n(\nu)$ according to Debye's approximation is constant, whereas from equation (94) of § 3.5 we see that it is equal to $2N_a\pi^{-1}(\nu_c^2-\nu^2)^{-1/2}$ where ν_c is the frequency corresponding to the value at the zone edge; these two frequency distributions are compared in Fig. 7.1. In order to obtain N_a normal modes it

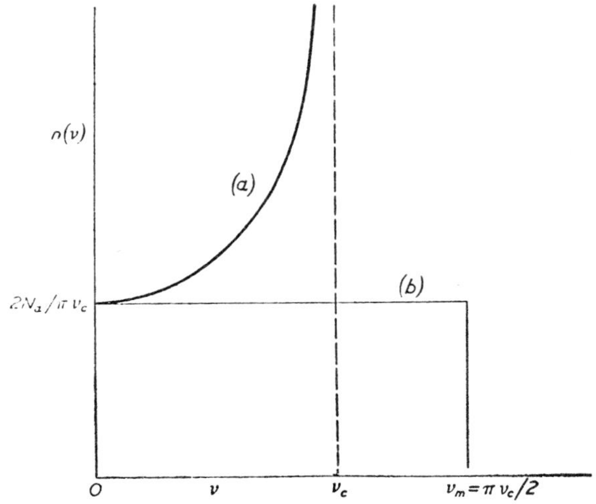

FIG. 7.1. Comparison of the frequency distribution $n(\nu)$ given by Debye's approximation with that given by the exact expression for a linear crystal; (a) exact, (b) Debye

will readily be seen that we must take ν_m, the Debye cut-off frequency equal to $\pi\nu_c/2$. At low temperatures, only frequencies near $\nu = 0$ will be effective, and the Debye approximation will give the same specific heat as the exact expression. At higher temperatures, however, all values of ν up to ν_c will be effective and it will be seen that the two distributions differ markedly. In particular, the real distribution has a high peak at $\nu = \nu_c$ so that an Einstein approximation with $\nu = \nu_c$ would be a better approximation. When we take both transverse and longitudinal modes into account as in equation (98) of § 3.5 we obtain a distribution of the form shown in Fig. 7.2. This has two sharp peaks and at high temperatures could be approximated by means of two Einstein distributions.

Similar results have been obtained by M. Blackman (*loc. cit.*) for square and simple cubic lattices and also show this phenomenon of peaking of the distribution at two frequencies. A number of calculations have been made for other lattices; for example, P. C. Fine* has studied the body-centred cubic lattice and R. B. Leighton† the face-centred cubic lattice. The frequency distribution found by Fine in his application of the calculation to the tungsten lattice is shown in Fig. 7.3 together with the distribution given by Debye's approximation. It will be seen that again we have marked peaks at two frequencies. Full discussions of the methods of obtaining the frequency distributions in terms of the force constants of the crystal by calculation have been given by J. de Launay‡ and by

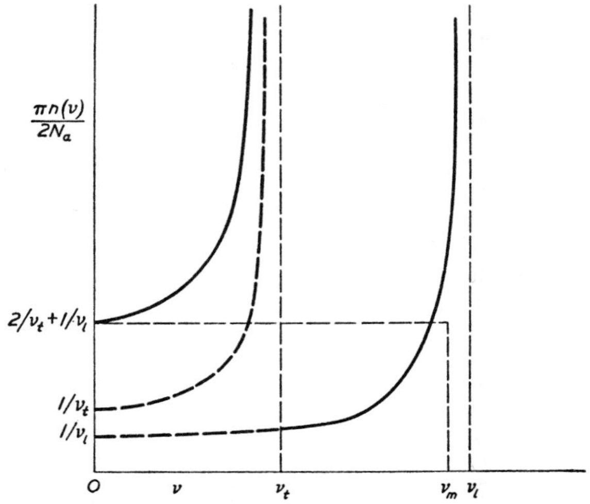

FIG. 7.2. Frequency distribution for a linear crystal taking both longitudinal and transverse modes into account

M. Blackman§. Until recently, these calculations have used only short-range forces between the atoms of the crystal but the validity of this procedure has recently been questioned, for example, by M. Lax**, mainly because of the failure of the $\omega/\mathbf{K}$ curves, calculated for the diamond lattice, to agree with the curves obtained from neutron scattering (see § 6.10).

Now that a start has been made in obtaining the $\omega/\mathbf{K}$ relationship experimentally (see § 6.10) it has become possible to derive specific heats from the experimental data. For example, F. A. Johnson and J. M. Lock††

* P. C. Fine, *Phys. Rev.*, **36**, 355 (1939).
† R. B. Leighton, *Rev. Mod. Phys.*, **20**, 165 (1948).
‡ J. de Launay, *Solid State Phys.*, **8**, 219 (1956).
§ M. Blackman, *Handb. Phys.*, **7**, (1), 325 (1955).
** M. Lax, *Phys. Rev. Letters*, **1**, 133 (1958).
†† F. A. Johnson and J. M. Lock, *Proc. Phys. Soc.*, **72**, 914 (1958).

have used the experimental $\omega/\mathbf{K}$ curves of B. N. Brockhouse and P. K. Iyengar* to derive the low-temperature specific heat for germanium and by this means have been able to explain a pronounced dip in the curve showing the Debye temperature Θ_D as a function of the temperature, which occurs at about 20°K. The experimental data give the $\omega/\mathbf{K}$ relationship along a few directions with special symmetry, namely $\langle 100 \rangle$, $\langle 110 \rangle$, $\langle 111 \rangle$; a simple method of interpolation has been used to obtain the full $\omega/\mathbf{K}$ relationship for the acoustic modes and so to obtain the frequency distribution. A more elaborate method of interpolation from the experimental data has been developed by J. C. Phillips† who has also included an optical mode in his calculation. The distribution

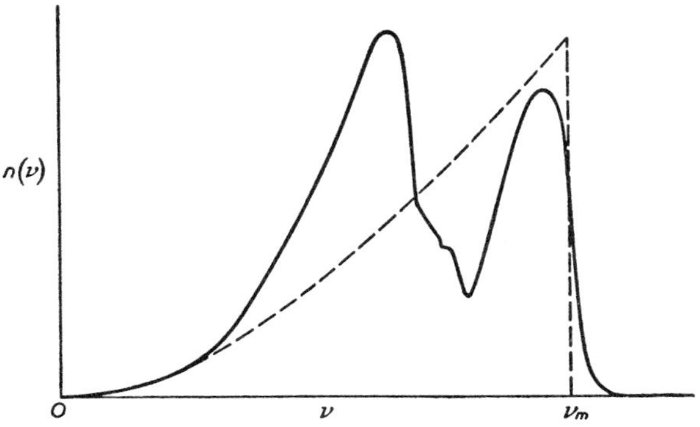

Fig. 7.3. Frequency distribution $n(\nu)$ for the body-centred cubic lattice (after P. C. Fine, *loc. cit.*). Dashed curve shows the distribution for the Debye model

shows three marked peaks corresponding to the two acoustic modes and the optical mode at the zone boundaries. The specific heat curve calculated from this frequency distribution agrees very well indeed with the experimental data over the whole range of temperature for which this is available. Using the same frequency distribution, but varying the values of the three frequencies at which it peaks, the experimental data for silicon and gray tin has also been fitted. This gives a method of obtaining some information about the $\omega/\mathbf{K}$ curves for these materials, in so far as it gives an indication of the position of the cut-off frequencies at the zone boundaries. The problem of deducing the $\omega/\mathbf{K}$ relationship from specific heat data is, however, a very difficult one since this data gives only an average over the zone.

* B. N. Brockhouse and P. K. Iyengar, *Phys. Rev.*, **111**, 747 (1958).
† J. C. Phillips, *Phys. Rev.*, **113**, 147 (1959).

7.5 Crystals with more than one atom per unit cell—effect of the optical modes

In our calculation of the specific heat using a Debye type of approximation we have considered, so far, only crystals with one atom per unit cell. Clearly, at high temperatures, where the law of Dulong and Petit holds, the presence of more than one atom per unit cell will make no difference to the value of C_v since we still have a contribution of $3k$ per atom to the specific heat. Again at low temperatures equation (22a) for C_v still holds, since the contribution from integration over branches of the curves other than the acoustic branches is negligible. If, however, we write C_v in terms of Θ_D we should note that the formulae (8) or (18) connecting ν_m with the volume V are of doubtful validity since they imply that integration over the acoustic branches gives a total of $3N_a$ modes whereas it should give only $3N_a/q$, q being the number of atoms per unit cell. In such formulae it would therefore be better to replace N_a by N_a/q.

At higher temperatures, the optical modes will begin to become excited before the classical region is reached for which $kT \gg h\nu$ for all values of ν. It is not a bad approximation to assume that for the optical modes ω is constant. We then have a contribution from each of the branches in the form of Einstein's formula for the specific heat. If, as is usual, all the optical branches lie fairly closely together it is again not a bad approximation to assume that we have a single frequency ν_0. The variation of the specific heat with temperature will then be given approximately by Einstein's formula (equation (49)); the acoustic modes having nearly all their states fully occupied except near their upper limits will also behave in a similar way. It has long been known that at high temperatures Einstein's formula fits the experimental data better than Debye's. The best approximation of this type is probably obtained from a function of the form

$$C_v = \frac{\mathbf{R}}{3q}\left[\phi_D\left(\frac{\Theta_1}{T}\right) + 2\phi_D\left(\frac{\Theta_2}{T}\right) + 3(q-1)\,\phi_E\left(\frac{\Theta_3}{T}\right)\right] \tag{24}$$

where ϕ_D and ϕ_E represent the functions given by the Debye and Einstein approximations. Even further elaboration may be attempted by introducing more than one optical frequency. The calculations referred to in § 7.4 have shown, however, that the dispersion laws on which these theories are based are poor approximations, and that the density of allowed states as a function of frequency differs markedly from that given by the simple approximations; this further elaboration is therefore hardly justified.

8

Motion of Electrons in a Perfect Crystal

8.1 Reduction to a problem involving a single electron

THE problem of determining exactly the motion of all the electrons in a crystal is much too difficult for solution by known methods. In principle, we should not only have to solve the wave equation for all the electrons but we should at the same time have to solve the equations for the motions of the atomic nuclei. We have already discussed this problem in Chapter 4 for a one-dimensional crystal and have seen how it is approached by various stages of approximation. First we determine the motions of the atoms regarded as single mass points, as discussed in Chapter 6, and then we assume that the atomic nuclei are fixed at the lattice points, and try to determine the motion of the electrons; finally we introduce interactions between the electrons and vibrating atomic ions. This approach to the problem is generally known as the 'adiabatic approximation'. In the present Chapter we shall assume that we have a perfect crystal and that the atoms of the crystal are fixed at the lattice points; our problem is to discuss the motion of the electrons in such a crystal.

To treat even this problem is very difficult if we regard it as a many-body problem, though this approach has been made in some recent work, account being taken of the mutual repulsions between the valence electrons as well as their interaction with the atomic ions. It is more usual, however, to start by seeking to reduce the problem to one involving a single electron; this approach is equivalent to the well-known Hartree method of finding the wave functions and charge density for the electrons in a heavy atom. We fix our attention on a single electron and try to determine its motion in the field of the atomic nuclei and all the other electrons, averaged in a suitable way over the motions of these electrons. The method used is one of *successive* approximation; using an approximate wave function for each electron the charge density is calculated and from this the potential energy of the average field is calculated; better wave functions are then found from the solution of the wave equation for this potential. By an iterative process the potential and wave functions are made self-consistent. Hartree's method has been used with great success when applied to atoms, but the complexity of solids is so much greater that, so far, only rather rough calculations have been made. The importance of the method lies in the fact that it shows us how, in principle at any rate, the problem may be reduced to that of studying the motion of one electron at a time in a field of force defined by means of an electrostatic potential. This problem has already been discussed in Chapter 4, and its relation to real crystals has been indicated, particularly in § 4.13. In spite of the rather drastic approximations made in such a one-electron treatment the results obtained are in surprisingly good agreement with a large amount of experimental data, and enable this data to be fitted into a consistent theory. Also, its limitations are obvious, so that when it fails the reason for its failure is usually fairly clear. Recent treatments of the many-body problem, particularly by D. Bohm and D. Pines* have shown why some of the approximations made, such as neglecting the strong mutual repulsion between the electrons except in its effect on the average potential, allow us to give much better descriptions of the motions of electrons in crystalline solids, by means of the one-electron treatment, than might have been expected.

It is clear that the potential energy which determines the static field in which each single electron is supposed to move has the same kind of periodicity and symmetry as the crystal lattice; this follows from the equivalence of each unit cell of the crystal. The fundamental problem of the one-electron treatment is therefore that of finding the wave functions and associated energy levels for an electron moving in a potential having the periodicity of the crystal lattice. This problem has already been discussed in some detail in Chapter 4 for a one-dimensional crystal, and we must now extend this treatment to deal with crystals in three dimensions.

8.2 Bloch's theorem

The fundamental theorem concerning motion of electrons in a periodic potential is Bloch's theorem, which we have discussed in its one-dimensional form (usually known as Floquet's theorem) in § 4.9. The generalisation of this theorem to three dimensions is fairly obvious; unfortunately the simple proof of the theorem valid for one dimension no longer can be applied in three dimensions. If the crystal lattice is defined by the three basic vectors $\mathbf{d}_1$, $\mathbf{d}_2$, $\mathbf{d}_3$ and the potential energy $V(\mathbf{r})$ is a periodic function, with periodicity of the lattice then we have

$$V(\mathbf{r}) = V(\mathbf{r}+l\mathbf{d}_1+m\mathbf{d}_2+n\mathbf{d}_3) \tag{1}$$

where l, m, n are any three integers. This strictly applies only for an infinite crystal, but for the moment we shall ignore surface effects and deal with an infinite or very large crystal. We then require solutions of Schrödinger's equation in the form

$$\nabla^2\psi+\frac{2m}{\hbar^2}[E-V(\mathbf{r})] = 0 \tag{2}$$

$V(\mathbf{r})$ being a function satisfying the condition (1).

* D. Bohm and D. Pines, *Phys. Rev.*, **80**, 903 (1950); *Ibid.*, **92**, 609 (1953).

The generalisation to three dimensions of solutions of the form given by equation (143) of § 4.9 will readily be seen to be

$$\psi = e^{i(\mathbf{k}.\mathbf{r})}\, U_{\mathbf{k}}(\mathbf{r}) \tag{3}$$

where we now use a real wave vector $\mathbf{k}$ and the functions $U_{\mathbf{k}}(\mathbf{r})$ have the periodicity of the crystal lattice, i.e.

$$U_{\mathbf{k}}(\mathbf{r}) = U_{\mathbf{k}}(\mathbf{r}+l\mathbf{d}_1+m\mathbf{d}_2+n\mathbf{d}_3) \tag{4}$$

where l, m, n are any three integers. The existence of such solutions in three dimensions was first demonstrated by F. Bloch* and this theorem is usually referred to as Bloch's theorem. These solutions are in the form of plane travelling waves, modulated by means of the function $U_{\mathbf{k}}(\mathbf{r})$, Single waves of this type are appropriate to infinite crystals, but combinations of such waves may be used just as for the other types of waves we have discussed, to provide solutions for finite crystals.

There is unfortunately no *simple* analytical proof of Bloch's theorem as in one dimension. The most direct proof depends on the fact that if $\psi(\mathbf{r})$ is a solution corresponding to energy E so also is $\psi(\mathbf{r}+\mathbf{d}_n)$ for all lattice vectors $\mathbf{d}_n$. $\psi(\mathbf{r})$ may therefore be expanded as a sum of these other solutions. By use of the theory of matrix transformations it may be shown that†, provided such solutions exist, they must be of the form given by equation (3). Such solutions do not exist for all values of E, as for the similar equation in one dimension. Many physicists are, however, unfamiliar with the advanced theory of matrices on which such a proof depends. An alternative approach is to use the theory of operators in quantum mechanics; again this involves rather advanced quantum theory and we shall only sketch the proof in brief outline‡. Schrödinger's equation may be written in operational form

$$H\psi = E\psi \tag{5}$$

where H is the operator

$$-\frac{\hbar^2}{2m}\nabla^2 + V(\mathbf{r}) \tag{6}$$

Solutions ψ of this equation, known as eigenfunctions, exist for certain values of E, known as eigenvalues. Now, if we have another operator L, we may form an operational equation

$$L\psi = l\psi \tag{7}$$

l being a constant, to determine the eigenvalues of the operator L. Only in very special circumstances can we find eigenfunctions ψ which satisfy

* F. Bloch, *Z. Phys.*, **52**, 555 (1928).

† See, for example, A. H. Wilson, *The Theory of Metals* (2nd Ed.). Cambridge University Press (1953), p. 29; or R. E. Peierls, *Quantum Theory of Solids*. Oxford University Press (1955), p. 73.

‡ See, for example, J. C. Slater, *Handb. Phys.*, **19**, 13 (1956).

both equations (5) and (7) and give eigenvalues *both* for the energy and the operator L. Well-known examples are one component of the angular momentum, and the total angular momentum, when $V(\mathbf{r})$ is the potential due to a single central field of force. The necessary and sufficient condition to be satisfied by the operator L is (see any advanced textbook on quantum mechanics) that L should commute with the energy operator H, i.e. that

$$LH = HL \tag{8}$$

Now let us define a translation operator $T_{\mathbf{n}}$ by means of the equation

$$T_{\mathbf{n}}\psi(\mathbf{r}) = \psi(\mathbf{r}+\mathbf{d}_{\mathbf{n}}) \tag{9}$$

Clearly this operator commutes with H when $V(\mathbf{r})$ is a function with periodicity of the lattice since it leaves H unchanged. Now if several operators have the property of having the same eigenfunctions as H it may be shown that they must mutually commute. This is clearly true of all the translation operators $T_{\mathbf{n}}$, since, if we make two successive lattice translations, the order is immaterial. It follows that the translation operators $T_{\mathbf{n}}$ have simultaneous eigenfunctions with H and we may write

$$T_n\psi(\mathbf{r}) = a_{\mathbf{n}}\psi(\mathbf{r}) \tag{10}$$

where $a_{\mathbf{n}}$ is a constant, this being appropriate to the equation of the form (7). Equation (10) may also be written in the form

$$\psi(\mathbf{r}+\mathbf{d}_{\mathbf{n}}) = a_{\mathbf{n}}\psi(\mathbf{r}) \tag{11}$$

Now, if $\psi(\mathbf{r})$ is normalised for an infinite crystal, we clearly must have

$$\int |\psi(\mathbf{r}+\mathbf{d}_{\mathbf{n}})|^2\,\mathrm{d}\mathbf{r} = \int |\psi(\mathbf{r})|^2\,\mathrm{d}\mathbf{r} \tag{12}$$

the translation having no effect on the value of the integral, since it is equivalent to a change of origin in the infinite crystal. It follows that $|a_{\mathbf{n}}|^2 = 1$ for all values of the three integers represented by $\mathbf{n}$. We note moreover that

$$T_{\mathbf{n}+\mathbf{m}}\psi = T_{\mathbf{n}}T_{\mathbf{m}}\psi \tag{13}$$

so that

$$a_{\mathbf{n}+\mathbf{m}} = a_{\mathbf{n}}a_{\mathbf{m}} \tag{14}$$

This is satisfied by taking $a_{\mathbf{n}}$ of the form

$$\ln a_{\mathbf{n}} = (\mathbf{a}\,.\,\mathbf{d}_{\mathbf{n}}) \tag{15}$$

where $\mathbf{a}$ is a constant vector. In order that $|a_{\mathbf{n}}|^2 = 1$, $\mathbf{a}$ must be purely imaginary so that we may write

$$a_{\mathbf{n}} = \mathrm{e}^{i\mathbf{k}.\mathbf{d}_{\mathbf{n}}} \tag{16}$$

We therefore have

$$\psi(\mathbf{r}+\mathbf{d}_{\mathbf{n}}) = \mathrm{e}^{i\mathbf{k}.\mathbf{d}_{\mathbf{n}}}\psi(\mathbf{r}) \tag{17}$$

Let us now consider the behaviour of the function $U_{\mathbf{k}}(\mathbf{r})$ which we shall *define* as equal to $\exp(-i\mathbf{k}.\mathbf{r})\psi(\mathbf{r})$; this function is clearly unchanged by the translation operator $T_{\mathbf{n}}$, i.e. it is a periodic function having the periodicity of the lattice. We may therefore write

$$\psi(\mathbf{r}) = e^{i\mathbf{k}.\mathbf{r}}\,U_{\mathbf{k}}(\mathbf{r})$$

which is the form required by Bloch's theorem.

The rather advanced quantum theory involving the use of operators may not be familiar to all readers and Bloch's theorem is sufficiently important to warrant yet another approach which will show the relationship between solutions of the form (3) and the descriptive waves we have already used to describe lattice vibrations. We start by using concepts already made familiar by the one-dimensional treatment. We have seen that in this case although we cannot find a solution in the form of a *single* plane wave, except for the case of a constant potential, we may find a solution in the form of a sum over an infinite set of plane waves, having wave numbers of the form $k_n = k + 2\pi n/d$. In three dimensions we should expect to use a triply infinite set of wave vectors and the natural set to use, in analogy with Fourier expansions, would be a set defined by an equation of the form (cf. § 5.4.3)

$$\mathbf{k}_{\mathbf{n}} = \mathbf{k} + 2\pi\mathbf{b}_{\mathbf{n}} \tag{18}$$

where $\mathbf{b}_{\mathbf{n}}$ represents a vector of the reciprocal lattice. We should then seek solutions* of the form

$$\psi(\mathbf{r}) = \sum_{\mathbf{n}} B_{\mathbf{n}} \exp\left[i(\mathbf{k} + 2\pi\mathbf{b}_{\mathbf{n}}).\mathbf{r}\right] \tag{19}$$

The constants $B_{\mathbf{n}}$ are not expected to be arbitrary but to be determinable apart from one. We are encouraged to use this form of expansion, since the periodic potential $V(\mathbf{r})$ may be expanded in the form

$$V(\mathbf{r}) = \sum_{\mathbf{m}} V_{\mathbf{m}} \exp\left[2\pi i\mathbf{b}_{\mathbf{m}}.\mathbf{r}\right] \tag{20}$$

(cf. § 5.4.3 equation (36)), where the constants $V_{\mathbf{m}}$ are such that $V_{-\mathbf{m}} = V_{\mathbf{m}}^{\star}$, since $V(\mathbf{r})$ is a real function. If we now substitute this form for $V(\mathbf{r})$ and also the expansion (19) of the wave function in Schrödinger's equation we obtain, writing $k_{\mathbf{n}}$ for $|\mathbf{k}_{\mathbf{n}}|$

$$\sum_{\mathbf{n}} B_{\mathbf{n}}\left(E - \frac{\hbar^2 \mathbf{k}_{\mathbf{n}}^2}{2m}\right) e^{2\pi i\mathbf{b}_{\mathbf{n}}.\mathbf{r}} = \sum_{\mathbf{m}}\sum_{\mathbf{n}} B_{\mathbf{m}}\, V_{\mathbf{n}}\, e^{2\pi i[(\mathbf{b}_{\mathbf{n}}+\mathbf{b}_{\mathbf{m}}).\mathbf{r}]} \tag{21}$$

In obtaining this equation we note that the factor $\exp(i\mathbf{k}r)$ goes out and this justifies its inclusion in (19); also we have used the result that

$$\nabla^2 \exp\left[i\mathbf{k}_{\mathbf{n}}.\mathbf{r}\right] = -k_{\mathbf{n}}^2 \exp\left[i\mathbf{k}_{\mathbf{n}}.\mathbf{r}\right] \tag{22}$$

* This method of approach is essentially the same as that due to L. Brillouin, *Wave Propagation in Periodic Structures*. Dover (1953), p. 140.

We now note that the vector $\mathbf{b_m}+\mathbf{b_n}$ is the same as the vector $\mathbf{b_{m+n}}$; we may therefore renumber the sum on the right-hand side of equation (21) by writing $\mathbf{n}'$ for $\mathbf{m}+\mathbf{n}$. When this is done we may write the double sum in the form

$$\sum_{\mathbf{n}'}\sum_{\mathbf{n}} B_{\mathbf{n}}\, V_{\mathbf{n}'-\mathbf{n}}\, e^{2\pi i\mathbf{b}_{\mathbf{n}'}.\mathbf{r}}$$

The summation indices $\mathbf{n}$ and $\mathbf{n}'$ may now be interchanged and equation (21) may be expressed in the form

$$\sum_{\mathbf{n}} B_{\mathbf{n}}\left(E-\frac{\hbar^2 k_{\mathbf{n}}^2}{2m}\right) e^{2\pi i\mathbf{b}_{\mathbf{n}}.\mathbf{r}} = \sum_{\mathbf{n}}\sum_{\mathbf{n}'} B_{\mathbf{n}'}\, V_{\mathbf{n}-\mathbf{n}'}\, e^{2\pi i\mathbf{b}_{\mathbf{n}}.\mathbf{r}} \tag{23}$$

We may now equate coefficients of $\exp(2\pi i\mathbf{b}_{\mathbf{n}}.\mathbf{r})$ to obtain the equations

$$\sum_{\mathbf{n}'} B_{\mathbf{n}'}\left[V_{\mathbf{n}-\mathbf{n}'}-\left(E-\frac{\hbar^2 k_{\mathbf{n}}^2}{2m}\right)\delta_{\mathbf{nn}'}\right] = 0 \tag{24}$$

(This procedure may be justified by multiplying in turn by $\exp(-2\pi i\mathbf{b}_{\mathbf{n}}.\mathbf{r})$ and integrating over all space.) These equations form a set of linear homogeneous equations similar to those obtained in many of the previous problems which we have treated using plane waves, only now we have an infinite set of equations. As before, however, these have only a non-zero solution provided the determinant of the coefficients is equal to zero, i.e. provided

$$\left\| V_{\mathbf{n}-\mathbf{n}'}-\left(E-\frac{\hbar^2 k_{\mathbf{n}}^2}{2m}\right)\delta_{\mathbf{nn}'}\right\| = 0 \tag{25}$$

This equation now has an infinite number of roots, giving, as we should expect, an infinite number of branches of the $E/\mathbf{k}$ curves. For each root we obtain a value $E(\mathbf{k})$ giving a band of allowed energies as $\mathbf{k}$ is varied, provided the value of E is real. In practice, this infinite determinantal equation is very difficult to solve numerically, but it enables us to deduce many of the important properties of the $E/\mathbf{k}$ curves and corresponding energy bands. We see at once that provided E is given by a real solution of equation (25) the wave function corresponding to this energy is of the form (19). For each value E_s of E the ratios of the constants $B_{\mathbf{n}}$ may be determined and, apart from an arbitrary constant the functions $U_s(\mathbf{r})$ given by

$$U_s(\mathbf{r}) = \sum_{\mathbf{n}} B_{\mathbf{n}}\, e^{2\pi i\mathbf{b}_{\mathbf{n}}.\mathbf{r}} \tag{26}$$

will be determined. The solutions of Schrödinger's equation corresponding for a *given* value of $\mathbf{k}$ to the energies E_s will therefore be of the form

$$\psi_s(\mathbf{r}) = U_s(\mathbf{r})\, e^{i\mathbf{k}.\mathbf{r}} \tag{27}$$

This is of the form required by Bloch's theorem since the functions $U_s(\mathbf{r})$ are clearly periodic with the periodicity of the lattice.

If we replace the wave vector $\mathbf{k}$ by any of the vectors $\mathbf{k_n}$ in equation (27) it will be clear that the form of the solution will not be changed, since, if we write $U'_s(\mathbf{r}) = U_s(\mathbf{r})\exp[2\pi i\mathbf{b_n}.\mathbf{r}]$, $U'_s(\mathbf{r})$ is also a periodic function with periodicity of the lattice. Moreover, it is clear from the form of equation (25) that the value of the energy is unchanged, since the replacement of $\mathbf{k}$ by $\mathbf{k_n}$ merely renumbers the wave vectors; $\mathbf{k}$ and $\mathbf{k_n}$ are therefore equivalent vectors. The vectors $\mathbf{k_n}/2\pi$ being derived from $\mathbf{k}/2\pi$ by a lattice translation in the reciprocal lattice, we may use a reduced representation, confining the vectors $\mathbf{k}$ to a zone in $\mathbf{k}$-space; this is just the first Brillouin zone introduced in our discussion of lattice vibrations. (The vector $\mathbf{k}/2\pi$ is restricted to the first Brillouin zone in the reciprocal lattice space.) It is interesting to see that the *same* zone is appropriate both for the wave vector $\mathbf{k}/2\pi$ describing the motion of electrons in the crystal and for the vector $\mathbf{K}/2\pi$ describing the lattice vibrations. As in one dimension we may give the vector $\mathbf{k_n}$ a physical interpretation (see § 4.5) in this type of problem, as distinct from the problem of lattice vibrations for which $\mathbf{k}$ is physically indistinguishable from $\mathbf{k_n}$; $\mathbf{k_n}$ is the wave number associated with a particular space harmonic in the expansion of the wave function as a sum of plane waves. The coefficients $B_\mathbf{n}$ being determined (apart from one), the amplitudes of these space harmonics are determined by the nature of the potential $V(\mathbf{r})$.

The energy E, being a periodic function of the wave vector $\mathbf{k}$, is usually only given for values of $\mathbf{k}$ in the first Brillouin zone. If we replace $\mathbf{k}$ by $-\mathbf{k}$ we do not change equation (25) since a change of i to $-i$ effectively changes $\mathbf{n}$ to $-\mathbf{n}$ throughout ($V_\mathbf{n} \to V^\star_\mathbf{n} = V_{-\mathbf{n}}$). We therefore obtain the same value for the energy, which is therefore an even function of $\mathbf{k}$.

When E is a single-valued function of the components k_r of the vector $\mathbf{k}$ it may therefore be expanded in a Taylor series near $\mathbf{k} = 0$ and we have

$$E = E_0 + \sum_{rs} A_{rs} k_r k_s \quad (r, s = 1, 2, 3) \tag{28}$$

where the constants A_{rs} are the components of a symmetrical tensor of rank 2, i.e. $A_{rs} = A_{sr}$. When we have degeneracy, and two $E/\mathbf{k}$ surfaces meet at $\mathbf{k} = 0$, the point $\mathbf{k} = 0$ will be a singular point on the combined surface. It is well known, from the theory of algebraic curves and surfaces, that a Taylor expansion may not be possible in such cases; it is then possible to have non-zero first derivatives at $\mathbf{k} = 0$. Some examples of such constant-energy surfaces will be discussed later.

There are, unfortunately, no examples in three dimensions corresponding to the Krönig–Penney problem or Mathieu's equation for which exact solutions may be found and we have to resort to approximate methods or to numerical analysis. When the potential $V(\mathbf{r})$ is nearly constant we may again obtain a solution by first-order perturbation analysis, and also when the overlap in the potential functions arising from individual atoms of the crystal is small—the tight-binding approximation. These we shall consider next and shall then discuss numerical methods for obtaining solutions for real crystals.

8.3 Motion in a small periodic potential

We shall now consider the motion when the kinetic energy is large compared with the potential energy. This problem bears no relation to the motion of the valence electrons in a real crystal for which the variable part of the potential energy and kinetic energy are comparable. It reveals some interesting general properties of the solutions for motion of electrons in a periodic potential in three dimensions and taken together with the tight-binding approximation gives us insight into the intermediate case which is generally met with in practice. The approximation would apply to the motion of very fast electrons injected into a crystal. In the wave equation we shall replace the potential energy $V(\mathbf{r})$ by the expression

$$V(\mathbf{r}) = V_0 + \epsilon W(\mathbf{r}) \tag{29}$$

where $W(\mathbf{r})$ is a periodic function with the periodicity of the lattice and ϵ is a small constant, whose square we shall neglect. If the energy E is equal to $E' + V_0$ we may write Schrödinger's equation in the form

$$\nabla^2 \psi + \frac{2m}{\hbar^2}[E' - \epsilon W(\mathbf{r})] = 0 \tag{30}$$

When $\epsilon = 0$ we have

$$\psi = A\,\mathrm{e}^{i\mathbf{k}_0 . \mathbf{r}} \tag{31}$$

where

$$E' = \hbar^2 k_0^2/2m$$

We now proceed as in the derivation of the equations subsequent to (20) taking for the wave function ψ an expansion of the form (19). From the form (31) of the zero-order approximation we may assume that all coefficients $B_\mathbf{n}$ are small except when $\mathbf{n} = 0$. We therefore shall write $\epsilon B'_\mathbf{n}$ for $B_\mathbf{n}$ except for $\mathbf{n} = 0$, for which we shall write $B'_0 = B_0$. On substituting E' for E and $\epsilon W(\mathbf{r})$ for $V(\mathbf{r})$ in equation (21) we obtain, neglecting powers of ϵ higher than the first,

$$B'_0\left(E' - \frac{\hbar^2 k^2}{2m}\right) + \epsilon \sum_\mathbf{n}{}' B'_\mathbf{n}\left(E' - \frac{\hbar^2 k_\mathbf{n}^2}{2m}\right)\mathrm{e}^{2\pi i \mathbf{b}_\mathbf{n} . \mathbf{r}} = \epsilon \sum_\mathbf{n} B'_0 W_\mathbf{n}\,\mathrm{e}^{2\pi i \mathbf{b}_\mathbf{n} . \mathbf{r}} \tag{32}$$

where $\sum'$ means, as usual, the sum over all values except 0. It is assumed that $W_0 = 0$, the constant term in the expansion of $W(\mathbf{r})$ being absorbed into the constant potential V_0. Multiplying in turn by $\exp(-2\pi i \mathbf{b}_\mathbf{n} . \mathbf{r})$ and integrating, we obtain the equations

$$E' = \hbar^2 k^2/2m \tag{33}$$

or

$$k = k_0 \tag{33a}$$

and

$$B'_\mathbf{n}/B'_0 = (2m/\hbar^2)\,W_\mathbf{n}/(k^2 - k_\mathbf{n}^2) \tag{34}$$

Equation (33) shows that to the first order in ϵ the energy is unchanged, and equation (34) gives the values of the coefficients $B'_{\mathbf{n}}$ in terms of the Fourier components $W_{\mathbf{n}}$ of the potential. We may proceed to a second-order approximation and obtain for the energy the value

$$E' = \frac{\hbar^2 k^2}{2m} + \frac{2m\epsilon^2}{\hbar^2} \sum_{\mathbf{n}}' \frac{|W_{\mathbf{n}}|^2}{(k^2 - k_{\mathbf{n}}^2)} \tag{35}$$

Both equations (34) and (35) indicate that the approximation breaks down when magnitude of the vector $\mathbf{k}$ is approximately equal to the magnitude of the vector $\mathbf{k}_{\mathbf{n}}$, i.e. when

$$|\mathbf{k}|^2 \simeq |\mathbf{k} + 2\pi \mathbf{b}_{\mathbf{n}}|^2 \tag{36}$$

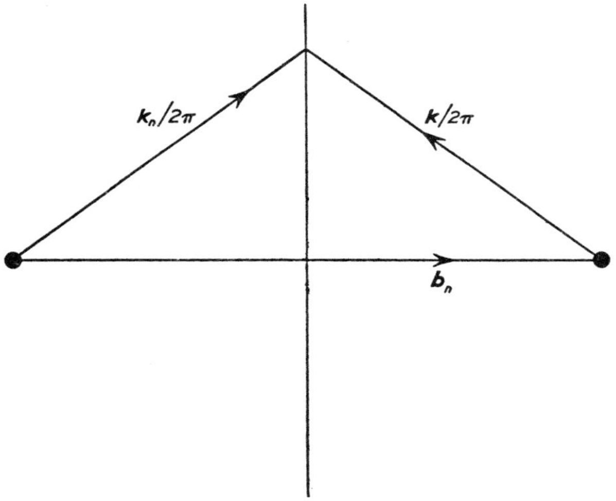

FIG. 8.1. Condition for Bragg reflection and connection with the Brillouin zones

since then the coefficient $B'_{\mathbf{n}}$ becomes very large, contrary to our assumption that it is small. Equation (36) may also be written in the form

$$(\mathbf{b}_{\mathbf{n}} . \mathbf{k}/2\pi) + \tfrac{1}{2}|\mathbf{b}_{\mathbf{n}}|^2 = 0 \tag{37}$$

This is just the equation for surface planes of the Brillouin zones and is also the condition for Bragg reflection (see § 5.4.6). When the condition (37) holds the electrons are strongly reflected from a set of planes in the crystal and a *single* travelling wave cannot propagate; we see that this happens whenever the wave vector $\mathbf{k}$ has its termination on the surface of a Brillouin zone, and the condition is illustrated in Fig. 8.1.

We have already discussed in § 4.11.1 a similar problem in one dimension. To obtain an approximate solution when $\mathbf{k} \simeq \mathbf{k}_{\mathbf{n}}$ we proceed in exactly the same way, taking as zero-order approximation a wave function having two large terms. Let $\mathbf{k}_1$ be taken as the particular vector of the set $\mathbf{k}_n$ for which $\mathbf{k} \simeq -\mathbf{k}_n$; we then take as our wave function the form

$$\psi = B_0 e^{i\mathbf{k}.\mathbf{r}} + B_1 e^{i(\mathbf{k}+2\pi\mathbf{b}_1.\mathbf{r})} + \epsilon \sum_{\mathbf{n}}'' B'_{\mathbf{n}} e^{i\mathbf{k}_\mathbf{n}.\mathbf{r}} \tag{38}$$

On substituting the appropriate values for B'_n in equation (32) we find on equating to zero the constant coefficient and the coefficient of $\exp(2\pi i\mathbf{b}_1.\mathbf{r})$ (justified by integration as before),

$$\left.\begin{aligned} B_0(k_0^2 - k^2) - 2B_1\, m\epsilon W_1/\hbar^2 &= 0 \\ -2B_0\, m\epsilon W_1/\hbar^2 + B_1(k_0^2 - k_1^2) &= 0 \end{aligned}\right\} \tag{39}$$

These two equations have only non-zero solutions if

$$(k_0^2 - k_1^2)(k_0^2 - k^2) = 4m^2\epsilon^2|W_1|^2/\hbar^4 \tag{40}$$

When we have $\mathbf{k}$ exactly equal to $\mathbf{k}_1$ we have

$$E' = \frac{\hbar^2 k_0^2}{2m} = \frac{\hbar^2 k^2}{2m} \pm \epsilon|W_1| \tag{41}$$

and we have two solutions separated by an energy gap $\Delta E = 2\epsilon|W_1|$.

Now suppose the value of $\mathbf{k}$ for which $\mathbf{k} = -\mathbf{k}_1$ is $\mathbf{k}'$, and we write $\mathbf{k} = \mathbf{k}' + \mathbf{u}$, where $\mathbf{u}$ is a small vector, then neglecting u^2 we have

$$\left.\begin{aligned} k^2 &= k'^2 + 2\mathbf{u}.\mathbf{k}' \\ k_1^2 &= k_1'^2 + 2\mathbf{u}.\mathbf{k}_1' \end{aligned}\right\} \tag{42}$$

where $\mathbf{k}_1' = \mathbf{k}' + 2\pi\mathbf{b}_1$. Also let

$$k_0^2 = k'^2 + \eta \tag{43}$$

and let us choose the direction of the vector $\mathbf{u}$ to be at right angles to the plane face of the Brillouin zone defined by the vector $\mathbf{b}_1$, i.e. parallel to $\mathbf{b}_1$; then we have $\mathbf{u}.\mathbf{k}' = -\mathbf{u}.\mathbf{k}_1'$, and on substituting in equation (40) we obtain

$$\eta^2 - 4(\mathbf{u}.\mathbf{k}')^2 = 4m^2\epsilon^2|W_1|^2/\hbar^4 \tag{44}$$

There is therefore no solution for which

$$|\eta| < 2m\epsilon|W_1|/\hbar^2$$

and we see that indeed we do have a forbidden energy gap $\Delta E = 2\epsilon|W_1|$. Moreover we see from equation (44) that η and hence the energy E' has a quadratic dependence on the magnitude of the vector $\mathbf{u}$ which shows that the normal derivative of E with respect to $\mathbf{k}$ at the zone face is equal to zero. A complexity arises if $|W_1| = 0$ and there is no forbidden energy gap; in this case we have a linear relationship between η and $|\mathbf{u}|$.

Thus we see that we may have a forbidden energy gap which occurs for values of $\mathbf{k}/2\pi$ corresponding to points on the surface of a Brillouin zone. When we use the reduced representation the various $E/\mathbf{k}$ surfaces are defined only for values of $\mathbf{k}/2\pi$ corresponding to the first zone and the discontinuities between the various surfaces then occur at the zone edge. If we plot the $E/\mathbf{k}$ curves for a particular direction of the vector $\mathbf{k}$ we shall obtain a set of curves similar to those obtained for one-dimensional problems and illustrated in Fig. 4.9; for the problem under discussion, however, it is more usual to use the expanded representation using a single Brillouin zone to represent one branch of the $E/\mathbf{k}$ surface. For a particular direction we then obtain $E/\mathbf{k}$ curves of the form illustrated in Fig. 4.8 and which approximate when $V(\mathbf{r})$ is small to the $E/\mathbf{k}$ surface for free electrons.

If we have more than one atom per unit cell we may obtain an interesting condition under which the discontinuity in energy at a zone boundary is small. As a first approximation the potential may be expressed as a sum of terms each of which comes from a single atom of the crystal. We then have

$$V(\mathbf{r}) = \sum_{\mathbf{n}j} V_j(\mathbf{r}+\mathbf{d}_\mathbf{n}+\mathbf{a}_j) \tag{45}$$

the jth atom in a unit cell being at vector distance $\mathbf{a}_j$ from the origin. The Fourier component $V_\mathbf{n}$ is then given by

$$V_\mathbf{n} = C\int e^{-2\pi i(\mathbf{b}_\mathbf{n}.\mathbf{r})} \sum_{\mathbf{n}j} V_j(\mathbf{r}+\mathbf{d}_\mathbf{n}+\mathbf{a}_j)\,d\mathbf{r} \tag{46}$$

where C is a constant. By changing the variable in each term of the sum it may readily be shown that $V_\mathbf{n}$ may be expressed in the form

$$V_\mathbf{n} = \sum_j V_{\mathbf{n}j} e^{2\pi i(\mathbf{b}_\mathbf{n}.\mathbf{a}_j)} \tag{47}$$

When all the atoms in the unit cell are identical, the expression for $V_\mathbf{n}$ may be further simplified and we have

$$V_\mathbf{n} = W_\mathbf{n} S_\mathbf{n} \tag{48}$$

where

$$S_\mathbf{n} = \sum_j e^{2\pi i(\mathbf{b}_\mathbf{n}.\mathbf{a}_j)} \tag{49}$$

The quantity $S_\mathbf{n}$ is known as the structure factor and plays an important part in determining the intensity of X-ray reflections from the particular planes associated with the reciprocal lattice vector $\mathbf{b}_\mathbf{n}$. In particular, if $S_\mathbf{n} = 0$, a reflection will be missing and, to the above approximation, $V_\mathbf{n}$ will be zero giving no energy gap at the zone edge. This, however, is only an approximate result since $V(\mathbf{r})$ cannot be accurately expressed in the form of equation (45); we may, however, expect a small energy gap if $S_\mathbf{n} = 0$.

The interpretation of the condition $S_\mathbf{n} = 0$ for no X-ray reflection is straightforward; the quantities $2\pi(\mathbf{b}_\mathbf{n}.\mathbf{a}_j)$ are the relative phases of the contributions from the various atoms in the unit cell, for the conditions of Bragg reflection from the planes to which the vector $\mathbf{b}_\mathbf{n}$ is normal. The relative phase of the incident wave at the jth atom is $(\mathbf{k}\,.\,\mathbf{a}_j)$ and of the scattered wave is $(\mathbf{k}'\,.\,\mathbf{a}_j)$ and the phase difference is $(\mathbf{k}'-\mathbf{k}\,.\,\mathbf{a}_j)$, which is equal to $2\pi(\mathbf{b_n}\,.\,\mathbf{a}_j)$ by equation (40) of § 5.5.

8.4 The tight-binding approximation

We shall now consider another extreme condition for which an approximate solution may readily be obtained; this is the tight-binding approximation which we have already discussed for one dimension in § 4.11.2. We shall first of all deal with a crystal with one atom per unit cell; the extension to polyatomic crystals and to crystals having a lattice with a basis is quite straightforward and will be left to the reader. Following exactly the method of § 4.11.2 we take as our approximate wave function

$$\psi(\mathbf{r}) = \sum_{\mathbf{n}} e^{i\mathbf{k}.\mathbf{d_n}}\,\psi_s(\mathbf{r}-\mathbf{d_n}) \tag{50}$$

where $\psi_s(\mathbf{r})$ is the normalised wave function for an electron in a stationary state with energy E_s of an isolated atom at the origin. This wave function is of the correct form since $\psi(\mathbf{r})\exp(-i\mathbf{k}\,.\,\mathbf{r})$ is clearly a periodic function with periodicity of the lattice. We use this wave function to calculate the 'average' value of the energy; proceeding exactly as in § 4.11.2 we obtain the equation

$$E = E_s + \sum_{\mathbf{n}} e^{-i\mathbf{k}.\mathbf{d_n}} \int \psi_s^\star(\mathbf{r}-\mathbf{d_n})\, W(\mathbf{r})\,\psi_s(\mathbf{r})\,d\mathbf{r} \tag{51}$$

where

$$W(\mathbf{r}) = V(\mathbf{r}) - U_0(\mathbf{r}) \tag{52}$$

$V(\mathbf{r})$ is the actual potential for the crystal and we suppose that it will not differ much from the potential function $U_0(\mathbf{r})$ for an isolated atom for values of $\mathbf{r}$ corresponding to points well inside the unit cell whose origin is at $\mathbf{r} = 0$. Two types of integral occur in equation (51); they may be defined as follows:

$$-I_s = \int \psi_s^\star(\mathbf{r})\, W(\mathbf{r})\,\psi_s(\mathbf{r})\,d\mathbf{r} \tag{53}$$

$$-J_{\mathbf{n}s} = \int \psi_s^\star(\mathbf{r}-\mathbf{d_n})\, W(\mathbf{r})\,\psi_s(\mathbf{r})\,d\mathbf{r} \tag{54}$$

(the negative signs are introduced since $W(\mathbf{r}) < 0$).

The integral I_s will be small since $W(\mathbf{r})$ will be small where $\psi_s(\mathbf{r})$ is appreciable; the integrals $J_{\mathbf{n}s}$ will also be small, since, although $W(\mathbf{r})$ will no longer be small where $\psi_s^\star(\mathbf{r}-\mathbf{d_n})$ is appreciable, the overlap of $\psi_s^\star(\mathbf{r}-\mathbf{d_n})$ and $\psi_s(\mathbf{r})$ will be small under the conditions we have assumed. In general it will be sufficient to evaluate the integrals $J_{\mathbf{n}}$ only for nearest neighbours.

Consider first the simple cubic lattice; the vectors $\mathbf{d_n}$ of the nearest neighbours will have components $(\pm d, 0, 0)$, $(0, \pm d, 0)$, $(0, 0, \pm d)$. The integrals J have all the same value for these vectors provided ψ_s is spherically symmetrical, e.g. an s-state wave function. We therefore obtain for E the value

$$E = E_s - I_s - 2J_s(\cos k_x d + \cos k_y d + \cos k_z d) \tag{55}$$

k_x, k_y, k_z being the components, along the cube axes, of the wave vector $\mathbf{k}$.

For a body-centred cubic lattice with nearest neighbours at the eight points $(\pm d/2, \pm d/2, \pm d/2)$ we have

$$E = E_s - I_s - 8J_s \cos\frac{k_x d}{2} \cos\frac{k_y d}{2} \cos\frac{k_z d}{2} \tag{56}$$

as may readily be seen by expressing the cosines as exponentials.

Exercise Show that, taking account only of the twelve nearest neighbours of the face-centred cubic lattice, the energy may be expressed in the form

$$E = E_s - I_s - 4J_s\left(\cos\frac{k_y d}{2}\cos\frac{k_z d}{2} + \cos\frac{k_x d}{2}\cos\frac{k_y d}{2} + \cos\frac{k_z d}{2}\cos\frac{k_x d}{2}\right) \tag{57}$$

It will be seen from the above that, since I_s and J_s are small, the energy of an electron is shifted only slightly from E_s, the value for a single atom, and the allowed values spread over a band whose width depends on the value of the integral J_s. When the overlap of the wave functions from neighbouring lattice sites is very small the energy band approximates closely to the atomic level; thus we see that for real crystals, as for one-dimensional crystals, the energy levels of the tightly bound electrons are hardly changed, but that no one electron can be allocated to a particular atom of the crystal. Near an individual atom the wave function (50), apart from a phase factor, is very nearly the same as for a single atom; this gives a very satisfactory description of the deep-lying atomic levels. Indeed it turns out that the bound electrons, i.e. those that do not take part in conduction processes, may be described quite well by this type of approximation even where there is some overlap of their wave functions. For the valence electrons, however, the overlap is so great that the approximation breaks down.

The form of the bands is quite well illustrated by equations (55)–(57). Near $\mathbf{k} = 0$ equation (55) may be expressed in the form

$$E = E_s - I_s - 2J_s + J_s d^2(k_x^2 + k_y^2 + k_z^2) \tag{58}$$

The centre of the first Brillouin zone at $\mathbf{k} = 0$ therefore corresponds to an energy minimum. Comparing equation (58) with that for the energy of a free particle of mass m_e,

$$E = E_0 + \frac{\hbar^2}{2m_e}(k_x^2 + k_y^2 + k_z^2)$$

we see that we may attribute to an electron near the bottom of the band an effective mass m_e given by

$$m_e^{-1} = 2J_s d^2/\hbar^2 \tag{59}$$

so that when J_s is small and the allowed energy band is narrow, the effective mass m_e is large. Similar expressions may be obtained for the face-centred and body-centred cubic lattices.

For the simple cubic lattice the maximum energy occurs at points corresponding to the corners of the first Brillouin zone when

$$\mathbf{k}/2\pi = d^{-1}(\pm\tfrac{1}{2}, \pm\tfrac{1}{2}, \pm\tfrac{1}{2})$$

For the body-centred cubic lattice the maximum energy also occurs at the outermost points of the first Brillouin zone, such as A in Fig. 6.9, for which $\mathbf{k}/2\pi = d^{-1}(0, 1, 0)$, and similarly for the face-centred cubic lattice at points such as D in Fig. 6.8 with $\mathbf{k}/2\pi = d^{-1}(0, \frac{1}{2}, 1)$.

Exercise Show that for the energy bands given by equations (55)–(57) the normal derivative $\partial E/\partial k_n$ in each case is zero at the planes bounding the first Brillouin zone.

For each atomic level having a wave function ψ_s we shall have an allowed energy band for the crystalline solid, the width of the band being determined by the extent of the overlap of the atomic wave functions from neighbouring sites. When the periodic potential energy may be regarded as a small perturbation we have seen that the forbidden energy gaps between the allowed bands are quite narrow, while for the tight-binding condition the allowed energy bands are narrow. For the outer electrons of the atoms of the crystal, the so-called valence electrons, neither of these conditions holds; the allowed energy bands are quite broad, and according to circumstances there may or may not be quite appreciable gaps between them. For real crystals, a numerical method must therefore be used to obtain the form of the allowed energy bands; we shall later discuss such methods, but first of all we may draw some important conclusions on the nature of the allowed and forbidden energy bands. If the atomic energy levels are well separated and lie deep in the atom, so that there is very little overlap of the wave functions, the bands will not overlap. If, however, two or more atomic wave functions ψ_s correspond to nearly equal values of the energy E_s the resulting bands for the solid may overlap, and a more detailed treatment will be necessary. In particular, if two or more of these wave functions are degenerate they will *both* have to be included in the equation (50) giving the wave function for the crystal. We have discussed a problem of this type for one dimension in § 4.11.3. We may, for example, have for the atomic wave function a p-state with three degenerate wave functions of the form $\psi_1 = xf(r)$, $\psi_2 = yf(r)$, $\psi_3 = zf(r)$. In this way we obtain three overlapping bands having energy of the form

$$E_1 = E_s - \alpha + \beta \cos k_x d - \gamma(\cos k_y d + \cos k_z d) \tag{60}$$

with two other similar equations, obtained by cyclic permutation, for E_2 and E_3. We shall not give the detailed analysis of this situation since the algebra is rather complex and no essentially new information is obtained beyond that already given by the study of the one-dimensional problem in § 4.11.3. If we consider *one particular direction* in the crystal then it is found that the E/k curves of the three bands do not intersect but behave like the two bands in Fig. 4.22(b). In three dimensions, however, this does *not* mean that the bands do not overlap. It is quite possible for the maximum value of E for a lower band in *one* direction to come above the minimum value of E for a higher band in *another* direction. This is a very important difference between one-dimensional and three-dimensional problems. Also in three dimensions it is not necessary for the maximum or minimum energy of a band to occur at $\mathbf{k} = 0$ or at the zone edge; well-established examples are known for which these occur at an interior point of the first Brillouin zone. We shall later discuss some actual examples taken from numerical calculations of the form of the bands.

8.5 The cellular method

If we know the wave function for an electron in any unit cell of a crystal the wave function for any other cell may be obtained by means of Bloch's theorem. Let us consider first of all a crystal with one atom per unit cell, and each atom having only a single valence electron. As a first approximation we might take as the potential field in the cell in which the atom is located the field of an atomic positive ion. We immediately meet with a difficulty here; an atom has spherical symmetry but the basic unit cells of the crystals we have discussed, in general, are not even centrally symmetrical. For example in the body-centred cubic lattice the basic unit cell is a parallelepiped with an atom at *one* corner, and if we take a cubic unit cell we have then two atoms per unit cell. The way out of this difficulty was first proposed by E. Wigner and F. Seitz*; it is very similar to that proposed by Brillouin (see § 6.5) for defining a zone in $\mathbf{k}$-space. The planes bisecting the lines joining a selected atom to its neighbours form a closed polyhedron; the smallest polyhedron formed in this way is used as a cell in which a solution of the wave equation is sought, and such cells are known as Wigner–Seitz cells. For a simple lattice with one atom per unit cell the Wigner–Seitz cell is exactly the same as the first Brillouin zone in a reciprocal lattice having the same structure as this lattice. For example, for the face-centred cubic lattice the Wigner–Seitz cell is a dodecahedron and for the body-centred cubic lattice it is a truncated octahedron. The atom in question lies at the centre of these cells and all space may be filled with them, each having one atom at its centre. An essential difference between Brillouin zones and Wigner–Seitz cells occurs, however, when we have a lattice with a basis. A good example is the diamond lattice, with two atoms per unit cell. The Wigner–Seitz cells for this lattice are of two types; they are similar in shape but are differently orientated. Each has volume equal to *half* that of a unit cell. Two

* E. Wigner and F. Seitz, *Phys. Rev.*, **43**, 804 (1933); *Ibid.*, **46**, 509 (1934).

taken together form an equivalent unit cell but this has no longer the symmetry properties of the crystal. For a *simple* lattice the Wigner–Seitz cell has all the symmetry properties of the crystal.

In principle, the programme to be followed in finding the electronic wave functions for a crystal is as follows. Starting with the potential for an atomic positive ion, the wave functions are calculated for a Wigner–Seitz cell. The boundary conditions are no longer those for an isolated atom but instead continuity of the wave function and its normal derivative are required at the boundary of the cell; these boundary conditions should determine the wave functions and energy levels. From the wave functions a charge distribution may be calculated, and this in turn, together with the field of the atomic core may be used to find a new potential function. This is now used to calculate new wave functions and the process is repeated till a self-consistent field is found as in Hartree's method for atomic fields. Such self-consistent calculations have been carried out in very few cases.

The method as used originally by Wigner and Seitz (*loc. cit.*) was very much simpler. The field was taken in the form of an analytical approximation to the field of the atomic ion (Na^+ in this case) and instead of using the proper Wigner–Seitz zone, a sphere of equivalent volume was used. The boundary conditions chosen were continuity of ψ, and $\partial\psi/\partial n$ equal to zero, at the surface of the equivalent sphere. The zero value of the normal derivative at the surface of two cells is required if the potential and wave functions are spherically symmetrical as was assumed. Spherically symmetrical wave functions are only possible for $\mathbf{k} = 0$. The method was extended by J. C. Slater* to apply to wave functions with values of $\mathbf{k}$ other than $\mathbf{k} = 0$; here the wave functions were fitted only at certain chosen points, generally the centres of the faces of the Brillouin zone. He applied the method to sodium and found that in the highest occupied energy band the form of the $E/\mathbf{k}$ relationship, except near the edges of the band (which is only half filled in any case), is very closely similar to that for free electrons, in spite of the fact that the potential $V(\mathbf{r})$ is by no means a small perturbation. This important result explains why the free-electron model serves so well for the alkali metals (see § 2.3).

For many purposes a wave function of the form $\psi(\mathbf{r},\mathbf{k}) = u(r)\exp(i\mathbf{k}.\mathbf{r})$ is quite a good approximation, and has been used for a number of calculations, $u(r)$ being the function obtained by the cellular method for $\mathbf{k} = 0$. It has been found in most of the numerical calculations that have been carried out that $u(r)$ is fairly constant except near the centre of the cell, so that over quite a large fraction of the volume of the cell the wave function behaves like a plane wave.

More elaborate methods for obtaining a better value of $V(\mathbf{r})$ without a full self-consistent calculation have been developed, and also better methods of applying the boundary conditions at the surface of the

* J. C. Slater, *Handb. Phys.*, **19**, 1 (1956).

Wigner–Seitz cell. It is beyond the scope of this book to go into these methods in detail; an excellent review of the subject, and an account of more modern methods of numerical analysis, such as the orthogonalised plane wave and augmented plane wave methods, has been given by J. C. Slater*, together with a full discussion of the results obtained not only for a number of metals but also for some important semiconductors and insulators. We shall later discuss the band structure of these near the maxima and minima of the valence and conduction bands.

A knowledge of the energy levels in a crystal enables a calculation to be made of the binding energy of the crystal. Some success has been had in this direction but recent work has raised doubts as to whether some of this has been fortuitous, and it is now well known that although calculations of the type we have discussed give a good indication of the *form* of the energy bands they do not give reliable values of the energy. This subject has also been discussed in some detail by J. C. Slater (*loc. cit.*).

8.6 Wave packets

In exactly the same way as for one dimension, Bloch-type wave functions for different values of the wave vector **k** (and hence for the energy $E(\mathbf{k})$) may be combined to form wave packets describing the motion of a single electron through the crystal. The form of such a wave packet would be

$$\Phi(\mathbf{r}, t) = \int A(\mathbf{k})\, U_{\mathbf{k}}(\mathbf{r})\, e^{i[\mathbf{k}.\mathbf{r}-\omega(\mathbf{k})t]}\, d\mathbf{k} \tag{61}$$

where $\hbar\omega(\mathbf{k}) = E(\mathbf{k})$. When $A(\mathbf{k})$ is appreciable over only a small range of values of **k** the function $U_{\mathbf{k}}(\mathbf{r})$ may be taken outside the integral and the wave packet behaves like a similar wave packet in free space, the dispersion relationship now being given by the $E/\mathbf{k}$ relationship for the crystal instead of by the relationship for free electrons; the discussion of the three-dimensional problem follows exactly that for one dimension as given in § 4.6. It will readily be seen that the velocity of the particle associated with the wave packet is given by the group velocity **U** where

$$\mathbf{U} = \nabla_k \omega \tag{62}$$

(cf. equation (74) of § 1.5). As we have seen, the average velocity $\bar{\mathbf{v}}$ of an electron represented by such a wave packet is just equal to the group velocity so that we have

$$\bar{\mathbf{v}} = \hbar^{-1} \nabla_k . E(\mathbf{k}) \tag{63}$$

As we saw in our previous discussion, this average is taken over a unit cell of the lattice, so that $\bar{\mathbf{v}}$ represents a macroscopic velocity from which the fluctuations over distances comparable to the atomic spacings are averaged out. We may note that, in general, for a value of **k** corresponding to a zone boundary, $\bar{\mathbf{v}} = 0$ since, as we have seen, $\partial E/\partial k_n = 0$.

* J. C. Slater, *Handb. Phys.*, **19**, 1 (1956).

Exercise Using the expression

$$\bar{\mathbf{v}} = \frac{\hbar}{mi}\int \psi^{\star}\nabla\psi\,d\mathbf{r}$$

for the average velocity $\bar{\mathbf{v}}$ of an electron, show by averaging over a unit cell that

$$\bar{\mathbf{v}} = \hbar^{-1}\nabla_k E(\mathbf{k})$$

(This gives a more rigorous proof of equation (63) since the quantity $\bar{\mathbf{v}}$ defined in this way is the same as the velocity of the centre of the wave packet (see § 1.4.1). Show first that

$$\bar{\mathbf{v}} = \frac{\hbar\mathbf{k}}{m} - i\frac{\hbar}{m}\int U_{\mathbf{k}}^{\star}\nabla U_{\mathbf{k}}\,d\mathbf{r} \tag{64}$$

if $\psi_{\mathbf{k}}$ is normalised to unity. Write down the equation satisfied by $U_{\mathbf{k}}(\mathbf{r})$, and operate on it with ∇_k; multiply by $U_{\mathbf{k}}^{\star}$ and integrate. On integrating by parts and making use of the equation satisfied by $U_{\mathbf{k}}^{\star}$ the required result is readily obtained.)

8.7 The crystal momentum vector P

Again, as in one dimension, we introduce a vector **P** called the crystal momentum, related to the wave vector **k** by means of the equation $\mathbf{P} = \hbar\mathbf{k}$. It should be noted that except for free electrons **P** is not equal to $m\bar{\mathbf{v}}$. The discussion we have already given of the properties of the crystal momentum in one-dimensional crystals in § 4.7 applies equally to the momentum vector **P**. In particular we see that **P** is not uniquely defined but may have added to it a constant vector of the form $h\mathbf{b}$, where **b** is a vector of the reciprocal lattice. Equation (63) for the average velocity may be expressed in terms of P in the form

$$\bar{\mathbf{v}} = \nabla_P E \tag{65}$$

Exercise (1) Show that $\bar{\mathbf{p}}$, the average over a unit cell of the true momentum **p** of an electron, defined by means of the equation

$$\mathbf{p}\psi = -i\hbar\nabla\psi$$

is equal to $m\bar{\mathbf{v}}$, where **v** is the average velocity, so that it is not equal, except for free electrons, to the crystal momentum **P**.

(2) For the region of an energy band for which

$$E(\mathbf{k}) = \frac{\hbar^2 k^2}{2m^{\star}}$$

show that $\bar{\mathbf{v}} = \hbar\mathbf{k}/m^{\star}$ and hence that $\mathbf{P} = m^{\star}\bar{\mathbf{v}} = m^{\star}\bar{\mathbf{p}}/m$.

(3) For the region of an energy band for which

$$E(\mathbf{k}) = E_0 - \frac{\hbar k^2}{2m^{\star}}$$

show that $\mathbf{P} = -m^{\star}\bar{p}/m$.

(4) If we define the momentum matrix element $\mathbf{p}_{nm}(\mathbf{k}, \mathbf{k}')$ by means of the equation

$$\mathbf{p}_{nm}(\mathbf{k}, \mathbf{k}') = -i\hbar\int \psi_n^{\star}(\mathbf{r}, \mathbf{k})\nabla\psi_m(\mathbf{r}, \mathbf{k}')\,d\mathbf{r}$$

where $\psi_n(\mathbf{r}, \mathbf{k})$ and $\psi_m(\mathbf{r}, \mathbf{k}')$ are two normalised Bloch wave functions belonging to different bands, show that

$$\mathbf{p}_{nm}(\mathbf{k}, \mathbf{k}') = 0 \quad \text{if } \mathbf{k} \neq \mathbf{k}'$$

[Express $\mathbf{p}_{nm}$ in the form

$$\left(\int_{\text{cell}} e^{i(\mathbf{k}'-\mathbf{k}).\mathbf{r}} u_n^\star[\nabla + i\mathbf{k}'] u_m \, d\mathbf{r}\right) \sum_{j=1}^{N} e^{i(\mathbf{k}'-\mathbf{k}).\mathbf{d}_j}$$

by using the periodic property of the function u_n.]

(5) If the wave functions ψ_n, ψ_m in exercise (4) are normalised so that when $\mathbf{k} = \mathbf{k}'$

$$\int_{\text{cell}} u_n^\star u_n \, d\mathbf{r} = 1$$

show that

$$\mathbf{p}_{nm}(\mathbf{k}) = -i\hbar \int_{\text{cell}} u_n^\star(\mathbf{r}, \mathbf{k}) \nabla u_m(\mathbf{r}, \mathbf{k}) \, d\mathbf{r}$$

(6) Show that the quantity $\mathbf{p}_{nm}(\mathbf{k})$ may also be expressed in the form

$$\mathbf{p}_{nm}(\mathbf{k}) = -m \frac{(E_n - E_m)}{\hbar} \int u_n^\star(\mathbf{r}, \mathbf{k}) \nabla_k u_m(\mathbf{r}, \mathbf{k}) \, d\mathbf{r}$$

(Proceed as in the exercise in § 8.6 but replacing u_n by u_m and retaining $u_n^\star$ as it stands. Use is then made of the equations satisfied by u_m and $u_n^\star$ having energy eigenvalues E_m and E_n.)

8.8 Equation of motion for an electron in a crystal

The existence of wave functions of the Bloch form and their formation into wave packets as discussed in § 8.6 shows that electrons may move freely through a perfect crystal. When we average the velocity over regions large compared with a unit cell we find that it will remain constant provided no *external* forces act on the electron—this follows from the fact that the distribution of the wave vectors $\mathbf{k}$ remains unchanged. There will be the usual spreading of the wave packet which is just an expression of the Uncertainty Principle as discussed in § 1.4. The only essential difference between the free motion of electrons in a perfect crystal and that of completely free electrons comes from the different dispersion relationship connecting the angular frequency ω (or energy) and the wave vector $\mathbf{k}$, and this may be expressed by saying that the electron no longer has its usual mass m. For a one-dimensional crystal we were able to define an effective mass $m^\star$ given by equation (93) of § 4.7; in three dimensions we can only define such a scalar effective mass under very restricted circumstances since the $E/\mathbf{k}$ relationship cannot normally be expressed, even near the minimum energy of a band by an expression of the form

$$E = E_0 + \frac{\hbar^2 k^2}{2m^\star} \tag{66}$$

When this is so, however, we may define an effective mass $m^\star$ by comparison with the free-electron relationship

$$E = \frac{\hbar^2 k^2}{2m}$$

Similarly when we have

$$E = E_0 - \frac{\hbar^2 k^2}{2m^\star} \tag{67}$$

near the maximum energy of a band we may define a scalar effective mass $m^\star$ of a particle which we shall later identify as a positive hole (see § 4.13.1). In general, the relationship between E and $\mathbf{k}$ involves the three components of the vector $\mathbf{k}$ and not simply its magnitude k. The latter restricted condition is usually referred to as that of 'spherical' energy bands since the constant-energy surfaces are spheres in $\mathbf{k}$-space. This condition occurs approximately for the alkali metals, but is much less common than was once thought. When we do not have 'spherical' energy bands we cannot define a scalar effective mass even at the extrema of the bands, and we must now see how to generalise this useful concept for all types of bands.

In order to do this we must discuss the motion of electrons in a crystal under the influence of an external field of force. Let the external force be denoted by the vector $\mathbf{F}$, and we shall first of all suppose that $\mathbf{F}$ is not always at right angles to the velocity $\mathbf{v}$; in this case we may proceed as in § 4.7, for one dimension, and write down the average rate at which the force $\mathbf{F}$ does work. This is given by $\mathbf{F}.\bar{\mathbf{v}}$ and is equal to the rate at which the energy E increases; we have therefore, using equation (65),

$$\frac{dE}{dt} = \nabla_P E.\frac{d\mathbf{P}}{dt} = \mathbf{F}.\bar{\mathbf{v}} = \nabla_P E.\mathbf{F} \tag{68}$$

hence we have

$$\frac{\hbar\, d\mathbf{k}}{dt} = \frac{d\mathbf{P}}{dt} = \mathbf{F} \tag{69}$$

Again we see that, if $\mathbf{F} = 0$, the crystal momentum $\mathbf{P}$, and the wave vector $\mathbf{k}$, are constant. When F is not equal to zero, equation (69) shows us how $\mathbf{P}$ and $\mathbf{k}$ change with time.

In three dimensions, the above treatment fails for a very important case, that in which the vector $\mathbf{F}$ is always at right angles to the velocity $\mathbf{v}$; an example of this condition occurs when we have an external magnetic field. Here the force $\mathbf{F}$ on an electron is equal to $-\boldsymbol{e}\mathbf{v}\times\mathbf{B}$, where $\boldsymbol{e}$ is the magnitude of the electronic charge and $\mathbf{B}$ the magnetic flux vector. In this case the force fluctuates rapidly as the electron moves over a single unit cell and we should form the average force $\bar{F}$ given by

$$\bar{F} = -\boldsymbol{e}[\bar{\mathbf{v}}\times\mathbf{B}] \tag{70}$$

We cannot now proceed as before, since the average work done by this force is equal to zero, the vector $\bar{\mathbf{F}}$ being at right angles to $\bar{\mathbf{v}}$; a more elaborate treatment is therefore required. There seems to be no simple way of proving that equation (69) still holds when **F** has the form (70). Indeed, neither equation (69) nor the more general result are *exactly* true. Equation (69) is only true for the force due to an electric field $\mathscr{E}$ provided this is not too great; in deriving it we have assumed that the wave function of the electron can be described using wave functions from a *single* energy band, and it has been shown by C. Zener* that in a very high electric field (generally of the order of 10^5 V/cm) electrons may be excited from one band to a higher band and equation (69) is no longer a good approximation (see § 8.13); for smaller fields it is, however, quite a good approximation.

When we have both external electric and magnetic fields equation (69) becomes

$$\frac{\hbar\,\mathrm{d}\mathbf{k}}{\mathrm{d}t} = \frac{\mathrm{d}\mathbf{P}}{\mathrm{d}t} = -e\mathscr{E} - e[\bar{\mathbf{v}} \times \mathbf{B}] \tag{71}$$

This equation holds, with the same restriction on $\mathscr{E}$, for values of **B** which are not too great; for larger values of **B** the paths of the electrons are bent into nearly circular orbits and the angular momentum becomes quantized; we shall later discuss this effect in more detail (see § 11.3.1). There seems to be no simple way of deriving equation (71) under the conditions for which it holds; a derivation of equation (69) using momentum eigenfunctions by J. C. Slater† may be extended to include magnetic fields; derivations by H. Jones‡ and A. H. Wilson§ are more straightforward but require somewhat lengthy mathematical analysis. A derivation based on the latter will be given in Appendix I.

Returning to equation (65) for the average velocity $\bar{\mathbf{v}}$, this may be differentiated with respect to the time to obtain the average acceleration; we have

$$\left.\begin{aligned}\frac{\mathrm{d}\bar{\mathbf{v}}}{\mathrm{d}t} &= \frac{\mathrm{d}}{\mathrm{d}t}\nabla_P E = \nabla_P\frac{\mathrm{d}E}{\mathrm{d}t}\\ &= \nabla_P\left(\nabla_P.\frac{\mathrm{d}\mathbf{P}}{\mathrm{d}t}\right)E\end{aligned}\right\} \tag{72}$$

Using equation (69) we may write equation (72) in the form

$$\frac{\mathrm{d}\bar{\mathbf{v}}}{\mathrm{d}t} = \nabla_P(\nabla_P E.\mathbf{F}) \tag{73}$$

The differential operator ∇_P acting on the scalar product $\nabla_P E.\mathbf{F}$ gives a vector; alternatively we may write equation (73) in the form

$$\frac{\mathrm{d}\bar{\mathbf{v}}}{\mathrm{d}t} = \mathbf{O}E.\mathbf{F} \tag{73a}$$

* C. Zener, *Proc. Roy. Soc.*, **A145**, 523 (1934).
† J. C. Slater, *Phys. Rev.* **76,** 1592 (1949).
‡ H. Jones, *Handb. Phys.*, **19,** 227 (1956).
§ A. H. Wilson, *The Theory of Metals*. Cambridge University Press (1954), p. 48.

where **O** is a tensor operator of rank 2 the symbol **O**.**F** representing a summation-type product which gives a vector derived from a tensor of rank 2 and a vector. (In Cartesian notation the x-component would be $O_{xx}F_x + O_{xy}F_y + O_{xz}F_z$ and similarly for the others). The tensor $\mathbf{O}E$, through the analogy of the classical equations of motion, is known as the effective-mass tensor. In Cartesian coordinates it has components $1/m_{rs}$ which may be written in the form

$$\frac{1}{m_{rs}} = \frac{1}{m_{sr}} = \frac{\partial^2 E}{\partial P_r \, \partial P_s} = \hbar^{-2} \frac{\partial^2 E}{\partial k_r \, \partial k_s} \tag{74}$$

equation (72) may then be expressed in the form

$$\frac{\mathrm{d}v_1}{\mathrm{d}t} = \frac{F_1}{m_{11}} + \frac{F_2}{m_{12}} + \frac{F_3}{m_{13}} \tag{75}$$

with similar equations for the components v_2, v_3, of $\mathbf{v}$, F_1, F_2, F_3, being the components of the force $\mathbf{F}$. If E may be expressed as a positive definite quadratic form in the vicinity of a band minimum, we have

$$E = E_0 + \tfrac{1}{2}\left[\frac{P_1^2}{m_{11}} + \frac{P_2^2}{m_{22}} + \frac{P_3^2}{m_{33}} + \frac{2P_1 P_2}{m_{12}} + \frac{2P_2 P_3}{m_{23}} + \frac{2P_3 P_1}{m_{13}}\right] \tag{76}$$

In general, we may make a transformation to axes for which $m_{rs}^{-1} = 0$, $s \neq r$, and write

$$E = E_0 + \tfrac{1}{2}\left[\frac{P_1^2}{m_1} + \frac{P_2^2}{m_2} + \frac{P_3^2}{m_3}\right] \tag{77}$$

so that we have only three non-zero components of the effective-mass tensor; in this case the constant-energy surfaces near the minimum are ellipsoids. If E is expressed in terms of the wave vector $\mathbf{k}$ in the form given by equation (28) we have

$$\frac{1}{m_{rs}} = \hbar^{-2} \frac{\partial^2 E}{\partial k_r \, \partial k_s} = \hbar^{-2} A_{rs} \tag{78}$$

A further simplification, which frequently occurs when an energy minimum lies on one of the axes of symmetry of a crystal, is that two of the quantities m_1, m_2, m_3 are equal. In this case it is usual to write $m_1 = m_l$, $m_2 = m_3 = m_t$ and to call m_l the longitudinal effective mass and m_t the transverse effective mass. We then have

$$E = E_0 + \tfrac{1}{2}\left[\frac{P_1^2}{m_l} + \frac{1}{m_t}(P_2^2 + P_3^2)\right] \tag{79}$$

and in this case the constant-energy surfaces are spheroids. Finally if $m_1 = m_2 = m_3$ we may write E in the form

$$E = \frac{P^2}{2m_e} = \frac{\hbar^2 k^2}{2m_e} \tag{80}$$

and we have a scalar effective mass m_e; the constant-energy surfaces are now spheres.

Near the maximum of a band, on the other hand, E may be expressed in the form

$$E = E_0 - \tfrac{1}{2}\left[\frac{P_1^2}{m_1'} + \frac{P_2^2}{m_2'} + \frac{P_3^2}{m_3'}\right] \tag{81}$$

In this case it is conventional to define the effective masses as the *negatives* of the reciprocals of the components of the tensor $\mathbf{O}E$, i.e.

$$\frac{1}{m_{rs}} = -\frac{\partial^2 E}{\partial P_r\, \partial P_s} \tag{82}$$

For an energy band given by equation (81) this makes the effective masses *positive* quantities; we shall later discuss the reason for this convention; we have therefore effective masses m_1', m_2', m_3'. For a 'spherical' band of this form, with

$$E \doteq E_0 - \frac{P^2}{2m_h} = E_0 - \frac{\hbar^2 k^2}{2m_h} \tag{83}$$

we have a scalar effective mass m_h. We shall later discuss a number of examples of energy bands having these forms near their extrema (see § 9.3).

Exercise Show that the effective-mass tensor may be expressed in the form

$$\frac{1}{m_{rr}} = \frac{1}{m} + \frac{i}{m}\int \left(\frac{\partial U_{\mathbf{k}}^{\star}}{\partial x_r}\frac{\partial U_{\mathbf{k}}}{\partial k_r} - \frac{\partial U_{\mathbf{k}}}{\partial x_r}\frac{\partial U_{\mathbf{k}}^{\star}}{\partial k_r}\right) d\mathbf{r} \tag{84}$$

$$\left.\begin{aligned} \frac{1}{m_{rs}} &= \frac{i}{m}\int \left(\frac{\partial U_{\mathbf{k}}^{\star}}{\partial x_r}\frac{\partial U_{\mathbf{k}}}{\partial k_s} - \frac{\partial U_{\mathbf{k}}}{\partial x_r}\cdot\frac{\partial U_{\mathbf{k}}^{\star}}{\partial k_s}\right) d\mathbf{r} \\ &= \frac{i}{m}\int \left(\frac{\partial U_{\mathbf{k}}^{\star}}{\partial x_s}\cdot\frac{\partial U_{\mathbf{k}}}{\partial k_r} - \frac{\partial U_{\mathbf{k}}}{\partial x_s}\cdot\frac{\partial U_{\mathbf{k}}^{\star}}{\partial k_r}\right) d\mathbf{r} \end{aligned}\right\} \tag{85}$$

x_r $(r = 1, 2, 3)$ being the components of the displacement vector and, k_r $(r = 1, 2, 3)$ the components of the wave vector, the wave function $\psi_{\mathbf{k}}(\mathbf{r}) = \exp(i\mathbf{kr})\, U_{\mathbf{k}}(\mathbf{r})$ being normalised to unity. (Start with equation (64) writing $\bar{v}_r$ as $\hbar^{-1}\,\partial E/\partial k_r$, etc., and differentiate with respect to k_s. After some partial integrations the required result will be obtained.)

8.9 Boundary conditions and the number of allowed energy levels

The actual values $E_{\mathbf{k}}$ of the energy E which are allowed for any particular crystal have not yet been determined; what we have found is the energy $E_{\mathbf{k}}$ associated with a particular solution $\psi_{\mathbf{k}}$ of Schrödinger's equation. We have treated $E_{\mathbf{k}}$ as a continuous function of $\mathbf{k}$, and indeed this is a good approximation for many purposes. The actual eigenvalues of E are determined by the boundary conditions in exactly the same way as the frequencies of the lattice vibrations were determined (see § 6.7). As before, boundary conditions for a finite crystal are difficult to apply and we use periodic boundary conditions. Thus we see that the allowed values of the wave vector $\mathbf{k}$ describing the motion of electrons in the crystal lattice are exactly the same as the allowed values of the wave vector $\mathbf{K}$ describing the lattice vibrations. They are uniformly spread and have density $V/8\pi^3$ in $\mathbf{k}$-space, where V is the volume of the crystal. For electrons, however, we have two particles occupying each level, one with each direction of the spin, according to the Pauli Principle (see § 2.4), so that for electrons the number of allowed levels $n(\mathbf{k})\,d\mathbf{k}$ with wave vector $\mathbf{k}$ between $\mathbf{k}$ and $\mathbf{k}+d\mathbf{k}$ is given by

$$n(\mathbf{k})\,d\mathbf{k} = V\,d\mathbf{k}/4\pi^3 \tag{86}$$

The total number of allowed values of $\mathbf{k}$ is N, as before, where N is the number of unit cells in the crystal; the number of energy levels for each allowed band which can be occupied by electrons is N, each level being capable of accepting two electrons.

For the deep-lying levels in an atom the electron spins are paired, and we have two electrons occupying each level defined apart from fine structure due to electron spin. Each such atomic level will give an energy band in the crystal and we see that each level in the band will be occupied by two electrons, i.e. there are $2N$ electrons to fill the N levels. When we have more than one atom per unit cell, each band will be split into s sub-levels, where s is the number of atoms per unit cell, and when we have paired electron spins each sub-band will be completely filled. We have already discussed such filled bands in § 4.13 and shown that they play no part in conduction processes. For the valence electrons we may either have an even number, with paired electron spins, or an odd number. In the latter case we shall have the highest band half-filled unless it overlaps with another band; this overlap is, as we have seen, quite possible. In any case, conduction will be possible and we have a metal. When the number of valence electrons is even, and the highest occupied band does not overlap another band we clearly have a completely filled band, and in this case we have an insulator or semiconductor (see § 4.13.2). If two bands overlap each will be only partially filled, and again we have a metal; we shall later discuss in more detail the various situations which may occur (see §§ 9.2, 9.3).

8.10 Expression for the electric current

The current density due to a charge q moving with velocity $\mathbf{v}$ in a volume V is equal to $q\mathbf{v}/V$. This may be seen in an elementary way as follows; suppose we have a large number N of electrons contained in a volume V, distributed uniformly throughout the volume and each moving with velocity $\mathbf{v}$. The average charge density will be qN/V and the charge crossing unit area normal to $\mathbf{v}$ per unit time will be $qN|\mathbf{v}|/V$; this will give the magnitude of the current density (charge flow per unit area per unit time). The vector current density will clearly be $qN\mathbf{v}/V$, having the same direction as the velocity vector $\mathbf{v}$. For a single charge the current density will therefore be $q\mathbf{v}/V$; if we have n charges per unit volume it will be $nq\mathbf{v}$ since $n = N/V$. For an electron with charge $-e$ the current density $\mathbf{I}$ will be given by

$$\mathbf{I} = -e\mathbf{v}/V \tag{87}$$

and if we have n electrons per unit volume with velocity $\mathbf{v}$ their contribution to the current density will be given by

$$I = -ne\mathbf{v} \tag{88}$$

The expression given by the quantum theory for the average velocity

$$\bar{v} = \frac{\hbar}{2im}\int (\psi^{\star}\nabla\psi - \psi\nabla\psi^{\star})\,d\mathbf{r} \tag{89}$$

(cf. exercise at end of § 8.7) is compatible with the expression for the current density given by the fundamental postulates of quantum mechanics. If the charge density ρ is given by $q\psi\psi^{\star}$ then the quantity $\mathbf{I}$ required to make ρ satisfy the continuity equation

$$\frac{\partial\rho}{\partial t} + \nabla.\mathbf{I} = 0 \tag{90}$$

is

$$\mathbf{I} = \frac{q\hbar}{2im}[\psi^{\star}\nabla\psi - \psi\nabla\psi^{\star}] \tag{91}$$

(see any textbook on elementary quantum mechanics). If we form the expression for the average current density due to an electron located somewhere in the volume V we have

$$\left.\begin{aligned}\bar{\mathbf{I}} &= \frac{1}{V}\int \mathbf{I}\,d\mathbf{r} \\ &= \frac{q\hbar}{2imV}\int (\psi^{\star}\nabla\psi - \psi\nabla\psi^{\star})\,d\mathbf{r}\end{aligned}\right\} \tag{92}$$

$$= q\bar{\mathbf{v}}/V \tag{92a}$$

When we take $q = -e$ and compare equation (92a) with equation (87) we see that the correct value to take for the velocity $\mathbf{v}$ in the expression for the average current density due to an electron is the quantum-mechanical average velocity $\bar{\mathbf{v}}$. This, as we have seen, is the velocity with which a wave packet representing an electron would move and is equal to $\hbar^{-1}\nabla_k E(\mathbf{k})$.

In a crystal, there will be a certain probability $f(\mathbf{k})$ that a particular level is occupied; this is given, in conditions of equilibrium, by the Fermi–Dirac function (see equation (55) of § 2.6) and the average number of electrons occupying the level will be equal to twice this value. The number of electrons in levels having values of $\mathbf{k}$ between $\mathbf{k}$ and $\mathbf{k}+d\mathbf{k}$ will be (cf. equation (86))

$$f(\mathbf{k})\,n(\mathbf{k})\,d\mathbf{k} = f(\mathbf{k})\,V\,d\mathbf{k}/4\pi^3 \tag{93}$$

The current density due to electrons in a particular energy band will therefore be given by

$$\mathbf{I} = \frac{-e}{4\pi^3} \int f(\mathbf{k})\,\bar{\mathbf{v}}(\mathbf{k})\,d\mathbf{k} \tag{94}$$

the integral being taken over the first Brillouin zone in **k**-space. (As usual, we have replaced a sum over the large but finite number of allowed values of **k** by an integral, and we note that the volume V has cancelled out.) The functions $f(\mathbf{k})$ and $\bar{\mathbf{v}}(\mathbf{k})$ are determined when the relationship between E and **k** is known, and we shall later use this expression to calculate the electrical conductivity.

For a completely filled band we have $f(\mathbf{k}) = 1$ for all allowed values of **k**, since each level is certainly occupied; this means that the Fermi level lies well above the top of the band (see § 2.6). We also note that $\bar{\mathbf{v}}(\mathbf{k}) = -\bar{\mathbf{v}}(-\mathbf{k})$, since $E(\mathbf{k})$ is an even function of **k**; the Brillouin zone being symmetrical about $\mathbf{k} = 0$, we clearly have $\mathbf{I} = 0$ for a completely filled band. Thus in calculating the current density we need only consider partially filled bands. When we have only one such band, the total current density is given by a single integral such as (94); when we have more than one partially filled band we must form an integral for each such band, and the total current density is given by the sum of these integrals.

8.11 Positive holes

A case of particular interest occurs when we have an energy band which is almost completely filled. Suppose first of all that there is only a single electron missing, and that the missing electron is from an energy level with wave vector $\mathbf{k}'$. The current density is then given by

$$\mathbf{I} = \frac{-e}{V} \sum_{\mathbf{k}}{}' \bar{\mathbf{v}}(\mathbf{k}) \tag{95}$$

where the sum $\sum'$ is taken over all electrons except the one having $\mathbf{k} = \mathbf{k}'$. We may also write equation (95) in the form

$$\mathbf{I} = \frac{-e}{V} \sum_{\mathbf{k}} \bar{\mathbf{v}}(\mathbf{k}) + \frac{e}{V} \bar{\mathbf{v}}(\mathbf{k}') \tag{96}$$

where the sum $\sum$ is taken over all electrons in the filled band, and so is equal to zero. We therefore have

$$\mathbf{I} = \frac{e}{V} \bar{\mathbf{v}}(\mathbf{k}') \tag{97}$$

This is just the current density we should have from a charge $+e$ moving with velocity $\bar{v}(\mathbf{k}')$. This suggests that a missing electron from a nearly filled band gives the same effect as a positive charge moving with the average velocity corresponding to the vector $\mathbf{k}'$ of the missing electron and equal to $\hbar^{-1}\nabla_k(E)$; we shall later give further reasons for this equivalence. Such a missing electron is therefore known as a *positive hole* or simply as a *hole*. Clearly if we have more than one electron missing, the current density is given by

$$\frac{e}{V}\sum_{\mathbf{k}} \bar{\mathbf{v}}(\mathbf{k}) \tag{98}$$

where the sum is taken over the values of $\mathbf{k}$ *for which electrons are missing*. In general, we shall not know definitely that an electron is missing, except for a completely empty band, but we shall know the *probability* of its being missing; this will be given by $f'(\mathbf{k}) = 1-f(\mathbf{k})$. Replacing the sum, as before by an integral, we have for the current density due to holes in a particular band

$$\mathbf{I}_h = \frac{e}{4\pi^3}\int [1-f(\mathbf{k})]\,\bar{\mathbf{v}}(\mathbf{k})\,d\mathbf{k} \tag{99}$$

When we have two bands, (1) nearly empty, and (2) nearly full, we have

$$I = \frac{-e}{4\pi^3}\int_{(1)} f(\mathbf{k})\,\bar{\mathbf{v}}(\mathbf{k})\,d\mathbf{k} + \frac{e}{4\pi^3}\int_{(2)} [1-f(\mathbf{k})]\,\bar{\mathbf{v}}(\mathbf{k})\,d\mathbf{k} \tag{100}$$

We must now consider how such a positive hole would move through a crystal. We have seen that it would have velocity $\bar{\mathbf{v}}(\mathbf{k})$, and we must ask how this changes in the presence of an external force. Suppose the electron which is missing had wave vector $\mathbf{k}$ at $t = 0$; then under a force $\mathbf{F}$ it would have wave vector $\mathbf{k}+\hbar^{-1}\mathbf{F}\Delta t$ a short time Δt later (see equation (69)), and corresponds to an electron with vector $\mathbf{k}+\hbar^{-1}\mathbf{F}\Delta t$ missing. The value of $\mathbf{k}$ for the positive hole therefore changes in the same way as for an electron. We therefore have

$$\dot{\bar{\mathbf{v}}}(\mathbf{k}) = \nabla_k \bar{\mathbf{v}}(\mathbf{k})\,\dot{\mathbf{k}} = \hbar^{-2}\nabla_k\nabla_k E\,.\,\mathbf{F} \tag{101}$$

When the force $\mathbf{F}$ is due to an electric field $\mathscr{E}$ and a magnetic field $\mathbf{H}$ we have

$$\mathbf{F} = -e\mathscr{E} - e[\bar{\mathbf{v}}\times\mathbf{B}] \tag{102}$$

Near the top of a band which is almost completely filled $\partial^2 E/\partial k^2$ is *negative* and we may make two interpretations of equation (102). This will perhaps be clearer if we apply the equation to a simple example, that of a 'spherical' band having $E(\mathbf{k})$ given by equation (83). In this case equation (102) becomes

$$-m_h\dot{\bar{\mathbf{v}}}(\mathbf{k}) = -e\mathscr{E} - e(\bar{\mathbf{v}}\times\mathbf{B}) \tag{103}$$

which may also be written as

$$m_h\dot{\bar{\mathbf{v}}}(\mathbf{k}) = e\mathscr{E} + e(\bar{\mathbf{v}}\times\mathbf{B}) \tag{104}$$

Equation (104) is equivalent to that of a particle with effective mass m_h and positive charge $+e$. Thus we see that we may either define the effective mass as before, in which case it will be negative and retain a negative charge for the hole or we may define the effective mass tensor *for a hole* by means of the equation

$$m_{rs}^{-1} = -\hbar^{-2}\,\partial^2 E/\partial k_r k_s \tag{105}$$

and interpret the hole as having a *positive* charge. A positive mass is a more familiar concept and a positive charge is in accordance with the interpretation of equation (97); this convention is therefore usually adopted. Thus we see that a positive hole behaves like a normal particle with a positive charge in an electric and magnetic field. We should not lose sight of the fact, however, that an individual electron near the top of a band does have a *negative* effective mass in the sense that it moves under the influence of an applied force in a direction opposite to that in which a free electron would move (see §§ 4.7, 4.13.1).

It is interesting to compare the different relationships between momentum and velocity for holes and electrons. We illustrate this by comparing an electron with scalar effective mass m_e and a hole with scalar effective mass m_h, $E(k)$ being given, respectively, by (80) and (83). For both, the crystal momentum $\mathbf{P}$ is equal to $\hbar\mathbf{k}$; for the electron

$$\mathbf{P} = m_e\,\bar{\mathbf{v}}(\mathbf{k}) \tag{106}$$

while for the hole

$$\mathbf{P} = -m_h\,\bar{\mathbf{v}}(\mathbf{k}) \tag{107}$$

Suppose we have an electron–hole pair created by absorption of a quantum of radiation, the electron being excited from a full band into an empty band above, with the *same* value of $\mathbf{k}$ (a so-called 'vertical' transition), then the electron and hole move off in *opposite* directions; a similar result may readily be obtained for more complex bands for which the effective mass has tensorial form.

We have seen that, for a single electron, we may obtain a wave function $\psi_\mathbf{k}$ in the form of a Bloch function which represents a stationary state with energy $E_\mathbf{k}$ in the absence of external fields. The so-called 'one-electron' approximation consists of treating the individual electrons in a crystal separately by means of such wave functions. We have also seen how such wave functions may be combined to form wave packets representing the motion of a single electron. Similar wave packets may, in principle, be formed to represent positive holes but are not generally useful in formulating the theory, equations such as (104) and its generalisation forming the basis of transport theory. It is, however, interesting to consider how such a wave packet may be formed.

The complete wave function for the valence electrons, which for simplicity we shall restrict to a single band, may be formed from linear combinations of products of wave functions of the type $\psi_\mathbf{k}$ for the N electrons. Spin wave functions may also be included if desired but we shall omit these and assume instead that only one electron may have a particular value of $\mathbf{k}$; we have then N values $\mathbf{k}_r$ for the N electrons, and, with this simplification, a full band is obtained when all these are occupied. It is well known (see any textbook on elementary quantum mechanics) that the proper linear combinations of n wave functions $\psi_1, \psi_2 \ldots \psi_n$ for n electrons whose position vectors are $\mathbf{r}_1, \mathbf{r}_2 \ldots \mathbf{r}_n$ is the determinant

$$D_n = \|\psi_s(\mathbf{r}_t)\| \tag{108}$$

This is an antisymmetrical wave function, as required for electrons, and satisfies the Pauli Principle since it vanishes if any two of the ψ_s are the same. If $n = N$ we may represent a full band by means of a wave function Ψ given by

$$\Psi = (N!)^{-1/2} D_N(\mathbf{r}_1, \mathbf{r}_2 \ldots \mathbf{r}_N) \tag{109}$$

the wave functions $\psi_s(\mathbf{r}_t)$ being in this case $\psi_{\mathbf{k}_s}(\mathbf{r}_t)$. From this wave function the charge density may be calculated in the usual way and also the current; it is easy to show from the orthogonal property of the wave functions that these are just equal to the sum of the charges and currents for the individual electrons. If we have a single electron missing we may form a wave function of the form

$$\Psi' = [(N-1)!]^{-1/2} D'_{N-1}(\mathbf{r}_1, \mathbf{r}_2, \ldots \mathbf{r}_{N-1}) \tag{110}$$

This represents a positive hole with $\mathbf{k} = \mathbf{k}_p$ where $\mathbf{k}_p$ is the value of $\mathbf{k}_s$ missing from (110). Forming the charge and current we again find that they have the values for the full band, *less* that for the missing electron, i.e. the behaviour is that of a positive hole. The analysis is straightforward and is left to the reader as an exercise.

We have seen that we may also form wave packets in the form

$$\phi_{s'}(\mathbf{r}_{t'}) = \sum_s A_{ss'} \psi_{\mathbf{k}_s}(\mathbf{r}_{t'}) \tag{111}$$

(We use the summation notation rather than the integral for convenience, to enable us to sum over a finite number of wave functions.) Suppose that the wave packet (111) has its sharp maximum in the constants $A_{ss'}$ corresponding to $\mathbf{k}_s = \mathbf{k}'_s$. We may form $N-1$ similar functions for the $(N-1)$ other values of $\mathbf{k}_s$. These represent wave packets with velocities $\bar{\mathbf{v}}_s$ corresponding to $\hbar^{-1}\nabla_{\mathbf{k}_s} E(\mathbf{k}_s)$. It may readily be shown that the constants $A_{ss'}$ may be chosen so as to make the functions $\phi_{s'}$ orthogonal and also normalised. We may now form the determinant

$$D'_{N-1} = \|\phi_{s'}(\mathbf{r}_{t'})\| \tag{112}$$

in which the value $\mathbf{k}''_s$ of $\mathbf{k}'_s$ is missing, and we may also form D'_N with all values of $\mathbf{k}'_s$ included. It will readily be seen that D'_N represents a full band and D'_{N-1} a wave packet of holes with $\mathbf{k}$ centred on $\mathbf{k} = \mathbf{k}''_s$.

8.12 Motion of an electron through a band

We must now return to consider in rather more detail the motion of an electron in a crystal under electric and magnetic forces. For simplicity, we shall consider first electrons near the bottom of a 'spherical' band for which $E(\mathbf{k})$ is given by equation (80). Suppose an electric field $-\mathscr{E}$ is applied in the direction of the lattice vector $\mathbf{d}_1$. Let us consider two electrons having values of k_1 equal to k_{10} and $-k_{10}$ at $t=0$; after a short time Δt these values will be, respectively, according to equation (69) $k_{10}+e\hbar^{-1}\mathscr{E}\Delta t$ and $-k_{10}+e\hbar^{-1}\mathscr{E}\Delta t$. If we suppose that only a limited number of energy levels with values of $\mathbf{k}$ near $\mathbf{k}=0$, and such that equation (80) holds, are occupied then it is clear that all the values of $\mathbf{k}$ in $\mathbf{k}$-space drift in the direction of the (1) axis. If before $t=0$ the system is in equilibrium the $\mathbf{k}$-values will be symmetrically distributed about the origin in $\mathbf{k}$-space; after a time Δt on the other hand the distribution will have drifted a distance $\hbar^{-1}e\mathscr{E}\Delta t$ in the direction of the (1) axis and will

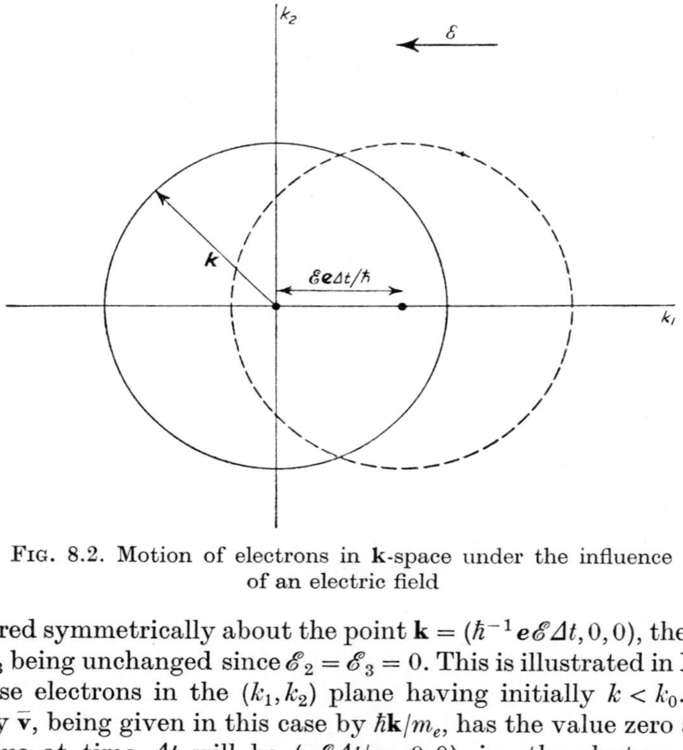

FIG. 8.2. Motion of electrons in $\mathbf{k}$-space under the influence of an electric field

be centred symmetrically about the point $\mathbf{k}=(\hbar^{-1}e\mathscr{E}\Delta t,0,0)$, the values of k_2, k_3 being unchanged since $\mathscr{E}_2=\mathscr{E}_3=0$. This is illustrated in Fig. 8.2 for those electrons in the (k_1,k_2) plane having initially $k<k_0$. If the velocity $\bar{\mathbf{v}}$, being given in this case by $\hbar\mathbf{k}/m_e$, has the value zero at $t=0$ the value at time Δt will be $(e\mathscr{E}\Delta t/m_e,0,0)$, i.e. the electron has an acceleration directed along the (1) axis equal to $e\mathscr{E}/m_e$. In a perfect crystal this acceleration would continue, the value of k_1 increasing; for larger values of k_1 however, the energy E would no longer be given by equation (80) and as $\mathbf{k}$ approached the edge of a Brillouin zone the effective mass would become negative and the electron would be slowed down, finally coming to rest when k_1 has the value corresponding to the zone edge ($\bar{v}=0$ at the zone edge). Further increase in k_1 would take it into the next zone; if we use the reduced representation we may say

instead that the electron now appears at the opposite edge of the first zone, with a negative value of k_1. The value of k_1 now increases till the electron is finally brought to rest at $k_1 = 0$, and the whole process is repeated; this is illustrated in Fig. 8.3. The electron therefore appears to oscillate backwards and forwards in the crystal having its motion reversed when k_1 changes sign. This corresponds of course to Bragg reflection at the

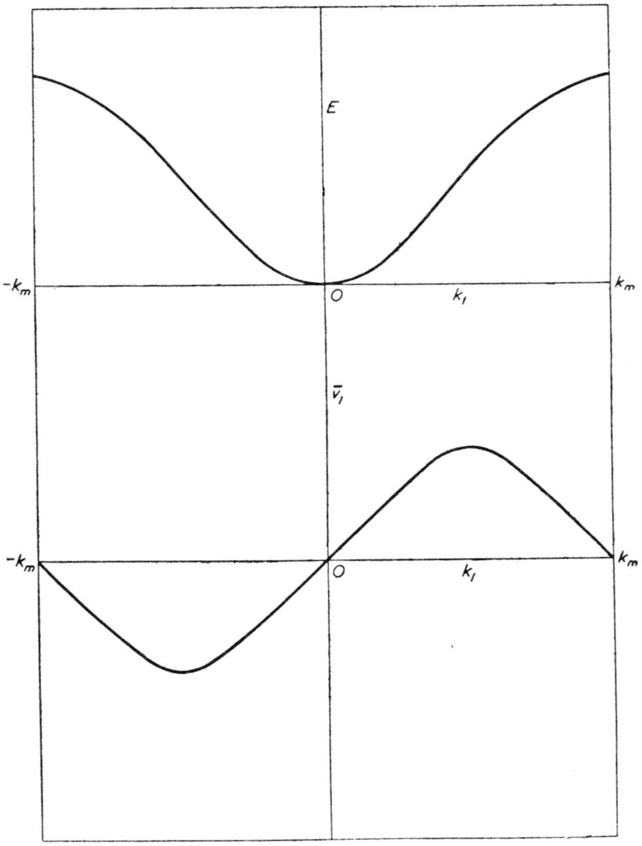

FIG. 8.3. Motion of an electron through a band

crystal planes, which as we have seen (see §§ 5.5 and 6.5), takes place whenever the vector **k** has its end on the surface of a zone. This behaviour does not take place in a real crystal since, except under very high values of $\mathscr{E}$, as we shall see, the value of the **k** vector is changed through scattering of the electron waves by lattice vibrations and by imperfections before it can increase its value by very much.

It is interesting to estimate how long an electron would take to complete one cycle; the time T for k to change from $-\pi/d$ to π/d is $h/e\mathscr{E}d$.

The cycle would be completed in about 2×10^{-9} sec. The total extent of the electron's orbit would be $\frac{1}{2}\bar{v}T$ where $\bar{v}$ is the average velocity equal to $h/4md$; if $d = 2 \times 10^{-8}$ cm the extent is 0·08 cm. In fact, as we shall see, such an electron would make a collision on the average about once every 10^{-12} sec and its velocity would never be much greater than its thermal velocity which at 300°K is about 10^7 cm/sec; between collisions it would therefore travel only a distance of about 10^{-5} cm. We shall later consider (§ 8.14) what happens when the field $\mathscr{E}$ is very large.

The shift of the mean value of k_1 as shown in Fig. 8.2 represents a drift velocity superimposed on the motion of the electrons, the value increasing with time. Except for very high fields, this drift velocity very quickly reaches a steady value, small compared with that which would give an appreciable shift in the Brillouin zone. When this is so the mean value of k^2 will only be changed to the second order so that the mean energy of the electrons is unchanged to the first order.

8.13 Cyclotron resonance

When we have a magnetic field alone acting on a crystal, the vector $\dot{\mathbf{k}}$ will be at right angles to $\bar{\mathbf{v}}(\mathbf{k})$; for the simple band structure under discussion $\dot{\mathbf{k}}$ will therefore also be at right angles to $\mathbf{k}$. Thus we see that for a magnetic field directed along the (3) axis, assumed to be at right angles to the (1) and (2) axes, the points representing the values of $\mathbf{k}$ for the electrons in $\mathbf{k}$-space are simply rotated about the (3) axis; this is illustrated in Fig. 8.4. In the crystal the electrons move in helices; this follows from the equations of motion

$$\left.\begin{aligned} m_e\dot{\bar{v}}_x &= -ev_y B \\ m_e\dot{\bar{v}}_y &= ev_x B \\ m_e\dot{\bar{v}}_z &= 0 \end{aligned}\right\} \tag{111}$$

If $\bar{v}_z = 0$ the motion is circular, being given by

$$\left.\begin{aligned} x - x_0 &= A\cos\omega t \\ y - y_0 &= A\sin\omega t \end{aligned}\right\} \tag{112}$$

where $\omega = eB/m_e$. The angular frequency ω is known as the cyclotron frequency, and its measurement gives a direct method of obtaining m_e (see § 11.3.2). If $m_e = m$ and $B = 0{\cdot}1$ Wb/m², $\omega/2\pi = 2{\cdot}8$ kMc/s, and the electron describes a circle in $3{\cdot}6 \times 10^{-10}$ sec. At room temperature many collisions per cycle would be made, but with very pure materials cooled with liquid helium the collision time may be long enough for several circles to be described without disturbance and the microwave resonance observed (see § 11.3.2).

We have already discussed briefly the effect of collisions in § 4.13.1. Apart from these, an electron would be accelerated according to the process we have described, under the influence of an electric field; an ideal crystal would therefore show no electrical resistance. Apart from the curious phenomenon of superconductivity all real crystals do show electrical resistance. Even for a perfect crystal, as we have seen in Chapter 6, the atoms are displaced from their positions on a perfect lattice by thermal vibrations, and this deviation from the ideal lattice leads, as we shall see, to electrical resistance, as does the presence of mechanical imperfections in the crystal. We defer a discussion of these topics to Chapter 11.

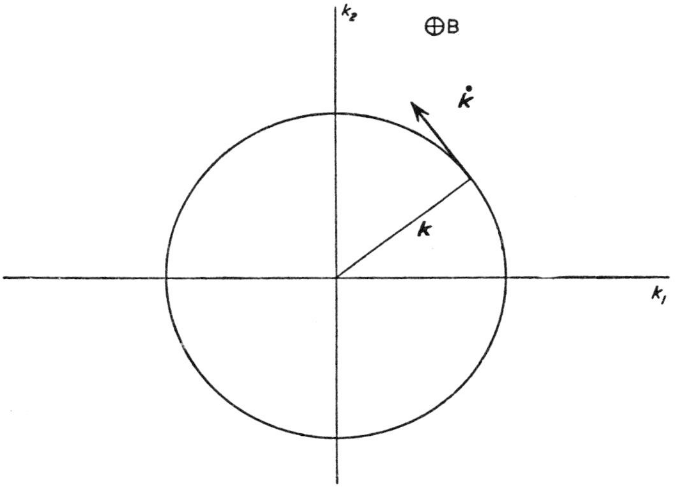

FIG. 8.4. Motion of electrons in **k**-space under the influence of a magnetic field

8.14 The Zener effect

We have already mentioned that when an electron suffers a Bragg reflection at a zone boundary there is a small but finite chance that instead of reflection it might make a transition into the next zone and so to a band of higher energy. The problem is most easily discussed in one dimension and has been treated by C. Zener*. An electron accelerated through a band by a high electric field $\mathscr{E}$ has its wave vector increased at a steady rate till $k = \pi/d$, d being the lattice spacing. Usually the wave vector will then revert to $-\pi/d$ corresponding to reflection at the zone boundary. On the other hand it is possible that the wave vector retains the value π/d and a transition is made to a higher band.

The elementary theory we have given does not take such transitions into account and we now wish to calculate the probability that they should take place. The exact treatment of the motion of electrons in a

* C. Zener, *Proc. Roy. Soc.*, **A145**, 523 (1934).

steady field presents some quite formidable theoretical problems, but an approximate solution may be found, at least in certain simple cases, without a great deal of trouble, and this illustrates the nature of the transitions which can take place. First of all we should note that this is not strictly a 'steady-state' problem. If, however, the field does not vary too much over a unit cell we may regard the electrons in the crystal as in quasi-stationary states, the states varying with position in the crystal. The energy levels can be regarded as 'local' levels; these may be represented on a diagram such as Fig. 8.5, the energies corresponding to the band edges being no longer 'horizontal' but having a slope $e\mathscr{E}$ due to the field. The energy of an electron will be conserved as it moves in the field and its energy measured from a band edge will change with time and

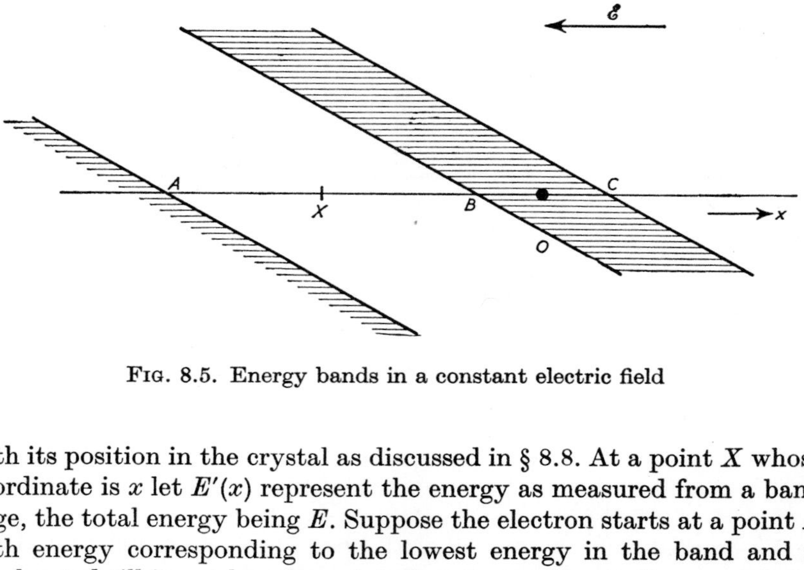

FIG. 8.5. Energy bands in a constant electric field

with its position in the crystal as discussed in § 8.8. At a point X whose coordinate is x let $E'(x)$ represent the energy as measured from a band edge, the total energy being E. Suppose the electron starts at a point A with energy corresponding to the lowest energy in the band and is accelerated till it reaches the point B corresponding to the upper band edge. It may then be reflected or may pass through the forbidden region (shown shaded) between the bands, emerging in the next band at the point C. We shall suppose that B and C correspond to the wave vector π/d. At each point along AC we could define a 'local' wave vector $k(x)$ by means of the equation giving the energy $E_0(k)$ of an electron in the crystal in terms of the wave vector in the absence of an external field; then we have $k(x)$ given by

$$E_0\{k(x)\} = E'(x) - E \tag{113}$$

The 'local' Bloch functions might be written in the form

$$b_k(x) = e^{ik(x)x}\, U_{k(x)}(x) \tag{114}$$

The phase of a wave, however, is an integrated quantity and a much better wave function is

$$b_k(x) = e^{i\int^x k(x)\,dx}\, U_{k(x)}(x) \tag{115}$$

In the region AB, $k(x)$ will be real, corresponding to an allowed band of energies; in BC, however, $k(x)$ will be complex, and our main problem is to obtain the value of $k(x)$ in this region. So far, we have neglected such values of $k(x)$, and it is interesting to find that they are of importance in some problems. It turns out that we need only be concerned with values of k near π/d. The variation of the periodic terms $U_k(x)$ will be quite small compared with the variation of the exponential term. Let us express $k(x)$ in the region BC in the form

$$k(x) = \alpha(x) + i\beta(x) \tag{116}$$

where α and β are real. We shall only be concerned with a *decreasing* wave in BC. If ψ_B and ψ_C represent the wave functions of an electron at B and C, respectively, then we see from equation (115) that

$$|\psi_C/\psi_B|^2 \simeq \exp\left[-2\int_B^C |\beta(x)|\,dx\right] \tag{117}$$

If we normalise the wave functions to represent one electron incident at the point B per second, the ratio $|\psi_C/\psi_B|^2$ gives the probability per unit time that such an electron will be transferred to C and so into the next band. To obtain the transition probability we must multiply by the number of times per second an electron is incident at B. We have seen in § 8.12 that the time taken for an electron to pass through the band is $\boldsymbol{h}/\boldsymbol{e}\mathscr{E}d$. The frequency is therefore $\boldsymbol{e}\mathscr{E}d/\boldsymbol{h}$ and the transition probability P per unit time is given by

$$P = \frac{\boldsymbol{e}\mathscr{E}d}{\boldsymbol{h}}\exp\left[-2\int_B^C |\beta(x)|\,dx\right] \tag{118}$$

We have already obtained the value of $\beta(x)$ for one problem, that of loose binding, although we have not expressed it in the form required; this problem has been treated in § 4.11.1. We expressed k in the form $k = \pi/d + k'$ and found that k' is given by the equation (217) of § 4.11.1, taking $n = 1$. This equation may be re-written in terms of the energy and the forbidden energy gap ΔE in the form (cf. equation (220) of § 4.11.1)

$$k'^2 = \eta^2 - \frac{4\pi^2 \boldsymbol{m}^2 d^2}{\boldsymbol{h}^4}(\Delta E)^2 \tag{119}$$

where η is connected with the energy E by means of the relationships $E = \boldsymbol{h}^2 k_0^2/8\pi^2 \boldsymbol{m}$, $k_0 = \pi/d + \eta$. Approximately we have therefore

$$E = \boldsymbol{h}^2/8\boldsymbol{m}d^2 + \boldsymbol{h}^2\eta/4\pi\boldsymbol{m}d \tag{120}$$

The two values of energy corresponding to $k = \pi/d$, i.e. to $k' = 0$ give

$$\eta = \pm\frac{2\pi \boldsymbol{m}d}{\boldsymbol{h}^2}\Delta E \tag{121}$$

the point $\eta = 0$ gives energy $\boldsymbol{h}^2/8\boldsymbol{m}d^2$ corresponding to the middle of the forbidden band. Now we may also write the energy in the form

$$E = E_0 + \boldsymbol{e}\mathscr{E}x \tag{122}$$

where E_0 is the energy at the mid-point O of the forbidden band. Thus we see that we may replace η by $(4\pi\boldsymbol{m}d/\boldsymbol{h}^2)(\boldsymbol{e}\mathscr{E}x)$ and write equation (119) in the form

$$k'^2 = \frac{16\pi^2\boldsymbol{m}^2d^2}{\boldsymbol{h}^4}\left[\frac{\Delta E^2}{4} - \boldsymbol{e}^2\mathscr{E}^2x^2\right] \tag{123}$$

The edges of the forbidden band correspond to $x = \pm b$ where $b = \Delta E/2\mathscr{E}\boldsymbol{e}$, and we note that at *both* edges $k' = 0$, so that $k = \pi/d$; also when $|x| > b$, k is real. We may therefore write the phase factor in the forbidden region in the form

$$k(x) = \frac{\pi}{d} \pm \frac{4\pi\boldsymbol{m}di}{\boldsymbol{h}^2}\left[\frac{\Delta E^2}{4} - \boldsymbol{e}^2\mathscr{E}^2x^2\right]^{1/2} \tag{124}$$

In order to obtain a decreasing solution we take

$$\left.\begin{aligned}\beta(x) &= \frac{4\pi\boldsymbol{m}d}{\boldsymbol{h}^2}\left[\frac{\Delta E^2}{4} - \boldsymbol{e}^2\mathscr{E}^2x^2\right]^{1/2}\\ &= \frac{4\pi\boldsymbol{m}d\boldsymbol{e}\mathscr{E}}{\boldsymbol{h}^2}[b^2 - x^2]^{1/2}\end{aligned}\right\} \tag{125}$$

We now substitute this expression in equation (118) to obtain the transition probability P, the limits of x being $\pm b$. We have

$$\int_{-b}^{b} (b^2 - x^2)^{1/2}\,\mathrm{d}x = \frac{\pi b^2}{2}$$

so that

$$P = \frac{\boldsymbol{e}\mathscr{E}d}{\boldsymbol{h}}\exp\left\{-\frac{\pi^2\boldsymbol{m}d\Delta E^2}{\boldsymbol{h}^2\boldsymbol{e}|\mathscr{E}|}\right\} \tag{126}$$

For small values of $\mathscr{E}$ this is extremely small but may become appreciable for large electric fields. This effect was originally proposed by Zener to explain breakdown at high electric fields in insulators, but in most cases it is now thought that breakdown in insulators and semiconductors is due to other causes, such as ionisation of impurities. A true Zener effect has, however, been recently observed in the high-field region of certain semiconductor diodes; careful observations are required to distinguish the phenomenon from other effects and in most practical cases transitions between adjacent bands due to the Zener effect can be neglected.

If we take $d = 2 \times 10^{-8}$ cm and $\Delta E = 1$ eV we find that

$$P \simeq 5 \times 10^4 \mathscr{E} \times 10^{-3\cdot5 \times 10^9/\mathscr{E}} \tag{127}$$

if $\mathscr{E}$ is expressed in V/m; fields of the order of a few times 10^7 V/m are therefore required to show the effect. Such fields can indeed occur at a junction between n-type and p-type semiconductors. For $\mathscr{E} = 5 \times 10^7$ V/m we have $P = 2\cdot5 \times 10^5$ or the probability per microsecond is 0·25. From our discussion in § 8.12 we see that the time required to perform an oscillation through the band is 4×10^{-13} sec which is comparable with the mean collision time. In our discussion we have neglected the effect of collisions; for the high fields which produce an appreciable probability of a transition to a higher band this will not seriously affect the calculation.

9

Distribution of Electrons between the Allowed Energy Levels in a Crystal

9.1 Form of the energy bands

We must now consider in rather more detail the actual form of the allowed energy bands for real crystals. Moreover, we must determine the distribution of the valence electrons in the highest occupied bands since this determines in a very fundamental way the properties of the crystal, for example, whether it is a metal, semiconductor or insulator. These three types of materials form convenient subdivisions of crystalline solids; we shall discuss them in turn and shall show how their characteristic properties are related to their electronic band structure. In particular, we shall see that the question as to whether neighbouring bands do or do not overlap is a very important one. The form of the bands may, in many cases, be deduced from very general considerations and, when this is so, has usually been verified when detailed calculations of the band structure have been carried out.

9.2 Metals

We shall first of all discuss briefly the properties of metals in terms of their electronic band structure. It is not proposed to treat the subject in detail since many excellent accounts are available which the reader should be able to appreciate, having mastered the theoretical considerations of the previous chapter.

9.2.1 The alkali metals

The simplest metals are the alkali metals, Li, Na, K, Rb, Cs; these crystallize, at room temperature, in the body-centred cubic form* and have one valence electron, i.e. they consist of closed shells plus one extra electron. The second ionisation potential, especially for the lighter elements of the group, is very much higher than the first, e.g. 75 V as compared with 5·36 V for Li, and 47 V as compared with 5·12 V for Na. All the electrons, apart from the valence electron are therefore in 'deep levels' which, in the crystal, become filled bands separated by a large energy gap from the band occupied by the valence electrons. Since each

* Most of the alkali metals have phase transformations at low temperatures.

atom contributes one electron to the latter band, it will be exactly half filled; as we have seen in § 8.9, if we have N_a atoms in the crystal the band will accommodate $2N_a$ electrons and we have exactly N_a electrons available, which at very low temperatures will occupy the lowest N_a levels in the band. We have already discussed the alkali metals in § 2.7 in terms of the free electron model and we now seek to justify its use. We have seen that, near the bottom of a band for which the minimum occurs at $\mathbf{k} = 0$, the energy bands may be expressed as a quadratic function of the components of the vector **k**. From the cubic symmetry of the crystal we might expect that the energy, for the alkali metals, should be expressible as a function of k, i.e. that the bands are 'spherical'. This has been verified in a band structure calculation by J. C. Slater* who has also shown that, except near the edges of the Brillouin zones, we may write the energy E in the form

$$E = \frac{\hbar^2 k^2}{2m^\star} \tag{1}$$

the zero of energy being taken at the bottom of the band. The actual form of the energy bands as calculated by J. C. Slater for Na is shown in Fig. 9.1 where they are plotted for two Brillouin zones, in the $\langle 100 \rangle$ directions. It will be seen that except near the zone edges equation (1) is a good approximation for the energy.

We must now verify that if we assume the form (1) and fill all the lowest levels with electrons, we do not have occupied levels having values of **k** near those corresponding to the surface of the first Brillouin zone. If the constant-energy surfaces are spherical all levels with k less than a maximum value k_m will be occupied at the absolute zero of temperature and those with $k > k_m$ will be empty (see § 2.7). The value of k_m may be obtained as follows. If we write $\kappa_m = k_m/2\pi$ we see (§ 8.9) that we must have the volume of the sphere $|\boldsymbol{\kappa}| = \kappa_m$ in reciprocal lattice space equal to half the volume of the first Brillouin zone (since the number of allowed levels is proportional to the volume of $\boldsymbol{\kappa}$-space, cf. § 8.9); the volume of the first zone is equal to $2/d^3$ (see § 5.4.4) where d is the edge of the unit cube of the crystal lattice. We therefore have

$$\frac{4\pi}{3}\kappa_m^3 = \frac{1}{d^3} \tag{2}$$

or

$$\kappa_m = 0{\cdot}62/d \tag{2a}$$

It will be seen from Fig. 6.9 that the shortest distance from the centre of the first Brillouin zone to a point on the surface is $1/\sqrt{2}d = 0{\cdot}71/d$, so that the sphere $|\boldsymbol{\kappa}| = \kappa_m$, sometimes known as the Fermi sphere (cf. § 2.7) lies well within the first Brillouin zone. The energy corresponding to κ_m is the Fermi energy E_F (see § 2.7); the energy corresponding to $|\boldsymbol{\kappa}| = 1/\sqrt{2}d$ would therefore be about $1{\cdot}4\ E_F$ if the form (1) held right

* J. C. Slater, *Phys. Rev.*, **45**, 794 (1934).

up to the zone boundary. As we have seen in § 2.4 the value of E_F for Na is about 3 eV so that the difference is of the order of 1·2 eV. Even at quite high temperatures, therefore, electrons will not be excited to levels near the zone boundary. We thus see that we are justified in using a free-electron model for the alkali metals, modified possibly by introducing an effective mass m^* in place of the free-electron mass $\boldsymbol{m}$. Since we have already discussed the alkali metals in terms of this model in Chapter 2

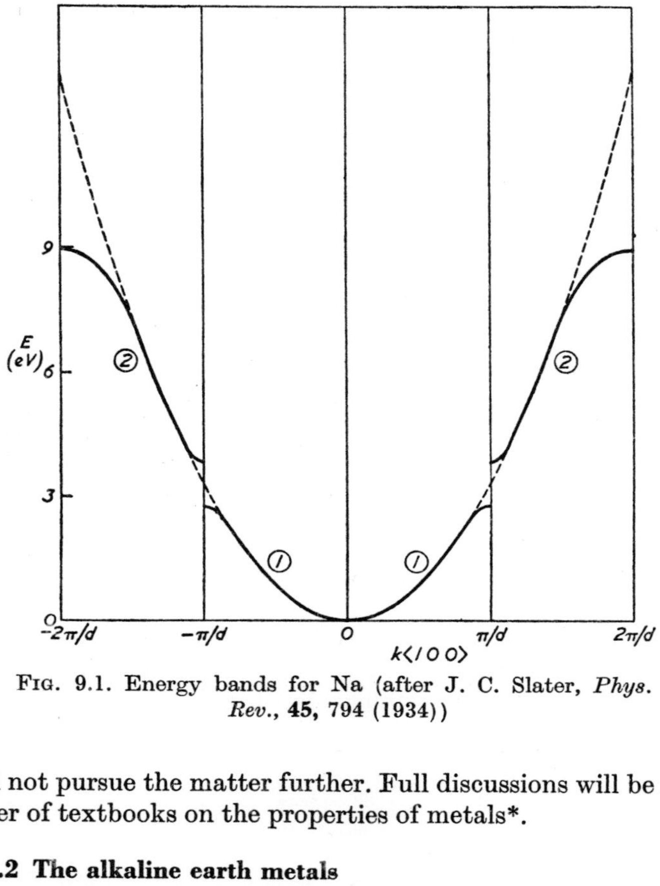

FIG. 9.1. Energy bands for Na (after J. C. Slater, *Phys. Rev.*, **45**, 794 (1934))

we shall not pursue the matter further. Full discussions will be found in a number of textbooks on the properties of metals*.

9.2.2 The alkaline earth metals

The alkaline earth metals such as Be and Mg have two valence electrons so that we should expect them to form filled bands if no overlapping occurs. In this case they would be insulators, so we may deduce from the fact that they are metals that overlapping of bands does occur. These metals crystallize in the close-packed hexagonal form and it is simpler to treat one of the cubic lattices.

* See, for example, N. F. Mott and H. Jones, *Properties of Metals and Alloys*. Oxford University Press (1936); J. C. Slater, *Handb. Phys.*, **19**, 1 (1956).

Overlapping of bands may readily be illustrated with the simple cubic lattice. If we have one electron per atom, the radius κ_m of the Fermi sphere is given by the equation

$$\frac{4\pi}{3}\kappa_m^3 = \frac{1}{2d^3} \tag{3}$$

or

$$\kappa_m = 0{\cdot}49/d \tag{3a}$$

It therefore just lies within the first Brillouin zone. When we have two electrons per atom we have

$$\kappa_m = 0{\cdot}62/d \tag{36}$$

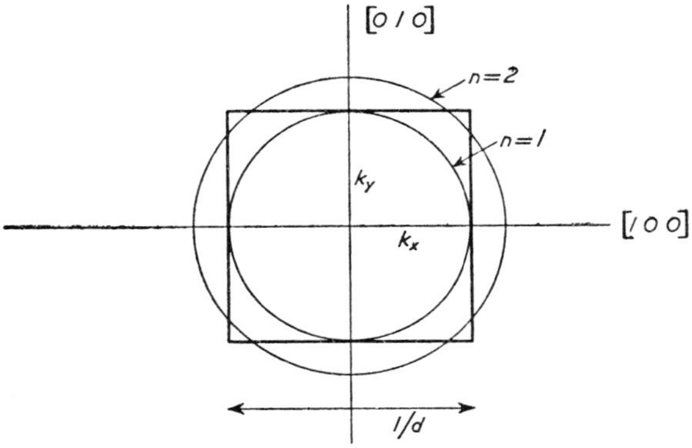

FIG. 9.2. Section of Fermi sphere and first Brillouin zone, in {100} plane, for simple cubic lattice with two electrons per atom

so that the Fermi sphere overlaps the first Brillouin zone which is a cube of side $1/d$; this is illustrated in Fig. 9.2. It will be seen that parts of the Fermi sphere lie outside the zone, and electrons in these parts would go into the next highest band provided the energy gap is small enough, otherwise they would be pushed into the corners of the zone, and the constant energy surfaces would no longer be approximately spherical.

Suppose we have the energy given by equation (1) apart from a very small energy gap at zone boundary. The lowest energy on the first Brillouin zone boundary will then be just less than $\boldsymbol{h}^2/8m^\star d^2$ and the highest just less than $3\boldsymbol{h}^2/8m^\star d^2$. If we represent the next highest energy band in the second zone, the lowest energy will be just greater than $\boldsymbol{h}^2/8m^\star d^2$ so that this band clearly overlaps the first. In order that no overlap should take place there should be a forbidden energy gap between the bands of the order of $2\boldsymbol{h}^2/8m^\star d^2$. Such a large energy gap would, of course, modify the form of the bands so much that equation (1) would no longer apply, but indicates that an appreciable gap is necessary if no overlapping is to take place. This is illustrated in Fig. 9.3 which shows two energy bands (marked (1) and (2)) plotted as E/k curves in the $\langle 100 \rangle$ and $\langle 111 \rangle$ directions, the energy being given by equation (1) except near the surface of the first Brillouin zone at which there is a forbidden energy gap ΔE; the reduced representation is used in this case, **k** being limited to the first Brillouin zone. The Fermi level E_F is also marked corresponding to two valence electrons per atom, the parts of the bands occupied at the absolute zero of temperature being shaded. It will be seen that band (2),

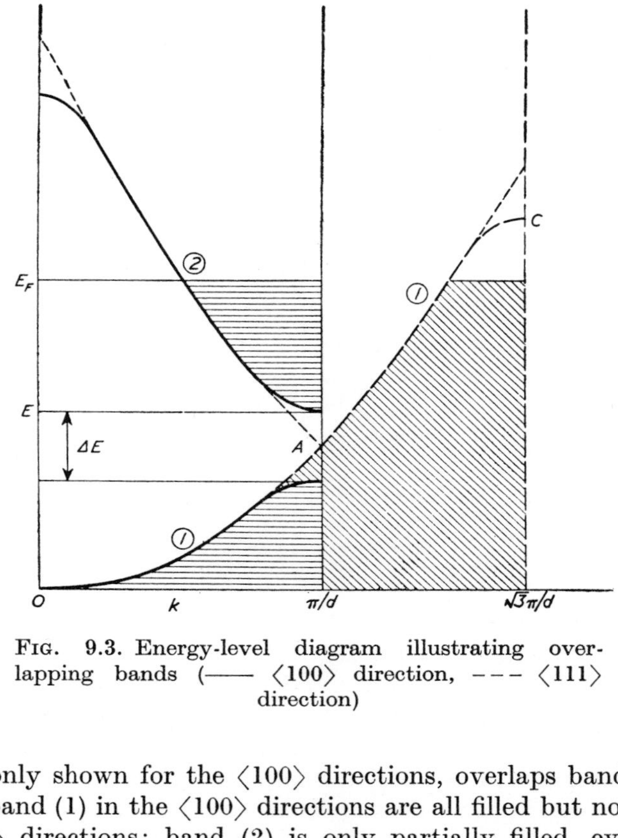

FIG. 9.3. Energy-level diagram illustrating overlapping bands (—— $\langle 100 \rangle$ direction, - - - $\langle 111 \rangle$ direction)

which is only shown for the $\langle 100 \rangle$ directions, overlaps band (1). The levels of band (1) in the $\langle 100 \rangle$ directions are all filled but not those in the $\langle 111 \rangle$ directions; band (2) is only partially filled, even in the $\langle 100 \rangle$ direction. If the Fermi level comes near the point C representing the highest point of band (1) only a few levels will be left unoccupied and these will behave like pockets of positive holes located at the corners of the Brillouin zone, giving two types of charge carriers, electrons in band (2) and holes in band (1).

It is interesting to see how the various energy bands in the crystal are associated with the energy levels of the isolated atoms. The valence electrons correspond to an s-state occupied by two electrons with opposite spins. In the isolated atoms the three p-states (degenerate apart from fine-structure) lie somewhat higher, but the energy separation is small and we should expect the p-states to be spread out into bands which overlap the bands arising from the s-states. This is verified by band-structure calculations such as that carried out by C. Herring and A. G. Hill* for Be. The bands arising from the s-state and the p-states could accommodate $8N$ electrons where N is the number of unit cells in the crystal. The alkaline earth metals crystallize in the close-packed hexagonal form and so have two atoms per unit cell and two electrons per atom; the number of electrons to be accommodated is therefore $4N$. On general grounds, therefore, we might expect the alkaline earth metals to have incompletely filled bands and to have the characteristic properties of metals.

9.2.3 Metals with three valence electrons

Metals such as Al have three valence electrons, two in s-states and one in a p-state in the isolated atoms. Al crystallizes in the face-centred cubic form and it is interesting to consider its properties assuming that the valence electrons behave essentially as if they were free, except that small energy gaps exist at the zone boundaries.

For the face-centred cubic lattice the radius κ_m of the Fermi sphere for one valence electron is given by

$$\frac{4\pi}{3}\kappa_m^3 = \frac{2}{d^3} \tag{4}$$

or

$$\kappa_m = 0{\cdot}78/d \tag{4a}$$

The Fermi sphere therefore lies well inside the first Brillouin zone, the nearest point of the surface being at a distance $\sqrt{3}/2d = 0{\cdot}86/d$ from the centre. When we have three valence electrons we have $\kappa_m = 1{\cdot}13/d$; the Fermi sphere does not intersect the surface of the first Brillouin zone (see Fig. 6.8), since the points of the zone boundary furthest from the centre are at a distance $1{\cdot}12/d$ from the centre; the first zone should therefore be fully occupied. It may be shown that the electrons in levels outside this zone go into the second and third zones. A detailed calculation of the band structure of Al by V. Heine† confirms these properties and it is suggested that the closeness of the Fermi surface to the points D (Fig. 6.8) may result in its being modified so that small pockets of positive holes exist at such points. Cyclotron resonance experiments (see § 8.13) by D. H. Langenberg and T. W. Moore‡ and by E. Fawcett§ have shown that at least two types of carriers are present in Al. Except near the zone boundaries, however, Heine has shown that the carriers behave very like free electrons.

* C. Herring and A. G. Hill, *Phys. Rev.*, **58**, 132 (1940).
† V. Heine, *Proc. Roy. Soc.*, A**240**, 340, 354, 361 (1957).
‡ D. H. Langenberg and T. W. Moore, *Phys. Rev. Letters*, **3**, 137 (1959).
§ E. Fawcett, *Phys. Rev. Letters*, **3**, 139 (1959).

9.2.4 The transition metals

The transition metals, of which Ni is typical, are of special interest since the band arising from the s- and p-type atomic states overlaps a narrow band arising from the lower-lying d-levels of the previous atomic shell, e.g. in Ni the band due to the two $4s$-electrons overlaps the band due to the eight $3d$-electrons; thus we have two partially filled bands. The properties of the transition metals have been discussed in terms of such a model by N. F. Mott* and by J. C. Slater† and a more recent calculation of the structure of the d-band of Ni has been made by G. C. Fletcher and E. P. Wohlfarth‡ using the tight-binding approximation, a problem for which this type of approximation is well suited. It appears that in Ni the d-band is not completely filled and that in addition to the conduction electrons in the s-band we have positive holes in the d-band.

9.2.5 The noble metals

The noble metals Cu, Ag, Au are in many respects similar to the alkali metals having a single valence electron in an s-state. Here, however, the electrons in the lower shells cannot be completely ignored and some of these have to be taken into account in band-structure calculations; for example, the ten $3d$-electrons of Cu must be taken into account as well as the $4s$-electron. Since the d-shell is, in a sense, complete, we might reasonably expect the d-band to be filled for Cu. This has been verified in a recent band-structure calculation by D. J. Howarth§, using the cellular method; nevertheless, the interaction between the d-band and the s-band cannot be ignored. Detailed discussion of the various groups of metals will be found in various accounts of the properties of metals referred to above; in particular, an interesting comparison has been drawn between the properties of the noble metals and of the transition metals by N. F. Mott and H. Jones**.

9.3 Semiconductors and insulators

In all the materials discussed in § 9.2 there existed overlapping bands which prevented the situation arising in which the highest occupied band was filled; at lower energies, however, filled bands making no contribution to conduction were found. We now consider materials for which the highest occupied band, at the absolute zero of temperature, is completely filled. We specify the absolute zero of temperature in order to eliminate, for the moment, thermal excitation of electrons from one band to another separated from it by a finite energy gap. We may see in a general way how such a condition may arise; it is almost always associated with closed

* N. F. Mott, *Proc. Phys. Soc.*, **47**, 571 (1935).

† J. C. Slater, *Phys. Rev.*, **49**, 537, (1936).

‡ G. C. Fletcher and E. P. Wohlfarth, *Phil. Mag.*, **42**, 106 (1951).

§ D. J. Howarth, *Phys. Rev.*, **99**, 469 (1955).

** N. F. Mott and H. Jones, *The Theory of the Properties of Metals and Alloys*. Oxford University Press (1936), p. 189.

shells of eight electrons in a way rather analogous to the formation of the rare gas atoms. The eight electrons may not come from a single atom but may be shared.

9.3.1 Semiconductors with the diamond-type lattice

The simplest materials showing this kind of behaviour are elements from the fourth column of the periodic table diamond, Si, Ge, and grey tin, all having the *diamond* crystal structure. Each atom has four valence electrons and there are two atoms per unit cell so that we have $8N$ electrons where N is the number of unit cells in the crystal. The atomic states occupied by the valence electrons are the s-state and three p-states. We should therefore expect to have in the crystal four overlapping bands, well separated from other bands, and we have exactly the number of electrons required to fill them all completely. We should therefore expect these materials, when pure, to be insulators at the absolute zero of temperature, and this is found to be so. Whether they continue to be insulators at higher temperatures depends on the value of ΔE, the forbidden energy gap between the valence band and the next highest band, usually called the conduction band, which originates in the d-levels of the constituent atoms. For diamond, ΔE is about 5 eV, and so it is an insulator; the probability of electrons being thermally excited into the higher band being very small at ordinary temperatures. For Si ($\Delta E \simeq 1{\cdot}1$ eV), of the purity available so far, conductivity at room temperature due to thermal excitation of the valence band electrons, generally called intrinsic conductivity, is so small as to be negligible compared with that due to electrons excited from impurities; intrinsic conductivity may, however, be induced at higher temperatures. Ge ($\Delta E \simeq 0{\cdot}65$ eV) of sufficient purity to show intrinsic conductivity at room temperature is commonplace.

Detailed calculations of the band structure for diamond, and for Si and Ge have been carried out by F. Herman*, who has shown that both the valence band and conduction band take the form of a complicated set of overlapping bands, the two being separated by a forbidden energy gap which varies considerably with the wave vector **k**. The form of the bands obtained by Herman is illustrated in Fig. 9.4, for the $\langle 100 \rangle$ and $\langle 111 \rangle$ directions in the crystal. The upper bands form together the conduction band and the lower bands the valence band. In addition to the three constituent bands of the valence band shown in Fig. 9.4 there is a still lower band, not shown, which joins up with the lowest of the three to form the full valence band. These four bands can each take $2N$ electrons so are occupied, at the absolute zero of temperature, with the $8N$ electrons in the crystal, N being, as before, the number of unit cells. The minimum at the zone boundary in a $\langle 111 \rangle$ direction is found to be the lowest point of the conduction band. This is very interesting. For

* F. Herman, *Phys. Rev.*, **93**, 1214 (1954); *Proc. Inst. Radio Engrs.*, **43**, 1703 (1955).

metals, we have usually found that, in the highest band, the electrons behaved very like free electrons near $\mathbf{k} = 0$; here this is certainly not so.

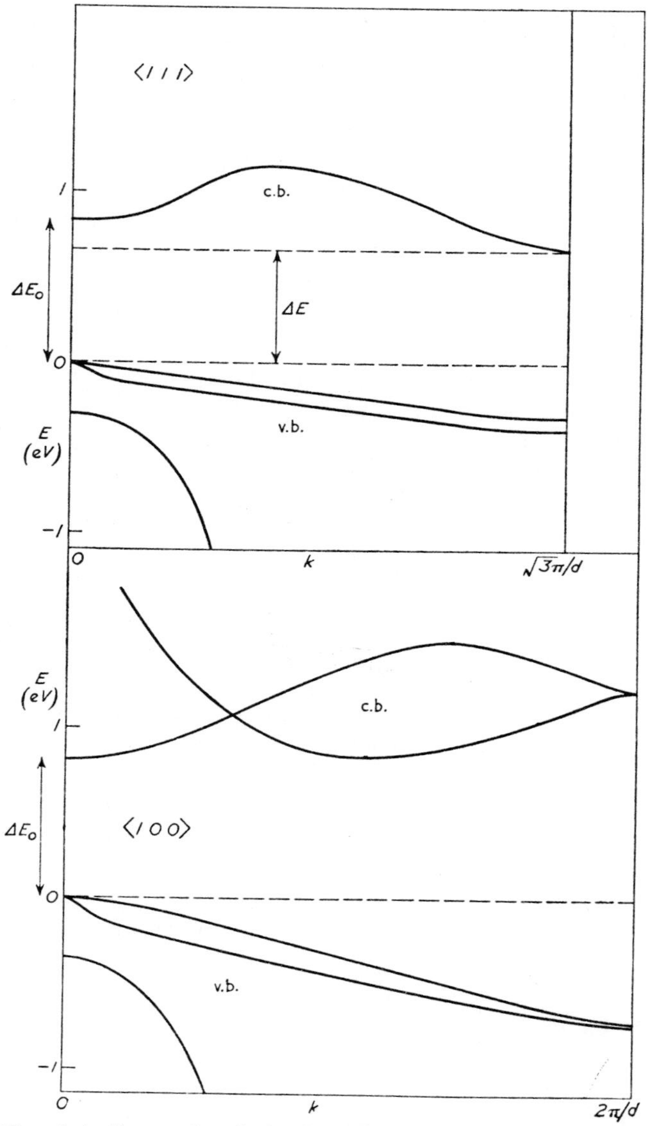

FIG. 9.4. Energy bands for Ge (after F. Herman, *Proc. Inst. Radio Engrs.*, **43**, 1703 (1955))

Normally, the conduction band will contain only a small number of electrons excited from the valence band (or, as we shall see, from impurities) and these will be located, not near $\mathbf{k} = 0$, but near the points

$\mathbf{k} = \pi/d\,(\pm 1,\ \pm 1,\ \pm 1)$. The new feature introduced by this type of band structure is that we have multiple minima in the conduction band. We shall discuss this feature in more detail later. The highest point of the valence band, which is completely filled by electrons at the absolute zero of temperature, is at $\mathbf{k} = 0$. Here again we have a new feature, which we shall later discuss in more detail, arising from the fact that the valence band is degenerate at $\mathbf{k} = 0$. The forbidden energy gap ΔE_0 ($> \Delta E$) at $\mathbf{k} = 0$ is also of importance; it has the value 0·8 eV.

A similar band structure is found for Si, but here the minimum which occurs in the conduction band for each of the $\langle 100 \rangle$ directions corresponds to the lowest energy of the band. Again we have multiple minima, and these occur at internal points of the first Brillouin zone.

These calculations have given the general form of the energy bands, which for both Si and Ge have now been verified by a great many experiments, including direct measurement of the effective-mass tensor by observation of cyclotron resonance (see § 11.3.2); they are not yet sufficiently accurate to predict the value of the forbidden energy gap with any degree of reliability but indicate its order of magnitude. Semiconductors such as Ge and Si are said to have valence binding, since the chemical bonds holding the atoms together are produced through the sharing of valence electrons between neighbouring atoms. A number of semiconductors in the form of intermetallic compounds such as InSb have similar binding, together with an ionic component of binding. If we write the compound in the form $In^-\ Sb^+$ it will be seen that each ion has four electrons and an electronic structure similar to that for Ge or Si may be built up. It is found that in this case the main contribution to the binding again comes from the closed groups of eight shared electrons, and that the properties of such semiconductors are in many respects similar to those of Ge and Si*. A band structure calculation for the so-called groups III–V semiconductors based on Herman's work modified by means of a perturbation to take account of the difference between the compound and the nearest element semiconductor, has been carried out by J. Callaway†. For InSb it appears that the minimum energy gap may occur at $\mathbf{k} = 0$, and there is now a good deal of experimental evidence to support this conclusion; indeed InSb is the only semiconductor for which it has been well established that the simple condition holds in which the lowest point of the conduction band and highest point of the valence band both occur at $\mathbf{k} = 0$.

9.3.2 Polar compound semiconductors

Another series of compounds for which it may be expected that the highest occupied band is completely filled at the absolute zero of tem-

* For a detailed account of these semiconductors see, for example, R. A. Smith, *Semiconductors*. Cambridge University Press (1959).

† J. Callaway, *J. Electron.*, **2**, 330 (1957).

perature are the ionic crystals, of which alkali halides such as NaCl are typical. Each of the halogen atoms has seven valence electrons and each of the metallic atoms only one. When we express the compound in the form $Na^+ Cl^-$ we see that the outer electrons of the Na^+ ion consist of a closed shell as do the outer electrons of the Cl^- ion. The closed shell of the Na^+ ion leads to a deep-lying narrow band and to the first order may be neglected so that we have again to fit eight electrons per molecule into the bands. For the NaCl structure we have one molecule per unit cell of the face-centred cubic lattice and so may again expect the valence band to consist of a series of overlapping bands all of which are completely filled. This has been verified in band structure calculations by W. Shockley* and by S. C. Tibbs†, and the fact that the forbidden energy gap between the full valence band and the next highest band is quite large is indicated by the fact that these materials (apart from ionic conduction) are insulators.

A considerable number of semiconductors in the form of polar compounds are also known, and their electronic band structure is similar except that they have much smaller forbidden energy gaps; of these PbS is typical. In ionic form this compound may be written as $Pb^{2+} S^{2-}$, again showing evidence of forming groups of eight electrons, although here the ionicity is not so complete as in the alkali halides, and there is evidence for some valence binding as well as ionic binding, i.e. electrons are shared between the Pb^{2+} and S^{2-} ions. The crystal structure is again of the sodium chloride form. One of the most difficult and sophisticated band-structure calculations attempted so far has been carried out for this material by Dorothy G. Bell *et al.*‡ The energy bands which they obtained are shown in Fig. 9.5 for one of the $\langle 110 \rangle$ directions, in which it was found that maxima in the energy in the valence band occurred. It will be seen that the valence band consists of three overlapping bands, derived from the $3p$-levels of the S atom; these three bonds are to include the four $3p$-electrons of S and the two $6p$-electrons of Pb. It was assumed that the two $3s$-electrons of S would form a deeper-filled band and it was verified that the two $6s$-electrons of Pb do so. The six p-electrons per unit cell just fill the three bands which form the complete valence band. The next highest band was found to originate from the $6p$-level of the Pb atom and was found to overlap with a still higher band having the properties of a band derived from s-states; only the lower band is shown in Fig. 9.5 and represents the conduction band. This also has its minima in the $\langle 110 \rangle$ directions.

This calculation gave the result, at the time very surprising, that the lowest minimum in the conduction band comes at the surface of the first

* W. Shockley, *Phys. Rev.*, **50**, 754 (1936).

† S. C. Tibbs, *Trans. Faraday Soc.*, **35**, 1471 (1939).

‡ Dorothy G. Bell, D. M. Hum, L. Pincherle, D. W. Sciama and P. M. Woodward, *Proc. Roy. Soc.*, A**217**, 71 (1953).

Brillouin zone in the ⟨110⟩ directions and that the maxima in the valence band also occur in this direction at interior points of the zone. These predictions have not yet had experimental verification, but the

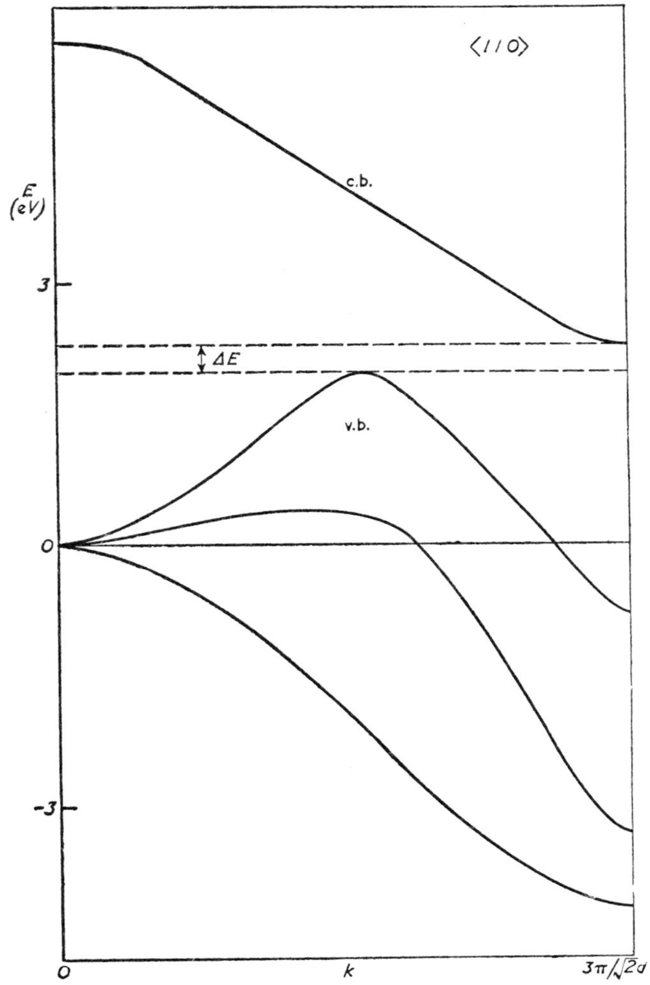

FIG. 9.5. Energy bands for PbS in ⟨110⟩ direction (after Dorothy G. Bell *et al.*, *Proc. Roy. Soc.*, **A217**, 71 (1953))

equally surprising later predictions by Herman for Ge and Si have been confirmed. The theoretical value of the forbidden energy gap of 0·3 eV is in excellent agreement with the experimental value, but this must be regarded as somewhat fortuitous.

9.3.3 Form of the energy bands near maxima and minima

It will be seen that, in general, the form of the energy bands for semiconductors is quite complex. Fortunately, since the conduction band is almost empty, and the valence band is almost full, we are nearly always only concerned with the form of the bands for values of **k** nearly equal to those corresponding to the lowest point in the conduction band and to the highest point in the valence band; these will be the same only in exceptional cases. We shall now consider in more detail the various

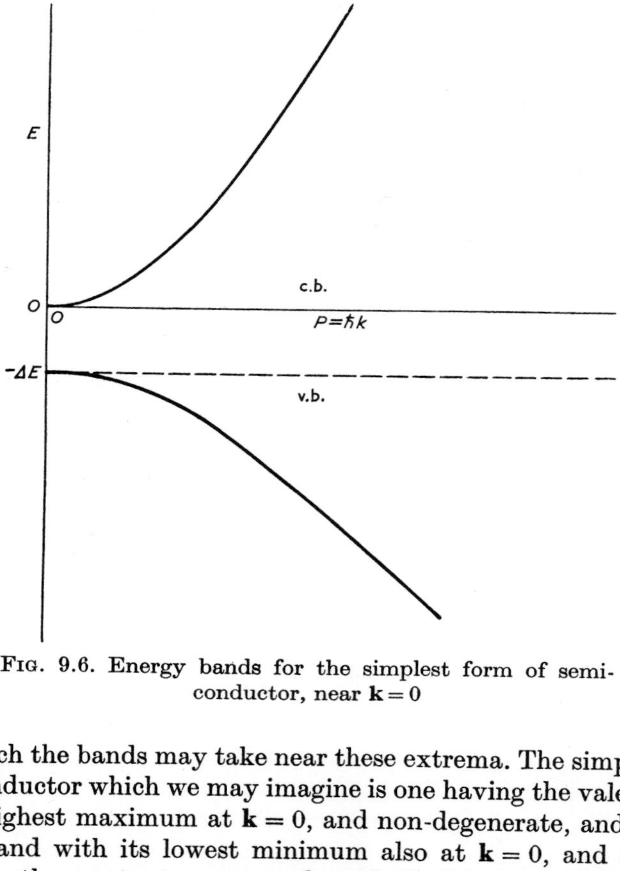

FIG. 9.6. Energy bands for the simplest form of semiconductor, near $\mathbf{k}=0$

forms which the bands may take near these extrema. The simplest form of semiconductor which we may imagine is one having the valence band with its highest maximum at $\mathbf{k} = 0$, and non-degenerate, and the conduction band with its lowest minimum also at $\mathbf{k} = 0$, and also non-degenerate, the constant-energy surfaces for both bands being spherical. Near $\mathbf{k} = 0$ the energy may be expressed in the forms (see § 8.8, equation (80))

$$E = \frac{\hbar^2 k^2}{2m_e} \tag{5}$$

for the conduction band, and

$$E = -\Delta E - \frac{\hbar^2 k^2}{2m_h} \tag{6}$$

for the valence band, m_e and m_h being, respectively, the scalar effective masses for electrons in the conduction band and for holes in the valence band; this form of the energy bands is illustrated in Fig. 9.6. Until quite recently nearly all discussions of the properties of semiconductors have been based on this form of band structure. Unfortunately no semiconductor is known to have bands of this simple form, although InSb comes near to it, but has almost certainly a degenerate valence band at $\mathbf{k} = 0$. When such degeneracy occurs in the valence band we have a band

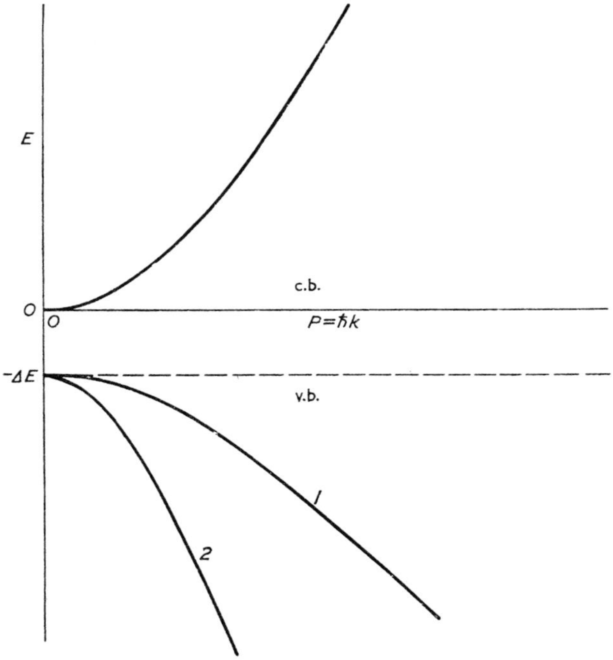

FIG. 9.7. Energy bands for a semiconductor having a degenerate valence band

structure such as that shown in Fig. 9.7. The two degenerate bands, however, perturb each other and it may be shown that this makes the constant-energy surfaces no longer quite spherical. We shall discuss this in more detail below in connection with structure of the valence bands of Si and Ge for which the form is known with high precision from cyclotron resonance experiments (see § 11.3.2). If, as a first approximation, we neglect this interaction, we have two 'spherical' bands with different effective masses m_{h_1}, m_{h_2} for the holes and both types of charge carrier will have to be taken into account. For crystals with cubic symmetry, having extrema at $\mathbf{k} = 0$ we should expect the energy bands near $\mathbf{k} = 0$ also to have cubic symmetry and if a quadratic expansion is permissible (it is not for degenerate bands) the bands should be 'spherical' and the effective mass scalar. For crystals not having cubic symmetry, for example, crystals with a hexagonal lattice, the constant-energy surfaces would be ellipsoidal and the energy would be given by equation (77) of § 8.8. Frequently there is an axis of symmetry (z-axis) and the energy would be given by an expression of the form

$$E = \frac{\hbar^2}{2}\left[\frac{k_x^2+k_y^2}{m_t}+\frac{k_z^2}{m_l}\right] \tag{7}$$

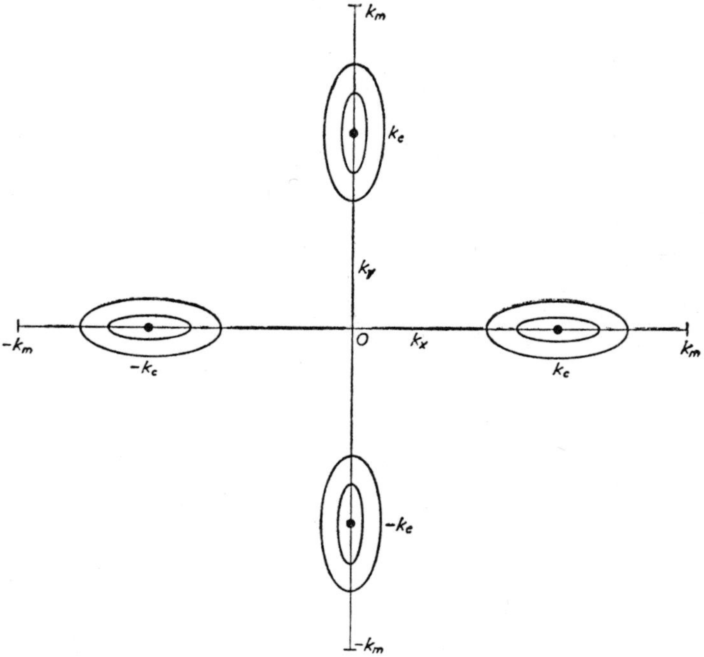

FIG. 9.8. Section of the constant-energy surfaces near the lowest minima in the conduction band of a cubic crystal, such as Si, for which these occur in the ⟨100⟩ directions at internal points of the first Brillouin zone

for the conduction band, and a similar expression for the valence band with appropriate changes of sign; the quantities m_e and m_t are known as the longitudinal and transverse effective masses.

9.3.4 Multiple equivalent maxima and minima

When the minimum value of the energy in the conduction band occurs for a value of **k** corresponding to an internal point of the first Brillouin zone, a very interesting situation occurs. As we have already seen, we shall now have multiple minima since the two points corresponding to **k** and $-\mathbf{k}$ give the same energy. For crystals with a high degree of symmetry more than two equivalent minima will exist. For example, with a crystal having cubic symmetry, and having a lowest minimum at an internal point in a $\langle 100 \rangle$ direction, equivalent minima will occur in the *six* directions $[\pm 100]$, $[0 \pm 10]$, $[00 \pm 1]$. Si is an example of such a crystal, and a section of the constant-energy surfaces by the (001) plane through the centre of the first Brillouin zone in $\mathbf{k}$-space is shown in Fig. 9.8. With this arrangement there is clearly no reason for the constant-energy surfaces to be spherical; indeed, it is more likely that they will have axial symmetry being ellipsoids of revolution with symmetry axes along the $\langle 100 \rangle$ directions. If the minimum in the (100) direction occurs for the value $(k_c, 0, 0)$ of the wave vector $\mathbf{k}$, the energy E near this minimum will be given by

$$E = \frac{\hbar^2}{2}\left[\frac{(k_x - k_c)^2}{m_l} + \frac{k_y^2 + k_z^2}{m_t}\right] \tag{8}$$

with corresponding equations for E at points near $(-k_c, 0, 0)$, $(0, \pm k_c, 0)$, $(0, 0, \pm k_c)$ in $\mathbf{k}$-space, obtained by cyclic permutation.

Near the energy minimum at $(k_c, 0, 0)$ we shall have for the components $\overline{v_x}$, $\overline{v_y}$, $\overline{v_z}$ of the average velocity of an electron (§ 8.6 equation (63))

$$\left.\begin{aligned} \overline{v_x} &= \hbar^{-1}\frac{\partial E}{\partial k_x} = \hbar(k_x - k_c)/m_l \\ \overline{v_y} &= \hbar^{-1}\frac{\partial E}{\partial k_y} = \hbar k_y/m_t \\ \overline{v_z} &= \hbar^{-1}\frac{\partial E}{\partial k_z} = \hbar k_z/m_t \end{aligned}\right\} \tag{9}$$

and we see that the velocity is equal to zero at the point $(k_c, 0, 0)$; similar equations apply near the other equivalent minima. The components P_x, P_y, P_z of crystal momentum are related to the components of the average velocity near the (100) minimum by means of the equations

$$\left.\begin{aligned} P_x &= P_c + m_l \overline{v_x} \\ P_y &= m_t \overline{v_y} \\ P_z &= m_t v_z \end{aligned}\right\} \tag{10}$$

where $P_c = \hbar k_c$. The effective-mass tensor has three non-zero components in this case, which must be used in the equations of motion.

An interesting case occurs when the minimum occurs at the surface of the Brillouin zone; for example if, in the above example k_c, is equal to k_m, the points in the $\langle 100 \rangle$ directions at the edge of the first Brillouin zone being $(\pm k_m, 0, 0)$, $(0, \pm k_m, 0)$, $(0, 0, \pm k_m)$, then the points $(k_m, 0, 0)$ and $(-k_m, 0, 0)$ are equivalent being at opposite ends of a diagonal of the zone and so separated by 2π times a vector of the reciprocal lattice.

Similarly, the remaining points are equivalent in pairs and we have only *three* minima instead of six, and the constant-energy surfaces for this case are illustrated in Fig. 9.9. A situation such as this occurs in Ge, but here the lowest minima in the conduction band occur in the ⟨111⟩ directions (see Fig. 9.4). There are eight such directions, but since the minima occur in equivalent pairs there are *four* equivalent minima in the conduction band. If the minima occur at internal points of the first

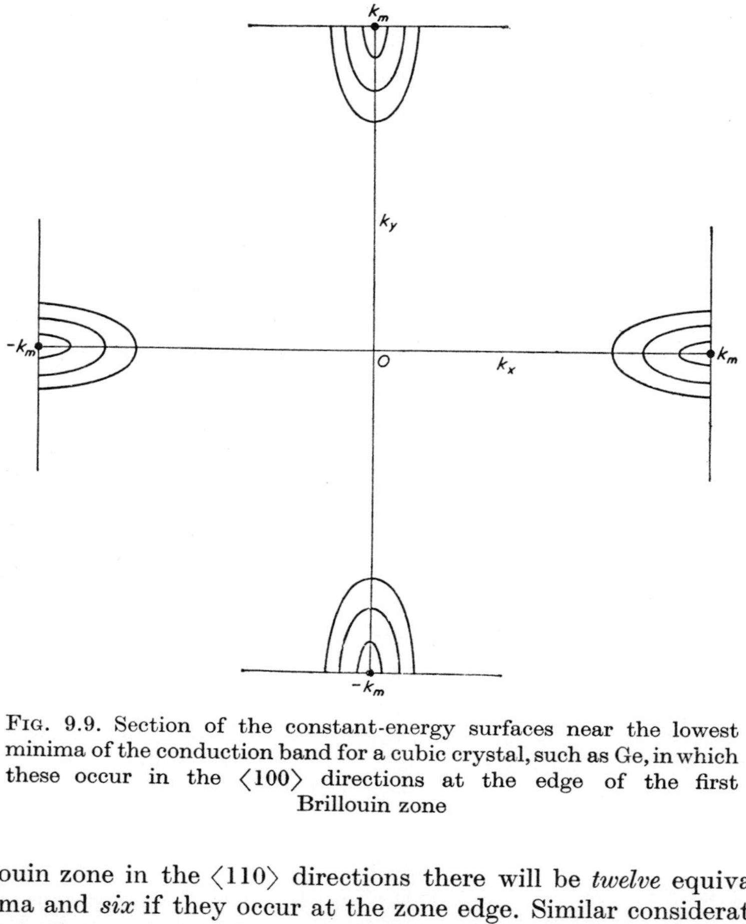

FIG. 9.9. Section of the constant-energy surfaces near the lowest minima of the conduction band for a cubic crystal, such as Ge, in which these occur in the ⟨100⟩ directions at the edge of the first Brillouin zone

Brillouin zone in the ⟨110⟩ directions there will be *twelve* equivalent minima and *six* if they occur at the zone edge. Similar considerations apply to the maxima of the valence band, and the band structure of PbS provides examples of both conditions, there being six equivalent minima in the conduction band and twelve equivalent maxima in the valence band (see Fig. 9.5).

Values of the longitudinal and transverse effective masses for Si and Ge have been determined accurately by means of cyclotron resonance experiments (see § 11.3.2) and the form of the bands near the minima established; a scalar effective mass corresponding to a single minimum at $\mathbf{k} = 0$ has also been found for one or two intermetallic semiconductors such as InSb and InAs. For most semiconductors, however, neither the form of the bands nor the numerical values of the components of the effective-mass tensor are known with any degree of certainty.

In addition to the lowest minima in the conduction band, there may be other minima at higher values of the energy. For example, in Ge there is a minimum at the centre of the first Brillouin zone ($\mathbf{k} = 0$) which lies 0·14 eV above the lowest minima at the zone edges in the $\langle 111 \rangle$ directions. Optical transition to this minimum have been observed (see § 13.3) and give rise to some very interesting phenomena. In general, the electrons in the conduction band will occupy the lowest minima, especially in the lower temperature range, but in some cases it may be necessary to take account of subsidiary minima. For Si it is not yet known with certainty whether there is a subsidiary minimum or a maximum in the energy at $\mathbf{k} = 0$.

9.3.5 Degenerate bands at k = 0

When the valence band has its maximum at $\mathbf{k} = 0$ and is non-degenerate we should expect, for a crystal with cubic symmetry, to have the energy E given in terms of $\mathbf{k}$, for small values of k, by an equation of the form

$$E = -\Delta E - Lk^2 \tag{11}$$

where L is a constant. When the valence band is degenerate at $\mathbf{k} = 0$ and the value $E = -\Delta E$ corresponds to points in two separate bands, as in Fig. 9.7, we should expect the energy to be given in a zero-order approximation by an equation of the form

$$(E - \Delta E - Lk^2)^2 = 0 \tag{12}$$

The next approximation would split the bands and we should expect to have an equation of the form

$$(E - \Delta E - Lk^2)^2 = \phi(\mathbf{k})$$

so that we have the two bands given by

$$E = -\Delta E - Lk^2 \pm [\phi(\mathbf{k})]^{1/2} \tag{13}$$

For a cubic crystal, neglecting orders of magnitude in $E - \Delta E$ greater than k_x^2, $k_x k_y$, etc., we should expect $\phi(k)$ to be a quartic expression with cyclic symmetry in k_x^2, k_y^2, k_z^2, since $\phi(\mathbf{k})$ must be an even function of $\mathbf{k}$. This would have the general form

$$\phi(\mathbf{k}) = M(k_x^4 + k_y^4 + k_z^4) + N(k_x^2 k_y^2 + k_y^2 k_z^2 + k_z^2 k_x^2) \tag{14}$$

which may be written in the form

$$\phi(\mathbf{k}) = Mk^4 + N'(k_x^2 k_y^2 + k_y^2 k_z^2 + k_z^2 k_x^2) \tag{15}$$

Writing A for $(2\boldsymbol{m}/\hbar^2)L$, B^2 for $(2\boldsymbol{m}/\hbar^2)^2 M$ and C^2 for $(2\boldsymbol{m}/\hbar^2)^2 N'$, we may express the energy E in the form, for the two bands,

$$E = -\Delta E - \frac{\hbar^2}{2\boldsymbol{m}}[Ak^2 \pm \{B^2 k^4 + C^2(k_x^2 k_y^2 + k_y^2 k_z^2 + k_z^2 k_x^2)\}^{1/2}] \tag{16}$$

This form has been confirmed theoretically, using the symmetry properties of the wave functions, determined from group theory, by G. Dresselhaus, A. F. Kip and C. Kittel* and by R. J. Elliott†, for Ge and Si. A third band split off from the degenerate pair has energy of the form

$$E = -\Delta E - \Delta E_3 - \hbar^2 Ak^2/2\boldsymbol{m} \tag{17}$$

the constant A being the same as in equation (16). The form of these bands has been verified experimentally for Si and Ge by Dresselhaus, Kip and Kittel, and also by others, using cyclotron resonance, and accurate values of the constants A, B, C determined.

If we neglect the constant C it will be seen that the degenerate bands split into two 'spherical' bands with scalar effective hole masses m_{h1}, m_{h2} given by

$$\left.\begin{aligned} \frac{\boldsymbol{m}}{m_{h1}} &= A - B \\ \frac{\boldsymbol{m}}{m_{h2}} &= A + B \end{aligned}\right\} \tag{18}$$

The split-off band, is also 'spherical' with scalar effective hole mass m_{h3} given by

$$\frac{\boldsymbol{m}}{m_{h3}} = A \tag{19}$$

For Ge and Si the constant C is not negligible and its inclusion gives rise to considerable warping of the constant-energy surfaces. The existence of two types of holes leads to some very interesting effects, but a discussion of these is beyond the scope of this Chapter.

9.3.6 Impurity levels

So far, we have considered only perfect crystals but we must now briefly consider the effect of imperfections in the crystal. Although these may occur only in small numbers they may have a profound effect on the electrical properties of a semiconductor. There are two main types of imperfection—mechanical imperfections and chemical impurities. It would take us too far afield from the main topic of this Chapter to consider in detail the various forms which both types of impurity may take. These have been discussed in considerable detail in a number of review articles and books on semiconductors‡. The effect of these imperfections

* G. Dresselhaus, A. F. Kip and C. Kittel, *Phys. Rev.*, **98**, 368 (1955).

† R. J. Elliott, *Phys. Rev.*, **96**, 268, 280 (1954).

‡ See, for example, R. A. Smith, *Semiconductors*. Cambridge University Press (1959); F. A. Kröger and H. J. Vink, *Solid State Phys.*, **3**, 307 (1956).

is to introduce discrete energy levels in addition to the levels found for the perfect crystal, provided they are far enough apart to be treated as isolated centres of potential in the crystal. When there is a high density of impurity centres the levels will interact and may form sub-bands. Such sub-bands, known as impurity bands, have a marked effect on the conductivity of certain semiconductors at very low temperatures. We shall not, however, discuss them in the present context and shall assume that each type of impurity centre produces an individual energy level. If we have N impurities of one particular type these will then give rise to N discrete levels (having each the same energy) in the crystal. If the levels occur above the bottom of the conduction band they will generally be empty and in any case will hardly be noticed among the many more levels due to the rest of the crystal. Similarly, if the levels occur below

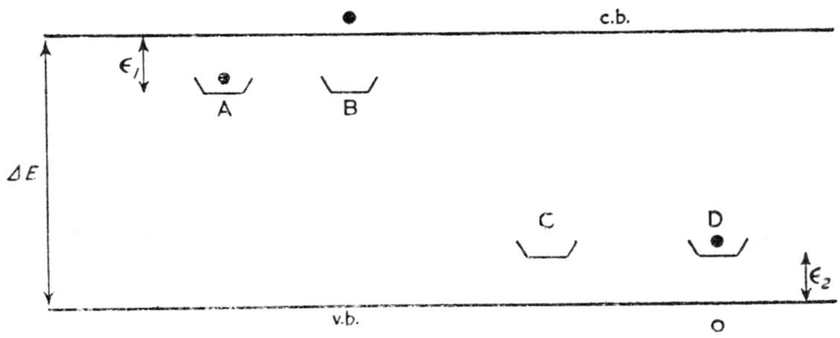

FIG. 9.10. Impurity energy levels in the forbidden energy gap of a semiconductor

the top of the valence band they will generally be unimportant. If, however, the levels occur in the forbidden energy band they will readily be noticed since there are no allowed levels from the perfect crystal there; in particular, if the impurity levels occur just below the bottom of the conduction band or just above the top of the valence band, their presence may greatly modify the electron concentration in the conduction band or the hole concentration in the valence band; the situation which we may envisage is illustrated in Fig. 9.10. The level denoted by A, at an energy ϵ_1 below the bottom of the conduction band is shown as occupied by an electron. If ϵ_1 is small, this condition will hold only at low temperatures; at ordinary temperatures the electron will be excited into the conduction band much more readily than electrons from the valence band and we shall have the situation illustrated by the level at B in Fig. 9.10. Here the impurity level will be left as an immobile positive charge and we have a free electron in the conduction band. It may readily be shown (see § 9.4.1) that a very small percentage of impurity will give far more electrons in the conduction band than would come from the valence band; for example, in Ge at room temperature with Sb impurity of one part in

10^9 most of the electrons in the conduction band will come from the Sb impurities. When electrons from such impurities, known as *n*-type impurities, are in considerable excess the semiconductor is known as an *n*-type semiconductor, and in this case we have relatively few positive holes. A similar situation occurs for impurities of types which give rise to levels which lie just above the valence band and can accept an electron. Except at very low temperatures an electron may be excited into the impurity level, as at *D* in Fig. 9.10, leaving behind a mobile positive hole in the valence band. When the holes arising from this type of impurity, known as a *p*-type impurity, are in considerable excess we have a *p*-type semiconductor. When the impurity content is so small that electrons and holes formed by thermal excitation of electrons from the valence band to the conduction band greatly outnumber those from impurities, we have what is known as an intrinsic semiconductor; in this case we have nearly equal numbers of holes and electrons.

We shall later consider in more detail how the occupancy of the allowed levels in the conduction and valence bands is modified by the presence of levels in forbidden energy gap. It will, however, already be evident that this modification can be quite drastic if $\epsilon_1 \ll \Delta E$ or $\epsilon_2 \ll \Delta E$, so that the presence of quite small concentrations of certain impurities can entirely change the properties of a semiconductor. It is for this reason that our detailed knowledge of these properties has been so inadequate until quite recently, when advances in technology have made possible the preparation of a few semiconductors such as Si, Ge, InSb in a very pure state. This has come about largely through the technological importance of semiconductors. The interesting and varied properties of these materials have been discussed in some detail by the author elsewhere* and also in a considerable number of books and review articles, so that there is no need to treat them in further detail.

Before concluding this section it would be of interest to consider a simple case in which it may readily be seen that an impurity level may be quite close to the conduction band or valence band. Suppose we have a semiconductor such as Ge and an impurity such as Sb. Ge occurs in the fourth column of the periodic table and so has four valence electrons which form valence bonds with the neighbouring atoms, giving a very stable structure. If an Sb atom replaces a Ge atom in the lattice there will be an extra electron left over from the bonding since the Sb atom has five valence electrons. An Sb^+ ion fits well into the bonding scheme and will cause little distortion of the lattice. If we take away one of the electrons and place it in the conduction band it will move in a field which is very nearly that of a perfect lattice plus that of a single positive charge located at the impurity. If we regard this as an *external* field we may now use the effective-mass concept to calculate the value of ϵ_1, the energy required to take an electron from the impurity level into the

* R. A. Smith, *Semiconductors*. Cambridge University Press (1959).

conduction band, as shown by N. F. Mott and R. W. Gurney*. If we have a scalar effective mass m_e for electrons in the conduction band, the problem simply reduces to finding the energy levels of a particle of charge $-\boldsymbol{e}$ and mass m_e in a Coulomb field modified by inclusion of a factor K^{-1} where K is the dielectric constant. The Coulomb field does not apply for values of the distance r from the impurity comparable with the atomic spacing, but, as we shall see, the main contribution, when the binding energy ϵ_1 of the electron to the impurity centre is small, comes from quite large values of r for which this is a good approximation. The value of ϵ_1 is then simply given by the expression

$$\epsilon_1 = m_e W_H / n^2 K^2 \boldsymbol{m} \tag{20}$$

where W_H is the energy of the ground state of the hydrogen atom (13·5 eV) and n is an integer. For most semiconductors K is quite large; for example for Ge $K = 16$, and, although in this case the effective mass is not scalar, an averaged value $m_e = 0{\cdot}2\boldsymbol{m}$ may be taken†; we have then $\epsilon_1 = 0{\cdot}01/n^2$ eV. Thus we see that we have a *series* of energy levels, the higher ones corresponding to excited states of the impurity centre. The ground state lies only 0·01 eV below the conduction band, whereas for Ge $\Delta E \simeq 0{\cdot}65$ eV; except at very low temperatures the electron will not be bound to the impurity centre and each impurity will give one electron in the conduction band.

The effective radius a of the impurity centre when occupied by an electron in the ground state is given by

$$a = (\boldsymbol{m}/m_e) K a_0 \tag{21}$$

where a_0 is the Bohr radius, $0{\cdot}53 \times 10^{-8}$ cm. For most semiconductors m_e is considerably less than the free electron mass $\boldsymbol{m}$ and K is also quite large; we therefore have $a \gg a_0$, and the wave function for an electron bound to an impurity centre extends over many lattice spacings; for excited states the effective radius is n^2 times larger. This justifies the approximate use of the Coulomb force between the electron and ionized impurity centre. Although equation (20) gives the impurity levels very well in Ge, deviations are found for other semiconductors; for Si, for example, although equation (20) gives the excited states fairly accurately it gives only approximate values for the ground states of impurities from column V of the periodic table. When accurate values of the impurity ionisation energies are required, the fact that the effective mass is tensorial in form must also be taken into account‡.

An exactly similar treatment may be given for impurities with valence one less than the atoms of the crystal, for example, elements from column

* N. F. Mott and R. W. Gurney, *Electronic Processes in Ionic Crystals*. Oxford University Press (1940), p. 83.

† See R. A. Smith, *Semiconductors*. Cambridge University Press (1959), p. 62.

‡ See, for example, W. Kohn and J. M. Luttinger, *Phys. Rev.*, **97**, 1721 (1955); *Ibid.*, **98**, 905 (1955).

III of the periodic table in Ge and Si. The impurity centre in this case may be regarded as a negative ion plus a positive hole, the binding energy ϵ_2 of the hole being given by equation (20) with m_e replaced by m_h. This binding energy is just equal to the height of the level above the top of the valence band (see Fig. 9.10). Such impurities therefore give energy levels close to the valence band.

Other types of impurity are known which give levels much nearer to the centre of the forbidden energy band; these generally differ in valency by more than one from one of the atoms of the crystal. They are known as deep-lying impurities and may have more than one energy level corresponding to the acceptance of more than a single electron, and are of importance in trapping electrons or holes as they move through the crystal and holding them bound for an appreciable time.

9.3.7 Excitons

We have shown that, on the basis of the one-electron theory of the band structure of crystalline solids, no energy levels exist in an infinite and perfect crystal in a forbidden energy gap. That there must exist certain states of the crystal corresponding to energies of electrons in the forbidden energy gap was shown very simply by N. F. Mott and R. W. Gurney*. Let us consider a crystal in which only one electron has been excited from the valence band to the conduction band leaving behind a positive hole which is free to move through the crystal. The electron and hole will attract each other and can form a series of bound states like a hydrogen atom; in this case they will move as a bound pair through the crystal. Such a pair was called by J. Frenkel, an exciton, since in its lowest state it corresponds to the first excited state of a crystal, the ground state corresponding to the situation when the electron returns to the valence band and the hole disappears. The binding energy of the exciton W_{ex}^n will be given by (cf. equation (20)),

$$W_{ex}^n = (m_r/\boldsymbol{m})\, W_H/n^2 K^2 \tag{22}$$

where m_r is a reduced effective mass for the hole and electron. The exciton has therefore a series of levels, the ground state being given by $n = 1$. The energy required to form an exciton in its ground state is therefore W_{ex}^1 *less* than the minimum energy ΔE required to form a free electron–hole pair. Thus we see that the first excited state of a crystal lies $(\Delta E - W_{ex}^1)$ above the lowest energy state, and corresponds to the condition that we have a single exciton in its ground state. Thus from the point of view of the single-electron model we can, in a sense, regard the exciton levels as lying just below the bottom of the conduction band in the forbidden energy gap and for many purposes this is rather convenient. However, it is not strictly correct; in addition to its binding energy the

* N. F. Mott and R. W. Gurney, *Electronic Processes in Ionic Crystals*. Oxford University Press (1940), p. 84.

exciton also has kinetic energy of its centre of gravity which can be expressed in terms of a wave vector $\mathbf{k}_{ex}$, and so should be represented on a separate energy-level diagram.

It is not difficult to see why the exciton levels do not appear in the one-electron theory of band structure. Their existence depends on an electron–hole interaction and this has been neglected (apart from an averaging process to get the correct potential); the exciton levels therefore represent a higher degree of approximation. Hydrogenic series of absorption lines have been observed for some time in a number of laboratories in certain high-energy gap semiconductors and these have been interpreted as being due to excitons. More recently, exciton lines have been observed in highly purified Ge for which there can be no doubt that they are not due to impurities which may also give a hydrogenic spectrum (cf. equation (20)). A discussion of the experimental data and of the part played by exciton formation in band-to-band transitions has been given by the author elsewhere*.

9.4 The Fermi level

We must now generalise the treatment given in §§ 2.4, 2.7 to determine the energy corresponding to the Fermi level in a crystal when the $E/\boldsymbol{k}$ relationship is no longer given by the form for free electrons and when we have to take account of more than one band and also of impurity levels.

9.4.1 Scalar effective mass

Let us consider first of all the simplest extension of the free-electron treatment which corresponds to the case of a single band for which we can represent the $E/\boldsymbol{k}$ relationship in terms of a single scalar effective mass m_e. The considerations of § 2.4 then apply provided we replace the free-electron mass $\boldsymbol{m}$ by m_e. In particular, the Fermi energy E_F is given in the degenerate case by

$$E_F = \left(\frac{3}{\pi}\right)^{2/3}\left(\frac{\boldsymbol{h}^2}{8m_e}\right)n^{2/3} \tag{23}$$

where n is the concentration of electrons in the band (cf. § 2.4 equation (31)). It will be noted that a given height of the Fermi level is attained for a smaller electron concentration if m_e is small. As we have seen, for a metal, the electrons are in the degenerate condition and E_F corresponds to the height of the Fermi level above the bottom of the partially filled band. For a semiconductor, as we shall see, the condition that the electrons are in a degenerate condition by no means always holds. However, for a strongly degenerate n-type semiconductor for which the Fermi level is above the bottom of the conduction band equation (23) will give the

* R. A. Smith, *Semiconductors*. Cambridge University Press (1959), pp. 68, 214.

height of the Fermi level above the bottom of the band provided E_F is several times greater than $\boldsymbol{k}T$ (see below).

We may treat similarly a single band of positive holes such as occurs in a strongly p-type semiconductor. The probability that a particular level having energy E is occupied is $F(x)$ where $F(x)$i s the Fermi–Dirac function and $x = (E - E_F)/\boldsymbol{k}T$; the probability that a particular level is *not* occupied is given by $F'(x)$ where

$$F'(x) = 1 - F(x) = F(-x) \tag{24}$$

If we now have a single band whose top is at $E = -\Delta E$ the probability of an energy level whose depth is E' below the top *not* being occupied is $F'(-x')$ where $x' = (E' - E_F')/\boldsymbol{k}T$, and $E_F' = -\Delta E - E_F$. Now $F'(-x')$ is equal to $F(x')$ and this gives the probability of occupation of a level at depth E' below the top of the band by a hole. Provided we replace E_F by E_F' we may now apply the same equations, so that equation (23) with m_e replaced by m_h gives E_F' the depth of the Fermi below the top of the band. Hence we have

$$E_F = -\Delta E - E_F' = -\Delta E - \left(\frac{3}{\pi}\right)^{2/3}\left(\frac{\boldsymbol{h}^2}{8m_h}\right)p^{2/3} \tag{25}$$

where p is the concentration of holes in the band. This again applies only when E_F' is several times greater than $\boldsymbol{k}T$ so that the holes are in a strongly degenerate condition.

When the bands overlap, as in some metals (see § 9.2.4), an integration must be carried out over both bands to obtain the Fermi level. This is also true for semiconductors except under the conditions discussed above. The subject has been treated in some detail by the author elsewhere* and we shall consider only two simple cases by way of illustration.

First let us consider an n-type semiconductor for which the ionization energy ϵ_1 of the impurities is much smaller than the forbidden-energy gap ΔE. We shall now assume that the electrons in the conduction band are non-degenerate and shall find an expression for the position of the Fermi level in this case. When this is so we may replace the Fermi–Dirac function $F(x)$ by the exponential e^{-x} (see § 2.7). We shall also assume that the impurities are fully ionised so that the concentration of electrons n in the conduction band is given equal to N_I the concentration of n-type impurities. To obtain an expression for the number of electrons in the conduction band in terms of the Fermi level we use the expression given by equation (24) of § 2.3 for $N(E)\,\mathrm{d}E$ the number of levels with energy between E and $E + \mathrm{d}E$ per unit volume, replacing the free-electron mass $\boldsymbol{m}$ by the scalar effective mass m_e. The concentration of electrons in the conduction band, equal to N_I to this approximation, is given by

$$n = N_I = 2\int_0^\infty N(E)\exp[(E_F - E)/\boldsymbol{k}T]\,\mathrm{d}E \tag{26}$$

* R. A. Smith, *Semiconductors*. Cambridge University Press (1959), Chapter 4.

A factor 2 has been included to take account of spin, the Fermi–Dirac function has been replaced by an exponential and the upper limit has been extended to ∞ instead of to E_m the energy corresponding to the top of the conduction band; since $(E_m - E_F)/kT$ will be quite large this is a legitimate extension. Inserting the value of $N(E)$ we have

$$n = N_I = 4\pi(2m_e)^{3/2}\,h^{-3}\int_0^{\infty} E^{1/2}\exp\left[(E_F - E)/kT\right]\mathrm{d}E \qquad (27)$$

$$= N_c\,\mathrm{e}^{E_F/kT} \qquad (28)$$

where
$$N_c = 2(2\pi m_e kT/h^2)^{3/2}$$

We therefore have

$$E_F = kT\ln(N_I/N_c) \qquad (29)$$

If $N_I < N_c$, E_F is negative and the Fermi level lies below the bottom of the conduction band, and clearly in the forbidden energy gap since the valence band is nearly full. We may also readily find an expression for the small number of holes in the valence band. If $N(E')\,\mathrm{d}E'$ is the number of levels with $E'\,(= -\Delta E - E)$ between E' and $E' + \mathrm{d}E'$ the concentration of holes p will be given by

$$p = 2\int_0^{\infty} N(E')\exp\left[(-E_F - \Delta E - E')/kT\right]\mathrm{d}E' \qquad (30)$$

$$= N_v\,\mathrm{e}^{-(\Delta E + E_F)/kT} \qquad (31)$$

where $N_v = 2(2\pi m_h kT/h^2)^{3/2}$. From equations (31) and (28) we note that the product np is constant being given by

$$np = N_c N_v\,\mathrm{e}^{-\Delta E/kT} \qquad (32)$$

N_c and N_v are of the order of 10^{19} cm^{-3}. It may be shown that the constancy of the product np holds for any value of N_I provided the semiconductor is non-degenerate (see R. A. Smith, *loc. cit.*). In particular, it holds when $N_I = 0$, i.e. for an intrinsic semiconductor, so that

$$np = n_i^2 \qquad (33)$$

where n_i is the equal electron and hole concentration for an intrinsic semiconductor, and hence from equation (32) is given by

$$n_i^2 = N_c N_v\,\mathrm{e}^{-\Delta E/kT} \qquad (34)$$

For an intrinsic non-degenerate semiconductor equation (28) will still hold provided we replace N_I by n_i, as will readily be seen from its derivation. A similar equation for the valence band will be obtained from equation (31) on replacing p by p_i. Since $n_i = p_i$ we obtain for the Fermi level the equation

$$E_F = -\frac{\Delta E}{2} + \tfrac{1}{2}kT\ln(N_v/N_c) \qquad (35)$$

$$= -\frac{\Delta E}{2} + \tfrac{3}{4}kT\ln(m_h/m_e) \qquad (36)$$

If $m_e = m_h$ the Fermi level lies half-way between the valence and conduction bands. An exactly similar treatment may be given for a p-type semiconductor. It will be seen that the electron or hole concentration of an intrinsic semiconductor depends markedly on ΔE and varies as $\exp[-\Delta E/2kT]$. A more accurate treatment shows that for a strongly n-type or p-type non-degenerate semiconductor (the condition being $N_I \gg n_i$ but also $N_I \ll N_c$ or N_v the electron concentration (n-type) varies as $\exp[-\epsilon_1/kT]$ or $\exp[-\epsilon_1/2kT]$, the hole concentration (p-type) as $\exp[-\epsilon_2/kT]$ or $\exp[-\epsilon_2/2kT]$, according to circumstances (see R. A. Smith, *loc. cit.* § 4.3). It will be clear that if $\epsilon_1 \ll \Delta E$ or $\epsilon_2 \ll \Delta E$ the temperature variation will be determined entirely by the impurities.

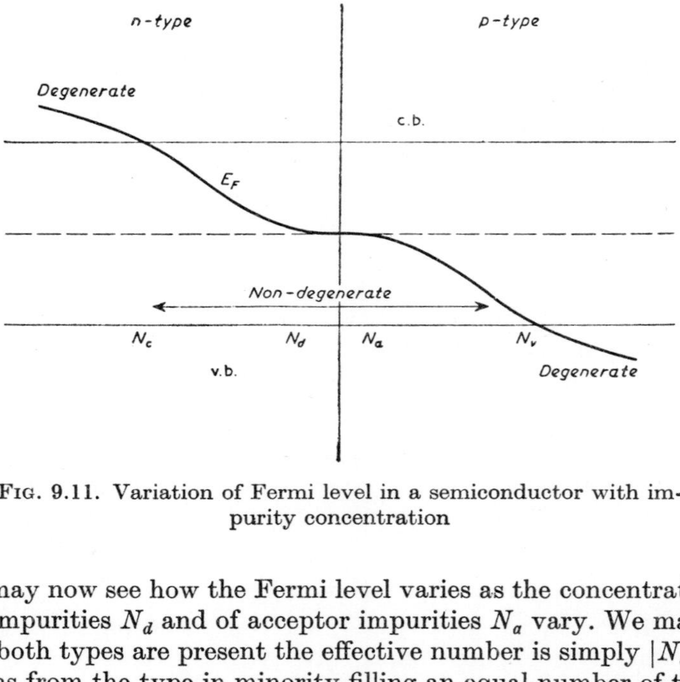

FIG. 9.11. Variation of Fermi level in a semiconductor with impurity concentration

We may now see how the Fermi level varies as the concentrations of donor impurities N_d and of acceptor impurities N_a vary. We may note that if both types are present the effective number is simply $|N_d - N_a|$, electrons from the type in minority filling an equal number of those in majority. The variation of the Fermi level is illustrated in Fig. 9.11. When $N_d \gg N_c$ the semiconductor is n-type and degenerate and the Fermi level lies above the bottom of the conduction band, as N_d decreases to a few times less than N_c the semiconductor becomes non-degenerate and the Fermi level is in the upper half of the forbidden energy gap. As N_d tends to zero the Fermi level approaches the middle of the forbidden energy gap crossing into the lower half when p-type impurities predominate to give a non-degenerate p-type semiconductor. Finally, if $N_a \gg N_v$ the Fermi level drops below the top of the valence band and the semiconductor is p-type and degenerate.

When we have two degenerate 'spherical' energy bands the integration to obtain the number of electrons or holes is taken over each band. There will be two integrals of the form appearing on the right-hand side of equation (27) each with its appropriate effective mass, which we shall denote by m_1 and m_2. It will be clear that equation (28) still applies provided we replace m_e in N_c or m_h in N_v by m_d where

$$m_d^{3/2} = m_1^{3/2} + m_2^{3/2} \tag{37}$$

m_d is known as the 'density-of-states effective mass'.

9.4.2 Ellipsoidal constant-energy surfaces

When the constant-energy surfaces are no longer spherical so that we do not have a scalar effective mass we can no longer use equation (24) of § 2.3 for $N(E)$. However, when we have ellipsoidal constant-energy surfaces with effective mass components m_1, m_2, m_3 the form of $N(E)$ is again readily obtained. By making the substitution $k'_x = (\boldsymbol{m}/m_1)^{1/2} k_x$, $k'_y = (\boldsymbol{m}/m_2)^{1/2} k_y$, $k'_z = (\boldsymbol{m}/m_3) k_z$ the energy may be expressed in 'spherical' form and we may proceed to carry out the integration over a sphere in $\mathbf{k}'$ space as before. We obtain the same result provided we replace m_e by a 'density-of-states' effective mass which will readily be seen to be given by

$$m_d = (m_1 m_2 m_3)^{1/3} \tag{38}$$

When we have M equivalent maxima or minima the integration must be carried over each and this introduces a factor M so that

$$N(E) = 2\pi \mathrm{M} (2m_d)^{3/2} E^{1/2} / \boldsymbol{h}^3 \tag{39}$$

In this case we may replace $m_e^{2/3}$ in equation (28) etc. by $\mathrm{M} m_d^{3/2}$.

9.4.3 General case

When the energy cannot be expressed as a quadratic function of the components of the wave vector $\mathbf{k}$ the determination of the function $N(E)$ is rather more difficult. Let $E(\mathbf{k}) = E$ represent a constant-energy surface in $\mathbf{k}$-space and let a neighbouring surface be given by $E(\mathbf{k}) = E + \delta E$. We wish to calculate the number of allowed values of $\mathbf{k}$ lying between the two surfaces. The density of allowed values in $\mathbf{k}$-space is equal to $V/8\pi^3$ (see § 8.9) where V is the volume of the crystal. The required number of energy levels is therefore equal to $(V/8\pi^3)\delta V_k$ where δV_k is the volume in $\mathbf{k}$-space between the two constant-energy surfaces. The normal distance between the surfaces is equal to $(\partial E/\partial k_n)^{-1} \delta E$ where $(\partial E/\partial k_n)$ is the normal derivative of E with respect to $\mathbf{k}$. Now we have

$$\frac{\partial E}{\partial k_n} = |\nabla_k E| = \left\{ \left(\frac{\partial E}{\partial k_x}\right)^2 + \left(\frac{\partial E}{\partial k_y}\right)^2 + \left(\frac{\partial E}{\partial k_z}\right)^2 \right\}^{1/2} \tag{40}$$

so that

$$N(E) = (V/8\pi^3) \iint |\nabla_k (E)|^{-1} \, \mathrm{d}S \tag{41}$$

This gives us a quite general expression for $N(E)$ for a single band, and may be used to calculate the number of electrons with energy less than E. This number $n(E)$ is given by

$$n(E) = 2\int_{E_0}^{E} N(E)\,F[(E-E_F)/\boldsymbol{k}T]\,\mathrm{d}E \tag{42}$$

where E_0 corresponds to the bottom of the band. By taking $E = E_m$ and equating this to the total number of electrons, the value E_F for the Fermi level may be obtained. When we have more than one partially filled band an integration must be carried out for each. In general, the integral (42) cannot be evaluated by analytical methods and must be calculated numerically. When we have strong degeneracy, however, the integral may be evaluated using the method given in § 2.7. We may take $E_0 = 0$ and extend the upper limit to ∞ without serious error; we then take as the function $G(E)$ in equation (77) of § 2.7

$$G(E) = \int_{0}^{E} N(E)\,\mathrm{d}E \tag{44}$$

If N is the total number of electrons we have

$$N = 2\int_{0}^{E_F} N(E)\,\mathrm{d}E + \frac{\pi^2}{3}\boldsymbol{k}^2T^2\left\{\frac{\mathrm{d}N(E)}{\mathrm{d}E}\right\}_{E=E_F} \tag{45}$$

If we define E_{F_0} by means of the equation

$$N = 2\int_{0}^{E_{F_0}} N(E)\,\mathrm{d}E \tag{46}$$

then E_{F_0} is the position of the Fermi level at the absolute zero of temperature; it is obtained by integration of equation (46) using equation (41) for $N(E)$. Using equation (46) an approximation to equation (45) may be written in the form

$$N(E_{F_0})(E_F-E_{F_0}) + \frac{\pi^2}{6}\boldsymbol{k}^2T^2\frac{\mathrm{d}N(E_{F_0})}{\mathrm{d}E_{F_0}} = 0$$

or

$$E_F = E_{F_0} - \frac{\pi^2\boldsymbol{k}^2T^2}{6N(E_{F_0})}\frac{\mathrm{d}N(E_{F_0})}{\mathrm{d}E_{F_0}} \tag{47}$$

9.5 Electronic specific heat

We have given expressions in § 2.7 for the electronic specific heat of a metal based on the free-electron model. These may be adapted at once to obtain expressions for metals in which we have a scalar effective mass m_e simply by replacing $\boldsymbol{m}$ by m_e. A discussion of the alkali metals was given in this way in § 2.7. The contribution of the free electrons and holes to the specific heat in semiconductors is generally quite small and the equations for the condition when we have a scalar effective mass meet almost all the requirements. It is, however, quite easy to derive a general expression for the electronic specific heat in the strongly degenerate condition, and this is the only condition in which it is appreciable. On comparison with equation (42) the total energy $\bar{E}$ of the electrons in the band will be given by

$$\bar{E} = 2\int_0^\infty EN(E)\,F[(E-E_F)/\boldsymbol{k}T]\,\mathrm{d}E \tag{48}$$

This integral may also be evaluated by the method given in § 2.7 on writing

$$G(E) = \int_0^E E'\,N(E')\,\mathrm{d}E' \tag{49}$$

giving

$$\bar{E} = 2\int_0^{E_F} EN(E)\,\mathrm{d}E + \frac{\pi^2}{3}\boldsymbol{k}^2\,T^2\frac{\mathrm{d}}{\mathrm{d}E}\{EN(E)\}_{E=E_F} \tag{50}$$

If the crystal corresponds to one mole, the atomic heat at constant volume C_v will be given by

$$C_v = \frac{\partial\bar{E}}{\partial T} = 2N(E_F)\,E_F\frac{\mathrm{d}E_F}{\mathrm{d}T} + \frac{2\pi^2}{3}\boldsymbol{k}^2\,T\left\{\frac{\mathrm{d}E_F N(E_F)}{\mathrm{d}E_F}\right\} \tag{51}$$

and on substituting in this equation for E_F we obtain

$$C_v = \frac{2\pi^2}{3}\boldsymbol{k}^2\,TN(E_F) \tag{52}$$

We note that the electronic specific heat is proportional to the absolute temperature T and to the density of states at the Fermi level. It will be seen on substituting the value of $N(E)$ for a 'spherical' band that equation (53) reduces to equation (92) of § 2.7.

At low temperatures we have seen that the specific heat due to the crystal lattice is proportional to T^3 (see § 7.3). The total specific heat may therefore, at low temperatures, be written in the form

$$C_v = AT + BT^3 \tag{53}$$

By plotting C_v/T against T^2 a linear plot should be obtained enabling the electronic component to be obtained from the intercept on the ordinate. This turns out to be so in many cases, but there are difficulties which must be avoided. These are discussed in most of the review articles referred to in Chapter 7.

10

Electrical and Thermal Transport Phenomena in Crystalline Solids

10.1 Collisions of electrons with crystalline imperfections

In the absence of external fields of force, an electron in a perfect crystal, at uniform temperature, moves with constant 'average' velocity $\bar{\mathbf{v}}$ through the lattice; this follows from the constancy of the wave vector $\mathbf{k}$ as discussed in Chapter 8. The term 'average' is used in the sense that $\bar{\mathbf{v}}$ represents an average taken over a unit cell of the crystal and is given in terms of the energy E by equation (63) of § 8.6. The variation of $\mathbf{k}$, and hence of $\bar{\mathbf{v}}$, in a perfect crystal, under the action of external forces, has been discussed in §§ 8.8 and 8.12. It was shown, for example, that, in a constant electric field, an electron in a perfect crystal would be accelerated so that its wave vector passed periodically right through a Brillouin zone. It was mentioned that this would not happen in real crystals because the electron would have made a collision with an imperfection before the value of the wave vector had changed appreciably. Even in a perfect crystal such collisions would take place, except at the absolute zero of temperature, with atoms of the crystal displaced from their lattice sites by the thermal vibrations. At the absolute zero of temperature such displacements will take place due to the 'zero-point' vibration but these are not random and it may be shown that they do not retard the electron's motion. At higher temperatures, however, the lattice vibrations play an important part in providing a damping force which rapidly prevents acceleration of an electron in a crystal and brings its average velocity to a steady value. In real crystals there are also sufficient imperfections to provide plenty of collisions, so that an electron remains 'free' only for a very short time, as we shall see. These imperfections may be due to displacements of the atoms of the crystal from their sites in the perfect lattice due to mechanical imperfections or to chemical impurities included in the crystal; these we shall class together as 'impurities'.

As we have also shown (see § 8.10), a perfect crystal at absolute zero would have no electrical resistance; resistance comes about through a damping force applied to electrons moving through the crystal under the influence of an electric field. At normal temperatures this damping force, in pure metals, comes mainly from scattering of electrons by the lattice vibrations, but at very low temperatures the so-called residual resistance is due to impurities. The much higher resistance of alloys, on the other hand, is due to scattering by the random arrangement of atoms on the lattice sites. In semiconductors, both lattice scattering and impurity scattering play an important part and we shall later discuss the effect of both types.

10.2 Mean free path and mean free time

In order to obtain an impression of the effect of collisions on the motion of electrons through a crystal under the influence of external fields of force we shall first of all consider a very much simplified model, and shall make some rather drastic assumptions which are not valid for most crystals with which we have to deal in practice; we shall later see how these assumptions have to be modified. We shall assume that an electron moves freely through a crystal, as in a perfect lattice, for a certain average time before making a collision, and that subsequent to a collision the direction of the electron's velocity bears no relationship to its previous velocity; we shall also assume that the collision is elastic; the latter assumption is indeed a very good approximation, the mass of the electrons being very much less than the mass of the atoms of the crystal. That it is not quite correct is clear from the fact that heating of the crystal by passage of an electric current corresponds to transfer of energy from the electrons to the atoms of the crystal; it turns out, however, to be a valid approximation except when we are dealing with very high electric fields leading to non-ohmic behaviour in semiconductors.

Since collisions correspond to a random process, the probability that an electron makes a collision at a time between t and $t+\mathrm{d}t$ subsequent to a former collision is independent of t and proportional to $\mathrm{d}t$; the probability per unit time P we shall also write as $1/\tau$, where τ has the dimensions of time. If we have n_0 electrons moving in a certain direction at $t = 0$ and n of these continue without making a collision at time t then we have

$$\mathrm{d}n = -Pn\,\mathrm{d}t = -(n/\tau)\,\mathrm{d}t \tag{1}$$

From equation (1) we obtain by integration

$$n = n_0\mathrm{e}^{-t/\tau} \tag{2}$$

The average time $\bar{t}$ between collisions is given by

$$\bar{t} = \frac{1}{n_0}\int_0^\infty \frac{nt\,\mathrm{d}t}{\tau} = \tau \tag{3}$$

The constant τ is therefore equal to the mean time between collisions, and the probability that an electron travels freely through the crystal for time t is equal to $e^{-t/\tau}$. If the magnitude of the velocity* of the electrons is v, we may define a mean free path l by means of the relationship $l = v\tau$.

Now let us suppose that the electrons are moving in the crystal under an applied electric field $\mathscr{E}$ which is in the x-direction. Let v_{x0} be the velocity of the electron in the x-direction at $t = 0$, the time at which it has just made a collision. Its x-component of velocity v_x at subsequent time t will then be given by

$$v_x = v_{x0} - \mathscr{E}\boldsymbol{e}t/m_e \tag{4}$$

provided the electron has *not* made a collision, where m_e is the effective mass, which we shall, for the moment, assume to be scalar; this follows from the equation of motion which takes the simple form $m_e\dot{v}_x = -\boldsymbol{e}\mathscr{E}$ when the effective mass is scalar. To obtain the average velocity $\overline{v_x}$ of all electrons in the x-direction, we must average over the times for which electrons are moving freely, so that equation (4) applies; we therefore have

$$\bar{v}_x = \bar{v}_{x0} - \frac{\boldsymbol{e}\mathscr{E}}{\tau m_e}\int_0^\infty t\,e^{-t/\tau}\,dt \tag{5}$$

The average velocity $\bar{v}_{x0}$ of electrons just subsequent to collisions is equal to zero in view of our assumption that the velocity is randomly directed. On performing the integration in equation (5), we therefore obtain

$$\bar{v}_x = -\boldsymbol{e}\mathscr{E}\tau/m_e \tag{6}$$

The effect of the electric field is therefore to set up a 'drift' velocity among the electrons. The field will not affect the y- and z-components of velocity and in this case $\overline{v_y} = \overline{v_{y0}} = \overline{v_z} = \overline{v_{z0}} = 0$.

10.3 Constant mean free time

The quantity τ will, in general, be dependent on the wave vector **k**, and to proceed further, we must make some assumptions about this dependence. To begin with we shall assume τ to be the same for all free electrons in the crystal. Equation (6) therefore applies to all free electrons and we may use it to obtain the current density J_x. We have

$$J_x = -\sum \boldsymbol{e}\bar{v}_x = n\boldsymbol{e}^2\mathscr{E}\tau/m_e \tag{7}$$

where n is the free electron concentration. The conductivity σ is therefore given by the equation

$$\sigma = n\boldsymbol{e}^2\tau/m_e \tag{8}$$

It is convenient to introduce a quantity μ_e called the electron mobility defined by means of the equation

$$\bar{v}_x = -\mathscr{E}\mu_e \tag{9}$$

* We now drop the term 'average', meaning average over a unit cell, to avoid confusion with an average with respect to time.

so that μ_e is the *magnitude* of the average drift velocity of electrons in unit electric field. We have then

$$\mu_e = e\tau/m_e \tag{10}$$

and equations (7) and (8) may be written in the forms

$$J_x = ne\mathscr{E}\mu_e \tag{11}$$

$$\sigma = ne\mu_e \tag{12}$$

When both holes and electrons are present, and the same assumptions apply, we obtain for the drift velocity $\bar{v}'_x$ of the holes

$$\bar{v}'_x = e\tau'\,\mathscr{E}/m_h$$

where τ' is the mean free time for the holes and m_h the (scalar) effective mass of the holes. The current density is therefore given by

$$J_x = ne^2\mathscr{E}\tau/m_e + pe^2\mathscr{E}\tau'/m_h \tag{13}$$

where p is the hole concentration. In terms of the hole mobility μ_h, given by

$$\mu_h = v'_x/\mathscr{E} = e\tau'/m_h \tag{14}$$

equation (14) may be written in the form

$$J_x = (ne\mu_e + pe\mu_h)\,\mathscr{E}$$

and the conductivity σ is given by

$$\sigma = ne\mu_e + pe\mu_h \tag{15}$$

It should be noted that the quantities μ_e, μ_h as defined above are *positive*; they are measured in M.K.S. units as $m^2/V\,sec$, but the unit $cm^2/V\,sec$ is frequently used.

Let us now try to obtain an estimate of the magnitude of the mean free time τ. If we know the order of magnitude of the carrier concentration we may obtain the value of the mobility μ. For normal metals the carrier concentration is of the order of the number of atoms per unit volume, and, as we shall see, there are, in any case, experimental methods for obtaining n or p. For a good semiconductor μ_e has a value of the order of $1000\ cm^2/V\,sec$ or $0{\cdot}1\ m^2/V\,sec$. If we take $m_e = m$, the free electron mass, we have $e/m \simeq 1{\cdot}75 \times 10^{11}$ C/kg, so that $\tau = 6 \times 10^{-13}$ sec. For a non-degenerate semiconductor for which the electron distribution is approximately Maxwellian the average velocity of an electron at room temperature is about 3×10^5 m/sec. The mean free path l is therefore about $2 \times 10^{-7}\ m = 2 \times 10^{-5}$ cm and so extends over many lattice spacings. In a time 10^{-5} sec the electron travels a distance of 3 m but drifts only a distance 10^{-2} cm in an electric field of 1 V/cm. The drift velocity is therefore very much less than the average velocity, which is virtually unchanged by the presence of the electric field.

It is also of interest to calculate the change $\Delta \boldsymbol{k}$ in the wave vector $\boldsymbol{k}$ during a collision time τ, under the influence of the electric field. We have

$$\dot{k}_x = \hbar^{-1} e\mathscr{E} \tag{16}$$

so that

$$\Delta k_x = \hbar^{-1} e\mathscr{E}\tau = m_e \hbar^{-1} \mathscr{E}\mu = -m_e \hbar \bar{v}_x \tag{17}$$

Again taking the above values for m_e and μ, and $\mathscr{E} = 1$ V/cm, giving $\bar{v}_x = 10$ m/s, we find that $\Delta k \simeq 1{\cdot}3 \times 10^2$ cm^{-1}. The change in Δk in crossing a Brillouin zone is of the order of $2\pi/d$ where d is the lattice spacing, or of the order of 10^8 cm^{-1}; the change in **k** between collisions is therefore extremely small and the process described in § 8.12, involving a large change in **k**, cannot take place except in very large electric fields.

10.3.1 Non-isotropic scattering

Scattering by lattice vibrations at normal temperatures is, as we shall see, mainly isotropic, so that our assumption, that the direction of motion of an electron after a collision is unrelated to its direction of motion just before the collision, is justified. For other important types of scattering, however, this is by no means so. For example, for scattering by ionized impurity centres in semiconductors there is a very high probability that the deflection of the electron from its direction before collision will be quite small. In order to deal with such a situation we introduce quantities known as the total and partial scattering cross-sections. The total scattering cross-section σ_t is related to the mean free path l by means of the equation

$$1/l = N\sigma_t \tag{18}$$

where N is the number of scattering centres per unit volume. It is therefore also related to the probability of scattering per unit time P and to the mean collision time τ by the equations

$$P = 1/\tau = Nv\sigma_t \tag{19}$$

The partial scattering cross-section $\sigma(\theta, \phi)$ enables us to obtain the probability per unit time that an electron will be scattered into a small solid angle in a direction marking an angle (θ, ϕ), in spherical polar coordinates, with its original direction of motion $\theta = 0$. The total cross-section σ_t is related to $\sigma(\theta, \phi)$ by the equation

$$\sigma_t = \int_0^{\pi} \int_0^{2\pi} \sigma(\theta, \phi) \sin\theta \, \mathrm{d}\theta \, \mathrm{d}\phi \tag{20}$$

so that $Nv\sigma(\theta, \phi)\,\mathrm{d}\Omega$ is probability of scattering into a small solid angle $\mathrm{d}\Omega$. The quantities σ_t and $\sigma(\theta, \phi)$ are commonly used in the theory of atomic collisions, and have the dimensions of area. Now let us consider an electron moving in the z-direction ($\theta = 0$) with velocity v_{z0} and making a collision in which it is scattered at an angle θ. The change in velocity δv_z in the z-direction will be given by

$$\delta v_z = (1-\cos\theta)\, v_{z0} \tag{21}$$

The average change in velocity in a collision $\overline{\delta v_z}$ will therefore be given by

$$\overline{\delta v_z} = \int_0^\pi \delta v_z \sigma(\theta) \sin\theta \,\mathrm{d}\theta \bigg/ \int_0^\pi \sigma(\theta)\sin\theta\,\mathrm{d}\theta \tag{22}$$

$$= \frac{2\pi v_{z0}}{\sigma_t}\int_0^\pi (1-\cos\theta)\,\sigma(\theta)\sin\theta\,\mathrm{d}\theta \tag{23}$$

if $\sigma(\theta,\phi)$ is independent of ϕ. If we define the cross-section σ_c by means of the equation

$$\sigma_c = 2\pi\int_0^\pi (1-\cos\theta)\,\sigma(\theta)\sin\theta\,\mathrm{d}\theta \tag{24}$$

we have

$$\overline{\delta v_z} = v_{z0}\,\sigma_c/\sigma_t \tag{25}$$

In order that the initial velocity v_{z0} be 'randomised' we must therefore have *on the average* σ_t/σ_c such collisions. The mean free time τ is related to the total cross-section σ_t by means of the relationship (19), so that for transport processes which depend on a change of velocity, the appropriate mean free time to use would therefore appear to be τ_c given by

$$\tau_c = \sigma_t\tau/\sigma_c \tag{26}$$

so that

$$\frac{1}{\tau_c} = 2\pi N v\int_0^\pi (1-\cos\theta)\,\sigma(\theta)\sin\theta\,\mathrm{d}\theta \tag{27}$$

The above argument can only be said to make this result plausible; we shall give later a proof valid under certain conditions (see § 10.9 equation (164)). The factor $(1-\cos\theta)$ is clearly introduced in order to give smaller weight to the small angle collisions which do not destroy the drift velocity. In terms of the probability $P(\theta,\phi)\,\mathrm{d}\Omega$ of scattering into a small solid angle $\mathrm{d}\Omega$ the transport mean free time τ_c may be expressed in the form

$$\frac{1}{\tau_c} = \iint (1-\cos\theta)\,P(\theta,\phi)\,\mathrm{d}\Omega \tag{28}$$

whereas the total collision mean free time τ_t is given by

$$\frac{1}{\tau_t} = \iint P(\theta,\phi)\,\mathrm{d}\Omega \tag{29}$$

and we see that $P(\theta,\phi) = Nv\sigma(\theta,\phi)$. The quantity τ_c is then the appropriate quantity to use in equations such as (10) and (14) for the mobility and for isotropic scattering clearly $\tau = \tau_c$.

10.3.2 Non-scalar effective mass

When τ (or τ_c) is the same for all free electrons it is very easy to see how to generalise the above results when the effective mass is no longer scalar. When we have, for example, a conduction band which is non-degenerate and has its lowest minimum at $\mathbf{k} = 0$, so that the energy E can be expressed in the form

$$E = \frac{\hbar^2}{2}\left[\frac{k_x^2}{m_1} + \frac{k_y^2}{m_2} + \frac{k_z^2}{m_3}\right] \tag{30}$$

the equations of motion of electrons in an external electric field whose components are $\mathscr{E}_x$, $\mathscr{E}_y$, $\mathscr{E}_z$ are

$$\left.\begin{aligned} m_1 \dot{v}_x &= -e\mathscr{E}_x \\ m_2 \dot{v}_y &= -e\mathscr{E}_y \\ m_3 \dot{v}_z &= -e\mathscr{E}_z \end{aligned}\right\} \tag{31}$$

Proceeding with each equation as before we have for the components J_x, J_y, J_z of the current density

$$\left.\begin{aligned} J_x &= ne\mu_1 \mathscr{E}_x \\ J_y &= ne\mu_2 \mathscr{E}_y \\ J_3 &= ne\mu_3 \mathscr{E}_z \end{aligned}\right\} \tag{32}$$

where $\mu_1 = e\tau/m_1$, $\mu_2 = e\tau/m_2$, $\mu_3 = e\tau/m_3$. The components of the conductivity tensor are given by

$$\left.\begin{aligned} \sigma_{xx} &= ne\mu_1 \\ \sigma_{yy} &= ne\mu_2 \\ \sigma_{zz} &= ne\mu_3 \\ \sigma_{xy} &= \sigma_{yx} = \sigma_{xz} = \sigma_{zx} = \sigma_{yz} = \sigma_{zy} = 0 \end{aligned}\right\} \tag{33}$$

Frequently we have an axis of symmetry, such as the c-axis in hexagonal crystals, and in this case we may take the z-axis along the axis of symmetry and write $\sigma_{xx} = \sigma_{yy} = \sigma_T$, $\sigma_{zz} = \sigma_L$. If the current density vector $\mathbf{J}$ is in the x–z plane, making an angle θ with the z-axis, we have

$$\left.\begin{aligned} J_x &= J \sin\theta = \sigma_T \mathscr{E}_x \\ J_y &= 0 \\ J_z &= J \cos\theta = \sigma_L \mathscr{E}_z \end{aligned}\right\} \tag{34}$$

The direction of the field and current are therefore not coincident. The resistivity σ^{-1} is equal in this case to $(\mathbf{J} . \mathscr{E})/J^2$ so that we have

$$\frac{1}{\sigma} = \frac{\sin^2\theta}{\sigma_T} + \frac{\cos^2\theta}{\sigma_L} \tag{35}$$

When we do not have the minimum of the band at $\mathbf{k} = 0$, and have several equivalent minima we must sum over the electrons at each minimum to obtain the current density. For a crystal with cubic symmetry we clearly must obtain $\sigma_{xx} = \sigma_{yy} = \sigma_{zz}$ and the other components of the conductivity tensor equal to zero; we have thus once again a scalar conductivity. This condition occurs for such materials as Si and Ge and has been discussed extensively in books on semiconductors*. For these, the assumption that τ is constant is far from correct and we shall not pursue the matter further in this section. We must now consider what happens when τ depends on the velocity (or energy) of the electrons.

10.4 Mean free time a function of the energy E

In order to simplify the problem and to obtain an insight into the effect of collisions we have, so far, regarded the mean free time τ as being the same for all electrons. This will not, in general, be so; τ will depend on the velocity of the electron or hole being scattered, and may also depend on the direction of motion through the crystal; this could be taken into account if we could express τ as a function of the wave vector **k**. We could then apply the above analysis to only those electrons having wave vectors in a narrow range and then sum over all the free electrons and holes in the crystal to obtain the current density. It must be admitted at once that this general treatment is quite difficult; first of all it has not proved possible to calculate $\tau(\mathbf{k})$ for non-isotropic conductors with any degree of accuracy, and secondly even if we *assume* a form for $\tau(\mathbf{k})$ the subsequent averaging process is complicated. It has been carried out for the element semiconductors Si and Ge assuming a form of τ which has the same symmetry as the effective-mass tensor. The problem is greatly simplified if we assume that τ may be expressed as a function of the velocity, or rather of the energy. We may then use the analysis of § 10.3 with $\tau = \tau(E)$ and sum over all the electrons (and holes) to obtain the conductivity.

Whenever we have to perform such a summation, or what is equivalent, when we average over groups of electrons with different energy E, we have to multiply by some weighting function which gives the relative importance of the different groups. At first sight this function might be supposed to be simply $N_e(E)\,\mathrm{d}E$, the number of electrons with energies between E and $E+\mathrm{d}E$. Further analysis of the problem, which we shall give later (see § 10.5), indicates, however, that the correct weighting function is $EN_e(E)\,\mathrm{d}E$ when we do not have degeneracy. In the degenerate condition the situation is more complex since we have to consider not only occupancy of the initial state but also occupancy of neighbouring states which may prevent transitions to them. If we accept that the correct average is given by use of the weighting function $EN_e(E)\,\mathrm{d}E$ we may define an average relaxation time $\langle\tau\rangle$ by means of the equation

$$\langle\tau\rangle = \int EN_e(E)\,\tau(E)\,\mathrm{d}E \Big/ \int EN_e(E)\,\mathrm{d}E \tag{36}$$

* See, for example, R. A. Smith, *Semiconductors*. Cambridge University Press (1959), p. 97 *et seq.*

The average drift velocity is then given by

$$\overline{v_x} = -e\mathscr{E}\langle\tau\rangle/m_e \tag{37}$$

and the electron mobility by

$$\mu_e = e\langle\tau\rangle/m_e \tag{38}$$

In deriving the expression for the appropriate weighting function we must consider the deviation of the electron distribution function from its equilibrium value; this is given by a well-known equation called Boltzmann's equation which we shall now proceed to obtain. Having obtained Boltzmann's equation it may be used to study the transport phenomena.

10.5 Boltzmann's equation

In thermal equilibrium, the distribution of electrons in energy in a crystal is given in terms of the density of states function $N(E)$ and the Fermi function $F[(E-E_F)/kT]$ as discussed in § 2.6. The probability that a state with wave vector $\mathbf{k}$ is occupied is the same as that of a state with wave vector $-\mathbf{k}$ so that no transport takes place. When we have transport of charge (electrical conduction) or transport of energy (thermal conduction) the distribution function must be modified by electric or magnetic fields or by temperature gradients. It is this modification of the distribution function which we now must calculate. Let $f(\mathbf{k},\mathbf{r},t)$ be the probability at time t of occupation of the state corresponding to the wave vector $\mathbf{k}$ at a point in the crystal given by the position vector $\mathbf{r}$. When we consider the variation of f with $\mathbf{r}$ we are concerned only with variations which are very small over a distance of the order of a lattice spacing, so that wave functions having a definite value of $\mathbf{k}$ may be specified locally in the crystal. The spatial variation is introduced mainly so as to be able to take account of inhomogeneity and variations of temperature throughout the crystal.

If we have a force $\mathbf{F}$, whose components are F_x, F_y, F_z, acting on the electrons, the value of $\mathbf{k}$ will change at a rate given by equation (69) of § 8.8, i.e. $\hbar\dot{\mathbf{k}} = \mathbf{F}$; at time $t+\mathrm{d}t$, $\mathbf{k}$ will have value $\mathbf{k}+\hbar^{-1}\mathbf{F}\,\mathrm{d}t$. An electron which at time t has position vector $\mathbf{r}$ and wave vector $\mathbf{k}$ will at time $t+\mathrm{d}t$ have position vector $\mathbf{r}+\mathbf{v}\,\mathrm{d}t$ and wave vector $\mathbf{k}+\hbar^{-1}\mathbf{F}\,\mathrm{d}t$. The function f is then given by $f(\mathbf{k}+\hbar^{-1}\mathbf{F}\,\mathrm{d}t, \mathbf{r}+\mathbf{v}\,\mathrm{d}t, t+\mathrm{d}t)$, so that the total rate of change of f is given by

$$\frac{\mathrm{D}f}{\mathrm{D}t} = \frac{\partial f}{\partial t}+\hbar^{-1}\left(F_x\frac{\partial f}{\partial k_x}+F_y\frac{\partial f}{\partial k_y}+F_z\frac{\partial f}{\partial k_z}\right)+v_x\frac{\partial f}{\partial x}+v_y\frac{\partial f}{\partial y}+v_z\frac{\partial f}{\partial z} \tag{39}$$

which may be expressed vectorially in the form

$$\frac{\mathrm{D}f}{\mathrm{D}t} = \frac{\partial f}{\partial t}+\hbar^{-1}\mathbf{F}\,.\,\nabla_k f+\mathbf{v}\,.\,\nabla_r f \tag{40}$$

This *total* rate of change in f must be brought about by collisions and we shall write the rate of change of f due to collisions in the form $\left.\frac{\mathrm{d}f}{\mathrm{d}t}\right]_{\mathrm{coll}}$ Thus we obtain the equation

$$\frac{\partial f}{\partial t} = \left.\frac{\mathrm{d}f}{\mathrm{d}t}\right]_{\mathrm{coll}} - \hbar^{-1}\mathbf{F}.\nabla_k f - \mathbf{v}.\nabla_r f \tag{41}$$

This equation is known as Boltzmann's equation and is the fundamental equation governing all transport phenomena. In steady-state conditions $\partial f/\partial t = 0$, and we have

$$\left.\frac{\mathrm{d}f}{\mathrm{d}t}\right]_{\mathrm{coll}} = \hbar^{-1}\mathbf{F}.\nabla_k f + \mathbf{v}.\nabla_r f \tag{42}$$

We must now consider the form of the collision term in equation (42). In the absence of external fields and temperature gradients the electron distribution will, if it is not already in the form corresponding to thermal equilibrium, relax back to this form in a certain time depending on the efficiency of the collision mechanism in bringing about equipartition of energy. In discussions of such processes in the dynamical theory of gases it is usual to assume that the rate of relaxation is proportional to the deviation of the function from equilibrium and to define a relaxation time τ by means of the equation

$$\left.\frac{\mathrm{d}f}{\mathrm{d}t}\right]_{\mathrm{coll}} = -\frac{f-f_0}{\tau} \tag{43}$$

where f_0 is the form of the function f corresponding to thermal equilibrium. A great deal of discussion has taken place as to whether this is justified. For small deviations from equilibrium such as we have here it would appear to be, but for non-isotropic crystals, as we shall see, τ is not necessarily a scalar quantity. If we assume the form given by equation (43) for the collision term, Boltzmann's equation becomes

$$f = f_0 - \tau\hbar^{-1}\mathbf{F}.\nabla_k f - \tau\mathbf{v}.\nabla_r f \tag{44}$$

In the absence of external fields and gradients when we do not have a steady state, the relaxation of the system is governed by the equation

$$\frac{\mathrm{d}f}{\mathrm{d}t} = -\frac{f-f_0}{\tau} \tag{45}$$

and the solution in this case is

$$f = f_0 + (f_1 - f_0)\,\mathrm{e}^{-t/\tau} \tag{46}$$

where f_1 is the value of f at $t = 0$. The system therefore relaxes exponentially towards its equilibrium value with time constant τ.

We shall first of all consider a homogeneous crystal at constant temperature so that f does not depend on the position vector $\mathbf{r}$. We have then for the equation to determine f

$$f - f_0 = -\tau\hbar^{-1}\mathbf{F}.\nabla_k f \tag{47}$$

We shall now assume, and later verify, that the deviation of f from its equilibrium value is small, so that f may be replaced by f_0 on the right-hand side of equation (47). We thus obtain for the modified distribution function f the value

$$f = f_0 - \tau\hbar^{-1}\mathbf{F}.\nabla_k f_0 \tag{48}$$

Since f_0 is a function of the energy E this equation may be written in the form

$$f = f_0 - \tau\hbar^{-1}\frac{\partial f_0}{\partial E}(\mathbf{F}.\nabla_k E) \tag{49}$$

Since $\nabla_k E = \hbar\mathbf{v}$ the equation may also be expressed in the form

$$f = f_0 - \tau\frac{\partial f_0}{\partial E}(\mathbf{F}.\mathbf{v}) \tag{50}$$

10.6 Electrical conduction

If the external force $\mathbf{F}$ is due to a constant electric field $\mathscr{E}$ in the x-direction we have for the disturbed distribution function f

$$f = f_0 + \tau e\mathscr{E}v_x\frac{\partial f_0}{\partial E} \tag{51}$$

Knowing the distribution function f, the current density J_x can be obtained by summing over all electrons with different values of the wave vector $\mathbf{k}'$; we have

$$J_x = -\frac{2e}{V}\int N(\mathbf{k}')fv_x\,d\mathbf{k}' \tag{52}$$

where $N(\mathbf{k}')$ is the density of allowed values of $\mathbf{k}'$ (excluding spin) and is equal to $V/8\pi^3$ where V, is the volume of the crystal (see § 8.9). The integral is to be taken over all partially filled bands. Inserting the value of f from equation (51) we obtain

$$J_x = -\frac{e^2\mathscr{E}}{4\pi^3}\int \tau v_x^2\frac{\partial f_0}{\partial E}\,d\mathbf{k}' \tag{53}$$

Let us consider first of all a single band, the bottom of which occurs at $E = 0$ and the top at $E = E_m$. The electron concentration n may then be obtained from the equation

$$nV = \frac{V}{4\pi^3}\int_0^{E_m} f_0\,d\mathbf{k}' \tag{54}$$

Using equation (54) the current density J_x may be expressed in the form

$$J_x = -\mathbf{e}^2\mathscr{E}n\int_0^{E_m}\tau v_x^2\frac{\partial f_0}{\partial E}\,\mathrm{d}\mathbf{k}'\bigg/\int_0^{E_m}f_0\,\mathrm{d}\mathbf{k}' \tag{55}$$

Since f_0 is given by the Fermi function (see § 2.6), we have

$$\frac{\partial f_0}{\partial E} = -f_0(1-f_0)/\mathbf{k}T$$

and an alternative form of equation (55) is

$$J_x = \frac{\mathbf{e}^2\mathscr{E}n}{\mathbf{k}T}\int_0^{E_m}\tau v_x^2 f_0(1-f_0)\,\mathrm{d}\mathbf{k}'\bigg/\int_0^{E_m}f_0\,\mathrm{d}\mathbf{k}' \tag{55a}$$

When τ is a function of the energy E and we have a scalar effective mass m_e, so that

$$E = \tfrac{1}{2}m_e v^2 = \hbar^2 k'^2/2m_e \tag{56}$$

equation (55) may be considerably simplified. Clearly we have

$$\int_0^{E_m} v_x^2\phi(E)\,\mathrm{d}\mathbf{k}' = \int_0^{E_m} v_y^2\phi(E)\,\mathrm{d}\mathbf{k}' = \int_0^{E_m} v_z^2\phi(E)\,\mathrm{d}\mathbf{k}' = \frac{2}{3m_e}\int_0^{E_m} E\phi(E)\,\mathrm{d}\mathbf{k}'$$

Moreover $\mathrm{d}\mathbf{k}' = \text{const.}\times E^{1/2}\mathrm{d}E$, so that the expression for J_x may be written as

$$J_x = \frac{2\mathbf{e}^2\mathscr{E}n}{3\mathbf{k}Tm_e}\int_0^{E_m}\tau E^{3/2}f_0(1-f_0)\,\mathrm{d}E\bigg/\int_0^{E_m}E^{1/2}f_0\,\mathrm{d}E \tag{57}$$

First let us suppose τ to be constant. We have

$$\int_0^{E_m}E^{3/2}f_0(1-f_0)\,\mathrm{d}E = -\mathbf{k}T\int_0^{E_m}E^{3/2}\frac{\partial f_0}{\partial E}\,\mathrm{d}E = \frac{3\mathbf{k}T}{2}\int_0^{E_m}E^{1/2}f_0\,\mathrm{d}E$$

the integrated term being negligible since we shall suppose that $E_m \gg \mathbf{k}T$. The expression for J_x therefore reduces to the simple form

$$J_x = \mathbf{e}^2\mathscr{E}n\tau/m_e \tag{58}$$

which is exactly the same as equation (7). We therefore see that the relaxation time τ is just the same as the mean free time between collisions.

When τ is not constant but varies with the energy E equation (57) may also be simplified under certain conditions. For a metal or highly degenerate semiconductor $f_0(1-f_0)$ or $\partial f_0/\partial E$ are only appreciable for a small band of energies of width of the order of $\boldsymbol{k}T$ near the Fermi level $E = E_F$. In this case

$$\frac{1}{\boldsymbol{k}T}\int_0^{E_m} \phi(E)\, f_0(1-f_0)\,\mathrm{d}E = -\int_0^{E_m} \phi(E)\frac{\partial f_0}{\partial E}\,\mathrm{d}E \simeq \phi(E_F) \tag{59}$$

and

$$\int_0^{E_m} E^{1/2} f_0\,\mathrm{d}E \simeq \int_0^{E_F} E^{1/2}\,\mathrm{d}E = \tfrac{2}{3}E_F^{3/2} \tag{60}$$

On substituting in equation (57) we therefore obtain

$$J_x = \boldsymbol{e}^2 \mathscr{E} n \tau_F / m_e \tag{61}$$

where τ_F is the value of τ when $E = E_F$, i.e. it is the relaxation time or mean collision time for electrons at the Fermi surface. Thus the formulae for the mobility and conductivity given in § 10.3 still hold provided we take $\tau = \tau_F$. It is interesting to note that although the contribution to the conduction comes only from electrons near the Fermi surface, the total free electron concentration n occurs in equation (61); all electrons have their wave vectors changed by the field, but only those near the Fermi surface can have their wave vectors changed by collisions.

Another extreme case, in which the expression for the current density may be greatly simplified, occurs when f_0 corresponds to the classical distribution function, for example in a non-degenerate semiconductor. In this case $f_0 = A\exp[-E/\boldsymbol{k}T]$ where A is a constant, and $f_0 \ll 1$. Equation (57) may then be expressed in the form

$$J_x = \frac{\boldsymbol{e}^2 \mathscr{E} n}{m_e} \int_0^{\infty} \tau E^{3/2} f_0\,\mathrm{d}E \bigg/ \int_0^{\infty} E^{3/2} f_0\,\mathrm{d}E \tag{62}$$

since we may extend the upper limit of integration to infinity without appreciable error, and we have made use of the relationship

$$\int_0^{\infty} E^{3/2} f_0\,\mathrm{d}E = \frac{3\boldsymbol{k}T}{2}\int_0^{\infty} E^{1/2} f_0\,\mathrm{d}E \tag{63}$$

The expression for the current density may therefore be written in the form

$$J_x = n\boldsymbol{e}^2 \mathscr{E} \langle\tau\rangle / m_e \tag{64}$$

where

$$\langle\tau\rangle = \int_0^{\infty} \tau E^{3/2}\,\mathrm{e}^{-E/\boldsymbol{k}T}\,\mathrm{d}E \bigg/ \int_0^{\infty} E^{3/2}\,\mathrm{e}^{-E/\boldsymbol{k}T}\,\mathrm{d}E \tag{65}$$

This now shows us the correct way to average for electrons of different energy. The appropriate weighting function is not $N_e(E)$ but $EN_e(E)$

(or a constant times $EN_e(E)$), as already discussed in § 10.4 (cf. equation 36)). The weighting function in this case may be taken as

$$E^{3/2}\exp[-E/\boldsymbol{k}T]$$

since $N_e(E)$, the number of electrons with energies between E and $E+\mathrm{d}E$ is proportional to $E^{1/2}\exp[-E/\boldsymbol{k}T]$. An average of any quantity q with this weighting function we shall write as $\langle q\rangle$.

From the expression (64) for the current density we see that for a non-degenerate semiconductor the electron mobility μ_e will be given by

$$\mu_e = e\langle\tau_e\rangle/m_e \tag{66}$$

similarly for holes we have

$$\mu_h = e\langle\tau_h\rangle/m_h \tag{67}$$

but the weighting function for holes is clearly

$$(-E-\Delta E)^{3/2}\exp[E/\boldsymbol{k}T] \tag{68}$$

the top of the valence band being at $E = -\Delta E$. Using these expressions for μ_e and μ_h the formulae given in § 10.3 for the current density and conductivity still hold.

The above formulae may also readily be modified to give the electrical conductivity when the constant-energy surfaces are ellipsoidal, the energy E being given by an expression of the form

$$E = \frac{\hbar^2}{2}\left[\frac{k_x^2}{m_1}+\frac{k_y^2}{m_2}+\frac{k_z^2}{m_3}\right] \tag{69}$$

We make the transformation

$$\left.\begin{aligned} k'_x &= (m_e/m_1)^{1/2}k_x \\ k'_y &= (m_e/m_2)^{1/2}k_y \\ k'_z &= (m_e/m_3)^{1/2}k_z \end{aligned}\right\} \tag{70}$$

so that

$$E = \frac{\hbar^2}{2m_e}[k_x'^2+k_y'^2+k_z'^2] \tag{71}$$

We then have, writing $v'_x = \hbar k'_x/m_e$, $v'_y = \hbar k'_y/m_e$, $v'_z = \hbar k'_z/m_e$,

$$\left.\begin{aligned} v_x^2 &= (m_e/m_1)\,v_x'^2 \\ v_y^2 &= (m_e/m_1)\,v_y'^2 \\ v_z^2 &= (m_e/m_1)\,v_z'^2 \end{aligned}\right\} \tag{72}$$

On substituting in the expression (55) for J_x we obtain the same expression in terms of the quantities k'_x, v'_x, etc., as before, apart from a factor (m_e/m_1). We therefore have

$$J_x = ne^2\mathscr{E}_x/m\bar{\tau}_1 = ne\mu_1\mathscr{E}_x \tag{73}$$

where

$$\mu_1 = e\bar{\tau}/m_1 \tag{74}$$

and $\bar{\tau}$ is the appropriate average of τ. Similarly if the field is directed along the y- and z-directions, we have

$$J_y = ne^2\mathscr{E}_y\bar{\tau}/m_2 = ne\mu_2\mathscr{E}_y \tag{75}$$

$$J_z = ne^2\mathscr{E}_z\bar{\tau}/m_3 = ne\mu_3\mathscr{E}_z \tag{76}$$

where

$$\mu_2 = e\bar{\tau}/m_2 \tag{77}$$

$$\mu_3 = e\bar{\tau}/m_3 \tag{78}$$

The conductivity tensor has therefore components

$$\left.\begin{aligned} \sigma_{xx} &= ne\mu_1 \\ \sigma_{yy} &= ne\mu_2 \\ \sigma_{zz} &= ne\mu_3 \\ \sigma_{xy} &= \sigma_{yz} = \sigma_{xz} = \sigma_{zx} = \sigma_{yz} = \sigma_{zy} = 0 \end{aligned}\right\} \tag{79}$$

If we have a semiconductor with cubic symmetry so that we have a number of *equivalent* minima in the conduction band, to obtain the current density we must sum over all the equivalent minima. When these lie on the x, y, z axes (as for Si, see § 9.3) the non-zero components of the conductivity tensor are clearly equal and are given by

$$\sigma_{xx} = \sigma_{yy} = \sigma_{zz} = \sigma = \tfrac{1}{3}ne(\mu_1+\mu_2+\mu_3) \tag{80}$$

$$= \tfrac{1}{3}ne^2\bar{\tau}\left(\frac{1}{m_1}+\frac{1}{m_2}+\frac{1}{m_3}\right) \tag{81}$$

The conductivity tensor therefore reduces to a scalar σ, and the conductivity is isotropic. This may be also shown to be true for any crystal having cubic symmetry*.

When the constant-energy surfaces are more complex, for example, for a semiconductor with a degenerate valence band (see § 9.3), the integration over the wave vector $\mathbf{k}'$ is more complicated but has been performed in a number of cases of practical interest*. It is not even justifiable to assume that τ is a scalar quantity and it has been proposed that τ may be taken as a tensor having the same symmetry properties as the effective-mass tensor. We note that the quantity which appears in our equations is usually m/τ, and it is this quantity which is given by a number of measurements rather than τ and m independently.

We shall later show how to calculate τ (see § 10.9), but it is of interest to see how a simple law for the variation of τ with E affects the mobility; we shall consider the condition of non-degeneracy first of all. Suppose that we may express the relaxation time τ in the form $\tau = AE^{-s}$; then we have

$$\langle\tau\rangle = A\int_0^\infty E^{3/2-s}\,\mathrm{e}^{-E/kT}\,\mathrm{d}E \Bigg/ \int_0^\infty E^{3/2}\,\mathrm{e}^{-E/kT}\,\mathrm{d}E \tag{82}$$

* See, for example, R. A. Smith, *Semiconductors*. Cambridge University Press (1959), § 5.3.

The integrals in equation (82) may be expressed in terms of the Γ-function

$$\Gamma(t) = \int_0^\infty x^{t-1} e^{-x} \, dx$$

If we make the transformation $x = E/kT$ we obtain

$$\langle \tau \rangle = A(kT)^{-s} \Gamma(\tfrac{5}{2} - s)/\Gamma \tfrac{5}{2} \tag{83}$$

When we have strong degeneracy we simply have

$$\bar{\tau} = A E_F^{-s} \tag{84}$$

We shall later give some examples of the variation of $\langle \tau \rangle$ and hence of the mobility μ with temperature (see § 10.9).

10.7 Heat conduction

We must now consider the effect on the distribution function f of macroscopic spatial variations within a crystal, and as an example we shall discuss the situation which arises when we have a temperature gradient. Such a gradient clearly will cause conduction of heat, but also gives rise to interesting thermoelectric effects.

It will now be necessary to include in Boltzmann's equation the second term on the right-hand side of equation (42). We shall suppose that we have a constant electric field $\mathscr{E}$ directed along the x-axis and a constant temperature gradient dT/dx, also in this direction; it is desirable to introduce *both* the electric field and the temperature gradient at the beginning of our discussion. Proceeding as before, and taking $f = f_0$ in the terms giving $(f - f_0)$ we obtain for the modified distribution function f the expression

$$f = f_0 + \tau e \mathscr{E} v_x \frac{\partial f_0}{\partial E} - \tau v_x \frac{\partial f_0}{\partial x} \tag{85}$$

We have

$$\frac{\partial f_0}{\partial x} = \frac{\partial f_0}{\partial T} \cdot \frac{dT}{dx} \tag{86}$$

and, since E_F must now be regarded as a function of T,

$$\frac{\partial f_0}{\partial T} = kT \frac{\partial f_0}{\partial E} \frac{d}{dT}\left(\frac{E - E_F}{kT}\right) \tag{87}$$

$$= -\left\{T \frac{d}{dT}\left(\frac{E_F}{T}\right) + \frac{E}{T}\right\} \frac{\partial f_0}{\partial E} \tag{87a}$$

Equation (85) may thus be expressed in the form

$$f = f_0 + \tau v_x \frac{\partial f_0}{\partial E}\left[e\mathscr{E} + \frac{dT}{dx}\left\{T \frac{d}{dT}\left(\frac{E_F}{T}\right) + \frac{E}{T}\right\}\right] \tag{88}$$

The electric current density J_x may again be obtained as in § 10.6 by summing over all values of the wave vector $\mathbf{k}'$. We have on comparison with equation (53)

$$J_x = -\frac{e}{4\pi^3}\int \tau v_x^2 \frac{\partial f_0}{\partial E}\left[e\mathscr{E}_x + \left\{T\frac{\mathrm{d}}{\mathrm{d}T}\left(\frac{E_F}{T}\right) + \frac{E}{T}\right\}\frac{\mathrm{d}T}{\mathrm{d}x}\right]\mathrm{d}\mathbf{k}' \tag{89}$$

The heat current W_x arises from the transport of energy E as distinct from the transport of charge $-e$ and so is given by

$$W_x = \frac{1}{4\pi^3}\int \tau v_x^2 \frac{\partial f_0}{\partial E}\left[e\mathscr{E}_x E + \left\{ET\frac{\mathrm{d}}{\mathrm{d}T}\left(\frac{E_F}{T}\right) + \frac{E^2}{T}\right\}\frac{\mathrm{d}T}{\mathrm{d}x}\right]\mathrm{d}\mathbf{k}' \tag{90}$$

10.7.1 Thermal conductivity

When we have a single band characterised by a scalar effective mass m_e, the integrals in (89) and (90) may be simplified by proceeding as in § 10.6. On comparing equations (89) and (90) with equation (53) we readily obtain the following equations for J_x and W_x,

$$J_x = \frac{ne}{m_e}\left[\left\{e\mathscr{E} + T\frac{\mathrm{d}}{\mathrm{d}T}\left(\frac{E_F}{T}\right)\frac{\mathrm{d}T}{\mathrm{d}x}\right\}\bar{\tau} + \frac{1}{T}\frac{\mathrm{d}T}{\mathrm{d}x}\,\overline{(E\tau)}\right] \tag{91}$$

$$W_x = -\frac{n}{m_e}\left[\left\{e\mathscr{E} + T\frac{\mathrm{d}}{\mathrm{d}T}\left(\frac{E_F}{T}\right)\frac{\mathrm{d}T}{\mathrm{d}x}\right\}\overline{(E\tau)} + \frac{1}{T}\frac{\mathrm{d}T}{\mathrm{d}x}\,\overline{(E^2\tau)}\right] \tag{92}$$

where $\bar{\tau}$, $\overline{(E\tau)}$, $\overline{(E^2\tau)}$ are appropriate averages of the quantities τ, $E\tau$, $E^2\tau$. For a non-degenerate semiconductor these averages are represented by $\langle\tau\rangle$, $\langle E\tau\rangle$, $\langle E^2\tau\rangle$ being taken with the weighting function

$$E^{3/2}\exp[-E/\boldsymbol{k}T]$$

For a metal or degenerate semiconductor $\bar{\tau} = \tau_F$, $\overline{(E\tau)} = E_F\tau_F$, $\overline{(E^2\tau)} = E_F^2\tau_F$.

If no electric current flows, we have $J_x = 0$, and we may eliminate $\mathscr{E}$ by means of equation (91). On substituting in equation (92) we obtain

$$W_x = -\frac{n}{m_e T}\cdot\left\{\frac{\bar{\tau}\overline{(E^2\tau)} - \overline{(E\tau)}^2}{\bar{\tau}}\right\}\frac{\mathrm{d}T}{\mathrm{d}x} \tag{93}$$

The part of the thermal conductivity due to electronic conduction, κ_e, is therefore given by

$$\kappa_e = -W_x\bigg/\left(\frac{\mathrm{d}T}{\mathrm{d}x}\right) = \frac{n\{\bar{\tau}\overline{(E^2\tau)} - \overline{(E\tau)}^2\}}{Tm_e\bar{\tau}} \tag{94}$$

For semiconductors κ_e, in general, represents only a small part of the total thermal conductivity, the lattice vibrations contributing the major part of the heat transfer. For a non-degenerate n-type semiconductor we have

$$\kappa_e = n\{\langle\tau\rangle\langle E^2\tau\rangle\langle E\tau\rangle^2\}/Tm_e\langle\tau\rangle \tag{95}$$

Since we may express the electrical conductivity σ in the form (cf. equation (64))

$$\sigma = ne^2\langle\tau\rangle/m_e$$

we may write equation (95) as

$$\kappa_e = L\sigma T = \mathscr{L}(\mathbf{k}^2/e^2)\,\sigma T \tag{96}$$

where $\mathscr{L}$ is a number given by

$$\mathscr{L} = \frac{\langle\tau\rangle\langle E^2\tau\rangle - \langle E\tau\rangle^2}{k^2 T^2\langle\tau\rangle^2} \tag{97}$$

The quantity L giving the ratio of κ_e to σT is known as the Lorentz ratio and $\mathscr{L}$ as the Lorentz number. Equation (96) expresses what is known as the Wiedemann–Franz law, which states that the ratio $\kappa_e/\sigma T$ is constant. The value of the Lorentz number $\mathscr{L}$ is of some interest. When $\tau = AE^{-s}$, the expressions for $\langle E\tau\rangle$ and $\langle E^2\tau\rangle$ may be evaluated in the same way as that for $\langle\tau\rangle$ (see equation (83)); we have

$$\langle E\tau\rangle = AkT\Gamma(\tfrac{7}{2}-s)/\Gamma(\tfrac{5}{2}) \tag{98}$$

$$\langle E^2\tau\rangle = Ak^2 T^2\Gamma(\tfrac{9}{2}-s)/\Gamma(\tfrac{5}{2}) \tag{99}$$

Since $\Gamma(1+s) = s\Gamma(s)$ we obtain

$$\mathscr{L} = \tfrac{5}{2}-s \tag{100}$$

Under certain circumstances, the value of s, when scattering of electrons is due to the lattice vibrations, is $\frac{1}{2}$ (see § 10.9.2), so that in this case $\mathscr{L} = 2$. (The theory of conduction by electrons treated as a classical gas of free particles gives $\mathscr{L} = 3$.)

For a p-type non-degenerate semiconductor an expression analogous to that given by equation (95) gives the contribution to the thermal conductivity due to the free positive holes, the average values of τ, τE, etc., being taken with the weighting function given by equation (68). For a semiconductor having appreciable numbers of both free electrons and holes, some interesting effects occur, the electronic thermal conductivity not being given by the sum of the separate contributions of the electrons and holes. The expressions for κ_e in this case are somewhat complex and will not be given here*.

For a metal, most of the thermal conductivity is due to the electrons and it is quite a good approximation to set the total thermal conductivity κ equal to κ_e. When we insert the values of $\bar{\tau}$, $(\overline{\tau E})$, etc., in equation (97) we obtain the surprising result that $\kappa = 0$. This is only because we have used too crude an approximation to obtain the average values $\bar{\tau}$, $(\overline{\tau E})$, etc., since the expression in equation (97) contains as a factor the quantity

* See, for example, R. A. Smith, *Semiconductors*. Cambridge University Press (1959), p. 169.

$(E_F/\boldsymbol{k}T)^2$ which is very large for a metal or highly degenerate semiconductor. When we expand the expressions for $\bar{\tau}$, $(\overline{\tau E})$, etc., as far as the second order in the small quantity $\boldsymbol{k}T/E_F$ we indeed obtain a non-zero contribution. If we write $J_x/\mathscr{E} = n\boldsymbol{e}^2\bar{\tau}/m_e$ we see from equation (57) that we may write

$$\bar{\tau} = -E_F^{-3/2}\int_0^\infty \tau E^{3/2}\frac{\partial f_0}{\partial E}\,\mathrm{d}E \tag{101}$$

Similarly we have

$$(\overline{E\tau}) = -E_F^{-3/2}\int_0^\infty \tau E^{5/2}\frac{\partial f_0}{\partial E}\,\mathrm{d}E \tag{102}$$

and

$$(\overline{E^2\tau}) = -E_F^{-3/2}\int_0^\infty \tau E^{7/2}\frac{\partial f_0}{\partial E}\,\mathrm{d}E \tag{103}$$

We now use equations (77), (81) of § 2.7 to obtain better approximations to these quantities than the first-order approximations quoted above. We obtain in this way

$$\bar{\tau} = \tau_F + \frac{\pi^2}{6}\frac{\boldsymbol{k}^2T^2}{E_F^{3/2}}\frac{\mathrm{d}^2}{\mathrm{d}E^2}(\tau E^{3/2}) \tag{104}$$

$$(\overline{E\tau}) = E_F\tau_F + \frac{\pi^2}{6}\frac{\boldsymbol{k}^2T^2}{E_F^{3/2}}\frac{\mathrm{d}^2}{\mathrm{d}E^2}(\tau E^{5/2}) \tag{105}$$

$$(\overline{E^2\tau}) = E_F^2\tau_F + \frac{\pi^2}{6}\frac{\boldsymbol{k}^2T^2}{E_F^{3/2}}\frac{\mathrm{d}^2}{\mathrm{d}E^2}(\tau E^{7/2}) \tag{106}$$

From these equations, neglecting higher powers of $\boldsymbol{k}T/E_F$ than the second, we obtain

$$\bar{\tau}(\overline{E^2\tau}) - (\overline{E\tau})^2 = \frac{\pi^2}{3}\boldsymbol{k}^2\boldsymbol{T}^2\tau_F^2 \tag{107}$$

and on substituting in equation (95) we have

$$\kappa = \frac{\pi^2 n\boldsymbol{k}^2 T\tau_F}{3m_e} \tag{108}$$

On taking the ratio $\kappa/\sigma T$ we obtain

$$\frac{\kappa}{\sigma T} = L = \frac{\pi^2}{3}\frac{\boldsymbol{k}^2}{\boldsymbol{e}^2} = \mathscr{L}\frac{\boldsymbol{k}^2}{\boldsymbol{e}^2} \tag{109}$$

For the condition of strong degeneracy we therefore have $\mathscr{L} = \pi^2/3$, which is very nearly equal to the classical value 3; a value very nearly equal to this is found for the simple metals.

10.7.2 The thermoelectric effect

We have already seen that, even when no electric current flows, we have an electric field developed in a crystal when we have a temperature gradient. If we let $J_x = 0$ in equation (91) we obtain

$$\mathscr{E} = -\frac{1}{\bar{\tau}e}\left\{T\frac{\mathrm{d}}{\mathrm{d}T}\left(\frac{E_F}{T}\right)\bar{\tau}+\frac{1}{T}\overline{(E\tau)}\right\}\frac{\mathrm{d}T}{\mathrm{d}x} \tag{110}$$

$$= T\frac{\mathrm{d}}{\mathrm{d}T}\left\{\frac{\overline{(E\tau)}-E_F\bar{\tau}}{eT\bar{\tau}}\right\}\frac{\mathrm{d}T}{\mathrm{d}x} \tag{110a}$$

Now the electric field produced by a temperature gradient in a homogeneous material is related to the temperature gradient by means of an equation of the form

$$\mathscr{E} = -\mathscr{T}\frac{\mathrm{d}T}{\mathrm{d}x} \tag{111}$$

the coefficient $\mathscr{T}$ being known as the Thomson coefficient. We therefore have

$$\mathscr{T} = -T\frac{\mathrm{d}}{\mathrm{d}T}\left[\frac{\overline{(E\tau)}-E_F\bar{\tau}}{eT\bar{\tau}}\right] \tag{112}$$

Moreover, the Thomson coefficient is related to the absolute thermoelectric power $\mathscr{P}$ through the relationship*

$$\mathscr{T} = T\frac{\mathrm{d}\mathscr{P}}{\mathrm{d}T} \tag{113}$$

We therefore have

$$\mathscr{P} = \frac{E_F\bar{\tau}-\overline{(\tau E)}}{eT\bar{\tau}} \tag{114}$$

For a non-degenerate n-type semiconductor we have

$$\mathscr{P} = \frac{E_F\langle\tau_e\rangle-\langle E\tau_e\rangle}{eT\langle\tau_e\rangle} \tag{115}$$

If $\tau = AE^{-s}$ we have $\langle E\tau_e\rangle/\langle\tau_e\rangle = (\frac{5}{2}-s)\,\boldsymbol{k}T$ so that

$$\mathscr{P} = -\frac{\boldsymbol{k}}{\boldsymbol{e}}\left\{(\tfrac{5}{2}-s)-\frac{E_F}{\boldsymbol{k}T}\right\} \tag{116}$$

Since $E_F < 0$, and s is always less than $\frac{5}{2}$, we see that the thermoelectric power is negative. For a p-type non-degenerate semiconductor with $\tau = A'E^{-s'}$ it may readily be shown that†

$$\mathscr{P} = \frac{\boldsymbol{k}}{\boldsymbol{e}}\left\{(\tfrac{5}{2}-s')+\frac{E_F}{\boldsymbol{k}T}+\frac{\Delta E}{\boldsymbol{k}T}\right\} \tag{117}$$

* See, for example, R. A. Smith, *The Physical Principles of Thermodynamics*. Chapman & Hall (1952), pp. 84–86.

† See, for example, R. A. Smith, *Semiconductors*. Cambridge University Press (1959), p. 172.

where, as usual, ΔE is equal to the forbidden energy gap. Moreover, it may also be shown* that for a mixed semiconductor

$$\mathscr{P} = \frac{\sigma_e \mathscr{P}_e + \sigma_h \mathscr{P}_h}{\sigma_e + \sigma_h} \tag{118}$$

where $\mathscr{P}_e$, σ_e and $\mathscr{P}_h$, σ_h are, respectively, the contributions to the thermoelectric power and electrical conductivity due to the electrons and holes taken separately.

These formulae show that the thermoelectric power of a non-degenerate semiconductor changes sign as the semiconductor changes from n-type to p-type, and a measurement of the thermoelectric power gives a simple indication of the type of semiconductor. The value of $\mathscr{P}$ may be of the order of 1 mV; if we take $s = \frac{1}{2}$ and $E_F = -0{\cdot}25$ eV (a fairly typical value for an n-type semiconductor) we have, for $T = 300°$K, $\mathscr{P} = -13\ \boldsymbol{k}/\boldsymbol{e}$; since $\boldsymbol{k}/\boldsymbol{e} = 8{\cdot}6 \times 10^{-5}$ V/°C we have $\mathscr{P} = 1{\cdot}1$ mV/°C.

For a metal or highly degenerate semiconductor equation (115) shows that if we use the first-order approximations for $\bar{\tau}$ and $(\overline{E\tau})$ we again obtain zero, and we must use the second-order approximations given by equations (104) and (105). In this way we obtain

$$\mathscr{P} = -\frac{\pi^2}{6}\cdot\frac{\boldsymbol{k}}{\boldsymbol{e}}\cdot\frac{\boldsymbol{k}T}{E_F}\left[3 + \frac{2E_F}{\tau_F}\left(\frac{d\tau}{dE}\right)_{E_F}\right] \tag{119}$$

When $\tau = AE^{-s}$ we have

$$\mathscr{P} = -\frac{\pi^2 \boldsymbol{k}}{3\boldsymbol{e}}\cdot\frac{\boldsymbol{k}T}{E_F}[\tfrac{3}{2} - s] \tag{119a}$$

We therefore see that for the degenerate condition the thermoelectric power is much smaller, being reduced by a factor of the order of $\boldsymbol{k}T/E_F$. For the intermediate condition the formulae are more complex and are given in textbooks dealing specifically with metals or semiconductors such as those to which we have already made reference.

For a metal with $E_F = 5$ eV and $T = 300°$K we have $E_F/\boldsymbol{k}T \simeq 200$ so that for $s = \frac{1}{2}$ we have $\mathscr{P} \simeq -0{\cdot}03\ \boldsymbol{k}/\boldsymbol{e} = -2{\cdot}6\ \mu$V/°C. Thermoelectric powers of the order of a few μV/°C are found for many metals at room temperature. Equation (118) shows that the thermoelectric power should vanish at low temperatures; indeed it becomes very small but some curious effects are observed which cannot be explained by the simple theory, including positive values for some metals for which the conductivity is thought to be due mainly to electrons. A discussion of these phenomena is, however, beyond the scope of this book.

10.8 Conduction in magnetic fields

When we have a magnetic field present we must insert in Boltzmann's equation the appropriate force term, given by equation (70) of § 8.8, in order to obtain the modified distribution function; we may then use this

* See, for example, R. A. Smith, *Semiconductors*. Cambridge University Press (1959), p. 172.

function to obtain expressions for the electric and heat currents as in § 10.7. An alternative approach to the problem is to derive an expression for the velocity from the equations of motion and average this over various collision times as in § 10.2 in order to obtain the average velocity for a particular group of electrons. We may then use the weighting function which we have already derived to obtain a weighted average over all groups of electrons. This method has been used by Harvey Brooks* and by the author† to discuss in some detail the properties of semiconductors in a magnetic field; it has the advantage of showing up rather clearly the physical processes taking place. We shall, however continue to use Boltzmann's equation which provides a rather more rigorous treatment of the problem.

We shall first suppose that the temperature is constant, and that we have a homogeneous crystal, so that we do not require to include the spatial variation in Boltzmann's equation. When we include the force term for the magnetic field, the total force **F** on an electron is

$$-e\mathscr{E}-e(\mathbf{v}\times\mathbf{B})$$

where **B** is the magnetic flux. Boltzmann's equation then becomes (cf. equation (47) of § 10.5)

$$f-f_0 = e\tau\hbar^{-1}[\mathscr{E}.\nabla_k f+(\mathbf{v}\times\mathbf{B}).\nabla_k f] \tag{120}$$

As a first approximation we may take, as before, $f=f_0$ in the terms on the right-hand side of equation (120). Since

$$\left[\hbar^{-1}\nabla_k f_0 = \mathbf{v}\frac{\partial f_0}{\partial E}\right]$$

the second term gives no contribution since $(\mathbf{v}\times\mathbf{B}).\mathbf{v}=0$; we therefore have as the first approximation

$$f = f_0+e\tau(\mathscr{E}.\mathbf{v})\frac{\partial f_0}{\partial E} \tag{121}$$

We now insert this approximation in the right-hand side of equation (120). We may neglect all terms in $\mathscr{E}^2$ since non-ohmic effects occur only for very large electric fields in homogeneous material; we retain, however, the terms involving products of $\mathscr{E}$ and **B** since these are linear in both $\mathscr{E}$ and **B**. We obtain

$$f = f_0+e\tau(\mathscr{E}.\mathbf{v})\frac{\partial f_0}{\partial E}+e^2\tau^2\hbar^{-1}(\mathbf{v}\times\mathbf{B}).\nabla_k(\mathscr{E}.\mathbf{v})\frac{\partial f_0}{\partial E} \tag{122}$$

Since we have

$$\hbar^{-1}(\mathbf{v}\times\mathbf{B}).\nabla_k\frac{\partial f_0}{\partial E} = (\mathbf{v}\times\mathbf{B}).\mathbf{v}\frac{\partial^2 f_0}{\partial E^2} = 0$$

* H. Brooks, *Advanc. Electron. Elect. Phys.*, **7**, 85 (1955).
† R. A. Smith, *Semiconductors*. Cambridge University Press (1959), Chapters 5 and 6.

equation (120) may be expressed in the form

$$f = f_0 + e\tau(\mathscr{E}\,.\,\mathbf{v})\frac{\partial f_0}{\partial E} + e^2\tau^2\hbar^{-2}\frac{\partial f_0}{\partial E}(\mathbf{v}\times\mathbf{B})\,.\,\nabla_k(\mathscr{E}\,.\,\nabla_k E) \tag{123}$$

For a band having scalar effective mass m_e we have

$$\hbar^{-2}\nabla_k(\mathscr{E}\,.\,\nabla_k E) = \frac{1}{m_e}\mathscr{E}$$

so that

$$f = f_0 + e\tau(\mathscr{E}\,.\,\mathbf{v})\frac{\partial f_0}{\partial E} + \frac{e^2\tau^2}{m_e}\frac{\partial f_0}{\partial E}(\mathbf{v}\times\mathbf{B})\,.\,\mathscr{E} \tag{124}$$

If the magnetic field is directed along the z-axis it will clearly not affect the motion of electrons in this direction and we shall assume that $\mathscr{E}_z = v_z = 0$. We have then

$$f = f_0 + e\tau(\mathscr{E}_x v_x + \mathscr{E}_y v_y)\frac{\partial f_0}{\partial E} + e\omega_c\tau^2(v_y\mathscr{E}_x - v_x\mathscr{E}_y)\frac{\partial f_0}{\partial E} \tag{125}$$

where we have written ω_c for eB/m_e.

We may now write down expressions for the current density as in § 10.6. On integrating over all values of $\mathbf{k}'$ we see that terms containing products such as $v_x v_y$ give no contribution. On comparing with equation (53) we see that

$$\left.\begin{aligned} J_x &= -\frac{e^2\mathscr{E}_x}{4\pi^3}\int \tau v_x^2\frac{\partial f_0}{\partial E}\,\mathrm{d}\mathbf{k}' + \frac{e^2\omega_c\mathscr{E}_y}{4\pi^3}\int \tau^2 v_x^2\frac{\partial f_0}{\partial E}\,\mathrm{d}\mathbf{k}' \\ J_y &= -\frac{e^2\mathscr{E}_y}{4\pi^3}\int \tau v_y^2\frac{\partial f_0}{\partial E}\,\mathrm{d}\mathbf{k}' - \frac{e^2\omega_c\mathscr{E}_x}{4\pi^3}\int \tau^2 v_y^2\frac{\partial f_0}{\partial E}\,\mathrm{d}\mathbf{k}' \end{aligned}\right\} \tag{126}$$

The integrals in equations (126) may be simplified in exactly the same way as in § 10.6. There is no need to carry out the detailed analysis, since we may see at once on comparison with equations (53) and (64) that

$$\left.\begin{aligned} J_x &= \frac{ne^2}{m_e}(\bar{\tau}\mathscr{E}_x - \omega_c\overline{\tau^2}\mathscr{E}_y) \\ J_y &= \frac{ne^2}{m_e}(\bar{\tau}\mathscr{E}_y + \omega_c\overline{\tau^2}\mathscr{E}_x) \end{aligned}\right\} \tag{127}$$

where $\bar{\tau}$ and $\overline{\tau^2}$ are average values of τ and τ^2 defined as in § 10.6. For a metal or degenerate semiconductor $\bar{\tau} = \tau_F$ and $\overline{\tau^2} = \tau_F^2$ so that $\overline{\tau^2} = (\bar{\tau})^2$. For a non-degenerate semiconductor $\overline{\tau^2} = \langle\tau^2\rangle$ and is not, in general, equal to $\langle\tau\rangle^2$. (The symbol $\langle\ \rangle$ refers as before to averaging with the weighting function $E^{3/2}\exp[-E/kT]$ or the corresponding function for holes.)

10.8.1 The Hall effect

If we have a filament with current flowing in the x-direction so that $J_y = 0$ we see that

$$\mathscr{E}_y = -\omega_c(\overline{\tau^2}/\bar{\tau})\mathscr{E}_x \tag{128}$$

so that in the presence of magnetic field a transverse voltage equal to $w\mathscr{E}_y$ is produced, where w is the width of the filament; this effect is known as the Hall effect. The angle θ which the resultant field makes with the current direction is known as the Hall angle and is given by the equation

$$\tan\theta = \mathscr{E}_y/\mathscr{E}_x = -\omega_c(\overline{\tau^2}/\bar{\tau}) = -\frac{eB\bar{\tau}}{m_e}\left(\frac{\overline{\tau^2}}{\bar{\tau}^2}\right) \tag{129}$$

If τ is constant or equal to τ_F we have

$$\tan\theta = -B\mu_e \tag{130}$$

where μ_e is the electron mobility; for a non-degenerate n-type semiconductor we have

$$\tan\theta = -rB\mu_e \tag{130a}$$

where

$$r = \langle\tau_e^2\rangle/\langle\tau_e\rangle^2 \tag{131}$$

The Hall coefficient R is defined by means of the equation

$$\mathscr{E}_y = RBJ_x \tag{132}$$

so that

$$R = -\frac{1}{ne}\left(\frac{\overline{\tau^2}}{\bar{\tau}^2}\right) \tag{133}$$

For a simple metal or a degenerate n-type semiconductor we have

$$R = -\frac{1}{ne} \tag{134}$$

A measurement of the Hall coefficient therefore gives a very direct method of obtaining the free electron concentration n. For conduction by positive holes we simply replace ω_c by $-\omega_c'$ where $\omega_c' = eB/m_h$. In this case we have

$$R = \frac{1}{pe} \tag{135}$$

where p is the hole concentration. For n- and p-type non-degenerate semiconductors we have respectively

$$R = -\frac{r}{ne} \tag{136}$$

and

$$R = \frac{r'}{pe} \tag{137}$$

where $r = \langle\tau_e^2\rangle/\langle\tau_e\rangle^2$ and $r' = \langle\tau_h^2\rangle/\langle\tau_h\rangle^2$; the sign of the Hall coefficient, like that of the thermoelectric power, therefore indicates the type of semiconductor. For mixed conduction it may be shown that*

$$R = -\frac{1}{e}\frac{rb^2n - r'p}{(bn+p)^2} \tag{138}$$

* R. A. Smith, *Semiconductors*. Cambridge University Press (1959), p. 123.

where b is the mobility ratio μ_e/μ_h. Clearly some care is required in the definition of n-type and p-type when $n \simeq p$ since the change over from positive to negative Hall coefficient does not occur for quite the same ratio of n/p as for the change of sign of the thermoelectric power.

When we have conduction predominantly by one type of carrier we have, for electrons

$$-R\sigma = \mu_H = r\mu_e \tag{139}$$

and for holes

$$R\sigma = \mu_H = r'\mu_h \tag{140}$$

The quantity μ_H is called the Hall mobility; it is not always carefully distinguished from μ_e or μ_h which are sometimes called the conductivity mobilities. When $\tau = AE^{-s}$ it may readily be shown by expressing the integrals for $\langle\tau\rangle$ and $\langle\tau^2\rangle$ as Γ-functions that

$$r = \Gamma(\tfrac{5}{2}-2s)\,\Gamma(\tfrac{5}{2})/\{\Gamma(\tfrac{5}{2}-s)\}^2 \tag{141}$$

for $s = \frac{1}{2}$ we have $r = 3\pi/8$. When we have the intermediate situation between non-degeneracy and full degeneracy the formulae for the Hall coefficient become quite complex; they are given in books on semiconductors and metals, such as those already referred to.

When the effective mass is no longer scalar, the formulae for the Hall coefficient also take a more complex form. They may be derived from Boltzmann's equation with the type of transformation used in § 10.6 to treat the electrical conductivity or may be derived from the equations of motion, averaged over all the free electrons and holes with the appropriate weighting function; they are of greatest importance in the study of semiconductors and are given in books on this subject such as those quoted. It may be shown, for example, that for a semiconductor with cubic symmetry the electric field $\mathscr{E}$ may be expressed in terms of the current density vector $\mathbf{J}$ by means of the vector equation

$$\mathscr{E} = \sigma^{-1}\mathbf{J} - R\mathbf{J}\times\mathbf{B} \tag{142}$$

The formula for the Hall coefficient R must now be modified to take account of the anisotropy of the effective mass tensor; if m_L and m_T are the longitudinal and transverse components of the effective mass it may be shown that* for an n-type semiconductor

$$R = -\frac{r}{ne}\frac{3K(K+2)}{(2K+1)^2} \tag{143}$$

where $K = m_L/m_T$, and similar formulae hold for p-type and mixed conductors.

* R. A. Smith, *Semiconductors*. Cambridge University Press (1959), p. 120.

10.8.2 Magneto-resistance

If we substitute the value of $\mathscr{E}_y$ given by equation (128) in equation (126) we obtain

$$J_x = \frac{ne^2}{m_e}\{\bar{\tau}\mathscr{E}_x + \omega_c^2(\overline{\tau^2})^2(\bar{\tau})^{-1}\mathscr{E}_x\} \tag{144}$$

This equation would appear to indicate that the conductivity increases quadratically with B; it is, however, incorrect since our approximation has neglected terms in B^2. In order to obtain the correct equation, as far as second-order terms in B, we must go back to Boltzmann's equation and take the approximation one step further. Clearly the only term to contribute further is the force term in $e\mathscr{E}$. When we do this we obtain (cf. equation (156))

$$J_x = \frac{ne^2}{m_e}\{\bar{\tau} + \omega_c^2(\overline{\tau^2})^2/\bar{\tau} - \omega_c^2\overline{\tau^3}\}\mathscr{E}_x \tag{145}$$

If we write $\rho = 1/\sigma$ and

$$\left.\begin{aligned} \sigma/\sigma_0 &= (1 - \Delta\sigma/\sigma_0) \\ \rho/\rho_0 &= (1 + \Delta\rho/\sigma_0) \end{aligned}\right\} \tag{146}$$

we have

$$-\frac{\Delta\sigma}{\sigma_0} = \frac{\Delta\rho}{\rho_0} = \frac{e^2B^2}{m_e^2}\frac{(\overline{\tau^3})\bar{\tau} - (\overline{\tau^2})^2}{\bar{\tau}^2} \tag{147}$$

Thus we see that for a metal or degenerate semiconductor, for which, $\bar{\tau}$, $\overline{\tau^2}$, $\overline{\tau^3}$ are equal to τ_F, τ_F^2, τ_F^3, respectively, the transverse component of magneto-resistances is zero, for a crystal whose constant energy surfaces have spherical symmetry; for a non-degenerate semiconductor, however, this is not so.

When the effective-mass tensor is no longer scalar the anisotropy introduced will produce a non-zero transverse magneto-resistance even in the degenerate condition. When the constant-energy surfaces are ellipsoids, formulae for the magneto-resistance components may be obtained in the same way as for the Hall coefficient, an approximation correct to the second order in B being required. The formulae are somewhat complex and will not be given here since they are given for a variety of conditions in books devoted particularly to metals or to semiconductors, such as those already referred to.

Under certain conditions we need not restrict our approximation to small powers of B; if we restrict our solution of Boltzmann's equation to powers of $\mathscr{E}$ not greater than the first, we may obtain, in principle, a solution of the equation for any value of B. When we no longer take $f = f_0$ in the magnetic term of equation (120), but only in the term involving $\mathscr{E}$ we obtain

$$f = f_0 + e\tau(\mathscr{E}.\mathbf{v})\frac{\partial f_0}{\partial E} + e\tau\hbar^{-1}(\mathbf{v}\times\mathbf{B}).\Delta_k f \tag{148}$$

We shall restrict our discussion to a crystal with scalar effective electron mass m_e. The term in $\mathscr{E}$ follows from the previous analysis; when the magnetic field is along the z-axis equation (148) reduces to

$$f = f_0 + e\tau(\mathscr{E}_x v_x + \mathscr{E}_y v_y)\frac{\partial f_0}{\partial E} + eB\tau\hbar^{-1}\left(v_y\frac{\partial f}{\partial k_x} - v_x\frac{\partial f}{\partial k_y}\right) \tag{149}$$

If we write $f = f_0 + f_1$ we have

$$f_1 = e\tau(\mathscr{E}_x v_x + \mathscr{E}_y v_y)\frac{\partial f_0}{\partial E} + eB\tau\hbar^{-1}\left(v_y\frac{\partial f_1}{\partial k_x} - v_x\frac{\partial f_1}{\partial k_y}\right) \tag{150}$$

since

$$v_y\frac{\partial f_0}{\partial k_x} = v_x\frac{\partial f_0}{\partial k_y}$$

When we examine the solution previously obtained for the modified distribution function when we have no magnetic field we see that f_1 has the form

$$f_1 = v_x\phi(E)$$

It is therefore reasonable now to seek a solution of the form

$$f_1 = v_x\phi(E) + v_y\psi(E) \tag{151}$$

We note that

$$\frac{\partial v_x}{\partial k_x} = \frac{\hbar}{m_e} = \frac{\partial v_y}{\partial k_y}$$

and that

$$\left(v_y\frac{\partial}{\partial k_x} - v_x\frac{\partial}{\partial k_y}\right)\phi(E) = 0$$

Hence we have, on substituting for f_1, in equation (150)

$$v_x\phi(E) + v_y\psi(E) = e\tau(\mathscr{E}_x v_x + \mathscr{E}_y v_y)\frac{\partial f_0}{\partial E} + \omega_c\tau[v_y\phi(E) - v_x\psi(E)] \tag{152}$$

where, as before $\omega_c = eB/m_e$.

We see that we obtain a solution if we take

$$\left.\begin{aligned}\phi(E) &= e\tau\mathscr{E}_x\frac{\partial f_0}{\partial E} - \omega_c\tau\psi(E)\\ \psi(E) &= e\tau\mathscr{E}_y\frac{\partial f_0}{\partial E} + \omega_c\tau\phi(E)\end{aligned}\right\} \tag{153}$$

These give

$$\phi(E) = e\tau\frac{\partial f_0}{\partial E}(\mathscr{E}_x - \omega_c\tau\mathscr{E}_y)/(1+\omega_c^2\tau^2) \tag{154}$$

$$\psi(E) = e\tau\frac{\partial f_0}{\partial E}(\mathscr{E}_y + \omega_c\tau\mathscr{E}_x)/(1+\omega_c^2\tau^2) \tag{155}$$

so that

$$f = f_0 + \frac{e\tau}{1+\omega_c^2\tau^2}(v_x\mathscr{E}_x + v_y\mathscr{E}_y)\frac{\partial f_0}{\partial E} + \frac{e\omega_c\tau^2}{1+\omega_c^2\tau^2}(v_y\mathscr{E}_x - v_x\mathscr{E}_y)\frac{\partial f_0}{\partial E} \tag{156}$$

On comparison with equation (125) it will be seen that we have simply introduced a factor $(1+\omega_c^2\tau^2)^{-1}$ into each of the terms giving $f-f_0$.

Using this distribution function we may readily obtain expressions for the current density; from equation (156) it will be seen that we get the same equations as before provided in taking the averages we replace $\bar{\tau}$ by the average value of $\tau/(1+\omega_c^2\tau^2)$, τ^2 by the average of $\tau^2/(1+\omega_c^2\tau^2)$, etc. We may also now readily obtain the correct value for the current density J_x as far as terms in B^2 by expanding the denominators in equation (156); this leads at once to equation (145).

At this stage, however, we must point out that although we have, in principle, obtained a solution for any value of B, this method of approach has implied an assumption that the force equation may be used to describe the motion of electrons for all values of the magnetic field. We shall see later, however, that even the motion of completely free electrons in a strong magnetic field is not in accordance with the classical equations of motion (see § 11.3). We should therefore expect our 'effective mass' approximation in which we regard the electrons as moving classically in the magnetic field with their appropriate effective mass to break down for high fields. The quantity ω_c which we have written for eB/m_e is just the angular cyclotron frequency, $2\pi/\omega_c$ being the time required for an electron to describe a circular orbit in the field, and as we shall see, such orbits are quantized (see § 11.3.1). We should therefore expect these quantum effects to become important when $\omega_c\tau \gg 1$, which is the same as the condition $\mu_e B \gg 1$. Fortunately it turns out that most of the transport phenomena are hardly affected by this quantum effect, but some very interesting effects are introduced when we study other phenomena such as the absorption of infra-red radiation, and the diamagnetism of the conduction electrons (see § 12.3).

10.8.3 Other magnetic effects

When we have a temperature gradient in the crystal, a number of transport phenomena take place in the presence of a magnetic field. When a current flows in a filament producing a transverse Hall voltage it also produces a transverse temperature gradient which we neglected in our previous treatment; we have therefore to distinguish between the adiabatic and isothermal Hall effects, but the difference between the two coefficients is usually small; the existence of the transverse temperature gradient is known as the Ettingshausen effect. Again, when we have a thermal current along the filament due to a temperature gradient, a transverse electric field is produced; this is known as the Nernst effect; a transverse thermal gradient is also produced, the effect being known as the Righi–Leduc effect. The magnitudes of these effects are, for not too large fields, all proportional to B, the magnetic flux at right-angles to the filament, and the voltages and temperature gradients occur in a direction at right-angles to the magnetic field. They have been studied most extensively for semiconductors. Formulae for the various coefficients specifying the magnitude of the effects may be derived in a similar way to that which we have used to derive the Hall coefficient, the temperature gradients in the x- and y-directions now being taken into account. The formulae for these coefficients are rather complex and involve averages such as $\overline{\tau^2}$, $(\overline{E^2\tau^2})$, etc.; we shall not derive them since they have been treated in detail in books on semiconductors and metals such as those to which we have already referred.

10.9 Calculation of the relaxation time τ

We must now see how to calculate the relaxation time τ. Suppose we have an electron initially in the state corresponding to the wave vector $\mathbf{k}$, and that a transition is made to a state with wave vector $\mathbf{k}'$. We shall assume that we are dealing with collisions in which no energy is lost by the electron so that the vector $\mathbf{k}'$ ends on the same constant-energy surface as $\mathbf{k}$. This is a reasonable assumption since the collision is likely to be elastic, the electron having insufficient energy to raise the scattering centre to an excited state and practically no energy being lost by direct transfer because of the large mass of the centre compared with that of the electron. Let the probability per unit time that the scattering takes place so that $\mathbf{k}'$ ends on a surface element $\mathrm{d}S'$ of the constant-energy surface be $P(\mathbf{k},\mathbf{k}')\,\mathrm{d}S'$. Let $\mathrm{d}\Omega$ be the solid angle in $\mathbf{k}$-space subtended by $\mathrm{d}S'$ at the origin, then the probability of scattering into the solid angle $\mathrm{d}\Omega$ will be equal to $k^2 P(\mathbf{k},\mathbf{k}')\,\mathrm{d}\Omega$; we may therefore relate $P(\mathbf{k},\mathbf{k}')$ to the partial cross-section $\sigma(\theta,\phi)$ defined in § 10.3.1. The total cross-section σ_t for scattering is given by (cf. equation (27))

$$\frac{1}{\tau_t} = Nv\sigma_t = \iint P(\mathbf{k},\mathbf{k}')\,k^2 \sin\theta\,\mathrm{d}\theta\,\mathrm{d}\phi \tag{157}$$

where θ is the angle between $\mathbf{k}$ and $\mathbf{k}'$ and ϕ is the azimuthal angle about $\theta = 0$. Comparing this equation with equation (20), giving the definition of the partial cross-section $\sigma(\theta,\phi)$, we see that $\sigma(\theta,\phi) = k^2 P(\mathbf{k},\mathbf{k}')$. If $P(\mathbf{k},\mathbf{k}')$ for a given value of $\mathbf{k}$ depends only on the angle of scattering θ we may write $k^2 P(\mathbf{k},\mathbf{k}') = \sigma(\theta)$ and we have

$$\sigma_t = 2\pi \int_0^\pi \sigma(\theta)\sin\theta\,\mathrm{d}\theta \tag{158}$$

The relaxation time τ_t defined in equation (157), as we shall see, is not equal to the relaxation time used in transport theory. In order to calculate the latter we must evaluate the rate of change of the distribution function f. The number of electrons leaving a volume element of $\mathbf{k}$-space $\mathrm{d}\mathbf{k}$ will be proportional to the number of occupied states in the volume element, namely $(V/4\pi^3)f(\mathbf{k})\,\mathrm{d}\mathbf{k}$. The function $P(\mathbf{k},\mathbf{k}')$ defined above gives the transition probability, *assuming that such a transition can take place*; in practice the transition probability must be multiplied by the probability that the state with wave vector $\mathbf{k}'$ is unoccupied, namely $1-f(\mathbf{k}')$. The total number of electrons transferred from the volume element $d\mathbf{k}$ per unit time is therefore

$$\frac{V\,d\mathbf{k}}{4\pi^3}\iint_S P(\mathbf{k},\mathbf{k}')\,f(\mathbf{k})[1-f(\mathbf{k}')]\,dS' \tag{159}$$

the integration to be taken over the constant-energy surface S. The number of electrons brought into the volume element $d\mathbf{k}$ per unit time by collisions is by a similar argument equal to

$$\frac{[1-f(\mathbf{k})]\,V\,d\mathbf{k}}{4\pi^3}\iint P(\mathbf{k}',\mathbf{k})f(\mathbf{k}')\,dS' \tag{160}$$

If the rate of change of f due to collisions is $\left.\frac{df}{dt}\right]_{\text{coll}}$ the total number of electrons leaving the volume element is clearly

$$-\frac{V}{4\pi^3}\left.\frac{df}{dt}\right]_{\text{coll}} d\mathbf{k}$$

It is well known in quantum mechanics that the transition probability between two states is the same for transitions taking place either way, so that $P(\mathbf{k},\mathbf{k}') = P(\mathbf{k}',\mathbf{k})$. We may therefore write

$$\left.\frac{df}{dt}\right]_{\text{col}} = \iint P(\mathbf{k},\mathbf{k}')\{[1-f(\mathbf{k})]f(\mathbf{k}')-f(\mathbf{k})[1-f(\mathbf{k}')]\}\,dS' \tag{161}$$

$$= \iint P(\mathbf{k},\mathbf{k}')\{f(\mathbf{k}')-f(\mathbf{k})\}\,dS' \tag{162}$$

It is interesting to note that this is just what we should have obtained if we had ignored the exclusion principle, which therefore does not affect the rate of change of f due to collisions.

In order to relate the transition probability $P(\mathbf{k},\mathbf{k}')$ to the relaxation time τ we must use one of the formulae we have derived for the modified distribution function f. The simplest case is that corresponding to an electric field along the x-axis and the function f is given by equation (51). We have therefore from the definition of $\left.\frac{df}{dt}\right]_{\text{coll}}$

$$\left.\frac{df}{dt}\right]_{\text{coll}} = \frac{f_0-f}{\tau} = -e\mathscr{E}v_x\frac{\partial f_0}{\partial E} \tag{163}$$

If we assume that we have a scalar effective mass m_e we have $m_e v_x = \hbar k_x$, and on substituting in equation (162) for $f(\mathbf{k})$ we obtain

$$\left.\frac{df}{dt}\right]_{\text{coll}} = \tau e\mathscr{E}\iint P(\mathbf{k},\mathbf{k}')\left\{\frac{\partial f_0}{\partial E}(v_x'-v_x)\right\}dS' \tag{164}$$

on equating the right-hand sides of equations (163) and (164) we obtain

$$\frac{1}{\tau} = \iint P(\mathbf{k},\mathbf{k}')\left(1-\frac{v'_x}{v_x}\right) \mathrm{d}S' \tag{165}$$

When $P(\mathbf{k},\mathbf{k}')$ depends only on θ the angle between $\mathbf{k}$ and $\mathbf{k}'$ we have, since $v'_x/v_x = \cos\theta$,

$$\frac{1}{\tau} = 2\pi \int P(\theta)\,(1-\cos\theta)\sin\theta\,\mathrm{d}\theta \tag{166}$$

where, as before, $P(\theta) = k^2 P(\mathbf{k},\mathbf{k}')$.

In terms of the partial cross-section $\sigma(\theta)$ which is equal to $P(\theta)/Nv$, and conductivity cross-section σ_c (cf. § 11.3.1), we have

$$\frac{1}{\tau} = Nv\sigma_c = 2\pi Nv \int_0^\pi \sigma(\theta)\,(1-\cos\theta)\sin\theta\,\mathrm{d}\theta \tag{167}$$

This proves the result previously quoted in § 10.3.1 equation (27).

When the effective mass is not scalar the calculation of the relaxation time becomes much more difficult. However, in some very important cases most of the scattering which contributes to the value of τ takes place at quite small angles, and in spite of the factor $(1-\cos\theta)$ which occurs in equation (167) the major contribution to the integral comes from such angles. When this is so it is quite a good approximation to use the effective mass associated with the direction of the vector $\mathbf{k}$. For example when the constant-energy surfaces are ellipsoids with principal axes along the x, y, z axes, the effective masses for the principal directions being m_1, m_2, m_3, then we may calculate $\sigma(\theta)$ for these principal directions assuming electron masses m_1, m_2, m_3. Suppose these cross-sections are σ_1, σ_2, σ_3; then we may define relaxation times τ_1, τ_2, τ_3 which give mobilities μ_1, μ_2, μ_3 in the principal directions where

$$\left.\begin{aligned} \mu_1 &= e\tau_1/m_1 \\ \mu_2 &= e\tau_2/m_2 \\ \mu_3 &= e\tau_3/m_3 \end{aligned}\right\} \tag{168}$$

For crystals with cubic symmetry, the conductivity mobility is obtained by averaging the contributions from the various ellipsoids.

When we have various scattering mechanisms the *probabilities* of collisions due to the different processes are added. If each taken alone leads to relaxation times τ_a, τ_b, ... then the relaxation time τ is clearly given by

$$\frac{1}{\tau} = \frac{1}{\tau_a} + \frac{1}{\tau_b} + \ldots \tag{169}$$

10.9.1 Scattering by impurities

We shall now illustrate the calculation of the relaxation time τ, from which the mobility is determined, by obtaining its value for scattering of electrons by isolated impurity centres in the crystal. Firstly we shall consider ionised impurities such as those which occur in a semiconductor with shallow donor or acceptor levels. We shall assume that there is sufficient separation between the impurities so that they may be treated as single scattering centres. Except at small values of the distance r, the field produced by an ionised impurity whose charge is Ze is derived from the potential V_I given (in M.K.S. units) by

$$V_I = \frac{Ze}{4\pi\epsilon r} \tag{170}$$

where ϵ is the permittivity of the crystal. The potential we require is really the difference between the potential in the actual crystal and that in an ideal crystal, and this is given quite accurately except for values of r smaller than the lattice spacing d by (170). It is well known that for a Coulomb field the main contribution to scattering comes from small angles, which for the energies with which we are concerned, correspond to closest distances of approach considerably greater than d. The value of $\sigma(\theta)$ for such a field is well known and may be calculated using classical mechanics, for an electron of mass m_e.

Here an interesting point arises: we have shown that in an external field of force an electron in a crystal behaves like a particle with a certain effective-mass tensor and, when the latter is scalar, like a classical particle of mass m_e; this strictly applies only when the force varies slowly with distance in the crystal. It is well known, however, that for treating problems involving the scattering of electrons by free atoms quantum mechanics must be used. For scattering in a Coulomb field both classical and quantum mechanics give the same answer but this is not generally true, the quantum treatment giving the correct answer through taking account of the wave properties of the electron. It would therefore appear that in dealing with *external* forces which vary rather more rapidly a better approximation would be to treat the electron as a particle of effective mass m_e whose motion in the *external* field is also governed by quantum mechanics, i.e. by a Schrödinger equation. This may readily be extended to include a tensoral effective mass since the Schrödinger differential operator may be derived from the expression for the energy in terms of the crystal momentum; we shall later derive this result formally (see § 11.2.4).

For scattering in the potential field given by equation (170) it is well known that*

$$\sigma(\theta) = 4R^2 \operatorname{cosec}^4(\theta/2) \tag{171}$$

where $$R = Ze^2/4\pi m_e \epsilon v^2$$

When we insert this value in equation (167) and perform the integration we find that $1/\tau$ is infinite, and this comes about through the preponderance of small angle scattering involving electrons which pass very far

* See, for example, N. F. Mott and H. S. W. Massey, *Theory of Atomic Collisions* (2nd Ed.). Oxford University Press (1949), Chapter 3.

from the scattering centre. Equation (170) no longer holds at such large distances; at a distance of about half the mean separation of the impurities the force arising from V_I must vanish, since the crystal is electrically neutral. If there are N_I impurities per unit volume we shall therefore assume that the field due to the potential V_I is 'cut off' at $r = r_m$ where $(2r_m)^{-3} = N_I$; r_m is therefore equal to half the mean distance between impurities. If an electron passes a centre at distance r greater than r_m we shall assume it to be undeflected by *that particular centre*. The angular deflection θ_m when r is just less than r_m is given by

$$\tan(\theta_m/2) = R/r_m \tag{172}$$

Thus we have

$$\frac{1}{\tau} = N_I v\sigma_c = 2\pi N_I v \int_{\theta_m}^{\pi} \sigma(\theta)(1-\cos\theta)\sin\theta\,d\theta \tag{173}$$

$$= -4\pi N_I v R^2 \ln(\sin\tfrac{1}{2}\theta_m) \tag{174}$$

$$= 2\pi N_I v R^2 \ln\left(1+\frac{r_m^2}{R^2}\right) \tag{175}$$

Since we have $E = \frac{1}{2}m_e v^2$ we may write $r_m/E = 2E/E_m$ where E_m is the potential energy of the electron when $r = r_m$; thus we may write equation (175) in the form

$$\tau = AE^{3/2}\Big/\ln\left(1+\frac{4E^2}{E_m^2}\right) \tag{176}$$

where

$$A = \frac{16\pi(2m_e)^{1/2}\epsilon^2}{Z^2 e^4 N_I}$$

For non-degenerate semiconductors, under certain circumstances $E \gg E_m$ for all important values of E, and we may write $\tau = AE^{3/2}$, so that the various formulae which we have deduced assuming $\tau = AE^{-s}$ may be used with $s = -3/2$. When this approximation does not apply τ must be averaged using the expression (176). This does not lead to an integral over E which is readily evaluated. However, the approximation is generally adopted of replacing E in the logarithmic part by $3kT$ the value of E for which the integrand in the expression for $\langle\tau\rangle$ has its maximum value. We then have

$$\langle\tau\rangle = \ln\left[\left(1+\frac{36k^2T^2}{E_m^2}\right)\right]^{-1}\frac{\int_0^\infty E^3\exp[-E/kT]\,dE}{\int_0^\infty E^{3/2}\exp[-E/kT]\,dE} \tag{177}$$

$$= \frac{64\pi^{1/2}\epsilon^2(2kT)^{3/2}m_e^{1/2}}{N_I Z^2 e^4}\left[\ln\left(1+\frac{144\pi^2\epsilon^2k^2T^2}{Z^2e^4N_I^{2/3}}\right)\right]^{-1} \tag{178}$$

From this mobility μ may be obtained as $e\langle\tau\rangle/m_e$ giving what is generally known as the Conwell–Weisskopf formula being first derived by Esther M.

Conwell and V. F. Weisskopf*. Neglecting the variation of the logarithmic term, we see that the mobility, when ionised impurity scattering predominates is proportional to $T^{3/2}$, so that this type of scattering is important at low temperatures. Other methods†, taking account of the screening effect of the electrons on the Coulomb field, have been used to determine r_m resulting in some modification of equation (178).

The scattering of electrons by neutral impurities has been discussed in a similar way by C. Erginsoy‡, using results of quantum mechanical calculations of the scattering cross-section for neutral hydrogen atoms. It is shown that in this case that we have approximately

$$\sigma_c = 20a_1 \hbar/m_e v \tag{179}$$

where $a_1 = (m/m_e)(\epsilon/\epsilon_0)a_0$, a_0 is the Bohr radius, and ϵ_0 the permittivity of free space; we therefore have, if N_n is the concentration of neutral impurities,

$$1/\tau = N_n v\sigma_c = 20a_1 \hbar N_n/m_e \tag{180}$$

so that τ is a constant, to this approximation. The use of these formulae is extensively discussed in books devoted to semiconductors.

10.9.2 Scattering by thermal vibrations of the lattice

In metals, and in pure semiconductors, except at low temperatures the most important contribution to the relaxation time τ comes from the scattering of electrons by the thermal vibrations of the lattice. We shall give in this section only an elementary treatment of the problem of calculating the relaxation time, deferring to § 13.4 a more complete analysis based on the quantum theory of the interaction of the electron waves and the waves representing the lattice vibrations. The latter type of wave will be travelling through the crystal with the speed of sound and so will have a velocity of the order of 10^5 cm/sec. At ordinary temperatures, even in a non-degenerate semiconductor, electrons having energies of the order of kT will have velocities of the order of 10^7 cm/sec, and in a metal those electrons with energies near the Fermi level will have much higher velocities still; thus to a first approximation a sound wave may be regarded as stationary. A plane compressional wave will cause periodic dilation and compression of the lattice, and will set up periodic potential changes in planes throughout the crystal; it is in this periodic potential that an electron may be scattered. The scattering planes may be regarded to a first approximation as being a wavelength λ_s apart and having their normals in the direction of $\mathbf{K}_s$ the wave vector representing the sound wave; in the term 'sound wave' we include all

* E. M. Conwell and V. F. Weisskopf, *Phys. Rev.*, **77**, 388 (1950). The formula usually quoted differs from (178), replacing $4\pi\epsilon$ by ϵ; the 4π occurs in (178) through the use of M.K,S. units.

† See; for example, R. B. Dingle, *Phil. Mag.*, **46**, 831 (1955).

‡ C. Erginsoy, *Phys. Rev.*, **79**, 1013 (1950).

the waves describing the lattice vibrations but, as we shall see later, the longitudinal acoustic waves are generally the most important, except in polar semiconductors in which the strong coupling to the electric dipolar field of the optical modes makes them also important in determining the probability of scattering.

We have discussed the scattering of X-rays from parallel planes in § 5.5, and have shown that if $\mathbf{k}$ is the wave vector of the incident wave and $\mathbf{k}'$ that of the scattered wave then we get a high probability of scattering only when the vector $\mathbf{k}-\mathbf{k}'$ is parallel to the normal to the scattering planes and has magnitude $2\pi/p$ where p is the separation between the planes (Bragg's law). In this case the separation is $\lambda_s = 2\pi/K_s$ and the normal is parallel to $\mathbf{K}_s$ so we have as the scattering condition

$$\mathbf{k}-\mathbf{k}' = \mathbf{K}_s \tag{181}$$

As we have seen in § 6.9.2 $\hbar\mathbf{K}_s$ is the momentum of the phonon associated with the sound wave so that equation (181) is simply the condition for conservation of crystal and phonon momentum, corresponding either to the creation of a phonon of momentum $\hbar\mathbf{K}_s$ or absorption of a phonon of momentum $-\hbar\mathbf{K}_s$. This relationship gives us an added justification for the terms crystal momentum and phonon momentum (see also § 6.11).

Since the reflecting planes are moving, the reflected waves will suffer a change in frequency due to the Doppler effect, and for electron waves, this corresponds to a change in energy, so that k' will not be quite equal to k. To obtain the Doppler shift we consider a fixed point on one of the reflecting planes; if $\mathscr{V}$ is the velocity of this point the reflection condition requires that the phase term in the reflected wave should be the same as that in the incident wave, i.e. that

$$(\mathbf{k}.\mathbf{r}-\omega t) = \mathbf{k}'.\mathbf{r}-\omega' t \tag{182}$$

where $\mathbf{r} = \mathscr{V}t$. Since $\mathscr{V} = u\mathbf{K}_s/K_s$ where $u = \omega_s/K_s$ is the velocity of the (long) sound wave we have

$$\omega-\omega' = \frac{u}{K_s}[(\mathbf{k}-\mathbf{k}').\mathbf{K}_s] \tag{183}$$

and, using equation (181), this reduces to

$$\omega-\omega' = uK_s = \omega_s \tag{184}$$

In terms of the energies E, E', of the incident and reflected electrons this becomes

$$E-E' = \hbar\omega_s \tag{185}$$

and this is the condition of conservation of energy when a phonon is created, $\hbar\omega_s$ being the phonon energy. Clearly we could also have for a wave having wave vector $-\mathbf{K}_s$

$$E-E' = -\hbar\omega_s \tag{186}$$

corresponding to absorption of a phonon. Except at low temperatures in non-degenerate conditions, the phonon energy is usually quite small compared with the kinetic energy of the electron, as we shall verify, so that $k' \simeq k$. If we take $k' = k$ we see that the angle of scattering θ is given by

$$K_s = |\mathbf{k}-\mathbf{k}'| = 2k \sin \tfrac{1}{2}\theta \tag{187}$$

and the greatest value of K_s is therefore $2k(\theta = \pi)$. The shortest wavelength of the acoustic waves which are effective is therefore $\lambda_e/2$ where λ_e is the electron wavelength; for electrons near the bottom of the conduction band this is long compared with the lattice spacing and so K_s corresponds to long acoustic waves. The change in energy $E-E'$ is then approximately given by

$$E-E' = \hbar\omega_s = \hbar u K_s = 2m_e uv \sin \tfrac{1}{2}\theta \tag{188}$$

The largest value (when $\theta = \pi$) of this is $2m_e uv$ so that

$$(E-E')/E = 4u/v \tag{189}$$

For electrons having a classical distribution at 300°K $v \simeq 10^7$ cm/sec whereas $u \simeq 10^5$ cm/sec so that the fractional change in energy is quite small. This will no longer be so at very low temperatures.

For a given *direction* of the vector $\mathbf{K}_s$ the angle of scattering is fixed and only phonons with the particular value of K_s given by equation (187) will cause scattering in this direction. If the probability of finding a phonon with a wave vector having magnitude K_s, between zero and $2k$ is the same for all such values of K_s then the scattering will be isotropic. The condition for this is that the maximum value of $\hbar\omega_s$ be such that $\hbar\omega_s \ll \boldsymbol{k}T$, i.e. that $2m_e uv \ll \boldsymbol{k}T$. For electrons with energy of the order of $\boldsymbol{k}T$ this becomes $4u/v \ll 1$ which certainly holds at ordinary temperatures. For metals, the electrons in question lie near the Fermi level and have much higher velocities, so that isotropic scattering by the lattice vibrations can generally be assumed; moreover the scattering is elastic to quite a high degree of approximation.

Although we require a quantum mechanical calculation to obtain the numerical value of τ, we may obtain its variation with temperature for a metal quite readily, since we do not require to know its variation with energy in the degenerate condition. The amplitude of the scattered electron wave will be proportional to the potential change, which in turn will be proportional to the amplitude of the wave describing the lattice vibrations. The probability of scattering will thus be proportional to the square of the amplitude of the acoustic wave. For temperatures well above the Debye temperature this will be proportional to the temperature (see § 7.3) so that we may write $\tau_F = BT^{-1}$ where B is a constant. The mobility and conductivity will therefore be proportional to $1/T$; this is found to be so for most metals except at low temperatures. As we shall see (§ 13.4.4), the mean free path l for lattice scattering at ordinary temperatures is independent of the velocity of the electron so that we may write

$$\tau = l/v \tag{190}$$

where $l = a/T$, a being a constant. When we have a scalar effective mass m_e we may express τ in the form

$$\tau = am_e^{1/2}/2^{1/2}\,TE^{1/2} \tag{191}$$

The index s in the equation $\tau = AE^{-s}$ is therefore $\frac{1}{2}$. For a non-degenerate semiconductor with $s = \frac{1}{2}$ we have from § 10.6 equation (83),

$$\left.\begin{aligned}\langle\tau\rangle &= A(\boldsymbol{k}T)^{-1/2}\Gamma(2)/\Gamma(\tfrac{5}{2})\\ &= 4A/3(\pi\boldsymbol{k}T)^{1/2}\end{aligned}\right\} \tag{192}$$

since $\Gamma 2 = 1$ and $\Gamma\frac{5}{2} = 3\pi^{1/2}/4$. Thus we have

$$\langle\tau\rangle = \frac{4am_e^{1/2}}{3(2\pi\boldsymbol{k}T)^{3/2}} \tag{193}$$

The mobility μ_e when we have only lattice scattering by the acoustical modes of vibration is therefore given by

$$\mu_e = \frac{4ea}{3(2m_e\,\boldsymbol{k}T^3)^{1/2}} \tag{194}$$

This gives the well known $T^{-3/2}$ law for the variation of mobility in pure semiconductors at high temperatures; it is unfortunately not very well obeyed for most semiconductors since both anisotropy and other scattering mechanisms tend to modify the mobility to an appreciable extent. We defer the calculation of the constant a to § 13.4.4.

10.10 Low-mobility semiconductors

Although the theory of electrical conductivity which we have given seems to be applicable to metals and to normal semiconductors having a high electron and hole mobility, we may readily show that it cannot be applied without serious modification to materials for which the mobility is low. There are many such materials which appear to have mobilities of about 1 cm^2/V sec. For such materials the relaxation time τ would have a value of about 6×10^{-16} sec if we have $m_e = \boldsymbol{m}$ and a smaller value still if $m_e < \boldsymbol{m}$. Now because of collisions the value of the energy cannot be precisely stated for a time much greater than τ so that, if δE is the uncertainty in energy, we must have (cf. § 1.4, equation (67))

$$\delta E\tau \sim \hbar \tag{195}$$

Thus if $\tau = 6 \times 10^{-16}$ sec, $\delta E \sim 1$ eV. The allocation of energy levels in a band therefore becomes meaningless, particularly if the energy spread of the carriers, as in a non-degenerate semiconductor, is only of the order of $\boldsymbol{k}T$. For a semiconductor like Ge, on the other hand, $\delta E \sim 10^{-3}$ eV so that the energy may be fairly closely specified.

For low-mobility semiconductors the band theory of conduction must be abandoned and we must regard conduction by electrons as a form of field-assisted tunnelling between adjacent atoms of the crystal. This process has been discussed extensively by A. F. Joffé* and has been applied by him to the study of liquid and amorphous semiconductors. The limitations of the band theory, particularly as applied to narrow bands with high effective mass have also been recently discussed by H. Fröhlich and G. L. Sewell†. The theory of conduction by 'jumping' from site to site has also been used by N. F. Mott‡ to discuss conduction by impurities in semiconductors at low temperatures, the so-called impurity band conduction. The full details of the theory of this type of conduction have not yet been worked out to anything like the same extent as for conduction in materials having a high electron or hole mobility.

* A. F. Joffé, *J. Phys. Chem. Solids*, **8**, 6 (1959).

† H. Fröhlich and G. L. Sewell, *Proc. Phys. Soc.*, **74**, 643 (1959).

‡ N. F. Mott, *Report of Conference on Semiconductors*. London Physical Society (1956), p. 5.

11

The Effective-mass Approximation

11.1 The quasi-classical approximation

We have shown in Chapter 8 that, to a high degree of approximation, an electron moving in a perfect crystalline lattice in an external field of force **F** may be regarded as a particle moving classically in the field, the particle having a tensorial effective mass; the equations of motion were derived in § 8.8. These may be expressed in terms of the wave vector **k** or crystal momentum **P** by means of the vector equation

$$\frac{\mathrm{d}\mathbf{P}}{\mathrm{d}t} = \hbar\frac{\mathrm{d}\mathbf{k}}{\mathrm{d}t} = \mathbf{F} \tag{1}$$

This equation may be transformed into an acceleration equation, giving the rate of change of the 'averaged' velocity $\mathbf{v} = \hbar^{-1}\nabla_k . E$, in the form

$$\frac{\mathrm{d}\mathbf{v}}{\mathrm{d}t} = \frac{1}{\mathbf{M}_e}.\mathbf{F} \tag{2}$$

where $1/\mathbf{M}_e$ represents the effective-mass tensor whose Cartesian components $1/m_{rs}$ are given by

$$\frac{1}{m_{rs}} = \hbar^{-2}\frac{\partial^2 E}{\partial k_r\,\partial k_s} = \frac{\partial^2 E}{\partial P_r\,\partial P_s} \tag{3}$$

When the energy E may be expressed in the quadratic form

$$E = E_0 + \sum_{r=1}^{3}\sum_{s=1}^{3} A_{rs}(P_r - P_{0r})(P_s - P_{0s}) \tag{4}$$

where the quantities A_{rs} are constants, and $A_{rs} = A_{sr}$, then $1/m_{rs} = A_{rs}$ and the components of the effective-mass tensor are constants. The various simplifications of equation (2) which may be made when some of the quantities A_{rs} are zero or equal have been discussed in § 8.8; in particular, when $A_{rs} = m_e^{-1}\delta_{rs}$ so that the effective mass is a scalar m_e, the equation of motion reduces to the simple classical form

$$m_e\frac{\mathrm{d}\mathbf{v}}{\mathrm{d}t} = \mathbf{F} \tag{5}$$

We shall refer to this, and the more general form (2), as the quasi-classical approximation. It should be clearly appreciated that these equations are in no sense based on classical mechanics—their derivation depends essentially on the quantum theory of electron waves in crystals as shown in § 8.8. The term quasi-classical is used to indicate that their form is classical. Once they have been derived, however, they may be used to describe the motions of the conduction electrons in the crystal by treating the electrons as classical particles. In the derivation of equation (1) we pointed out that there were certain restrictions under which it could be applied. In particular, the force **F** must be 'slowly varying', i.e. it must change very slightly between neighbouring cells in the crystal.

We have applied equations of this form to discuss the motion of electrons under external electric and magnetic fields and have found that this description leads to results in excellent agreement with experiment when the fields are not too strong. We have also used the idea of effective mass in § 9.3.6 to discuss the motion of an electron in the Coulomb field of an impurity atom in a semiconductor. Here, however, we have a rather paradoxical state of affairs in that, while we regarded the electron as a *particle* of mass m_e, we used *wave mechanics* to derive the energy levels of the impurity centre, quoting the well-known result for a hydrogen atom. Indeed we may readily see that the quasi-classical approximation only holds provided the wavelength λ_e of the quasi particle is short compared with the distance over which the field varies appreciably; this is the well-known criterion for the application of classical mechanics to the motion of a particle in a field of force.

For the motion of a free particle of mass m in a field of force given by a potential function $V(\mathbf{r})$ the classical equation of motion

$$m\frac{\mathrm{d}\mathbf{v}}{\mathrm{d}t} = -\nabla V(\mathbf{r}) \tag{6}$$

is replaced by Schrödinger's equation

$$\nabla^2\psi+\frac{2m}{\hbar^2}[E-V(\mathbf{r})]\psi = 0 \tag{7}$$

or more generally by the equation

$$H[-i\hbar\nabla,\mathbf{r}]\psi = E\psi \tag{8}$$

where $H[\mathbf{p},\mathbf{r}]$ is the classical Hamiltonian expressed in terms of the momentum $\mathbf{p}$. Equations (7) or (8) determine the stationary-state wave function ψ associated with the energy E. Because of the similarity of equations (5) and (6) it would seem not unreasonable to replace equation (5), when we are dealing with an 'external' field of force in a crystal to which classical mechanics cannot be applied, by an equation of the form

$$\nabla^2\psi+\frac{2m_e}{\hbar^2}[E-V(\mathbf{r})]\psi = 0 \tag{9}$$

where $V(\mathbf{r})$ is the potential which determines the force. This is effectively what we have done in discussing the energy levels of an impurity centre in a semiconductor in § 9.3.6. We shall devote most of the present Chapter to proving that such an equation can indeed be used to describe the motion. Some thought will have to be given to the interpretation of the wave function ψ. It is clearly not the same as the wave functions used to describe the motion of the electron in the perfect crystal; as we shall see, it is not the whole wave function but may be interpreted as a slowly varying amplitude.

The extension of equation (9) to the case when the effective mass is tensorial may be expected to follow in the same way as equation (8) is an extension of equation (7), the Hamiltonian H being the sum of the energy of the electron in the crystal as a function of the crystal momentum $\mathbf{P}$ (which we should expect to replace the momentum $\mathbf{p}$ of a free particle) and the potential energy $V(\mathbf{r})$ being derived from the *external* force. We might reasonably therefore expect the equation which determines the motion of an electron in a crystal under an external force to be

$$[E_p(-i\hbar\nabla)+V(\mathbf{r})]\psi = E\psi \tag{10}$$

where $E_p(\mathbf{P})$ is the energy of an electron in the perfect crystal given as a function of the crystal momentum $\mathbf{P}$. In terms the wave vector $\mathbf{k}$, equation (10) may be written in the form

$$[E_k(-i\nabla)+V(\mathbf{r})]\psi = E\psi \tag{10a}$$

where $E_k(\mathbf{k})$ is the energy of electrons in the perfect crystal as a function of the wave vector $\mathbf{k}$. For slowly varying fields it is well known that equations such as (7) and (8) give the same results as classical mechanics; similarly, equations (9) and (10) will give the same results as the quasi-classical approximation when this is applicable. Equations (10) and (10a) clearly reduce to equation (9) when we have a scalar effective mass m_e; they represent a higher degree of approximation than equation (2). So far, we have only given plausible arguments for their form; we shall now proceed to derive them using the quantum theory of the motion of electrons in a crystal.

11.2 Quantum theory of the effective-mass approximation

The wave equation describing the motion of electrons in a crystal in a perturbing field of force may be derived in a number of ways. An elegant derivation, which also shows up well the physical principles involved, originally given by G. Wannier*, has been developed by J. C. Slater†, and we shall first of all follow this method of derivation. In order to use Wannier's method we shall have to introduce some wave functions which he used and which are generally known as Wannier functions; they are

* G. Wannier, *Phys. Rev.*, **52**, 191 (1937).
† J. C. Slater, *Handb. Phys.*, **19**, 1 (1956).

built up from the Bloch wave functions which we have already used in our discussion of the motion of electrons in a perfect lattice. These functions are particularly well suited to this kind of problem, whereas for many other problems the Bloch functions are to be preferred. As we shall see, the Wannier functions are localised, whereas, the Bloch functions are spread throughout the whole crystal; the latter are therefore appropriate for the discussion of problems in which we do not require to specify the *position* of an electron closely, while the former are useful when discussing problems associated with a definite point in the crystal such as an isolated impurity centre. It was indeed in order to obtain a localised wave function that the Wannier functions were first introduced.

We know that the Bloch functions $b_{\mathbf{k}}(\mathbf{r})$ defined by

$$b_{\mathbf{k}}(\mathbf{r}) = u_{\mathbf{k}}(\mathbf{r})\, e^{i\mathbf{k}.\mathbf{r}} \tag{11}$$

are solutions of the wave equation for the perfect crystal and hence that a wave function representing a solution of the wave equation may be expanded as a series of such functions. If we restrict the values of $\mathbf{k}$ to the first Brillouin zone there will be N such allowed values corresponding to each energy band, where N is the number of unit cells in the crystal. In order to obtain an exact expansion of the wave function we should require to use Bloch functions $b_{\mathbf{k}n}(\mathbf{r})$ corresponding to all bands. However, when we have a substantial gap between the bands it appears that we may obtain, under certain conditions, a good approximation by using Bloch functions only from the band in which we are interested, and these we shall denote by $b_{\mathbf{k}}(\mathbf{r})$. We then have for the expansion of the wave function ψ

$$\psi = \sum_n \sum_{\mathbf{k}} A_n(\mathbf{k})\, b_{n\mathbf{k}}(\mathbf{r}) \tag{12}$$

11.2.1 The Wannier functions

The expansion given in equation (12) is not very easily interpreted physically if a number of coefficients $A_n(\mathbf{k})$ are required to give an accurate description of the wave function representing the motion of an electron in the perturbing field of force. To overcome this difficulty Wannier (*loc. cit.*) introduced a new set of wave functions, derived from the Bloch wave functions, which have the property of being localised. Consider the wave function

$$\chi_n(\mathbf{r}, \mathbf{k}) = \sum_{\mathbf{k}} \alpha_n(\mathbf{k})\, b_{n\mathbf{k}}(\mathbf{r}) \tag{13}$$

where the constants α_n are at our disposal, the sum being taken over the N allowed values of $\mathbf{k}$. In the first unit cell of the crystal we may choose the constants α_n to make all the functions $b_n(\mathbf{k})$ add. We shall assume that the Bloch functions are normalised for a volume V containing N unit cells, and we have already seen that they are orthogonal, so that we have

$$\int_V b_n^{\star}(\mathbf{k})\, b_{n'}(\mathbf{k}')\, d\mathbf{r} = N \int_{\text{cell}} u_n^{\star}(\mathbf{k})\, u_{n'}(\mathbf{k}')\, d\mathbf{r} = \delta_{nn'}\, \delta_{\mathbf{k}\mathbf{k}'} \tag{14}$$

In the definition of the Bloch functions there is an arbitrary phase term and we use the constants α_n which may be written in the form $\exp(i\beta_n)$ to take out this phase term. Indeed we may assume that the Bloch functions are so *defined* that they add to give the *maximum* contribution in the first unit cell so that we may take $\alpha_n = N^{-1/2}$ for all values of n. When this is done we write the function $\chi_n(\mathbf{r}, \mathbf{k})$ as $a_{n0}(\mathbf{r}, \mathbf{k})$, and we have

$$a_{n0}(\mathbf{r}, \mathbf{k}) = N^{-1/2} \sum_{\mathbf{k}} b_{n\mathbf{k}}(\mathbf{r}) = N^{-1/2} \sum_{\mathbf{k}} u_{n\mathbf{k}}(\mathbf{r})\, e^{i\mathbf{k}.\mathbf{r}} \tag{15}$$

Similarly for the jth unit cell we may define a similar function a_{nj} by taking $\beta_n = -\mathbf{k}.\mathbf{R}_j$ where $\mathbf{R}_j$ is the position vector defining the jth unit cell, so that we have

$$a_{nj}(\mathbf{r}, \mathbf{k}) = N^{-1/2} \sum_{\mathbf{k}} e^{-i\mathbf{k}.\mathbf{R}_j} b_{n\mathbf{k}}(\mathbf{r}) \tag{16}$$

$$= N^{-1/2} \sum_{\mathbf{k}} u_{n\mathbf{k}}(\mathbf{r})\, e^{i\mathbf{k}.(\mathbf{r}-\mathbf{R}_j)} \tag{17}$$

Since $u_{n\mathbf{k}}(\mathbf{r})$ is a function having the periodicity of the lattice, we may write equation (17) in the form

$$a_{nj}(\mathbf{r}, \mathbf{k}) = N^{-1/2} \sum_{\mathbf{k}} u_{n\mathbf{k}}(\mathbf{r}-\mathbf{R}_j)\, e^{i\mathbf{k}.(\mathbf{r}-\mathbf{R}_j)} \tag{18}$$

We therefore see that if we substitute $\mathbf{r}+\mathbf{R}_j$ for $\mathbf{r}$ in a_{nj} we obtain the function a_{n0}. The function a_{nj} in the jth unit cell therefore behaves exactly in the same way as the function a_{n0} does in the first cell for which we may take $R_j = 0$. The function a_{nj} is a function of $\mathbf{r}-\mathbf{R}_j$ and may be written as $a_n(\mathbf{r}-\mathbf{R}_j)$; it is a function of *position only*, unlike the Bloch functions b_{nk} which depend on both $\mathbf{r}$ and $\mathbf{k}$, and in this lies the chief merit of the Wannier functions, as the functions $a_n(\mathbf{r}-\mathbf{R}_j)$ are called. For the jth unit cell the function a_{nj} will have its maximum value, the phase terms $\exp[i\mathbf{k}.(\mathbf{r}-\mathbf{R}_j)]$ being equal to unity when $\mathbf{r} = \mathbf{R}_j$, so that the functions $u_{n\mathbf{k}}(\mathbf{r})$ add constructively. As we move away from this cell the phase terms will cause destructive interference and the value of $|a_{nj}|$ will decrease rapidly. The function a_{nj} is therefore localised about the jth cell rather like the atomic wave functions $\psi_n(\mathbf{r}-\mathbf{R}_j)$ which we used in the tight-binding approximation in § 8.4. They have the advantage over the atomic wave functions that they are built up from solutions of the wave equation for the perfect crystal.

Since we have N Bloch functions for each band we have also N Wannier functions, given by the N values of j. These two sets of functions may be regarded as two sets of variables connected by means of a linear transformation, the typical element α_{rs} of the transformation matrix being

$$\alpha_{rs} = N^{-1/2} e^{-i\mathbf{k}_r.\mathbf{R}_s} \tag{19}$$

We have already had this type of transformation in the theory of lattice vibrations; it is normalised (we chose $\alpha_n = N^{-1/2}$ above so that this would be so), and the quantities α_{rs}, $\alpha^{\star}_{rs}$ from an orthogonal set, so that we have

$$\sum_{r} \alpha_{rs} \quad \alpha^{\star}_{rs'} = \delta_{ss'} \tag{20}$$

$$\sum_{s} \alpha_{rs} \quad \alpha^{\star}_{r's} = \delta_{rr'} \tag{21}$$

The transformation may therefore be written in reverse in the form

$$b_{n\mathbf{k}_r}(\mathbf{r}) = N^{-1/2} \sum_{s} \alpha^{\star}_{rs} a_{ns} \tag{22}$$

or

$$b_{n\mathbf{k}}(\mathbf{r}) = N^{-1/2} \sum_{j=1}^{N} \mathrm{e}^{i\mathbf{k}.\mathbf{R}} \, a(\mathbf{r}-\mathbf{R}_j) \tag{23}$$

From equation (23) it will be clear that by substituting in equation (12) we may expand the general wave function ψ in terms of the Wannier functions as well as in terms of the Bloch functions. We shall now show that the Wannier functions from a normalised orthogonal set and so are equally good as functions for expansion.

Let us consider the integral

$$\int a^{\star}_{n}(\mathbf{r}-\mathbf{R}_j)\, a_{n'}(\mathbf{r}-\mathbf{R}_{j'})\, \mathrm{d}\mathbf{r} \tag{24}$$

taken over the volume V. It is clear that the integral will be zero unless $n = n'$ since Bloch functions with different n are orthogonal, since they correspond to different eigenvalues, if, as we shall assume, we are not dealing with overlapping bands. We then have

$$\int a^{\star}_{n}(\mathbf{r}-\mathbf{R}_j)\, a_{n}(\mathbf{r}-\mathbf{R}_{j'})\, \mathrm{d}\mathbf{r} = N^{-1} \sum_{\mathbf{k},\mathbf{k}'} \mathrm{e}^{i(\mathbf{k}.\mathbf{R}_j - \mathbf{k}'\mathbf{R}_{j'})} \int b^{\star}_{n\mathbf{k}}(\mathbf{r})\, b_{n\mathbf{k}'}(\mathbf{r})\, \mathrm{d}\mathbf{r} \tag{25}$$

By the orthogonal property of the Bloch functions for different values of $\mathbf{k}$ we see that all the integrals in (25) vanish unless $\mathbf{k} = \mathbf{k}'$. We therefore have, using equation (14) for the normalisation of the Bloch functions,

$$\int a^{\star}_{n}(\mathbf{r}-\mathbf{R}_j)\, a_{n}(\mathbf{r}-\mathbf{R}_j)\, \mathrm{d}\mathbf{r} = N^{-1} \sum_{\mathbf{k}} \mathrm{e}^{i\mathbf{k}.(\mathbf{R}_j - \mathbf{R}_{j'})} = \delta_{jj'} \tag{26}$$

Equation (26) follows from the normality and orthogonality of the transformation matrix given by equation (21). We therefore see that the Wannier functions are orthogonal and normalised to unity for the volume V.

11.2.2 Some examples of Wannier functions

In general, the Wannier functions are a good deal more difficult to compute than the Bloch functions, since we require Bloch functions corresponding to all values of the wave vector $\mathbf{k}$. For many problems Bloch functions corresponding to a limited range of values of $\mathbf{k}$ are sufficient; for metals, for example, we only require values of $\mathbf{k}$ lying near the Fermi surface, and for semiconductors values which nearly correspond to a maximum or minimum of the energy bands. As we shall see, however, the Wannier functions enable us to derive one of the most important equations in the whole of the theory of the motion of electrons in crystalline solids and it is at least interesting to study some examples in order to become more familiar with their properties.

If the Bloch functions may be written in the form

$$b(\mathbf{r},\mathbf{k}) = u(\mathbf{r})\,e^{i\mathbf{k}.\mathbf{r}} \tag{27}$$

$u(\mathbf{r})$ being independent of $\mathbf{k}$, then the Wannier functions take a particularly simple form (we drop the suffix n for the moment). Since, for a large crystal, the allowed values of the wave vector $\mathbf{k}$ lie very close together we may replace the sum over $\mathbf{k}$ in the definition of $a(\mathbf{r}-\mathbf{R}_j)$ by an integral, the density of allowed values being $V/8\pi^3$ (see § 8.9), and we then have

$$a(\mathbf{r}-\mathbf{R}_j) = \frac{N^{-1/2}\,Vu(\mathbf{r})}{8\pi^3}\int_{\text{zone}} e^{i\mathbf{k}.(\mathbf{r}-\mathbf{R}_j)}\,d\mathbf{k} \tag{28}$$

For example, when the first Brillouin zone is a cube of side d we readily obtain if $\mathbf{R}_j = (X_j, Y_j, Z_j)$

$$\left.\begin{aligned} a(\mathbf{r}-\mathbf{R}_j) &= \frac{N^{-1/2}\,Vu(\mathbf{r})}{8\pi^3}\int_{-\pi/d}^{\pi/d}\int_{-\pi/d}^{\pi/d}\int_{-\pi/d}^{\pi/d}\exp\{ik_x(x-X_j)+\\ &\qquad +ik_y(y-Y_j)+ik_z(z-Z_j)\}\,dk_x\,dk_y\,dk_z\\ &= N^{1/2}u(\mathbf{r})\frac{\sin\{\pi(x-X_j)/d\}\sin\{\pi(y-Y_j)/d\}\sin\{\pi(z-Z_j)/d\}}{\pi^3(x-X_j)(y-Y_j)(z-Z_j)/d^3}\end{aligned}\right\} \tag{29}$$

If we take $u(\mathbf{r}) = (Nd^3)^{-1/2}$ then $b_k(\mathbf{r})$ corresponds to the normalised wave function for a completely free electron and we have

$$a(r-R_j) = \frac{\sin\{\pi(x-X_j)/d\}\sin\{\pi(y-Y_j)/d\}\sin\{\pi/(y-Y_j)/d\}}{\pi^3(x-X_j)(y-Y_j)(z-Z_z)/d^{3/2}} \tag{30}$$

It may readily be verified that $a(\mathbf{r}-\mathbf{R}_j)$ is normalised since

$$\int_{-\infty}^{\infty}\frac{\sin^2\alpha x}{x^2}\,dx = \pi\alpha \tag{31}$$

The value of $|a(r)|^2$ therefore falls off from $r = 0$ like the intensity in a diffraction pattern, i.e. along the x-axis like $\sin^2(\pi x/d)/(\pi x/d)^2$ having its peak at $x = 0$. The Wannier functions therefore, so to speak, give us a 'look' at each unit cell, the resolution being limited by the diffraction pattern of the cell.

Another type of Bloch function for which the corresponding Wannier functions are readily calculated is that given by the tight-binding approximation

$$b_{\mathbf{k}}(\mathbf{r}) = B\sum_j \phi(\mathbf{r}-\mathbf{R}_j)\,e^{i\mathbf{k}.\mathbf{R}_j} \tag{32}$$

where B is a normalising constant. If we neglect all the overlap integrals of the form

$$\alpha(R_j) = \int \phi^\star(\mathbf{r})\,\phi(\mathbf{r}-\mathbf{R}_j)\,\mathrm{d}\mathbf{r}, \quad \mathbf{R}_j \neq 0 \tag{33}$$

the normalising constant B is simply $N^{-1/2}$. We then have

$$a(\mathbf{r}-\mathbf{R}_j) = N^{-1}\sum_{\mathbf{k}}\sum_{j'}\phi(\mathbf{r}-\mathbf{R}_{j'})\,\mathrm{e}^{i\mathbf{k}.(\mathbf{R}_j-\mathbf{R}_{j'})} \tag{34}$$

When we perform the sum over $\mathbf{k}$ with $R_j \neq R_{j'}$ we get zero, and when $R_j = R_{j'}$ we get $N\phi(r-R_j)$; we therefore have simply

$$a(\mathbf{r}-\mathbf{R}_j) = \phi(\mathbf{r}-\mathbf{R}_j) \tag{35}$$

and the Wannier functions are the same as the 'atomic' functions ϕ. When we do not neglect the overlap integrals $\alpha(\mathbf{R}_j)$ the Wannier functions may also be readily calculated when the overlap integrals are small (see exercises).

Exercises (1) Show that, when the overlap integrals defined by equation (33) are small, the normalising factor B for the wave function (32) is given by

$$B = N^{-1/2}\left\{1-\tfrac{1}{2}\sum_j \alpha(\mathbf{R}_j)\,\mathrm{e}^{-i\mathbf{k}.\mathbf{R}_j}\right\}$$

Hence show that the Wannier functions $a(\mathbf{r}-\mathbf{R}_j)$ are given by

$$a(\mathbf{r}-\mathbf{R}_j) = \phi(\mathbf{r}-\mathbf{R}_j)-\tfrac{1}{2}\sum_{j'\neq 0}\alpha(\mathbf{R}_{j'})\,\phi(\mathbf{r}-\mathbf{R}_j-\mathbf{R}_{j'})$$

(2) Show that, if we neglect $\mathrm{e}^{-\alpha_0 d}$ in comparison with unity, the normalised Bloch wave function for a series of deep, narrow, one-dimensional potential wells (§ 4.2) is given by

$$b_k(x) = \left(\frac{\alpha_0}{N}\right)^{1/2}[\mathrm{e}^{-\alpha_0(x-rd)}+\mathrm{e}^{ikd}.\mathrm{e}^{-\alpha_0(rd+d-x)}]\,\mathrm{e}^{ikrd} \quad rd \leqslant x \leqslant (r+1)d$$

Hence show that the Wannier function for the rth cell is given by

$$a(x-rd) = \alpha_0^{1/2}\,\mathrm{e}^{-\alpha_0|x-rd|}$$

which is just the 'atomic' wave function for the rth potential well.

Show that if we include terms involving higher powers of $\mathrm{e}^{-\alpha_0 d}$ the normalising factor of $b_k(x)$ is altered and extra terms are introduced into the expression for the Wannier functions.

11.2.3 Expansion in Wannier functions

If we substitute for the Bloch functions in equation (12) we see that the perturbed wave function can be expanded using Wannier functions in the form

$$\psi = \sum_{nj} A_n\,F_n(\mathbf{R}_j)\,a_n(\mathbf{r}-\mathbf{R}_j) \tag{36}$$

This reminds us of the wave function used in the tight-binding approximation in § 8.4 where the functions $a(\mathbf{r}-\mathbf{R}_j)$ were replaced by the atomic wave functions $\phi(\mathbf{r}-\mathbf{R}_j)$. When we are not concerned with the variation of ψ within a cell but rather with the average value over quite a number of cells in problems in which we know that the constants $F_n(\mathbf{R}_j)$ do not vary rapidly from cell to cell, we may replace the constants $F_n(\mathbf{R}_j)$ by a continuous function $F_n(\mathbf{r})$.

For the moment we shall restrict the expansion by using Bloch and Wannier functions from a single band and drop the index n; the wave function ψ may then be written in the form

$$\psi = (V/N)^{1/2} \sum_j F(\mathbf{R}_j)\, a_n(\mathbf{r}-\mathbf{R}_j) \tag{37}$$

V/N being the volume of a unit cell. The factor $(V/N)^{1/2}$ is introduced, for purposes of normalisation, and we have

$$\int |\psi|^2 \,\mathrm{d}\mathbf{r} = (V/N) \sum_j |F(\mathbf{R}_j)|^2 = 1 \tag{38}$$

If the wave function ψ represents a Bloch function corresponding to an electron moving in a perfect crystal with wave vector $\mathbf{k}$ we see from equation (23) that $F(\mathbf{R}_j) = V^{-1/2}\exp(i\mathbf{k}.\mathbf{R}_j)$. In this case $|F(\mathbf{R}_j)|^2$ is constant and equal to V^{-1}. If $|F(\mathbf{R}_j)|$ is a slowly varying function of $\mathbf{R}_j$ it may be regarded as an amplitude determining the magnitude of the wave function ψ in the jth unit cell, since $a(\mathbf{r}-\mathbf{R}_j)$ falls off rapidly outside this cell. Suppose we have a volume ΔV, small compared with the total volume V of the crystal but containing many unit cells, and that $|F(\mathbf{R}_j)|$ does not vary much over the volume ΔV, which is centred on the cell defined by $\mathbf{R}_j$. The probability ΔP that an electron, whose motion is described by the wave function (37), is in the volume ΔV is given by

$$\Delta P = \int_{\Delta V} |\psi|^2 \,\mathrm{d}\mathbf{r} \simeq VN^{-1}|F(\mathbf{R}_j)|^2 \int_{\Delta V} \sum_{jj'} a^\star(\mathbf{r}-\mathbf{R}_j)\, a(\mathbf{r}-\mathbf{R}_{j'}) \,\mathrm{d}\mathbf{r} \tag{39}$$

From the orthogonal property of the Wannier functions we see that the integral of the sum which occurs in equation (39) is just equal to the number of unit cells in the volume ΔV which is $N\Delta V/V$; we therefore have

$$\Delta P = |F(\mathbf{R}_j)|^2 \Delta V \tag{40}$$

Thus we see that if we replace the constants $F(\mathbf{R}_j)$ by a continuous function $F(\mathbf{r})$ this function behaves like a normalised wave function, the probability density being given by $|F(\mathbf{r})|^2$; equation (38) is then equivalent to the normalising condition

$$\int_V |F(\mathbf{r})|^2 \,\mathrm{d}\mathbf{r} = 1 \tag{41}$$

We shall now see how the constants $F(\mathbf{R}_j)$ may be determined and how, under certain circumstances, a differential equation for $F(\mathbf{r})$ may be found, which turns out to be equation (10) or (10a).

11.2.4 Differential equation for the amplitude function

We now seek to solve the wave equation

$$[H_0 + V(\mathbf{r})]\psi = E\psi \tag{42}$$

where H_0 is the Hamiltonian for the perfect crystal and $V(\mathbf{r})$ is a slowly varying perturbing potential. The Bloch functions satisfy the equations

$$H_0 b_{\mathbf{k}}(\mathbf{r}) = E(\mathbf{k})\, b_{\mathbf{k}}(\mathbf{r}) \tag{43}$$

where $E(\mathbf{k})$ is the energy function for the band with which we are concerned. From the definition of the Wannier functions we see that

$$H_0 a(\mathbf{r}-\mathbf{R}_j) = N^{-1/2} \sum_{\mathbf{k}} e^{-i\mathbf{k}.\mathbf{R}_j} E(\mathbf{k})\, b_{\mathbf{k}}(\mathbf{r}) \tag{44}$$

In order to obtain the equations which determine the N quantities $F(\mathbf{R}_j)$ we substitute the wave function ψ given by equation (37) in the wave equation (42) changing the summation index to j'; we then multiply by $a^\star(\mathbf{r}-\mathbf{R}_j)$ and integrate over the volume V. In this way we obtain the N equations

$$\sum_{j'} (H_0)_{jj'} F(\mathbf{R}_{j'}) + \sum_{j'} V_{jj'} F(\mathbf{R}_{j'}) = EF(\mathbf{R}_j) \tag{45}$$

where

$$(H_0)_{jj'} = \int a^\star(\mathbf{r}-\mathbf{R}_j)\, H_0\, a(\mathbf{r}-\mathbf{R}_{j'})\, \mathrm{d}\mathbf{r} \tag{46}$$

and

$$V_{jj'} = \int a^\star(\mathbf{r}-\mathbf{R}_j)\, V(\mathbf{r})\, a(\mathbf{r}-\mathbf{R}_{j'})\, \mathrm{d}\mathbf{r} \tag{47}$$

Since the operator H_0 is periodic it will be seen that the integral in equation (46) may be re-written by means of a transformation in the form

$$(H_0)_{jj'} = \int a^\star(\mathbf{r})\, H_0\, a(\mathbf{r}-\mathbf{R}_{j'}+\mathbf{R}_j)\, \mathrm{d}\mathbf{r} \tag{48}$$

This shows that $(H_0)_{jj'}$ is a function of $\mathbf{R}_j-\mathbf{R}_{j'}$ and may be written in the form

$$(H_0)_{jj'} = h_0(\mathbf{R}_j-\mathbf{R}_{j'}) \tag{49}$$

In terms of the quantities $h_0(\mathbf{R}_j)$ equation (45) may be re-written in the form

$$\sum_{j'} [h_0(\mathbf{R}_j-\mathbf{R}_{j'}) + V_{jj'}]\, F(\mathbf{R}_{j'}) = EF(\mathbf{R}_j) \tag{50}$$

These are N simultaneous equations to determine the N quantities $F(\mathbf{R}_j)$.

So far, apart from assuming that we require only functions from a single band we have made no approximations; to solve these equations, however, we must do so. This is necessary in order to evaluate the matrix elements $V_{jj'}$. It is clear that $V_{jj'}$ will only be appreciable when there is considerable overlap of the functions $a(\mathbf{r}-\mathbf{R}_j)$ and $a(\mathbf{r}-\mathbf{R}_{j'})$; this will only be so when $\mathbf{R}_j-\mathbf{R}_{j'}$ is small. Now if $V(\mathbf{r})$ is a slowly varying function we may regard it as constant over the region where the product

$a^{\star}(\mathbf{r}-\mathbf{R}_j)a(\mathbf{r}-\mathbf{R}_{j'})$ is appreciable and take it outside the integral for $V_{jj'}$, which is then equal to zero when $j \neq j'$. We therefore see that $V_{jj'}$ will be very small when $j \neq j'$ and $V_{jj} \simeq V(\mathbf{R}_j)$. For such a function $V(\mathbf{r})$, the equations (50) may therefore be reduced to the form

$$\sum_{j'} h_0(\mathbf{R}_j-\mathbf{R}'_{j'})\,F(\mathbf{R}'_j)+V(\mathbf{R}_j)\,F(\mathbf{R}_j) = EF(\mathbf{R}_j) \tag{51}$$

By re-ordering the index j these may also be written in the form

$$\sum_{j'} h_0(\mathbf{R}_{j'})F(\mathbf{R}_j-\mathbf{R}_{j'})+V(\mathbf{R}_j)F(\mathbf{R}_j) = EF(\mathbf{R}_j) \tag{52}$$

As we shall see, this set of equations for the N quantities $F(\mathbf{R}_j)$ is equivalent to the differential equation (10) in which ψ is replaced by $F(\mathbf{r})$, the values of $F(\mathbf{R}_j)$ being obtained by putting $\mathbf{r} = \mathbf{R}_j$ in the function $F(\mathbf{r})$.

To show this we must examine the properties of the operator $E(-i\nabla)$ obtained by replacing $\mathbf{k}$ by $-i\nabla$ in $E(\mathbf{k})$. Let us first of all consider the operator $\exp[-i\nabla.\mathbf{R}_j]$. We have

$$\{\exp[-\nabla.\mathbf{R}_j]\}\phi(\mathbf{r}) = \phi(\mathbf{r})-\mathbf{R}_j.\nabla\phi(\mathbf{r})+(\mathbf{R}_j.\nabla)[\mathbf{R}_j.\nabla\phi(\mathbf{r})]/2+\dots$$

On comparing this with the Taylor expansion of $\phi(\mathbf{r}-\mathbf{R}_j)$ we see that the two expressions are the same so that we have

$$\{\exp[-\nabla.\mathbf{R}_j]\}\phi(r) = \phi(\mathbf{r}-\mathbf{R}_j) \tag{53}$$

The exponential operator is therefore just equivalent to the translation operator, which turns $\phi(\mathbf{r})$ into $\phi(\mathbf{r}-\mathbf{R}_j)$.

We may expand the function $E(\mathbf{k})$ in terms of the quantities $h_0(\mathbf{R}_j-\mathbf{R}_{j'})$ as follows: we have from equation (43)

$$E(\mathbf{k}) = \int b^{\star}_{\mathbf{k}}(\mathbf{r})\,H_0\,b_{\mathbf{k}}(\mathbf{r})\,\mathrm{d}\mathbf{r} \tag{54}$$

and, on substituting for the Bloch functions in terms of Wannier functions, we obtain

$$E(\mathbf{k}) = N^{-1}\sum_{jj'} \mathrm{e}^{-i\mathbf{k}.(\mathbf{R}_j-\mathbf{R}_{j'})}\,h_0(\mathbf{R}_j-\mathbf{R}_{j'}) \tag{55}$$

If we sum for a series of fixed values of $\mathbf{R}_{j'}$ we get in each case the same value so that we may write

$$E(\mathbf{k}) = \sum_{j} \mathrm{e}^{-i\mathbf{k}.\mathbf{R}_j}\,h_0(\mathbf{R}_j) \tag{56}$$

and this is the required expansion.

On replacing $\mathbf{k}$ by $-i\nabla$ we see that

$$E(-i\nabla)\,F(\mathbf{r}) = \sum_{j'} \mathrm{e}^{-\nabla.\mathbf{R}_{j'}}\,h_0(\mathbf{R}_{j'})\,F(\mathbf{r}) \tag{57}$$

$$= \sum_{j'} h_0(\mathbf{R}_{j'})\,F(\mathbf{r}-\mathbf{R}_{j'}) \tag{58}$$

The equation

$$-E(-i\nabla)\,F(\mathbf{r})+[E-V(\mathbf{r})]\,F(\mathbf{r}) = 0 \tag{59}$$

may therefore be written in the form

$$-\sum_j h_0(\mathbf{R}_j)\, F(\mathbf{r}-\mathbf{R}_j)+[E-V(\mathbf{r})]\, F(\mathbf{r}) = 0 \tag{60}$$

and on setting successively $\mathbf{r}=\mathbf{R}_j$ we obtain the set of equations (52) which are therefore seen to be equivalent to equation (59). The function $F(\mathbf{r})$, as we have seen, gives the probability density in the form $|F(\mathbf{r})|^2$ if normalised according to equation (41); it is therefore equivalent to the 'averaged' wave function for the motion of the electron in the crystal. Equation (59) may therefore be used to determine the value of the energy E of states associated with impurities or may be used to obtain scattering cross-sections of impurities by applying the usual theory of atomic collisions (see § 13.2.2); we shall later consider some examples. Before doing so, however, we must examine the conditions under which wave functions from a single band may be used.

When we expand the wave function ψ in terms of Wannier functions derived from several bands as in equation (36), we obtain, proceeding in exactly the same way as before, equations for the quantities $F_n(\mathbf{R}_j)$ of the form

$$\sum_{n'j'} (H_0)_{nn',jj'}\, F_{n'}(\mathbf{R}_{j'})+\sum_{n'j'} V_{nn',jj'}\, F'_n(\mathbf{R}_{j'}) = EF_n(\mathbf{R}_j) \tag{61}$$

where
$$(H_0)_{nn',jj'} = \int a_n^\star(\mathbf{r}-\mathbf{R}_j)\, H_0\, a_{n'}(\mathbf{r}-\mathbf{R}_{j'})\, d\mathbf{r} \tag{62}$$

and
$$V_{nn',jj'} = \int a_n^\star(\mathbf{r}-\mathbf{R}_j)\, V(\mathbf{r})\, a'_n(\mathbf{r}-\mathbf{R}_{j'})\, d\mathbf{r} \tag{63}$$

It will readily be seen that $(H_0)_{nn',jj'}$ is zero unless $n=n'$, and we may write

$$(H_0)_{nn,jj'} = h_{0n}(\mathbf{R}_j-\mathbf{R}_{j'}) \tag{64}$$

It will be clear also that $V_{nn',jj'}$ is very small unless $n=n'$, and therefore unless $R_j \simeq R_{j'}$ as we have already seen. When $n=n'$ we therefore have

$$V_{nn',jj'} \simeq V(\mathbf{R}_j)\, \delta_{jj'} \quad n = n' \tag{65}$$

$$\simeq V_{nn'}(\mathbf{R}_j)\, \delta_{jj'} \quad n \neq n' \tag{66}$$

where
$$V_{nn'} = \int a_n^\star(\mathbf{r})\, V(\mathbf{r})\, a_{n'}(\mathbf{r})\, d\mathbf{r} \tag{67}$$

We therefore obtain the equations

$$\sum_j h_{0n}(\mathbf{R}_j-\mathbf{R}_{j'})\, F_n(\mathbf{R}_{j'})+V(\mathbf{R}_j)\, F_n(\mathbf{R}_j)+\sum_{\substack{n' \\ (n\neq n')}} V_{nn'}\, F_{n'}(\mathbf{R}_j) = EF_n(\mathbf{R}_j) \tag{68}$$

The operators $E_n(-i\nabla)$ may be used as before to express these equations in differential form to determine the functions $F_n(\mathbf{r})$ from which the constants $F_n(\mathbf{R}_j)$ are derived. We obtain in this way the set of coupled differential equations

$$E_n(-i\nabla)\, F_n(\mathbf{r})+V(\mathbf{r})\, F_n(r)+\sum_{\substack{n' \\ (n\neq n')}} V_{nn'}\, F_{n'}(r) = EF_n(\mathbf{r}) \tag{69}$$

Let us consider the case when we have only two bands, 1 and 2, and we are primarily concerned with states in band 1. The zero-order approximation to $F_1(\mathbf{r})$ is obtained from the equation

$$E_1(-i\nabla)\,F_1(\mathbf{r})+V(r)\,F_1(\mathbf{r}) \;=\; EF_1(\mathbf{r}) \tag{70}$$

If we assume for the moment that $|F_2(r)| \ll |F_1(r)|$ we may use this zero-order approximation to obtain an equation for $F_2(r)$; we have then

$$\{E_2(-i\nabla)+V(\mathbf{r})-E\}\,F_2(\mathbf{r}) \;=\; -V_{12}F_1(\mathbf{r}) \tag{71}$$

Suppose we are concerned with two bands separated by an amount ΔE and we take the zero of energy at the bottom of band 1; then we may write

$$E_2(-i\nabla) \;=\; -\Delta E+E_2'(-i\nabla) \tag{72}$$

If we are concerned with bands for which the separation is large so that $\Delta E \gg E$, since clearly $|-\Delta E-E_2(\mathbf{k})| \leqslant \Delta E$ and the average value of $V(\mathbf{r})$ is of the order of E, we see that the order of magnitude of $F_2(\mathbf{r})$ is given by

$$|F_2(\mathbf{r})| \;\sim\; \frac{V_{12}}{\Delta E}|F_1(\mathbf{r})| \tag{73}$$

Again V_{12} will not be greater in order of magnitude than E so we may write

$$|F_2(\mathbf{r})| \;<\; \left|\frac{E}{\Delta E}\right|\;|F_1(r)| \tag{74}$$

Thus, provided we are concerned with states whose energy E above the bottom of the band 1 is small compared with the separation of the bands, the approximation in which we use only wave functions from band 1 is justified. When this condition is not satisfied, wave functions from both bands are likely to be required in order to obtain an accurate description of the perturbed state. For example, if we are considering the bound states of an impurity centre in an n-type semiconductor we may obtain a good description of these states using only wave functions of the conduction band provided the ionisation energy W_I is small compared with the forbidden energy gap ΔE. This will be true for the so-called 'shallow' impurity levels given by group V elements in silicon or germanium, but not for the deep-lying impurity levels having energy levels near the centre of the conduction band. For the latter the differential equation will be invalid in any case since $V(\mathbf{r})$ is likely to vary too rapidly to enable us to use the approximate method of calculating $V_{jj'}$. In this case, the set of linear algebraic equations must be used and not the wave equation for $F(\mathbf{r})$. In general, the solution of the problem becomes very difficult and has not been carried out in many cases of interest. We shall now consider some examples in which the wave equation for $F(\mathbf{r})$ may be used. The analysis we have given has been directed towards the study of the motions of electrons, but clearly an exactly similar effective-mass formalism may be set up to describe the motion of holes, the function $E(\mathbf{k})$ referring to the band in which the holes occur. In particular, when the hole mass is a scalar m_h the wave equation for the hole motion is

$$\nabla^2 F(\mathbf{r})+\frac{2m_h}{\hbar^2}[E-V(\mathbf{r})]\,F(\mathbf{r}) = 0 \qquad (75)$$

with an obvious extension when the effective hole mass is tensorial.

In order to use the differential equation

$$-E(-i\nabla)\,F(\mathbf{r})+[E-V(\mathbf{r})]\,F(\mathbf{r}) = 0 \qquad (76)$$

to obtain the function $F(\mathbf{r})$ the operator $E(-i\nabla)$ must reduce to a simple form; when we are concerned with values of $\mathbf{k}$ near a minimum we have seen that a Taylor expansion to second-order terms may be used, giving us a second-order differential equation. However, we have seen that in order to define the Wannier functions *all* allowed values of $\mathbf{k}$ in the first Brillouin zone are used. We must now show that when $F(\mathbf{r})$ is a slowly varying function of $\mathbf{r}$, only Bloch functions corresponding to values of $\mathbf{k}$ near the minimum of $E(\mathbf{k})$ are of importance in determining the wave function. Let us first suppose that we have a minimum in $E(\mathbf{k})$ at $\mathbf{k} = 0$ corresponding to the bottom of the conduction band. If we expand the wave function given by equation (37) in terms of Bloch functions we obtain,

$$\psi = \left(\frac{V}{N}\right)^{1/2} \sum_j \sum_{\mathbf{k}} F_n(\mathbf{R}_j)\, e^{-i\mathbf{k}.\mathbf{R}_j}\, b_{\mathbf{k}}(\mathbf{r}) \qquad (77)$$

$$= V^{1/2} \sum_{\mathbf{k}} G(\mathbf{k})\, b_k(\mathbf{r}) \qquad (78)$$

where

$$G(\mathbf{k}) = N^{-1/2} \sum_j F(\mathbf{R}_j)\, e^{-i\mathbf{k}.\mathbf{R}_j} \qquad (79)$$

The relationship between $G(\mathbf{k}_r)$ and $F(\mathbf{R}_j)$ will be seen to be the same as that between $b(\mathbf{k}_r)$ and $a(\mathbf{r}-\mathbf{R}_j)$; we may therefore invert the transformation and obtain

$$F(\mathbf{R}_j) = N^{-1/2} \sum_{\mathbf{k}} G(\mathbf{k})\, e^{i\mathbf{k}.\mathbf{R}_j} \qquad (80)$$

Equation (78) may also be written in the form

$$\psi = V^{1/2} \sum_{\mathbf{k}} G(\mathbf{k})\, u_k(\mathbf{r})\, e^{i\mathbf{k}.\mathbf{r}} \qquad (81)$$

Equation (79) for $G(\mathbf{k})$ is rather like a Fourier transform; it may be written approximately as

$$G(\mathbf{k}) = \frac{N^{1/2}}{V} \int_V F(\mathbf{r})\, e^{-i\mathbf{k}.\mathbf{r}}\, d\mathbf{r} \qquad (82)$$

and we see that $G(\mathbf{k})$ will only be appreciable if $\mathbf{k}$ is small. In this case we may obtain an approximate solution in rather simple form. We may assume that $u_{\mathbf{k}}(\mathbf{r})$ will not vary much with $\mathbf{k}$ over a small range of values of $\mathbf{k}$ near $\mathbf{k} = 0$; we may therefore replace $u_{\mathbf{k}}(\mathbf{r})$ in equation (80) by $u_0(\mathbf{r})$ the value of $u_{\mathbf{k}}(\mathbf{r})$ when $\mathbf{k} = 0$, and obtain

$$\psi = V^{1/2} u_0(\mathbf{r}) \sum_{\mathbf{k}} G(\mathbf{k})\, e^{i\mathbf{k}.\mathbf{r}} \tag{83}$$

Using equation (80) this may be written in the form

$$\psi = V^{1/2} u_0(\mathbf{r})\, F(\mathbf{r}) \tag{84}$$

In this case therefore the function $F(\mathbf{r})$ acts as a modulating function for the periodic function $u_0(\mathbf{r})$ which may vary rapidly in a single cell. If $E(\mathbf{k})$, near $\mathbf{k} = 0$, may be written in the form

$$E(\mathbf{k}) = \frac{\hbar^2}{2} \sum_{rs} \frac{\mathbf{k}_r \mathbf{k}_s}{m_{rs}} \tag{85}$$

$r, s = 1, 2, 3$, then $F(\mathbf{r})$ is obtained from the equation

$$-\frac{\hbar^2}{2} \sum_{r,s} \frac{1}{m_{rs}} \frac{\partial^2 F}{\partial x_r\, \partial x_s} + V(\mathbf{r})\, F = EF \tag{86}$$

When the lowest energy in the conduction band occurs at $\mathbf{k} = \mathbf{k}_0$ instead of at $\mathbf{k} = 0$ an expansion of the energy function may be made about $\mathbf{k} = \mathbf{k}_0$. If we let $\mathbf{k} = \mathbf{k}_0 + \mathbf{k}'$ the previous analysis may be repeated using $\mathbf{k}'$ as variable. We obtain in this way a wave function ψ given by (cf. equation (78))

$$\psi = V^{1/2} \sum_{\mathbf{k}'} G(\mathbf{k}')\, b_{\mathbf{k}'}(\mathbf{r}) \tag{87}$$

where $G(\mathbf{k}')$ is only appreciable when $|\mathbf{k}'|$ is small, and $b_{\mathbf{k}'}(\mathbf{r})$ is the Bloch function for which $\mathbf{k} = \mathbf{k}_0 + \mathbf{k}'$; the function $F(\mathbf{r})$ now satisfies the equation

$$E(-i\nabla + \mathbf{k}_0)\, F(\mathbf{r}) + V(\mathbf{r})\, F(\mathbf{r}) = EF(\mathbf{r}) \tag{88}$$

As before, we may assume that the part $u_{\mathbf{k}}(\mathbf{r})$ of the Bloch functions do not vary much for small values of $\mathbf{k}'$ and we may write

$$b_{\mathbf{k}'}(r) = u_{\mathbf{k}_0}(\mathbf{r})\, e^{i(\mathbf{k}_0 + \mathbf{k}').\mathbf{r}} \tag{89}$$

On substituting in equation (87) we obtain

$$\begin{aligned} \psi &= V^{1/2} u_{\mathbf{k}_0}(\mathbf{r})\, e^{i\mathbf{k}_0.\mathbf{r}} \sum G(\mathbf{k}')\, e^{i\mathbf{k}'.\mathbf{r}} \\ &= V^{1/2} b_{\mathbf{k}_0}(\mathbf{r})\, F(\mathbf{r}) \end{aligned} \tag{90}$$

where $b_{\mathbf{k}_0}(\mathbf{r})$ is the Bloch function associated with the energy minimum.

When the energy $E(\mathbf{k})$ may be expressed near the minimum in the form

$$E(\mathbf{k}) = \frac{\hbar^2}{2} \sum_{r,s} \frac{(k_r - k_{0r})(k_s - k_{0s})}{m_{rs}} \tag{91}$$

$r, s = 1, 2, 3$, and $k_0 \equiv (k_{01}, k_{02}, k_{03})$, the equation satisfied by the function $F(\mathbf{r})$ is also given by equation (86).

When we have a number of equivalent minima the wave function ψ will have to be a linear combination of the wave functions for each minimum, since each corresponds to a degenerate state; thus we may write the complete wave function in the form

$$\psi = \sum_{r=1}^{M} \alpha_r F_r(\mathbf{r})\, b_{\mathbf{k}_{0r}}(\mathbf{r}) \tag{92}$$

where $\mathbf{k}_{0r}$ is the value of $\mathbf{k}$ corresponding to the rth minimum. Such wave functions have been used by W. Kohn* to discuss impurity levels in Si and Ge.

An alternative method of deriving the fundamental equations of the effective-mass approximation has been given by W. Kohn and J. M. Luttinger†; in this they used the set of functions $u_0(\mathbf{r})\exp(i\mathbf{k}.\mathbf{r})$ for purposes of expansion and obtained essentially the same equations as those given above. For the case of an impurity centre having a Coulomb field a simple treatment is possible as has been shown by W. Kohn*; the expansion is made straight away in terms of Bloch functions.

We shall now consider some examples of the application of the above theory and these will give us an opportunity to discuss in a more practical manner the conditions of applicability of the effective-mass approximation.

11.2.5 Shallow hydrogenic impurity levels

We have already discussed in § 9.3.6 the shallow impurity levels associated with the Coulomb field of a donor or acceptor impurity differing by one unit of valence from the atoms of the host lattice. The perturbing potential in this case is given by (M.K.S. units)

$$V(\mathbf{r}) = \frac{-e^2}{4\pi\epsilon r} \tag{93}$$

ϵ being the permittivity of the crystal. We first shall assume that we have a scalar effective mass m_e and are concerned with electrons near the minimum of the conduction band of a semiconductor; an exactly similar analysis can be given for holes, the energy levels lying just above the top of the valence band. In this case equation (10a) reduces to

$$\nabla^2 F + \frac{2m_e}{\hbar^2}\left[E + \frac{e^2}{4\pi\epsilon r}\right] F = 0 \tag{94}$$

which is the same as the wave equation for a hydrogen atom with effective nuclear charge $\epsilon^{-1/2}$. The allowed energy levels corresponding to bound electrons are then given by

$$E_n = -\frac{e^4 m_e}{8\epsilon^2 h^2 n^2} \tag{95}$$

* W. Kohn, *Solid State Phys.* **5**, 258 (1957).
† W. Kohn and J. M. Luttinger, *Phys. Rev.*, **97**, 869 (1955).

where n is integer; equation (95) may also be expressed in the form

$$E_n = -\frac{W_H}{n^2}\left(\frac{m_e}{m}\right)\left(\frac{\epsilon_0}{\epsilon}\right)^2 \tag{95a}$$

Where W_H is the ionization energy of the hydrogen atom ($W_H \sim 13{\cdot}5$ eV); ϵ_0 is the permittivity of free space, so that the dielectric constant K is equal to ϵ/ϵ_0. The normalised function $F(\mathbf{r})$ for the lowest state, corresponding to $n = 1$ is given by

$$F(\mathbf{r}) = (\pi a^3)^{-1/2}\,\mathrm{e}^{-r/a} \tag{96}$$

where

$$a = a_0\left(\frac{\epsilon}{\epsilon_0}\right)\left(\frac{m}{m_e}\right) \tag{97}$$

a_0 being the Bohr radius $\boldsymbol{h}^2\epsilon_0/\pi m\boldsymbol{e}^2$. For many semiconductors $m_e/\boldsymbol{m} \sim 0{\cdot}1$ and $\epsilon/\epsilon_0 \sim 10$ so that $a \sim 100\,a_0$ and $E_1 \sim 10^{-3}\,W_H$. Thus we see that the energy level lies quite close to the bottom of the conduction band and the function $F(\mathbf{r})$ extends over many unit cells.

It is of interest to calculate some of the quantities which we have used in the derivation of the wave equation for $F(\mathbf{r})$. The coefficient $G(\mathbf{k})$ in the expansion of the complete wave function in Bloch functions (equation (81)) is given by (cf. equation (82))

$$G(\mathbf{k}) = C\int_V \mathrm{e}^{-r/a-i\mathbf{k}.\mathbf{r}}\,\mathrm{d}\mathbf{r} \tag{98}$$

where $C = N^{1/2}V^{-1}(\pi a^3)^{-1/2}$. To evaluate this integral we may take the z-axis as the direction of the vector $\mathbf{k}$; moreover, since the function $\exp(-r/a)$ falls off rapidly for $r \gg a$ we may extend the integration over the volume V to all space without introducing appreciable error. We then have

$$\left.\begin{aligned} G(\mathbf{k}) &= 2\pi C\int_0^\infty\int_0^\pi \mathrm{e}^{-r/a-ikr\cos\theta}\,r^2\sin\theta\,\mathrm{d}\theta \\ &= \frac{2\pi C}{ik}\int_0^\infty r[\mathrm{e}^{-r/a-ikr}-\mathrm{e}^{-r/a+ikr}]\,\mathrm{d}r \\ &= \frac{8\pi Ca^3}{(1+k^2a^2)^2} \end{aligned}\right\} \tag{99}$$

Thus we see that $G(\mathbf{k})$ falls of quite rapidly when $k > 1/a$, and is thus confined to a small part of the first Brillouin zone, at the edge of which k has a magnitude of the order of $1/a_0$.

The simple hydrogen-like model for an impurity level is likely to fail in a number of respects. We may verify that for the very shallow levels with $E \sim -10^{-2}$ eV very little contribution comes from the potential given by equation (93) for values of r of the order of the lattice spacing; for such small values of r the Coulomb form is clearly not applicable. For larger values of $|E|$ of the order of 0·1 eV this is no longer true and the value of $V(\mathbf{r})$ in the cell in which the impurity is located has an appreciable effect on the value of E. Thus while the prediction of the simple hydrogenic model, indicating the same value for the ionization energy of group V donors in Ge and Si, is well verified for Ge it is not for Si; for the latter, a considerable variation of the ionization energy is found, the values, in this case, being considerably greater than for Ge. For excited states, however, the agreement, as would be expected from the fact that these have bigger 'orbits', is much better. For Si and Ge there is, however, a more serious defect of the simple hydrogenic model—the effective-mass tensor is by no means scalar, and the wave function $F(\mathbf{r})$ is changed appreciably through the dissimilarity of the transverse and longitudinal components of the effective-mass tensor; we shall now discuss this effect.

11.2.6 Shallow impurity levels with non-scalar effective mass

When the constant-energy surfaces are in the form of spheroids the energy near the bottom of the conduction band may be expressed in the form

$$E(\mathbf{k}) = \frac{\hbar^2}{2}\left[\frac{k_x^2+k_y^2}{m_T}+\frac{k_z^2}{m_L}\right] \tag{100}$$

The differential equation for $F(\mathbf{r})$ for an impurity with potential given by equation (93) is now

$$\frac{1}{m_T}\left(\frac{\partial^2 F}{\partial x^2}+\frac{\partial^2 F}{\partial y^2}\right)+\frac{1}{m_L}\frac{\partial^2 F}{\partial z^2}+\frac{2}{\hbar^2}[E-V(r)]\,F = 0 \tag{101}$$

This equation cannot be separated into three ordinary differential equations; however, it may be reduced to a second-order differential equation if we let

$$\left.\begin{aligned} x &= \rho\cos\phi \\ y &= \rho\sin\phi \end{aligned}\right\} \tag{102}$$

The equation becomes

$$\frac{1}{m_T}\left[\frac{\partial^2 F}{\partial \rho^2}+\frac{1}{\rho}\frac{\partial F}{\partial \rho}+\frac{1}{\rho^2}\frac{\partial^2 F}{\partial \phi^2}\right]+\frac{1}{m_L}\frac{\partial^2 F}{\partial z^2}+\frac{2}{\hbar^2}\left[E+\frac{e^2}{4\pi\epsilon(\rho^2+z^2)^{1/2}}\right]F = 0 \tag{103}$$

We may separate out the variable ϕ by writing

$$F(\mathbf{r}) = f(\rho, z)\,\mathrm{e}^{im\phi} \tag{104}$$

and we obtain the equation

$$\frac{1}{m_T}\left[\frac{\partial^2 f}{\partial \rho^2}+\frac{1}{\rho}\frac{\partial f}{\partial \rho}-\frac{m^2}{\rho^2}\right]+\frac{1}{m_L}\frac{\partial^2 f}{\partial z^2}+\frac{2}{\hbar^2}\left[E+\frac{e^2}{4\pi\epsilon(\rho^2+z^2)^{1/2}}\right]f = 0 \tag{105}$$

In order to solve this equation numerical methods based on the variation principle must be used; this has been carried out for Si and Ge by various authors* and astonishingly good agreement with experiment has been found. In order to label the levels, the magnetic quantum number m is used and two other quantum numbers, l and n; l is no longer the 'angular momentum' quantum number since this is not now a 'good' quantum number, but the states can be referred to as $s, p, d \ldots$ states if they reduce to these states as $m_T \rightarrow m_L$. The states we have now are, in the order found for their energy, $1s$, $2p$ $(m = 0)$, $2s$, $2p$ $(m = \pm 1)$, $3p$ $(m = 0)$, $3p$ $(m = \pm 1)$, etc.; the states with $m = \pm 1$ are doubly degenerate in m.

11.3 Quantum theory of motion in a magnetic field

One very important problem to which we can apply the theory we have given of the quantum mechanical treatment of the motion of electrons in the effective-mass approximation is that of motion in a magnetic field. We have already mentioned that the quasi-classical approach is likely to be applicable only when $\omega_c \tau < 1$, where ω_c is the cyclotron resonance angular frequency and τ the mean collision time; this condition we have also expressed in the form $B\mu < 1$ where μ is the mobility and B the magnetic flux.

11.3.1 Magnetic quantization

As long ago as 1930 it was shown by L. Landau† that the classical theory of the motion of free electrons in a magnetic field is inadequate, even for energies for which motion in an electric field could be quite well described by means of classical theory. The reason for this is not hard to see. The projection of the motion of a free electron in a magnetic field, in the absence of an electric field, on a plane at right angles to the direction of the field, is a circle of radius r given by the equation

$$mv^2/r = Bev \tag{106}$$

where v is the magnitude of the electron's velocity. Such a motion, according to crude quantum theory, should have its angular momentum quantized, only values $n\hbar$ being allowed, and this would give

$$mvr = n\hbar \tag{107}$$

The frequency $\nu_c = 2\pi/\omega_c$ of motion round the circle, known as the cyclotron resonance frequency, is given by $\omega_c = Be/m$ and the energy $\frac{1}{2}mv^2$ by $\frac{1}{2}nh\nu_c = \frac{1}{2}n\hbar\omega_c$ (see § 8.13). The energy of the motion at right angles to the direction of the field is therefore quantized, although the total energy may take on any value being given by $\frac{1}{2}n\hbar\omega_c + \frac{1}{2}mv_z^2$. As we shall see, a proper quantum-mechanical treatment gives a slightly different

* See, for example, C. Kittel and A. H. Mitchell, *Phys. Rev.*, **96**, 1488 (1954); M. A. Lampert, *Ibid.*, **97**, 352 (1955); W. Kohn, *Solid State Phys.*, **5**, 258 (1957).

† L. Landau, *Z. Phys.*, **64**, 629 (1930).

answer, which coincides with the above only when n is odd. We see that the lowest allowable energy is not equal to zero but to $\frac{1}{2}\hbar\omega_c$. It is interesting to inquire over what range of electron energies the classical treatment is approximately correct. As is usual in such discussions, we shall suppose that the classical motion is approached when the fractional change in energy between allowed levels, corresponding to a change of one in the quantum number n, is small compared with the total energy, i.e. that n is large. If E is the energy of the electron, then we must have

$$\hbar\omega_c \leqslant E \tag{108}$$

If we express the energy in eV and B in Wb/m², we have

$$\hbar\omega_c \simeq 1{\cdot}15 \times 10^{-4}\, B$$

so that except for electrons with quite small energies equation (108) will hold for fields for which B is of the order of 1 Wb/m² (10 kG). It should be noted that it is for slow electrons, for which the spacing of the quantized levels, which is independent of the electron's energy, is large compared with the electron's total energy that the classical theory is inadequate. For free electrons, the classical motion giving helical orbits in a magnetic field is usually adequate. The above considerations also apply to the quasi-free electrons in crystalline solids and here we may expect to have some interesting quantum effects, particularly at low temperatures. Another condition, however, comes into operation; if an electron makes a collision before completing a number of circles its angular momentum, and hence its energy, will not be quantized and its motion should be correctly described by a quasi-classical treatment. The condition for the full quantum theory to be valid is therefore $\tau \gg \omega_c^{-1}$ or $B\mu \gg 1$, since $\mu = e\tau/m_e$; thus it is only for high fields and low temperatures that we expect to observe marked quantum effects.

11.3.2 The wave equation for motion in a magnetic field

When we come to consider the wave equation for the function $F(\mathbf{r})$, giving the averaged probability density for motion in a magnetic field, we run into a slight difficulty; the perturbing term in the Hamiltonian can no longer be expressed as a potential function $V(\mathbf{r})$. The modification of the Schrödinger wave equation for free electrons to take account of the presence of a magnetic field is, however, well known; it is obtained by replacing the momentum operator $\mathbf{p} = -i\hbar\nabla$ by the operator $\mathbf{p}' = \mathbf{p} - e\mathbf{A}$ (usually written as $\mathbf{p} - e\mathbf{A}/c$ in Gaussian units)*; $\mathbf{A}$ is the vector potential describing the magnetic field, so that the flux $\mathbf{B}$ is given by $\mathbf{B} = \nabla \times \mathbf{A}$. Thus, for an electron having a scalar effective mass m_e we should expect the wave equation to become, in the absence of other perturbing fields

$$\frac{1}{2m_e}(i\hbar\nabla + e\mathbf{A})^2 F(\mathbf{r}) = EF(\mathbf{r}) \tag{109}$$

* See, for example, L. I. Schiff, *Quantum Mechanics* (5th Ed.). McGraw-Hill (1955), p. 137.

The extension of this equation to a band having ellipsoidal constant-energy surfaces is straightforward and we have

$$\sum_{r=1}^{3} \frac{1}{2m_r}(i\hbar\nabla_r + eA_r)^2 F(\mathbf{r}) = EF(\mathbf{r}) \tag{110}$$

The extension to a band of arbitrary form, is, however, quite difficult and no simple equation such as (10) or (10a) seems possible; the problem has been discussed at some length by W. Kohn*.

We shall solve equation (109) for a magnetic field along the z-axis, following essentially the treatment of Landau (*loc. cit.*) for free electrons. It turns out to be more convenient to use a vector potential **A** with only one component rather than the more usual form symmetrical in x and y; we therefore take $\mathbf{A} = (0, Bx, 0)$. On substituting in equation (109) we obtain

$$\frac{\partial^2 F}{\partial x^2} + \frac{\partial^2 F}{\partial z^2} + \left(\frac{\partial}{\partial y} - \frac{ieBx}{\hbar}\right)^2 F + \frac{2m_e EF}{\hbar^2} = 0 \tag{111}$$

The variables in this equation may be separated by taking a solution of the type

$$F = \phi(x)\exp[ik_y y + ik_z z] \tag{112}$$

giving for $\phi(x)$ the equation

$$\frac{d^2\phi}{dx^2} + \left\{\frac{2m_e E}{\hbar^2} - k_z^2 - \left(k_y - \frac{eBx}{\hbar}\right)^2\right\} F = 0 \tag{113}$$

Let us write

$$E' = E - \frac{\hbar^2 k_z^2}{2m_e}$$

$$x = x' + \frac{\hbar k_y}{eB} = x' + x_0$$

Equation (113) may then be expressed in the form

$$\frac{d^2\phi}{dx'^2} + \frac{2m_e}{\hbar^2}\{E' - \tfrac{1}{2}m_e\omega_c^2 x'^2\}\phi = 0 \tag{114}$$

where now we have written ω_c for eB/m_e. This equation we recognise at once as the wave equation for a harmonic oscillator of frequency ω_c, the classical Hamiltonian being made up of the sum of the kinetic energy $\frac{1}{2}m_e\dot{x}'^2 = \frac{1}{2}p_{x'}^2/m_e$ and the potential energy $\frac{1}{2}m_e\omega_c^2 x'^2$. The solutions are well known and may be written in the form†

$$\phi(x') = H_n(\alpha x')\,e^{-\frac{1}{2}\alpha^2 x'^2} \tag{115}$$

* W. Kohn, *Phys. Rev.*, **115**, 1460 (1959).

† See, for example, L. I. Schiff, *Quantum Mechanics* (5th Ed.). McGraw-Hill (1955), p. 64.

where $\alpha^2 = m_e\omega_c/\hbar$, and $H_n(x)$ is a polynomial of degree n. The allowed values E'_n of E' are

$$E'_n = (n+\tfrac{1}{2})\,\hbar\omega_c \tag{116}$$

where n is a positive integer, and do *not* depend on k_y; the energy E is given by the equation

$$E = (n+\tfrac{1}{2})\,\hbar\omega_c + \frac{\hbar^2 k_z^2}{2m_e} \tag{117}$$

Equation (117) shows us that the energy is made up of the quantized energy of the motion in the x,y-plane together with the kinetic energy of translation parallel to the direction of the field. If we take $m_e = \boldsymbol{m}$ we obtain the energy of motion of completely free electrons in a magnetic field. If we take $k_z = 0$, we see that equation (117) reduces to equation the value obtained in § 11.3.1 only for odd values of n; the crude theory gives too many levels. We note that the motion in the y- and z-directions is unrestricted but that motion in the x-direction is centred on

$$x = x_0 = \hbar k_y/\boldsymbol{e}B$$

and restricted by the rapid fall-off of the function $\phi(x')$. The function $|\phi(x')|^2$ does not extend much beyond $x'_m = (2n)^{1/2}/\alpha = (2n\hbar/\boldsymbol{e}B)^{1/2}$; this is approximately equal to $(2E/\boldsymbol{e}B\omega_c)^{1/2}$ when n is large which is also equal to $m_e v/B\boldsymbol{e}$ the radius of the classical orbit; for an electron with $E = 1$ eV and $B = 1$ Wb/m^2 the value of x'_m is about 4×10^{-4} cm. The cyclotron frequency $\nu_c = \omega_c/2\pi$ is equal to $28(\boldsymbol{m}/m_e)\,B$ kMc/sec if B is expressed in Wb/m^2. Clearly the x-direction has no special significance and could have been chosen in any direction at right angles to the z-axis; the energy levels are therefore highly degenerate and we shall later calculate the number in a small energy interval dE (see § 12.3.1); this is allowed for in the solution we have obtained by the independence of energy of the quantity k_y. To fix the actual number of allowed energy levels we shall have to introduce boundary conditions, and shall do this when we come to use these results to discuss the diamagnetism of the free carriers in a crystal (see § 12.3).

If k_z is kept fixed we note that a transition between neighbouring energy levels corresponds to an energy change $\hbar\omega_c$. Now according to the well-known selection rules for the harmonic oscillator n can only change by ± 1; a change from n to $n+1$, with absorption of energy $\hbar\omega_c$, corresponds to cyclotron resonance absorption, and we get the same value $\omega_c/2\pi$ for the resonant frequency as we should obtain from a quasi-classical treatment* (see § 11.3.1).

If we have the constant energy surfaces in the form of ellipsoids we must use equation (110). Suppose that the magnetic field is directed along the z-axis, one of the principal axes of the ellipsoids, and we take

* See also, R. A. Smith, *Semiconductors*. Cambridge University Press (1959), p. 233.

the same representation as before for the vector potential and also the form (112) for the wave function. The equation which we now have to solve for $\phi(x)$ is

$$\frac{1}{m_1}\frac{d^2\phi}{dx^2}+\left\{\frac{2E}{\hbar^2}-\frac{k_z^2}{m_3}-\frac{1}{m_2}\left(k_y-\frac{eB_x}{\hbar}\right)^2\right\}\phi = 0 \tag{118}$$

If we now write $E' = E-\hbar^2k_z^2/2m_3$ this may be expressed in the form

$$\frac{d^2\phi}{dx^2}+\frac{2m_1}{\hbar^2}\left[E'-\frac{e^2B^2}{m_1m_2}(x-x_0)^2\right]\phi = 0 \tag{119}$$

and the energy E is given by

$$E = \hbar\omega_3(n+\tfrac{1}{2})+\hbar^2k_z^2/2m_3 \tag{120}$$

where

$$\omega_3 = \frac{eB}{(m_1m_2)^{1/2}} \tag{121}$$

Similarly, when the magnetic field is directed along the x- and y-axes the energy is given by expressions similar to equation (120) with ω_3 replaced by ω_1 and ω_2, respectively, where

$$\omega_1 = \frac{eB}{(m_2m_3)^{1/2}} \tag{122}$$

and

$$\omega_2 = \frac{eB}{(m_1m_3)^{1/2}} \tag{123}$$

Generally, if the magnetic field is in a direction defined with respect to the principal axes of a constant-energy ellipsoid by the direction cosines (l, m, n), the energy E' due to the quantized motion is given by

$$E' = \hbar\omega(n+\tfrac{1}{2}) \tag{124}$$

where

$$\omega^2 = l^2\omega_1^2+m^2\omega_2^2+n^2\omega_3^2 \tag{125}$$

(see exercise).

The analysis we have given for electrons may readily be extended to deal with positive holes, the only change being that the effective-mass tensor appropriate to holes is used to obtain the equation for the wave function $F(\mathbf{r})$. When the valence band is degenerate at its maximum value, as for Si and Ge, the situation becomes very complex; however, when the interaction between the degenerate bands is small, the two bands may be treated as separate to the first approximation and the condition for resonance obtained for each. For Si and Ge the interaction between the bands is so strong that this is a poor approximation and the quantum theory for this situation has been given by J. M. Luttinger*. The use of cyclotron resonance has played a large part in the elucidation of the band structure of semiconductors and is discussed in detail in books dealing with these materials†.

* J. M. Luttinger, *Phys. Rev.*, **102**, 1030 (1956).

† See, for example, R. A. Smith, *Semiconductors*. Cambridge University Press (1959), § 7.8.

Exercise Prove the result given by equation (125). (Transform equation (110) by means of the linear transformation

	x'	y'	z'
x	l_1	m_1	n_1
y	l_2	m_2	n_2
z	l_3	m_3	n_3

taking the z'-axis along the magnetic field so that $(l, m, n) \equiv (n_1, n_2, n_3)$. Take the vector potential components in the (x', y', z') axes as $(0, Bx', 0)$. Use a wave function of the form

$$F = \phi(x') \exp[ik_2 y' + ik_3 z']$$

to obtain an equation for $\phi(x')$ in the form

$$\frac{d^2\phi}{dx'^2} + (\alpha + \beta x)\frac{d\phi}{dx'} + (\gamma + \delta x + \epsilon x^2)\phi = 0$$

Now remove the term in $d\phi/dx'$ by means of the substitution

$$\phi(x') = \psi(x') \exp(-\tfrac{1}{2}\alpha x - \tfrac{1}{4}\beta x^2)$$

This gives an equation for $\psi(x)$

$$\frac{d^2\psi}{dx^2} + (A + Bx + Cx^2)\psi = 0$$

which can be expressed in the form

$$\frac{d^2\psi}{dx'^2} + \frac{2M}{\hbar^2}[E' - \tfrac{1}{2}M\Omega^2(x' - x_0)^2]\psi = 0$$

where
$$E = E' + G(k_2, k_3)$$

We therefore obtain

$$E = (n + \tfrac{1}{2})\hbar\Omega + G(k_2, k_3)$$

with
$$\Omega^2 = n_1^2\omega_1^2 + n_2^2\omega_2^2 + n_3^2\omega_3^2)$$

12

Magnetic Phenomena due to Free Electrons and Holes in Crystals

12.1 Sources of magnetism

THE phenomenon of magnetism is an extremely complex one, since magnetic effects are due to a number of causes. In a crystalline solid we may separate, to some extent, the effects due to the current carriers and those due to ions at the lattice points of the crystal. For the latter, magnetic effects arise from two main causes: (1) uncompensated electron spins giving the ions a magnetic moment; (2) the diamagnetic effect of the ions. The first effect normally leads to paramagnetism, but when there is strong coupling between the spins of neighbouring ions we have cooperative phenomena giving ferromagnetism or antiferromagnetism. A full discussion of these phenomena would require a whole book, and fortunately a number of books on the subject are available*. The diamagnetic effect of the ions is usually similar to that for the isolated ions but may be modified by the interaction between them. If we treat the valence electrons in the highest full band in an insulator or semiconductor as 'free' the diamagnetism of the residual ions is certainly given to a good approximation as that of the isolated ions. For a metal this is also a good approximation if we regard the ions as those left after the removal of the conduction electrons and holes, but the diamagnetism due to the ions is in this case generally much less than the contributions due to the free carriers; it is only the latter that we shall discuss in this Chapter.

The contribution to the magnetic moment of the crystal, when placed in a magnetic field, from the free carriers also arises from two main causes, paramagnetism due to their spin and diamagnetism due to their motion. For simplicity we shall discuss electrons, but holes may be treated in exactly the same way. The absence of an electron with magnetic moment μ in a nearly full band will give rise to a hole with magnetic moment $-\mu$, so that the hole may be regarded as a particle with a magnetic moment having the same magnitude as that of the electron.

* See, for example, J. H. Van Vleck, *The Theory of Electronic and Magnetic Susceptibilities*. Oxford University Press (1932); L. F. Bates, *Modern Magnetism*. Cambridge University Press (1951).

12.2 Spin paramagnetism of free electrons and holes in a crystal

The electron has a magnetic moment $\boldsymbol{\beta}$ equal to one Bohr magneton; this is given (in M.K.S. units) by

$$\boldsymbol{\beta} = \frac{e\hbar}{2m} = 9{\cdot}4 \times 10^{-24}\,\mathrm{A\,m^2} \tag{1}$$

(in Gaussian units $\boldsymbol{\beta} = e\hbar/2mc$).

If s is the component of angular momentum due to the spin of the electron in a fixed direction we have*

$$s = \pm\tfrac{1}{2}\hbar \tag{2}$$

so that the magnetic moment M is given by

$$M = \pm\frac{es}{m} \tag{3}$$

i.e. the spin and magnetic moment when measured in a fixed direction can each take only two values. If the fixed direction is defined by means of a magnetic field then the component of the spin and angular momentum parallel to the field can each take only the two values given by equations (2) and (3), respectively. If we introduce the gyromagnetic ratio $e/2m$ which gives the ratio of the magnetic moment to the angular momentum in classical electromagnetic theory (see § 12.3), we have

$$M = 2\left(\frac{e}{2m}\right)s \tag{4}$$

It is customary to write a magnetic moment in the form

$$M = g\left(\frac{e}{2m}\right)p_m \tag{5}$$

where p_m is the angular momentum. For the moment due to the spin of the electron we see that, $g = 2$†.

The magnetic energy arising from the presence of a vector magnetic moment $\mathbf{M}$ in a magnetic field whose flux is $\mathbf{B}$ is equal to $-\mathbf{M}.\mathbf{B}$; the energy due to the spin of the electron therefore takes the two values $\pm\boldsymbol{\beta}B$. If transitions are induced between two states with the same value of $\mathbf{k}$ but with different values of the spin the energy absorbed or emitted will be equal to $2\boldsymbol{\beta}B$. The resonant frequency ν_s for such absorption (known as spin resonance) will be given by

$$h\nu_s = \hbar\omega_s = 2\boldsymbol{\beta}B = Be\hbar/m$$

so that

$$\omega_s = Be/m \tag{6}$$

* This is shown in most textbooks on quantum mechanics.

† Modern developments have shown that the value of g is not exactly equal to 2 but is 2·0023.

The spin resonance frequency is therefore equal to the cyclotron frequency when $m_e = m$ (cf. § 11.3.1). For an electron in an atom, the spin resonance frequency also corresponds to absorption due to the turning over of the spin; because of coupling between the spin and orbital angular momentum, however, the effective magnetic moment is modified and it is customary to write

$$h\nu_s = g'\beta B \tag{7}$$

so that for a free electron* $g' = 2$. Another frequency which is sometimes used is the Larmor frequency ν_L; a charge system describing closed orbits, according to classical mechanics, rotates with angular velocity $\omega_L = 2\pi\nu_L$ about the direction of a magnetic field corresponding to flux B, where, for electrons

$$\omega_L = eB/2m = \tfrac{1}{2}\omega_s \tag{8}$$

When we apply a magnetic field to a crystal there will be a tendency for the electron spins to be orientated so that the magnetic moment points in the direction of the field; due to collisions, on the other hand, a number will be made to point in the opposite direction. The problem of calculating the magnetic moment due to all the spins of the free electrons consists in determining the average number pointing in each direction in thermal equilibrium.

12.2.1 Non-degenerate condition

Let us first of all consider the case of an n-type semiconductor in which the conduction electrons are in a non-degenerate condition, so that we may apply classical statistics. For an electron of moment μ whose wave vector $\mathbf{k}'$ has a particular value, the probabilities P_+ and P_- that the spin is parallel and anti-parallel to the field are proportional, respectively, to $\exp[\beta B/kT]$ and $\exp[-\beta B/kT]$, as given by the Boltzmann factor. Since this is independent of $\mathbf{k}'$, and there being no question as to whether such energy levels are or are not occupied, the probabilities that any electron is parallel or anti-parallel to the field are also proportional to P_+ and P_-. Since $P_+ + P_- = 1$, we have therefore

$$P_+ = \frac{e^{\beta B/kT}}{e^{\beta B/kT} + e^{-\beta B/kT}} \tag{9}$$

$$P_- = \frac{e^{-\beta B/kT}}{e^{\beta B/kT} + e^{-\beta B/kT}} \tag{10}$$

The total magnetic moment M, per unit volume, due to the conduction electrons is therefore given by

$$M = \beta n(P_+ - P_-) \tag{11}$$

where n is the electron concentration; thus we have

$$M = \beta n \tanh(\beta B/kT) \tag{12}$$

* g and g' are not quite the same quantity.

The saturation value of M for strong magnetic fields is therefore equal to βn, and this value is obtained when $\beta B \gg kT$. For fields so small, on the other hand, that $\beta B \ll kT$ we have

$$M = n\beta^2 B/kT \tag{13}$$

Let us consider the numerical significance of these conditions. The energy $E_m = \beta B$ may be expressed in the form

$$E_m = \beta B = 5{\cdot}85 \times 10^{-5}\, B \text{ eV}$$

if B is in Wb/m^2; since, for $T = 300°\text{K}$, $kT = 0{\cdot}025$ eV we see that even for a field of 1 Wb/m^2 (10 kG) $\beta B \ll kT$, and the expression (13) gives the magnetic moment. Only for very high fields and very low temperatures will the condition of saturation be approached.

The magnetic susceptibility χ is sometimes defined for unit volume, for unit mass, or for one mole; we shall adopt the definition for unit volume. In this case we define χ by means of the equation

$$\mathbf{M} = \chi \mathbf{H} \tag{14}$$

where $\mathbf{H}$ is the magnetic field corresponding to the flux $\mathbf{B}$. The magnetic moment is also related to the vectors $\mathbf{B}$ and $\mathbf{H}$ by means of the equation

$$\mathbf{M} = \mu_0^{-1}\mathbf{B} - \mathbf{H} \tag{15}$$

where μ_0 is the permeability of free space $4\pi \times 10^{-7}$ H/m, so that we have

$$\chi = (\mu/\mu_0) - 1 \tag{16}$$

When $\mu \simeq \mu_0$, as is usual except for ferromagnetic or antiferromagnetic materials, we see that χ is a small dimensionless quantity. When χ is small we may write approximately $\mathbf{B} = \mu_0 \mathbf{H}$ so that χ is given in terms of the flux $\mathbf{B}$ by the equation

$$\chi = \mu_0 M/B \tag{17}$$

When equation (13) is applicable we therefore have for χ_s, the contribution to the susceptibility from the spins of the free electrons,

$$\chi_s = n\mu_0 \beta^2/kT \tag{18}$$

We see that $\chi > 0$ so that we have paramagnetism and that $\chi \propto T^{-1}$. Equation (18) may be written in numerical form as

$$\chi_s = 8{\cdot}1 \times 10^{-31}\, n/T \tag{19}$$

with n expressed in m^{-3}. Thus we see that, with $n = 10^{16}$ cm^{-3} and $T = 10°\text{K}$, $\chi_s \simeq 10^{-9}$, so that for a non-degenerate semiconductor the contribution of χ_s to the magnetic susceptibility would be quite small. The total susceptibility of such materials is of the order of 10^{-7}, so that values of n of the order of 10^{18} cm^{-3} would be required to make an appreciable contribution except at very low temperatures. In this case equation (18) would not apply, since the semiconductor would no longer be non-degenerate. We shall now discuss the condition of strong degeneracy, applicable to metals and highly degenerate semiconductors.

12.2.2 Degenerate condition

When the electron concentration is so high that we have a condition of degeneracy, as in a metal or very impure semiconductor, we must use the Fermi function to obtain the distribution of electrons in energy. For each allowed value of the wave vector $\mathbf{k}'$ we now have *two* distinct values of the energy E, i.e. the spin degeneracy is removed by the magnetic field. These values we shall denote by E_+ and E_- and correspond, respectively, to electrons with spins parallel and anti-parallel to the magnetic field; we have

$$E_+ = -\boldsymbol{\beta}B + E_{\mathbf{k}'}$$

$$E_- = \boldsymbol{\beta}B + E_{\mathbf{k}'}$$

where $E_{\mathbf{k}'}$ is the energy given as a function of $\mathbf{k}'$ in the absence of the magnetic field. The probability of occupation of the state having wave vector $\mathbf{k}'$ and energy E_+ is

$$\frac{1}{\exp[(-\boldsymbol{\beta}B + E_{\mathbf{k}'} - E_F)/kT] + 1}$$

which may be written as $f_0(E_{\mathbf{k}'} - \boldsymbol{\beta}B)$ where f_0 is the distribution function in the absence of a magnetic field. Similarly the probability of the occupation of a state with wave vector $\mathbf{k}'$ and energy E_- is $f_0(E_{\mathbf{k}'} + \boldsymbol{\beta}B)$. The magnetic moment M is obtained by summing over the allowed values of $\mathbf{k}'$ for unit volume and is given by

$$M = \frac{\boldsymbol{\beta}}{8\pi^3} \int \{f_0(E_{\mathbf{k}'} - \boldsymbol{\beta}B) - f_0(E_{\mathbf{k}'} + \boldsymbol{\beta}B)\}\, d\mathbf{k}' \tag{20}$$

In the highly degenerate condition it is clear that the main contribution to equation (20) will come from values of $\mathbf{k}'$ corresponding to energies near the Fermi energy E_F since, if $E_{\mathbf{k}'}$ is much less than E_F,

$$f_0(E_{\mathbf{k}'} - \boldsymbol{\beta}B) \simeq f_0(E_{\mathbf{k}'} + \boldsymbol{\beta}B) \simeq 1$$

provided $\boldsymbol{\beta}B \ll E_F$, as will generally be so; also if $E_{\mathbf{k}'}$ is much greater than E_F both $f_0(E_{\mathbf{k}'} - \boldsymbol{\beta}B)$ and $f_0(E_{\mathbf{k}'} + \boldsymbol{\beta}B)$ will be small. We may therefore expand $f_0(E_{\mathbf{k}'} + \boldsymbol{\beta}B)$ and $f_0(E_{\mathbf{k}'} - \boldsymbol{\beta}B)$ as

$$\left.\begin{aligned} f_0(E_{\mathbf{k}'} + \boldsymbol{\beta}B) &= f_0(E_{\mathbf{k}'}) + \boldsymbol{\beta}B\frac{df_0}{dE_{\mathbf{k}'}} \\ f_0(E_{\mathbf{k}'} - \boldsymbol{\beta}B) &= f_0(E_{\mathbf{k}'}) - \boldsymbol{\beta}B\frac{df_0}{dE_{\mathbf{k}'}} \end{aligned}\right\} \tag{21}$$

since it will be a good approximation to assume that $\beta B \ll E_{\mathbf{k}'}$ if $\beta B \ll E_F$. Substituting in equation (20) we obtain

$$M = -\frac{2\beta^2 B}{8\pi^3}\int \frac{\mathrm{d}f_0}{\mathrm{d}E_{\mathbf{k}'}}\,\mathrm{d}\mathbf{k}' \tag{22}$$

If we transform as in § 2.4 to an integration with respect to the energy E we obtain

$$M = -2\beta^2 B\int \frac{\mathrm{d}f_0}{\mathrm{d}E} N(E)\,\mathrm{d}E \tag{22a}$$

where $N(E)$ is the number of states having each spin value with energy between E and $E+\mathrm{d}E$ in the absence of a magnetic field. (The total number of electrons with either spin may be written either as

$$\frac{1}{8\pi^3}\int f_0\,\mathrm{d}\mathbf{k}'$$

or as

$$\int f_0 N(E)\,\mathrm{d}E$$

The limits of integration of E in the case of strong degeneracy can be taken as 0 and ∞, without serious error.)

The integral (22a) may be evaluated approximately as in § 2.7 since the integrand is large only near $E = E_F$, and we may take $N(E)$ outside the integral sign as $N(E_F)$; we therefore have, since

$$\int_0^\infty \frac{\mathrm{d}f_0}{\mathrm{d}E}\,\mathrm{d}E = -1$$

$$M = 2\beta^2 BN(E_F) \tag{23}$$

When the effective mass is a scalar m_e we have (cf. equation (24) of § 2.3),

$$N(E) = CE^{1/2} \tag{24}$$

the constant C being determined by the equation

$$n = 2\int_0^{E_F} CE^{1/2}\,\mathrm{d}E = \tfrac{4}{3}CE_F^{3/2} \tag{25}$$

Equation (23) may therefore be written in the form

$$M = 3n\beta^2 B/2E_F \tag{26}$$

The spin susceptibility χ_s is therefore given by

$$\chi_s = 3n\mu_0\beta^2/2E_F \tag{27}$$

and if we write $E_F = \boldsymbol{k}T_F$ this becomes

$$\chi_s = 3n\mu_0\beta^2/2\boldsymbol{k}T_F \tag{27a}$$

This should be compared with equation (18). It will be seen that the contribution per electron is reduced by the factor $3T/2T_F$ which for a metal will be of the order of 10^2 or more; on the other hand n will be much greater than for the non-degenerate case. For example if we take $n = 10^{23}$ cm^{-3} and $E_F = 2$ eV, so that $T_F = 2{\cdot}4 \times 10^4$ °K, we have $\chi_s = 3 \times 10^{-6}$.

We may also express equation (27) in terms of the effective mass m_e, when this is scalar; using equation (23) of § 9.4.1 we obtain

$$\chi_s = \frac{4\pi m_e \mu_0 \beta^2}{h^2}\left(\frac{3n}{\pi}\right)^{1/3} \tag{28}$$

For ellipsoidal constant-energy surfaces the effective mass m_e is simply replaced by $\mathrm{M}^{2/3} m_d$, where m_d is the density-of-states effective mass given by

$$m_d = (m_1 m_2 m_3)^{1/3} \tag{29}$$

and M is the number of equivalent minima (cf. § 9.4.2 equation (39)).

Expressions for the magnetic susceptibility in the intermediate region between the highly degenerate and non-degenerate conditions may be obtained by integration of equation (22) without approximation; these have been given by E. C. Stoner* for the case of a scalar effective mass and may be extended to deal with the situation in which we have a non-scalar effective mass by using the appropriate expression for the Fermi energy (see exercise (2)).

Exercise (1) Show that a higher approximation to the spin susceptibility, to the first order in kT/E_F is given by

$$\chi_s = \frac{3n\mu_0 \beta^2}{2kT_F}\left[1 - \frac{\pi^2}{12}\left(\frac{T}{T_F}\right)^2\right]$$

(Use equations (27) and (81) of § 2.7 to obtain a higher approximation to the integral in equation (22)).

(2) Show that the expression given by equation (22) may be expressed in the form

$$M = 4\pi\mu_0 \beta^2 B(2m_e/h^2)^{3/2}(kT)^{1/2} F'_{1/2}(E_F/kT)$$

where

$$F_{1/2}(x_0) = \int_0^\infty \frac{x^{1/2}\,\mathrm{d}x}{\mathrm{e}^{x-x_0}+1}$$

and $F'_{1/2}(x_0)$ is its first derivative with respect to x_0. The value of E_F is then obtained from the equation

$$n = 4\pi(2m_e kT/h^2)^{3/2} F_{1/2}(E/kT)$$

12.3 Diamagnetism of free electrons and holes in a crystal

In addition to the contribution to the magnetic moment from the spin of an electron or hole, we should also expect a contribution due to the orbital angular momentum associated with the circular motion having

* E. C. Stoner, *Proc. Roy. Soc.*, **A152**, 672 (1935).

cyclotron angular frequency $\omega_c = eB/m$ (see § 11.3.1) in a magnetic field, when $\omega_c \tau \gg 1$, τ being the mean free time (see § 10.2). Let us first of all try to calculate this in a very elementary way. An electron moving in a circle of radius r with angular velocity ω_c is equivalent to a current i of magnitude $e\omega_c/2\pi$. Classical electromagnetic theory gives for the magnetic moment M of the current the value iA where A is the area enclosed by the orbit; thus we have

$$M = er^2 \omega_c/2 \tag{30}$$

$$= E/B \tag{31}$$

where E is the kinetic energy of the electron. One might therefore be tempted to suppose that the magnetic moment in the non-degenerate case is of the order of nkT/B. This is clearly incorrect since it is inconceivable that the magnetic moment should increase indefinitely with T and be *inversely* proportional to B. It was shown as long ago as 1911 by N. Bohr* that in fact on classical theory the diamagnetism of electrons in free space is exactly zero and the same argument should apply to electrons in a crystal. The trouble with the above argument is that incomplete orbits at the edge of the space confining the carriers, in this case the crystal, give a contribution which just cancels the contribution due to the closed orbits†. We shall later calculate the diamagnetic susceptibility using wave mechanics and show that this tends to zero if we let h tend to zero, thus confirming the result that a proper treatment of the susceptibility based on classical mechanics should give χ equal to zero; such a treatment was first given for electrons in free space by L. Landau‡.

It is also of interest to calculate the magnitude of the magnetic moment due to a circular orbit using elementary quantum theory. The condition of quantization for an electron with scalar effective mass m_e is (cf. § 11.3.1)

$$m_e r^2 \omega_c = n\hbar \tag{32}$$

and comparing this with equation (30) we see that

$$M = \frac{en\hbar}{2m_e} = n\left(\frac{m}{m_e}\right)\beta \tag{33}$$

The component M_z along a particular direction (the z-axis) is given by

$$M_z = m\left(\frac{m}{m_e}\right)\beta \tag{34}$$

* N. Bohr, Dissertation, University of Copenhagen (1911).

† See, for example, J. H. Van Vleck, *The Theory of Electric and Magnetic Susceptibilities*. Oxford University Press (1932), p. 94.

‡ L. Landau, *Z. Phys.*, **64**, 629 (1930).

where m is an integer ranging from $-n$ to $+n$. The potential energy E_p of such a moment in a magnetic field of flux B will be given by

$$\left.\begin{aligned} E_p &= m\left(\frac{m}{m_e}\right)\boldsymbol{\beta}B \\ &= \tfrac{1}{2}m\hbar\omega_c \end{aligned}\right\} \tag{35}$$

The kinetic energy we have already calculated using this elementary theory in § 11.3.1 as $\frac{1}{2}n\hbar\omega_c$; the total energy is therefore $\frac{1}{2}(n+m)\,\omega_c$ and we get the same values as before. These have been shown to be different from the values given by a proper wave-mechanical treatment and it is clear that to obtain a correct expression for the magnetic moment we must use such a treatment. We therefore return to the calculation of the energy levels based on the quantum theory of the effective-mass approximation as given in § 11.3.2.

12.3.1 Density of states in a strong magnetic field

In § 11.3.2 we have obtained the wave functions describing the motion of an electron in a strong magnetic field together with the allowed energy levels. The slowly varying part of the wave function $F(\mathbf{r})$ is given by

$$F(\mathbf{r}) = \phi_s(x-x_0)\,\mathrm{e}^{ik_y y+ik_z z} \tag{36}$$

where $\phi_s(x)$ is a harmonic oscillator wave function of degree s and $x_0 = \hbar k_y/eB$. When the effective mass m_e is scalar, the energy E corresponding to this wave function is

$$E_s(k_z) = (s+\tfrac{1}{2})\,\hbar\omega_c + \tfrac{1}{2}\hbar^2 k_z^2/m_e \tag{37}$$

where s is an integer. In order to determine the density of states corresponding to allowed values of the energy we must, as usual, specify the boundary conditions to be satisfied by the wave functions. For the variables y and z we may, as before, apply periodic boundary conditions; since the wave functions in x are limited to a distance of the order of $(2s)^{1/2}/\alpha$ from the points $x = x_0$ $(\alpha^2 = m_e\omega_c/\hbar)$, periodic boundary conditions are inappropriate. It is convenient to take a crystal bounded by planes perpendicular to the x-, y-, z-axes, the dimensions being L_1, L_2, L_3; then k_y and k_z have allowed values given by

$$\left.\begin{aligned} k_y &= \frac{2\pi r}{L_2} \\ k_z &= \frac{2\pi t}{L_3} \end{aligned}\right\} \tag{38}$$

r and t being integers. We shall assume that $F(\mathbf{r})$ is zero on the planes $x = 0$ and $x = L_1$, and this assumption will modify the wave functions within a distance of the order of $(2s)^{1/2}/\alpha$ of the boundary. We have seen, however, that for electrons of energies of a few electron volts and fairly strong fields this distance is quite small and we shall neglect these boundary electrons when we come to calculate the *total* energy of the system. (It is just these electrons which cannot be neglected if we try to calculate the magnetic moment directly.) The quantity x_0 therefore takes values from 0 to L_1 so that k_y extends from 0 to $\boldsymbol{e}BL_1/\hbar$. The allowed values of k_y are given by (38) and we see that there are $L_2/2\pi$ per unit extent of k_y, there are therefore altogether $\boldsymbol{e}BL_1L_2/\boldsymbol{h}$ allowed values of k_y, each corresponding to the same energy E given by (37). For a given value of k_z, each level is therefore now N-fold degenerate ($2N$-fold if we neglect spin separation) where $N = \boldsymbol{e}BL_1L_2/\boldsymbol{h}$. When the magnetic field is strong so that the interval $\hbar\omega_c$ between the levels for a given value of k_z is large, there is a strong bunching of the allowed values around the levels given by $E = (s+\frac{1}{2})\hbar\omega_c$. This is directly observable, as we shall see, as periodic variations of the magnetic susceptibility, known as the de Haas–van Alphen effect*, and also in periodic variations of infra-red absorption, known as the oscillatory magneto-absorption effect†. When the magnetic field is weak, however, we shall show that the density of states is practically unchanged.

Let us consider an interval of energy ΔE small enough so that $\Delta E \ll E$ yet which, for a given value of k_z, contains many values of $(s+\frac{1}{2})\hbar\omega_c$. The number of allowed energy values in the range is $\Delta E/\hbar\omega_c$ and the degeneracy of each is $\boldsymbol{e}BL_1L_2/\boldsymbol{h}$; the total number is therefore $m_e L_1 L_2 \Delta E/2\pi\hbar^2$. Now in the absence of a magnetic field we have

$$E = \frac{\hbar^2}{2m_e}(k_x^2+k_y^2+k_z^2) \tag{39}$$

If we keep k_z fixed, to find the number of levels in an interval ΔE, we first calculate the number of levels with $E' = E - \hbar^2 k_z^2/2m_e$ less than a fixed value E'_m. Proceeding as in § 2.4 we see that this value N_m is given by

$$N_m = \frac{L_1L_2}{4\pi^2}\iint \mathrm{d}k_x\,\mathrm{d}k_y \tag{40}$$

the limits of integration being such that $E' < E_m$. We have clearly

$$N_m = \frac{L_1L_2}{4\pi^2}\int_0^{k_m} 2\pi k\,\mathrm{d}k \tag{41}$$

where
$$E_m = \hbar^2 k_m^2/2m_e$$

Performing the integration we obtain

$$\left.\begin{aligned} N_m &= L_1L_2k_m^2/4\pi \\ &= \frac{L_1L_2}{4\pi}\left(\frac{2m_eE_m}{\hbar^2}\right) \end{aligned}\right\} \tag{42}$$

* See, for example, D. Shoenberg, *Philos. Trans.*, **A245**, 1 (1952); *Progr. Low Temp. Phys.*, **2**, 226 (1957).

† S. Zwerdling, B. Lax, and Laura M. Roth, *Phys. Rev.*, **108**, 1402 (1958).

The number of levels in an interval ΔE is obtained by differentiating N_m with respect to E_m and putting $E_m = E'$; the result is

$$N'(E')\,\Delta E' = L_1 L_2 m_e/2\pi\hbar^2 \tag{43}$$

which is the same as calculated above when we have a magnetic field. When, however, we evaluate the density of states for an energy interval ΔE small compared with the separation $\hbar\omega_c$ between the quantized levels we find a considerable difference between the two conditions. This simply means that when we have to examine the fine structure of due bunching of the levels to see a change in the density of the levels, no change being evident when we average over a large number of quantized levels. Moreover, due to collisions, the quantized levels will be broadened so that the bunching effect will be smeared out. For most purposes therefore the effects of quantization may be neglected when the magnetic field is weak.

Let us now calculate the density of states associated with each quantized level. Let us write

$$E_k = E - \tfrac{1}{2}(s+\tfrac{1}{2})\,\hbar\omega_c = \hbar^2 k_z^2/2m_e \tag{44}$$

The number of allowed values of k_z such that $|k_z| < k_{z0}$ is $L_3 k_{z0}/\pi$. The number of allowed values of E_k such that $E_k < E_{k0}$ is therefore also equal to $L_3 k_{z0}/\pi$ if $E_{k0} = \hbar^2 k_{z0}^2/2m_e$; the number is therefore equal to

$$2L_3(2m_e E_{k0})^{1/2}/\boldsymbol{h}$$

The number of allowed levels in the interval $\mathrm{d}E_k$, $N_k(E_k)\,\mathrm{d}E_k$ is therefore given by

$$N_k(E_k)\,\mathrm{d}E_k = \frac{L_3(2m_e)^{1/2}}{\boldsymbol{h}E_k^{1/2}} \tag{45}$$

We therefore have the interesting result that near each quantized level the density of states becomes very high and varies as $E_k^{-1/2}$; we have already had a similar result for a one-dimensional problem, but here we have it for a three-dimensional situation which occurs in practice.

To obtain the distribution of all energy levels we have now to sum distribution functions such as given by equation (45) over all the quantized states, remembering that each has degeneracy $\boldsymbol{e}BL_1L_2/\boldsymbol{h}$. In this way we obtain for the density of states $N(E)$

$$N(E) = \frac{\boldsymbol{e}BV(2m_e)^{1/2}}{\boldsymbol{h}^2}\sum_s \{E-(s+\tfrac{1}{2})\,\hbar\omega_c\}^{-1/2} \tag{46}$$

where V is the volume of the crystal; the sum is to be taken over values of s from zero to the maximum value which makes $E \geqslant (s+\frac{1}{2})\,\hbar\omega_c$. We also note that the lowest allowed value of E is no longer zero but $\frac{1}{2}\hbar\omega_c$, so that the bottom of the band is shifted up by an amount proportional to the magnetic field. The form of the function $N(E)$ is shown in Fig. 12.1. When the energy resolution is high enough to detect differences of the order of $\hbar\omega_c$ ($2{\cdot}9 \times 10^{-5}B$ eV if B is expressed in Wb/m^2) we see that the function $N(E)$ differs very considerably from the function of the form $AE^{1/2}$ giving the density of states in the absence of a magnetic field. For much lower energy resolution we may replace the sum in equation (46) by an integral and obtain the usual function $2\pi V(2m_e)^{3/2}E^{1/2}/h^3$ (cf. equation (24) of § 2.3). In this case the bunching of the states at the energy values $(s+\frac{1}{2})\hbar\omega_c$, clearly evident in Fig. 12.1, is smoothed out.

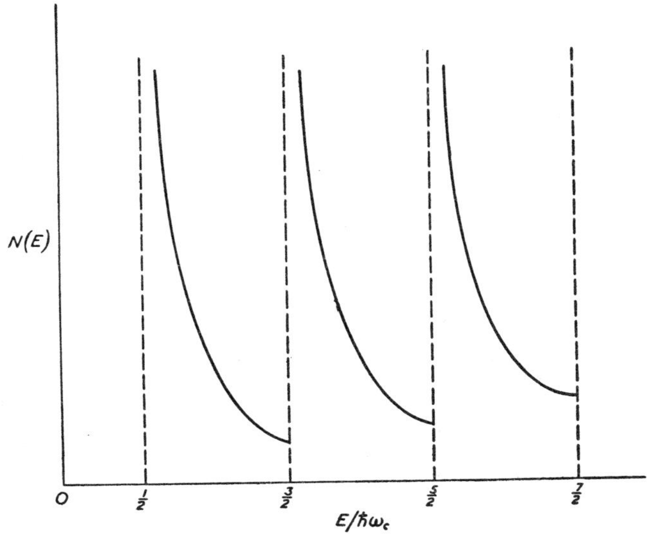

FIG. 12.1. The function $N(E)$ giving the density of allowed energy values in the presence of a magnetic field

12.3.2 The free energy

We have already discussed the difficulty in calculating the magnetic moment due to the orbital motion of the electrons associated with the 'orbits' which occur near the edge of the crystal; this difficulty arises both with the classical theory and with wave mechanics. Fortunately, the relatively small number of electrons having orbits near the edge does not materially affect the total energy of all the electrons and there is a method of deducing the magnetic moment from the variation of this energy with the magnetic field. The total energy U of the free electrons, (usually called the 'internal energy' in thermodynamics), is given by*

$$U = \sum_r E_r f(E_r) \tag{47}$$

where the sum is taken over all allowed states of energy E_r and $f(E_r)$ is the distribution function, given in this case by the Fermi function. The magnetic moment M is not given in simple terms by the variation of U

* See, for example, R. A. Smith, *The Physical Principles of Thermodynamics*. Chapman & Hall (1952), p. 154.

with the magnetic field **H** but by the variation of an associated function F, the free energy, equal to $U - TS$, where S is the entropy. This relationship, proved in books on thermodynamics*, is usually written (for unit volume) as

$$M = -\left.\frac{\partial F}{\partial H}\right)_T \tag{48}$$

It is, however, more correct to write

$$M = -\left.\frac{\partial F}{\partial B}\right)_T \tag{49}$$

The difference is negligible in c.g.s. units if the susceptibility χ is small, but equation (48) is dimensionally incorrect in M.K.S. units, while (49) is correct, and is based on the expression $-\mathbf{M}.\mathbf{B}$ for the energy† rather than $-\mathbf{M}.\mathbf{H}$ from which equation (48) is usually derived. In order to calculate the free energy F we do not require to have an expression for the entropy S since we may use a thermodynamic differential relationship connecting F and U namely‡

$$\frac{\partial}{\partial T}\left(\frac{F}{T}\right)_V = -\frac{U}{T^2} \tag{50}$$

If, therefore, we know U as a function of T, F may be obtained, apart from an irrelevant constant, by integration with respect to T.

12.3.3 Non-degenerate condition

We may readily obtain an expression for the function U in the non-degenerate condition. In this case $f(E_s)$ is proportional to $\exp[-E_s/\boldsymbol{k}T]$ and we have

$$U/n = \sum_r E_r \mathrm{e}^{-E_r/\boldsymbol{k}T} / \sum \mathrm{e}^{-E_r/\boldsymbol{k}T} \tag{51}$$

where n is the electron concentration. If we use the expression for the energy $E_s(k_z)$ given by equation (37), the sum with respect to r represents a sum over all values of the integer s together with a sum over the allowed values of k_z; the latter may clearly be replaced by an integral and we have

$$U = \frac{n \displaystyle\int_{-\infty}^{\infty} \sum_s \left\{(s+\tfrac{1}{2})\,\hbar\omega_c + \frac{\hbar^2 k_z^2}{2m_e}\right\} \exp\left\{-(s+\tfrac{1}{2})\,\hbar\omega_c - \frac{\hbar^2 k_z^2}{2m_e}\right\} \mathrm{d}k_z}{\displaystyle\int_{-\infty}^{\infty} \sum_s \exp\left\{-(s+\tfrac{1}{2})\,\hbar\omega_c - \frac{\hbar^2 k_z^2}{2m_e}\right\} \mathrm{d}k_z} \tag{52}$$

* See, for example, R. A. Smith, *The Physical Principles of Thermodynamics*. Chapman & Hall (1952), p. 154.

† J. A. Stratton, *Electromagnetic Theory*. McGraw-Hill (1941), p. 130.

‡ R. A. Smith, *loc. cit.*, p. 271.

We may readily show that

$$\int_{-\infty}^{\infty} k_z^2 \exp\left\{-\frac{\hbar^2 k_z^2}{2m_e}\right\} dk_z \Bigg/ \int_{\infty}^{\infty} \exp\left\{-\frac{\hbar^2 k_z^2}{2m_e}\right\} dk_z = \frac{m_e kT}{\hbar^2}$$

so that

$$U = \frac{n \sum_s (s+\frac{1}{2})\, \hbar\omega_c \, e^{-(s+\frac{1}{2})\hbar\omega_c/kT} + \frac{nkT}{2} \sum_s e^{-(s+\frac{1}{2})\hbar\omega_c/kT}}{\sum_s e^{-(s+\frac{1}{2})\hbar\omega_c/kT}} \tag{53}$$

Now we have

$$\sum_{s=0}^{\infty} e^{-\alpha s} = (1-e^{-\alpha})^{-1}$$

$$\sum_{s=0}^{\infty} s\, e^{-\alpha s} = -\frac{\partial}{\partial \alpha} \sum_{s=0}^{\infty} e^{-\alpha s} = e^{-\alpha}(1-e^{-\alpha})^{-2}$$

Using these results to evaluate the various sums in equation (53) we obtain

$$U = \tfrac{1}{2} nkT + \tfrac{1}{2} n\hbar\omega_c + n\hbar\omega_c (e^{\hbar\omega_c/kT} - 1)^{-1} \tag{54}$$

Since we have, in most practical conditions, $\hbar\omega_c \ll kT$ we may expand the exponential in equation (54) in powers of $\hbar\omega_c/kT$ and obtain

$$U = \tfrac{3}{2} nkT + \frac{1}{12} \frac{\hbar^2 \omega_c^2}{kT} + \ldots \tag{55}$$

We note that as $B \rightarrow 0$, U reduces to the classical value $\frac{3}{2}nkT$, and that the change in U is second order in B.

The free energy F is obtained by inserting the value for U given by equation (55) in equation (50) and integrating with respect to T. We obtain

$$F = -\tfrac{3}{2} nkT \ln T + \frac{n}{24} \frac{\hbar^2 \omega_c^2}{kT} \tag{56}$$

On inserting the value of ω_c and introducing the Bohr magneton β we have

$$F = -\tfrac{3}{2} nkT \ln T + \frac{nB^2 \beta^2}{6kT} \left(\frac{m}{m_e}\right)^2 \tag{57}$$

The magnetic moment M is obtained by using equation (49); we have

$$M = -\left.\frac{\partial F}{\partial B}\right)_T = -\frac{1}{3} \frac{n\beta^2 B}{kT} \left(\frac{m}{m_e}\right)^2 \tag{58}$$

We thus obtain a diamagnetic effect the magnitude of which is one-third of that obtained for the spin paramagnetism if the effective mass m_e is equal to the free electron mass $\boldsymbol{m}$, a result obtained for electrons in free space by Landau (*loc. cit.*). We also see that if we let $\boldsymbol{h} \to 0$ there is no magnetic moment, verifying the correct classical result.

From equation (58) the diamagnetic susceptibility χ_d may be obtained and is given by

$$\chi_d = -\frac{n}{3}\frac{\boldsymbol{\beta}^2\mu_0}{\boldsymbol{k}T}\left(\frac{\boldsymbol{m}}{m_e}\right)^2 \tag{59}$$

The total contribution to the susceptibility due to the conduction electrons χ_c is obtained by adding to χ_d the spin susceptibility χ_s as given by equation (18), and we have

$$\chi_c = \frac{n\mu_0\boldsymbol{\beta}^2}{\boldsymbol{k}T}\left(1-\frac{\boldsymbol{m}^2}{3m_e^2}\right). \tag{60}$$

We see that if $m_e < \boldsymbol{m}/\sqrt{3}$ the total contribution from the free electrons is diamagnetic; quite large negative values for $1-\boldsymbol{m}^2/3m_e^2$ are in fact found for a number of semiconductors and the observed susceptibility due to the conduction electrons agrees quite well with that predicted by equation (60). When we also have positive holes as current carriers they too make a contribution to the diamagnetic susceptibility given by equation (59) with appropriate effective mass.

When the constant-energy surfaces are ellipsoids the modification to equation (59) may be readily found. If the magnetic field is along the z-axis we have ω_c in equation (37) replaced by $\boldsymbol{e}B/(m_1m_2)^{1/2}$ and m_e by m_3 in the term involving k_z (see equation (120) of § 11.3.2). The mass m_3 disappears from the subsequent analysis and we obtain in place of equation (59)

$$\chi_d = -\frac{n}{3}\frac{\boldsymbol{\beta}^2\mu_0}{\boldsymbol{k}T}\frac{\boldsymbol{m}^2}{m_1m_2} \tag{61}$$

When we have ellipsoids with all symmetrical orientations such as in Si we clearly obtain

$$\chi_d = -\frac{n\boldsymbol{\beta}^2\mu_0\boldsymbol{m}^2}{9\boldsymbol{k}Tm_1m_2m_3}(m_1+m_2+m_3) \tag{62}$$

If, as for Si, $m_1 = m_2 = m_T$ and $m_3 = m_L$ we have

$$\chi_d = -\frac{n\boldsymbol{\beta}^2\mu_0\boldsymbol{m}^2(2m_T+m_L)}{9\boldsymbol{k}Tm_T^2m_L} \tag{63}$$

Equation (63) may also be shown to be true for Ge for which the axes of the constant-energy ellipsoids are not at right angles.

Exercises (1) Show, using the energy function

$$E = \pm\boldsymbol{\beta}B+\frac{\hbar^2k^2}{2m_e}$$

that the internal energy U is given by

$$U = \tfrac{3}{2}nkT - n\beta B \tanh(\beta B/kT)$$

Hence show that the free energy F is given by

$$F = -\tfrac{3}{2}nkT\ln T - nkT\ln\cosh(\beta B/kT)$$

Derive the expression given by equation (12) for the magnetic moment due to the spins of the conduction electrons.

(2) Show that the results of exercise (1) are unchanged for constant-energy surfaces in the form of ellipsoids.

(3) Show, using the energy function

$$E = \pm\beta B + (s+\tfrac{1}{2})\hbar\omega_c + \hbar^2 k_z^2/2m_e$$

to calculate the functions U and F, that the contributions to the magnetic susceptibility due to the spin paramagnetism and the diamagnetism of the free carriers are added.

(4) Show that when we do not make the approximation $B\beta \ll kT$, but use equation (54) for U, the free energy F is given by

$$F = -\tfrac{1}{2}nkT\ln T + \tfrac{1}{2}n\hbar\omega_c + nkT\ln(1-e^{-\hbar\omega_c/kT}) - nkT\ln(\hbar\omega_c/\mathbf{k})$$

(N.B. The last term is added as an integration constant so that $F \to -\tfrac{3}{2}nkT\ln T$ as $B \to 0$). Hence show that the magnetic moment M is given by

$$M = -n\beta\left(\frac{m}{m_e}\right)\left(1+\frac{kTm_e}{B\beta m}\right) + \frac{2n\beta m}{m_e(1-e^{-2\beta Bm/kTm_e})}$$

and verify that this reduces to equation (58) when $B\beta \ll kT$.

(5) Show that the internal energy function U in the form

$$U = \sum_s \frac{E_s}{e^{(E_s-E_F)/kT}-1}$$

may be derived from the free energy function F given by

$$F = nE_F - kT\sum_s \ln(1+e^{(E_F-E_s)/kT})$$

12.3.4 Degenerate condition

Before obtaining an expression for the diamagnetism of the free carriers in the degenerate condition let us consider the effect of magnetic quantization on the spin susceptibility. Since the distribution function $N(E)$ is drastically changed it is not at all clear that the spin susceptibility is not affected; we shall show, however, that this is not so provided

$$B\beta \ll kT(m_e/m)$$

It may readily be shown by considering the expression for the internal energy U that the spin and diamagnetic contributions may be evaluated separately and added. The magnetic moment due to the spins may therefore be obtained directly from the distribution of allowed energy states as before and is given by equation (20). The integration may be carried out in terms of the energy provided we now use equation (46) for $N(E)$. We have (cf. equation (22a))

$$M = \beta \int_0^\infty N(E)[f(E-\beta B)-f(E+\beta B)]\,\mathrm{d}E \tag{64}$$

where, as before, $f(E)$ is the Fermi-Dirac distribution function. Provided $\beta B \ll \boldsymbol{k}T$, $\boldsymbol{k}T$ being the range of E over which $f(E)$ is varying rapidly, we may write equation (64) in the form

$$M = -2\beta^2 B \int_0^\infty N(E)\frac{\partial f}{\partial E}\,\mathrm{d}E \tag{65}$$

The function $\partial F/\partial E$ is only appreciable over a range of values of E of the order of $\boldsymbol{k}T$ centred on the Fermi energy E_F. Provided $\boldsymbol{k}T \gg \beta B\boldsymbol{m}/m_e$ the function $N(E)$ will have many oscillations in the interval $\boldsymbol{k}T$ and we may replace it by its average value $\overline{N}(E)$ which, as we have seen in § 12.3.1, is just the density of states function in the absence of a magnetic field. Equation (65) therefore reduces, under these conditions, to equation (22) of § 12.3.4, and the spin susceptibility χ_s is given by equation (27). When we have very strong magnetic fields and very low temperatures so that $\boldsymbol{k}T$ is comparable with $\beta B\boldsymbol{m}/m_e$ the situation becomes very complex and we get variations in the susceptibility with field.

Let us now consider the diamagnetic susceptibility when we have a strongly degenerate condition. The function U is given by

$$U = \frac{\boldsymbol{e}B}{\pi\boldsymbol{h}}\sum_{s=0}^{\infty}\int_{-\infty}^{\infty}\frac{[\hbar\omega_c(s+\frac{1}{2})+\hbar^2k_z^2/2m_e]\,\mathrm{d}k_z}{\exp\phi(s+\frac{1}{2},k_z)+1} \tag{66}$$

where $$\boldsymbol{k}T\phi(s+\tfrac{1}{2},k_z) = \hbar\omega_c(s+\tfrac{1}{2})+\hbar^2k_z^2/2m_e-E_F \tag{67}$$

The factor $\boldsymbol{e}B/\pi\boldsymbol{h}$ gives the number of states, including spin degeneracy, per unit range of k_z, for unit volume (see § 12.3.1).

If we replace the sum with respect to s by an integral, we may write U in the form

$$U = \frac{\boldsymbol{e}B}{\pi\boldsymbol{h}}\int_{-\infty}^{\infty}\int_0^{\infty}\frac{[x\hbar\omega_c+\hbar^2k_z^2/2m_e]\mathrm{d}k_z\,\mathrm{d}x}{\exp\phi(x,k_z)+1} \tag{68}$$

Performing the integration first with respect to x and then with respect to k_z, assuming the exponential function to be zero in the second integration, we obtain

$$U = \frac{8\pi}{5}\frac{(2m_e)^{3/2}}{\boldsymbol{h}^3}E_F^{5/2} \tag{69}$$

$$= \tfrac{3}{5}nE_F \tag{69a}$$

and this is just the value of U which we expect when $B = 0$ (see § 2.4). To obtain a higher order approximation including terms in B^2 we must use a better approximation to the sum (66). When $f(x)$ is a not too rapidly varying function of x we have* approximately

$$\sum_{s=0}^{\infty} f(s+\tfrac{1}{2}) = \int_0^{\infty} f(x)\,\mathrm{d}x + \tfrac{1}{24} f'(0) \tag{70}$$

In this case

$$f(x) = \frac{eB}{\pi h} \int_{-\infty}^{\infty} \frac{[x\hbar\omega_c + \hbar^2 k_z^2/2m_e]\,\mathrm{d}k_z}{\exp\phi(x, k_z) + 1} \tag{71}$$

On performing the differentiation to obtain $f'(x)$ we find that when $E_F \gg kT$ the only term which gives an appreciable contribution is

$$f'(x) = \frac{eB}{\pi h} \int_{-\infty}^{\infty} \frac{\hbar\omega_c\,\mathrm{d}k_z}{\exp\phi(x, k_z) + 1} \tag{72}$$

When $x = 0$ we have

$$f'(0) = \frac{eB}{\pi h} \int_{-\infty}^{\infty} \frac{\hbar\omega_c\,\mathrm{d}k_z}{\exp\left(\dfrac{\hbar^2 k_z^2}{2m_e} - E_F\right)\Big/ kT + 1} \tag{73}$$

When $E_F \gg kT$ we have approximately

$$f'(0) = \frac{eB}{\pi h}\left[\frac{(2m_e)^{1/2}}{h}(\hbar\omega_c)\,E_F^{1/2}\right] \tag{74}$$

The internal energy U is therefore equal to

$$U = \tfrac{3}{5} n E_F + \frac{e^2 B^2}{12\pi h}\left(\frac{2E_F}{m_e}\right)^{1/2} \tag{75}$$

It will readily be seen that terms in the internal energy U which are independent of T give exactly the same terms in the free energy. We therefore obtain for the magnetic moment M,

$$M = -\frac{e^2 B}{6\pi h}\left(\frac{2E_F}{m_e}\right)^{1/2} \tag{76}$$

and for the diamagnetic susceptibility χ_d

$$\chi_d = -\frac{e^2 \mu_0}{6\pi h}\left(\frac{2E_F}{m_e}\right)^{1/2} \tag{77}$$

* This is a modified form of Euler's summation formula, correct to the approximation that we neglect $f'''(0)$; see, for example, T. J. I. Bromwich, *Theory of Infinite Series*. Macmillan (1908), p. 239.

this may also be written in the form

$$\chi_d = -\frac{2\beta^2\mu_0 N(E_F)}{3}\left(\frac{m}{m_e}\right)^2 \tag{78}$$

$$= -\frac{n\mu_0\beta^2}{2E_F}\left(\frac{m}{m_e}\right)^2 \tag{79}$$

Thus once again we obtain the result that the diamagnetic susceptibility has magnitude equal to $\frac{1}{3}(m/m_e)^2$ times the spin susceptibility as given by equation (27), a result we found also for the non-degenerate case. For the condition of strong degeneracy we may therefore write for the total susceptibility χ_c due to the free carriers

$$\chi_c = \frac{3n\mu_0\beta^2}{2E_F}\left[1-\frac{1}{3}\left(\frac{m}{m_e}\right)^2\right] \tag{80}$$

When the constant energy surfaces are ellipsoidal this result may be generalised as for the spin paramagnetism. This will be seen by carrying through the above analysis with the appropriate energy function, the effective mass which appears being $(m_1m_2)^{1/2}$ when the field is directed along the z-axis. The mass which occurs in E_F, however, is the 'density-of-states' effective mass given by equation (29). For crystals with constant-energy surfaces in the form of symmetrically placed ellipsoids the mass m_e in equation (80) is replaced by the average mass given in equation (63), so that

$$\chi_c = \frac{3n\beta^2\mu_0 m^2}{2E_F}\left[1-\frac{m^2(2m_T+m_L)}{3m_T^2m_L}\right] \tag{81}$$

Exercise Show by expressing the integral in equation (73) as a Fermi integral that the diamagnetic susceptibility χ_d may be expressed as

$$\chi_d = -\frac{4\pi}{3}\mu_0\beta^2(2m_e/h^2)^{3/2}(kT)^{1/2}F'_{1/2}(E_F/kT)(m/m_e)^2$$

and hence that quite generally, provided $B\beta \ll kT$

$$\chi_c = \chi_s\left[1-\frac{1}{3}\left(\frac{m}{m^\star}\right)^2\right]$$

where $m^\star$ is the appropriate effective mass.

12.3.5 The de Haas–van Alphen effect

The above calculation breaks down when the magnetic field is very high; the approximate method of summation is not really valid in this case and a more extended calculation must be made in order to obtain the expression for the magnetic moment. At high fields we must also take into account the second-order variation of the Fermi energy E_F

with field. If we consider only the spin paramagnetism we have under conditions of strong degeneracy

$$\left.\begin{aligned} n &= \int_0^{E_F} \{N(E-B\beta)+N(E+B\beta)\}\,\mathrm{d}E \\ &= 2\int_0^{E_F} N(E)\,\mathrm{d}E + \beta^2 B^2 \int_0^{E_F} \frac{\mathrm{d}^2 N}{\mathrm{d}E^2}\,\mathrm{d}E \\ &= 2\int_0^{E_F} N(E)\,\mathrm{d}E + B^2\beta^2 \left.\frac{\mathrm{d}N}{\mathrm{d}E}\right)_{E_F} \end{aligned}\right\} \tag{82}$$

For a scalar effective mass m_e we have

$$N(E) = \frac{2\pi(2m_e)^{3/2}}{h^3} E^{1/2}$$

so that

$$n = \frac{8\pi}{3}\frac{(2m)^{3/2}}{h^3} E_F^{3/2} + \frac{\pi(2m_e)^{3/2}}{h^3}\frac{B^2\beta^2}{E_F^{1/2}} \tag{83}$$

To the second order in B we therefore obtain

$$E_F = E_{F_0}\left[1 - \frac{1}{4}\frac{B^2\beta^2}{E_{F_0}^2}\right] \tag{84}$$

If $E_{F_0} \gg B\beta$ we see that the change in E_F is quite small.

For very strong magnetic fields the peaks in the density of states function given by equation (46) become well separated and a periodic variation of the susceptibility results. It requires a very high field to make $B\beta \gg kT$ even at quite low temperatures, and an average over a number of peaks in $N(E)$ occurs even at the highest fields quite apart from broadening of the peaks due to the uncertainty in the energy of the order of h/τ due to collisions, where τ is the mean relaxation time. Nevertheless, we should expect to have a maximum in $-\chi_d$ when the Fermi level coincides with one of the values $(s+\frac{1}{2})\,\hbar\omega_c$. If we neglect the small quadratic variation of the Fermi level with B, the condition for a maximum is

$$E_F = (s+\tfrac{1}{2})\,\hbar\omega_c$$

or

$$\frac{1}{B} = (s+\tfrac{1}{2})\,\hbar e/m_e E_F \tag{85}$$

$$= (s+\tfrac{1}{2})\left(\frac{m}{m_e}\right)\left(\frac{2\beta}{E_F}\right) \tag{85a}$$

The susceptibility therefore shows oscillations with constant separation equal to $(2\beta/E_F)(m/m_e)$ if plotted on a scale of $1/B$. The amplitude of the oscillations is quite difficult to calculate*, but they have been observed experimentally for a number of metals†, and the effect is known as the de Haas–van Alphen effect.

12.4 Transport phenomena in a strong magnetic field

We have already discussed the application of the Boltzmann equation to the study of electrical conduction in a strong magnetic field in § 10.8.2, and indicated that such analysis may not be strictly applicable when $\omega_c\tau \gg 1$. We have been concerned in the present Chapter with motion of electrons when $\omega_c\tau \gg 1$ and we must see how the transport equations are modified for this condition. If we apply the equation we have derived in § 10.8.2 (equation (156)), to calculate the components of the electric current due to electrons in a crystal for which we have a scalar effective mass m_e, we find that, if the magnetic field is in the z-direction,

$$J_x = \frac{ne^2}{m_e}\left\{\frac{\tau\mathscr{E}_x}{1+\omega_c^2\tau^2} - \frac{\omega_c\tau^2\mathscr{E}_y}{1+\omega_c^2\tau^2}\right\} \tag{86}$$

$$J_y = \frac{ne^2}{m_e}\left\{\frac{\tau\mathscr{E}_y}{1+\omega_c^2\tau^2} + \frac{\omega_c\tau^2\mathscr{E}_x}{1+\omega_c^2\tau^2}\right\} \tag{87}$$

$$J_z = \frac{ne^2}{m_e}.\tau\mathscr{E}_z \tag{88}$$

From equation (88) we see that such a treatment shows that there is no longitudinal magneto-resistance; if we average, as usual, over values of E we obtain

$$\bar{J}_z = \frac{ne^2}{m_e}\bar{\tau}\mathscr{E}_z = \sigma_0\mathscr{E}_z \tag{89}$$

where σ_0 is the conductivity in the absence of a magnetic field. If the x-direction is chosen as the direction of the component of current in the (x,y) plane, so that $\bar{J}_y = 0$, equation (87) gives, when $\omega_c\tau \gg 1$,

$$\mathscr{E}_x = -\frac{\mathscr{E}_y}{\omega_c}\overline{\left(\frac{1}{\tau}\right)} \tag{90}$$

and equation (86) reduces to

$$J_x = -\frac{ne^2\mathscr{E}_y}{m_e\,\omega_c} = -\frac{ne}{B}\mathscr{E}_y \tag{91}$$

The Hall constant R is thus given by

$$R = \mathscr{E}_y/BJ_x = -1/ne \tag{92}$$

This result, which we had previously for the degenerate condition in a small magnetic field, would therefore appear to hold for any condition

* R. E. Peierls, *Quantum Theory of Solids*. Oxford University Press (1955).

† D. Shoenberg, *Philos. Trans.*, **245**, 1 (1952); *Progr. Low Temp. Phys.*, **2**, 226 (1957).

in a very strong magnetic field. If we substitute for $\mathscr{E}_y$ from equation (90) in equation (92) we obtain

$$J_x = \frac{ne^2 \mathscr{E}_x}{m_e} \Big/ \overline{\left(\frac{1}{\tau}\right)} \tag{93}$$

$$= \sigma_{\infty T} \mathscr{E}_x \tag{93a}$$

where $\sigma_{\infty T}$, the conductivity for a large transverse magnetic field, is given by

$$\sigma_{\infty T} = \frac{ne^2}{m_e} \Big/ \overline{\left(\frac{1}{\tau}\right)} = \sigma_0 \Big/ \overline{\left(\frac{1}{\tau}\right)} \bar{\tau} \tag{94}$$

For a metal or degenerate semiconductor $\overline{(1/\tau)} = 1/\bar{\tau}$, so that $\sigma_{\infty T} = \sigma_0$ and again we have no magneto-resistance if the effective mass is scalar; for a non-degenerate semiconductor we have when $\tau = A\,\mathrm{e}^{-s}$ (see § 10.6)

$$\sigma_0/\sigma_{\infty T} = \Gamma(\tfrac{5}{2}+s)\,\Gamma(\tfrac{5}{2}-s)/(\Gamma(\tfrac{5}{2}))^2 \tag{95}$$

If $s = \frac{1}{2}$, as for scattering by acoustic modes of the lattice vibrations, we have $\sigma_0/\sigma_{\infty T} = \frac{32}{9}\pi$; if $s = \frac{3}{2}$, as is approximately true for ionised impurity scattering we have $\sigma_0/\sigma_{\infty T} = \frac{32}{3}\pi$. For all reasonable values of s the conductivity reaches a saturation value independent of the field, the value being less than that of the zero-field conductivity, i.e. the magneto-resistance coefficient is always positive. Of the above predictions only that given by equation (92) for the Hall coefficient is verified in practice.

That a calculation based on Boltzmann's equation and on the undisturbed distribution function f_0 is incorrect is hardly surprising in view of our discussion of the motion of electrons in a high magnetic field in the previous section. To begin with, the equation of motion giving $\dot{\mathbf{k}}$ based on the force $[e\mathbf{v} \times \mathbf{B}]$ is likely to be invalid when the motion is quantized; secondly, and even more important, is the fact that the distribution function f does not tend to f_0, the value for no electric and no magnetic field, when $\mathscr{E} = 0$, but to the value f_{0B} given by the quantized energy function. To obtain an exact treatment of this very difficult problem we should therefore abandon Boltzmann's equation and use a wave-mechanical treatment for the motion of the electrons. Such a treatment requires rather advanced quantum mechanical technique and will not be given here; a treatment based on the so-called density matrix has been given recently by P. N. Argyres* and by E. N. Adams and T. D. Halstein†. In particular, they find that equation (92) for the high-field value of the Hall coefficient is verified as a first approximation, but that small oscillatory terms are introduced, as for the magnetic susceptibility, in the degenerate condition.

When the magnetic field and electric current vectors are in the same direction a somewhat simpler treatment is possible, since Boltzmann's

* P. N. Argyres, *Phys. Rev.*, **109**, 1115 (1958).

† E. N. Adams and T. D. Halstein, *J. Phys. Chem. Solids*, **10**, 254 (1959).

equation in which f_0 is replaced by f_{0B} is probably a reasonably good approximation when $\hbar\omega_c \gg kT$. In this case only the quantized level with $s = 0$ is populated and f_{0B} is a function of k_z only; moreover, we have $\hbar\dot{k}_z = e\mathscr{E}_z$ and this is likely to be correct. Thus we obtain for the equation for the disturbed distribution function f

$$e\hbar^{-1}\frac{\partial f}{\mathrm{d}k_z}\mathscr{E}_y = \frac{f - f_{0B}}{\tau} \tag{96}$$

If we write the energy E in the form

$$E = \tfrac{1}{2}\hbar\omega_c + \hbar^2 k_z^2/2m_e = \tfrac{1}{2}\hbar\omega_c + E' \tag{97}$$

equation (96) may be written in the form

$$ev_z\mathscr{E}_z\frac{\partial f}{\partial E'} = \frac{f - f_{0B}}{\tau} \tag{98}$$

where the velocity v_z is given by $v_z = \hbar k_z/m_e$. As before, we may replace f by f_{0B} in the left-hand side of equation (98) and obtain

$$f = f_{0B} + e\tau v_z\mathscr{E}_z\frac{\partial f_{0B}}{\partial E'} \tag{99}$$

The function f_{0B} is given by

$$f_{0B} = \frac{1}{\mathrm{e}^{(E' + \frac{1}{2}\hbar\omega_c - E_F)/kT} + 1} \tag{100}$$

and, in the non-degenerate condition, may be replaced by

$$f_{0B} = A\,\mathrm{e}^{-E'/kT} = A\,\mathrm{e}^{-\hbar^2 k_z^2/2m_e kT} \tag{101}$$

The number of allowed states in the interval E' to $E' + \mathrm{d}E'$ is $\mathscr{C}E'^{-1/2}$ (see § 12.3.1) so that on comparison with the analysis given in § 10.6 we see that the current density J_z is given by

$$J_z = \frac{ne^2\mathscr{E}_z\bar{\tau}}{m_e} \tag{102}$$

where
$$\bar{\tau} = \int_0^\infty \tau(E')\,E'^{1/2}\mathrm{e}^{-E'/kT}\,\mathrm{d}E' \bigg/ \int_0^\infty E'^{1/2}\mathrm{e}^{-E'/kT}\,\mathrm{d}E' \tag{103}$$

We see that in calculating the average value of τ, due to the redistribution of the allowed energy values because of magnetic quantization, we replace the weighting function $E^{3/2}\exp(-E/kT)$ by $E^{1/2}\exp(-E/kT)$. We therefore see that unless τ is independent of the energy E, both with and without magnetic field, we have a change in conductivity in a longitudinal magnetic field. The prediction based on the quasi classical method of calculation that no such change takes place when the effective mass is scalar is therefore incorrect.

In order to calculate the value of the conductivity $\tau_{\infty L}$ in a large longitudinal magnetic field, given by

$$\sigma_{\infty L} = ne^2\bar{\tau}/m_e \tag{104}$$

we must know the variation of τ with E'. The average velocity of an electron whose quantized energy is in the zero-point mode is $(2E'/m_e)^{1/2}$, so it seems not unreasonable to assume that if $\tau = AE^{-s}$ in the absence of a magnetic field, then $\tau = AE'^{-s}$ in a strong field. If we make this assumption we find that

$$\bar{\tau} = A(\boldsymbol{k}T)^{-s}\Gamma(\tfrac{3}{2}-s)/\Gamma\tfrac{3}{2} \tag{105}$$

We therefore have

$$\frac{\sigma_{\infty L}}{\sigma_0} = \frac{\Gamma(\frac{3}{2}-s)\Gamma\frac{5}{2}}{\Gamma(\frac{5}{2}-s)\Gamma\frac{3}{2}} \tag{106}$$

and

$$\frac{\sigma_{\infty L}}{\sigma_0} = \frac{3}{3-2s} \tag{107}$$

If $s = \frac{1}{2}$ as for scattering by the acoustic modes then we have $\sigma_{\infty L}/\sigma_0 = \frac{3}{2}$. Thus we see that a longitudinal magnetic field *increases* the conductivity and we have a *negative* magneto-resistance coefficient. On the other hand, if $s \simeq -\frac{3}{2}$ as for ionised impurity scattering $\sigma_{\infty L} \gg \sigma_0$ and indeed the condition $\omega_c\tau \gg 1$ is not satisfied for small values of E'. We see, however, that the type of scattering has a marked effect on the high-field magneto-resistance.

Before leaving this subject we might make one further remark on the difficulty associated with an exact treatment. Although the wave equation based on the effective-mass approximation may be solved exactly when we have electric and magnetic fields at right angles, and the quantum-theoretical treatment referred to above is based on such a solution, we must recognise the limitation of even such a treatment. The quantity α which appears in the wave function of equation (115) of § 11.3.2 is equal to $(m_e\omega_c/\hbar)^{1/2}$, the extent of the wave function being of the order of $\alpha^{-1} = (\hbar/eB)^{1/2}$. We have also seen that the condition $\omega_c\tau \gg 1$ is equivalent to $B\mu_e \gg 1$ where μ_e is the electron mobility. For a semiconductor such as InSb, at low temperatures, it is found that $\mu_e \sim 10$ m²/V sec, so that with $B = 10$ Wb/m² (10^5 G) we have

$$B\mu_e = \omega_c\tau = 100$$

the condition $\omega_c\tau \gg 1$ may therefore readily be satisfied in this case. However, when $B = 10$ Wb/m² we have $\alpha^{-1} \simeq 10^{-6}$ cm. While such a distance includes quite a number of lattice spacings, an increase of B by a factor of 10 to 100 Wb/m², which would be required to give $\omega_c\tau \simeq 100$ for materials with mobilities of the order of 1 m²/V sec, would lead to a serious breakdown of the effective-mass approximation, since the wave function $F(\mathbf{r})$ could no longer be regarded as 'slowly varying'. In classical terms, the radius of the electron's orbit would then be of the order of a lattice spacing and the motion of the electron would be confined to the neighbourhood of a single lattice site.

13

Interaction of Electrons with Photons and Phonons

13.1 Interaction of electrons and positive holes in a crystal with the lattice vibrations

WE now come to the final phase of our treatment of the motion of electrons and positive holes in a real crystalline solid, based on the one-electron approximation. We have studied the motion of electrons in an ideal crystal in which we have assumed the atoms to be at rest at their lattice sites, and have discussed the vibrations of the atoms of the lattice, ignoring the effect of the very light conduction electrons. We must now take the approximation one stage further and consider in more detail the effect of the lattice vibrations on the motion of the electrons. We have already seen in § 10.9.2 how these vibrations lead to scattering of the electrons and are one cause of the frictional force which gives rise to electrical resistance. We shall be concerned only with the conduction electrons and with positive holes; there is also an interaction between the lattice vibrations and the electrons whose atomic energy levels are not too deep, giving rise to the thermal broadening of atomic spectral lines in solids, but we shall not discuss this effect.

Since the theory of the interaction of electrons with lattice vibrations, or phonons, is very similar to the theory of their interaction with light waves, or photons, we shall include in this Chapter a discussion of the latter as well. The study of the optical absorption due to excitation of electrons from the valence band to the conduction band of semiconductors has, in recent years, led to a much better understanding of their fundamental properties; absorption of infra-red radiation due to transitions of electrons and holes between different states within a band gives information on the density of allowed energy levels and has provided data on the form of the degenerate valence band of a number of important semiconductors. A study of absorption in the far infra-red due to transitions between different states of the lattice vibrations is beginning to give information on the lattice vibration spectrum (see § 6.11). We shall begin with a brief discussion of the quantum-mechanical formalism by means of which the transition probabilities, which determine the magnitude of these effects, may be determined.

13.2 Transition probabilities

The quantum theory on which the calculation of transition probabilities is based is given in full in textbooks on quantum mechanics*. We shall, however, give a brief discussion of those aspects of the theory which are particularly applicable to the problems we wish to discuss. The general problem may be expressed as follows: suppose we have a system described by a Hamiltonian H_0 and we have determined the stationary-state wave functions ψ_n and allowed values E_n of the energy E by solving the Schrödinger equation

$$H_0\psi = E\psi \tag{1}$$

so that the functions ψ_n satisfy the equations

$$H_0\psi_n = E_n\psi_n \tag{2}$$

Now suppose that, in addition to the forces included in the Hamiltonian H_0, we have additional forces acting on the system and described by an operator H_1; for example, H_1 may be the potential from which these forces may be derived but may also be a differential operator. We shall suppose that the magnitude of H_1 is small compared with the magnitude of H_0 so that H_1 may be regarded as a perturbation of the system. The effect of H_1 will be to render the states described by the wave functions ψ_n no longer stationary, and the energy E will no longer be exactly determined. Indeed, if at a certain time we know that the system is in a particular state ψ_m at a later time it will no longer be in this state but will require a wave function to describe it which contains all the wave functions ψ_n; in other words, we may say that transitions have been made from the state (m) to the other states (n). Our main problem is to calculate the probability per unit time of such transitions taking place under the influence of the forces described by H_1.

The starting point of this theory is the expansion of a wave function which satisfies the time-dependent wave equation

$$i\hbar\frac{\partial\Psi}{\partial t} = (H_0+H_1)\Psi \tag{3}$$

in a series of the form

$$\Psi = \sum a_n\psi_n e^{-iE_n t/\hbar} \tag{4}$$

We may note that the function $\Phi_n = \psi_n e^{-iE_n t/\hbar}$ satisfies the equation

$$i\hbar\frac{\partial\Phi_n}{\partial t} = H_0\Phi_n \tag{5}$$

so therefore does the wave function given by equation (4) for *all* values of a_n, if these are constant. They are in general, however, assumed to be functions of the time t and our problem is to find the particular functions

* See, for example, L. I. Schiff, *Quantum Mechanics* (2nd Ed.). McGraw-Hill (1955), Chapter 8.

for which the wave function (4) is a solution of equation (3). The expansion (4) is a generalisation of the type of expansion we have had previously, for example, the Fourier series. The requirement of the functions used for the expansion is that they form an orthogonal set and are 'complete', i.e. that they provide all the functions necessary; the set of eigenfunctions of an equation such as (1) may be shown to be such a 'complete' set and when multiplied by the time-dependent term as in (4) may be used for an expansion of this type. In many problems of quantum mechanics, for example, that of the hydrogen atom in free space, both discrete and continuous sets of allowed energy values are possible. We shall, however, be concerned with crystals of large but finite size and as we have seen all our allowed energy values are discrete, given for example by the periodic boundary conditions; we shall, therefore, assume that the sum in equation (4) is taken over a set of integral values of a summation index n, and in general we shall be concerned with a large but finite number of values of n. We shall frequently replace the sum by an integral, using the density of allowed values of E, but it should never be forgotten that this is an approximation and that we are really summing over a large number of discrete values.

We now substitute the wave function Ψ, given by equation (4), in equation (3); using equation (5) we obtain

$$i\hbar \sum \frac{\mathrm{d}a_n}{\mathrm{d}t} \psi_n \mathrm{e}^{-iE_n t/\hbar} = \sum a_n H_1 \psi_n \mathrm{e}^{-iE_n t/\hbar} \tag{6}$$

If we now multiply by $\psi_m^\star$ and integrate over the spatial coordinates, which we shall denote by the single symbol $\mathbf{r}$, we have, in view of the orthogonality of the functions ψ_n, which we also assume to be normalised,

$$i\hbar \frac{\mathrm{d}a_m}{\mathrm{d}t} = \sum a_n H_{mn} \mathrm{e}^{i(E_m - E_n)t/\hbar} \tag{7}$$

where the matrix element H_{nm} is given by

$$H_{mn} = \int \psi_m^\star H_1 \psi_n \, \mathrm{d}\mathbf{r} \tag{8}$$

So far, we have made no approximations and the equations in (7) are equivalent to the wave equation (3); they were first obtained by P. A. M. Dirac*.

To proceed further we make the assumption that H_1 is small and neglect products of H_1 with all the coefficients a_n except a_0. We assume that at $t = -\infty$ $a_n = 0$, $n \neq 0$ and $a_0 = 1$, i.e. that the system starts in the stationary state (0) with energy E_0. To this approximation we have what is usually called a first-order perturbation theory; higher orders of approximation may, of course, also be obtained. With this approximation we obtain from equation (7), writing $(E_m - E_n)/\hbar = \omega_{mn}$,

$$i\hbar \frac{\mathrm{d}a_m}{\mathrm{d}t} = a_0 H_{m0} \mathrm{e}^{i\omega_{m0} t} \tag{9}$$

* P. A. M. Dirac, *Proc. Roy. Soc.* **A112**, 661 (1926); *Ibid.*, **114**, 243 (1927).

If we write $a_0 = 1 + a_0'$, a_0' will clearly be of the same order of magnitude as a_n, $n \neq 0$, so that equation (9) may be written as

$$i\hbar \frac{\mathrm{d}a_m}{\mathrm{d}t} = H_{m0}\,\mathrm{e}^{i\omega_{m0}t} \tag{10}$$

We may now integrate equation (10) and obtain

$$a_m = -\frac{i}{\hbar}\int_{-\infty}^{t} H_{m0}\,\mathrm{e}^{i\omega_{m0}t}\,\mathrm{d}t, \quad m \neq 0 \tag{11}$$

13.2.1 Perturbation independent of the time

If the perturbation operator is constant, subsequent to a time which we shall take as $t = 0$, and zero for $t < 0$ we have for $t > 0$, from equation (11),

$$a_m(t) = -\frac{H_{m0}}{\hbar}\frac{\mathrm{e}^{i\omega_{m0}t} - 1}{\omega_{m0}}, \quad m \neq 0 \tag{12}$$

According to the general principles of quantum mechanics regarding the interpretation of the wave function Ψ, the probability of finding the system in state (m) at time t is

$$|a_m(t)|^2 = \frac{4|H_{m0}|^2 \sin^2(\frac{1}{2}\omega_{m0}t)}{\hbar^2\,\omega_{m0}^2} \tag{13}$$

This result is somewhat surprising at first sight since we might expect the probability to be proportional to t. The interpretation of equation (13) is, however, that under the perturbation the system makes transitions not only from state (0) to state (m) but also from state (m) to state (0) and the probability of occupation of state (m) varies periodically with t. In the conditions with which we shall be mainly concerned we have to consider not the transition to a single state but to a group of neighbouring states and we obtain a transition probability proportional to t when we average over these states. (When we are dealing with two isolated states as in microwave absorption, and highly monochromatic radiation, equation (13) is of great importance.)

13.2.2 Cross-section for scattering by an isolated impurity

Let us, for example, consider an electron in a crystal initially in a state (0) defined by a wave vector $\mathbf{k}_0$ and let us calculate, using the wave equation given by the effective-mass approximation (equation (7) of § 11.1), the probability of scattering by a potential $V(\mathbf{r})$ due to an isolated scattering centre. We shall assume for simplicity that we have a scalar effective mass m_e so that $E(\mathbf{k}') = \hbar^2 k'^2/2m_e$. We wish to calculate the probability that the electron will be scattered so that its wave vector will be changed from $\mathbf{k}_0$ to any one of a group of allowed values ending in a small volume in $\mathbf{k}$-space. This volume may be taken as the region between the two constant-energy surfaces corresponding to values E and $E+\mathrm{d}E$ of the energy, and bounded by the surface element $\mathrm{d}S$. To obtain the probability of scattering into a small solid angle subtended by $\mathrm{d}S$ we must integrate equation (13) over the corresponding volume of **k**-space, since the probabilities add. We may note that because of the denominator $\hbar\omega_{m0} = E(k')-E(k_0)$ the integrand peaks very sharply near $E = E(k_0)$. If $N(E)\,\mathrm{d}E\,\mathrm{d}S$ is the number of allowed states in the volume element, we may anticipate the final result by writing the probability as $\mathrm{d}Pt$ where $\mathrm{d}P$ is the probability per unit time. We have therefore, taking the state (m) in equation (13) as that corresponding to $\mathbf{k}'$

$$\mathrm{d}Pt = \frac{4}{\hbar^2}\int_{-\infty}^{\infty} \frac{N(E)\sin^2(\frac{1}{2}\omega_{k'0}t)|H_{k'0}|^2\,\mathrm{d}E\,\mathrm{d}S}{\omega_{k'0}^2} \tag{14}$$

where

$$\hbar\omega_{k'0} = E(k')-E(k_0) \tag{15}$$

We have extended the limits of integration to $\pm\infty$ in (14) since the integrand is very large when $\omega_{k0} \simeq 0$ and falls off rapidly when $E(k')$ differs very much from $E(k_0)$, i.e. unless $\mathbf{k}'$ lies very close to the constant-energy surface on which $\mathbf{k}_0$ lies. This is simply an expression of the conservation of energy. We note that it is not an exact expression unless $t \to \infty$; for finite values of t, the central peak of the function $\sin^2(\frac{1}{2}\omega_{k'0}t)/\omega_{k'0}^2$ has width ΔE of the order of $\hbar/t$ which is the uncertainty predicted by the Uncertainty Principle. When t is large enough to make the peak very narrow we take $N(E)|H_{k'0}|^2\,\mathrm{d}S$ outside the integral so that, writing $\mathrm{d}E = \hbar\,\mathrm{d}\omega_{k'0}$, we have

$$Pt = \frac{4N(E_0)|H_{k'0}|^2}{\hbar}\int_{-\infty}^{\infty}\frac{\sin^2(\frac{1}{2}\omega_{k'0}t)\,\mathrm{d}\omega_{k'0}}{\omega_{k'0}^2} \tag{16}$$

Now

$$\int_{-\infty}^{\infty}\frac{\sin^2 x\,\mathrm{d}x}{x^2} = \pi$$

so that the integral in (16) is equal to $\frac{1}{2}\pi t$. We now see that the total probability is indeed proportional to t, and we may write

$$\mathrm{d}P = \frac{2\pi}{\hbar}N(E_0)|H_{k'0}|^2\,\delta\{E(k_0)-E(k')\}\mathrm{d}S \tag{17}$$

The wave functions ψ_0 and ψ_k are in this case

$$\psi_0 = \frac{1}{V^{1/2}}\mathrm{e}^{i\mathbf{k}_0.\mathbf{r}} \tag{18}$$

$$\psi_k = \frac{1}{V^{1/2}}\mathrm{e}^{i\mathbf{k}'.\mathbf{r}} \tag{19}$$

where V is the volume of the crystal, so that

$$H_{k'0} = \frac{1}{V}\int V(\mathbf{r})\,e^{i\mathbf{K}.\mathbf{r}}\,d\mathbf{r} \tag{20}$$

where

$$\mathbf{K} = \mathbf{k}_0 - \mathbf{k}' \tag{21}$$

If we restrict the wave vector $\mathbf{k}'$ to end on a surface element dS corresponding to a small solid angle $\sin\theta\,d\theta\,d\phi$ we have (see § 8.9)

$$N(E)\,dE\,dS = \frac{V}{8\pi^3}k'^2\,dk'\sin\theta\,d\theta\,d\phi \tag{22}$$

$$= \frac{Vk'\,m_e}{8\pi^3\,\hbar^2}\sin\theta\,d\theta\,d\phi\,dE \tag{23}$$

so that, since $k' = k_0$,

$$dP = \frac{k_0\,m_e\,V}{4\pi^2\,\hbar^3}|H_{k0}|^2\sin\theta\,d\theta\,d\phi \tag{24}$$

If we introduce the partial cross-section $\sigma(\theta,\phi)$ we must remember that it gives the probability of scattering for unit particle flux. If the wave function ψ_0 is to represent unit flux it must be divided by $(v/V)^{1/2}$ where $v = \hbar k'/m_e$ is the velocity of the electron before and after scattering. To obtain $\sigma(\theta,\phi)\sin\theta\,d\theta\,d\phi$ we must therefore multiply P by $V/v = m_e V/\hbar k_0$ (cf. § 10.2 equation (18) with $N = 1/V$, since we are dealing with a single scattering centre). We therefore obtain

$$\sigma(\theta,\phi) = \left(\frac{2\pi m_e}{h^2}\right)^2\left|\int V(\mathbf{r})\,e^{i\mathbf{K}.\mathbf{r}}\,d\mathbf{r}\right|^2 \tag{25}$$

This is the well-known Born approximation for scattering in a potential $V(\mathbf{r})$. The conditions for the validity of the approximation have been extensively discussed in connection with the theory of atomic collisions*. Equation (25) has been widely used to calculate relaxation times due to impurity scattering; in particular it may be used to obtain the cross-section for a screened Coulomb potential as discussed in connection with scattering by ionised impurities in semiconductors in § 10.9.1. The modifications required when the effective mass is not scalar are indicated in the exercises.

Exercises (1) Show that the partial cross-section σ is given by

$$\sigma(\theta,\phi) = \frac{k_0^3}{4\pi^2\hbar(\mathbf{k}.\nabla_k E)_0\,v_0}\left|\int V(\mathbf{r})\,e^{i\mathbf{K}.\mathbf{r}}\,d\mathbf{r}\right|^2$$

where $\mathbf{K} = \mathbf{k}_0 - \mathbf{k}'$. (Use the result of § 9.4.3 equation (41) that $d\mathbf{k} = dE\,dS/|\nabla_k E|$ and that $v_0 = \hbar^{-1}|\nabla_{k_0}E|$; then replace dS by $k^2\sec\psi\sin\theta\,d\theta\,d\phi$ where ψ is the angle between the normal to the constant-energy surface and $\mathbf{k}$, so that

$$\cos\theta = (\mathbf{k}.\nabla_k E)/k|\nabla_k E|)$$

* See, for example, N. F. Mott and H. W. S. Massey, *Theory of Atomic Collisions* (2nd Ed.). Oxford University Press (1949).

(2) Show that if E is a quadratic function of the components of k'

$$\sigma(\theta,\phi) = \frac{k_0^3}{8\pi^2 E v_0 \hbar}\left|\int V(\mathbf{r})\,e^{i\mathbf{K}.\mathbf{r}}\,d\mathbf{r}\right|^2$$

(3) Show that if the constant energy surfaces are ellipsoids and $\sigma(\theta,\phi)$ is only appreciable when $\mathbf{k}$ makes a small angle with $\mathbf{k}_0$ then $\sigma(\theta,\phi)$ is given by equation (25) with m_e replaced by

$$\frac{(l^2+m^2+n^2)^{3/4}}{\left[\dfrac{l^2}{m_1}+\dfrac{m^2}{m_2}+\dfrac{n^2}{m_3}\right]^{1/2}\left[\dfrac{l^2}{m_1^2}+\dfrac{m^2}{m_2^2}+\dfrac{n^2}{m_3^2}\right]^{1/4}}$$

where m_1, m_2, m_3 are the effective masses and l, m, n are the direction cosines of $\mathbf{k}_0$ referred to the principal axes of the ellipsoids.

13.2.3 Periodic perturbation

Let us next consider a periodic perturbation which may be expressed in the form

$$H_1 = H_{10}(\mathbf{r})\cos\omega t \tag{26}$$

$$= \tfrac{1}{2}H_{10}(\mathbf{r})[e^{i\omega t}+e^{-i\omega t}] \tag{26a}$$

On substituting in equation (10) we obtain

$$i\hbar\frac{da_m}{dt} = H_{m0}[e^{i(\omega_{m0}+\omega)t}+e^{i(\omega_{m0}-\omega)t}] \tag{27}$$

where now

$$H_{m0} = \tfrac{1}{2}\int \psi_m^\star H_{10}(\mathbf{r})\,\psi_0\,d\mathbf{r} \tag{28}$$

On integrating equation (21) from $t = 0$ we obtain

$$a_m = -\frac{H_{m0}}{\hbar}\left[\frac{e^{i(\omega_{m0}+\omega)t}-1}{\omega_{m0}+\omega}+\frac{e^{i(\omega_{m0}-\omega)t}-1}{\omega_{m0}-\omega}\right] \tag{29}$$

We now see that a_m is only appreciable provided

$$\omega_{m0} \simeq \omega \tag{30}$$

or

$$\omega_{m0} \simeq -\omega \tag{31}$$

Equation (30) corresponds to the condition

$$E_m - E_0 \simeq \hbar\omega \tag{30a}$$

and equation (31) to the condition

$$E_0 - E_m \simeq \hbar\omega \tag{31a}$$

We shall suppose that $\omega > 0$ and consider first of all the case when $E_m > E_0$. In this case only the condition (30a) can be approximately satisfied and we may ignore the first term in the right-hand side of equation (29); if, on the other hand $E_m < E_0$ only the first term will give an appreciable contribution. For the moment we shall continue to assume that $E_m > E_0$; we have therefore

$$|a_m|^2 = \frac{4|H_{m0}|^2 \sin^2\{\frac{1}{2}(\omega_{m0}-\omega)t\}}{\hbar^2(\omega_{m0}-\omega)^2} \tag{32}$$

In this case we again do not get a probability proportional to t unless we integrate over a band of frequencies as in the theory of absorption of radiation by atoms*, or over a group of neighbouring states.

13.3 Absorption due to direct optical transitions between bands

An important application of the theory of transitions due to periodic perturbations is the calculation of the strength of the optical absorption due to excitation of electrons from one band to a higher band. The most striking example of such absorption is the marked absorption edge due to the optical excitation of electrons from the filled valence band to the conduction band in insulators and semiconductors; for the former this occurs in the visible or ultra-violet region of the spectrum and for the latter it usually occurs in the near infra-red. For such materials the optical absorption is small for frequencies ν such that $h\nu < E_m$ where E_m is approximately equal to the forbidden energy gap ΔE. As we shall see, in some cases E_m is not quite equal to ΔE, since under certain circumstances phonons as well as photons are involved in the transition. Generally speaking, however, when $h\nu$ is appreciably less than ΔE a photon has insufficient energy to raise an electron from the valence band to the conduction band and absorption is due to transitions *within* a band. In a nearly empty band there are comparatively few electrons to take part in such intra-band transitions and in a nearly full band there are comparatively few holes or vacant levels to which such transitions may take place. This absorption due to intra-band transitions, known as the free-carrier absorption is therefore weak compared with that due to the inter-band transitions, known as the fundamental absorption. The onset of this absorption when $h\nu \simeq \Delta E$ is known as the fundamental absorption edge and a study of the spectral form of this edge gives, as we shall see, a lot of valuable information about the form of the energy bands as well as the value of the important parameter ΔE.

We shall suppose that the top of the valence band is at $\mathbf{k} = 0$ and to begin with we shall assume it to be non-degenerate and specified by a scalar effective mass m_h so that in terms of the wave vector $\mathbf{k}'$ the energy is given by $E_v(\mathbf{k}')$ where

$$E_v(\mathbf{k}') = -\Delta E - \hbar^2 k'^2/2m_h \tag{33}$$

We shall suppose that the conduction band has a minimum, not necessarily the lowest minimum, at $\mathbf{k} = 0$ and that near this minimum we have the energy given by $E_c(\mathbf{k}'')$ where

$$E_c(\mathbf{k}'') = \hbar^2 k''^2/2m_e \tag{34}$$

* See, for example, L. I. Schiff, *Quantum Mechanics* (2nd Ed.). McGraw-Hill (1955), Chapter 10.

If we have a maximum in the energy of the conduction band at $\mathbf{k} = 0$ rather than a minimum we simply replace m_e by $-m_e$ in the subsequent analysis. We wish to calculate the probability of a transition from a state (0) corresponding to an electron in the valence band with wave vector $\mathbf{k}'$ and energy $E_v(\mathbf{k}')$ to a state (m) in the conduction band with wave vector $\mathbf{k}''$ and energy $E_c(\mathbf{k}'')$ under the influence of a light wave of frequency $\nu = \omega/2\pi$. Conservation of energy would therefore imply that

$$h\nu = \hbar\omega = E_c(\mathbf{k}'') - E_v(\mathbf{k}') \tag{35}$$

We shall not assume this relationship to be true to begin with but shall deduce it as a consequence of our study of the transition probability. Such a transition creates an electron with wave vector $\mathbf{k}''$ and velocity $\hbar^{-1}\nabla_{k''} E_c(\mathbf{k}'') = \hbar\mathbf{k}''/m_e$ in the conduction band, and a hole with wave vector $\mathbf{k}'$ and velocity $\hbar^{-1}\nabla_{k'} E_v(\mathbf{k}') = -\hbar\mathbf{k}'/m_h$ in the valence band.

The light wave is most conveniently represented by means of a vector potential $\mathbf{A}$; the electric and magnetic vectors $\mathscr{E}$ and $\mathscr{H}$ are given by* (M.K.S. units)

$$\mathscr{E} = -\frac{\partial \mathbf{A}}{\partial t} \tag{36}$$

$$\mathscr{H} = \mu^{-1}\nabla \times \mathbf{A} \tag{37}$$

where μ is the permeability, provided we have

$$\nabla . \mathbf{A} = 0 \tag{38}$$

The energy flux is given by the Poynting vector $\mathbf{S} = \mathscr{E} \times \mathscr{H}$ and the quantum flux F_q, equal to the number of quanta crossing unit area per unit time, for nearly monochromatic radiation, by $|\mathbf{S}|/\hbar\omega$. The vector potential $\mathbf{A}$ may be expressed in the form

$$\mathbf{A} = A\mathbf{a}_0 \cos(\mathbf{K}.\mathbf{r} - \omega t) \tag{39}$$

$$= \tfrac{1}{2}A\mathbf{a}_0 e^{i(\mathbf{K}.\mathbf{r}-\omega t)} + \tfrac{1}{2}A\mathbf{a}_0 e^{-i(\mathbf{K}.\mathbf{r}-\omega t)} \tag{39a}$$

where $\mathbf{a}_0$ is a unit polarisation vector and $\mathbf{K}$ is the wave vector giving the direction of propagation of a plane-polarised light wave, so that

$$\mathbf{a}_0 . \mathbf{K} = 0 \tag{40}$$

The magnitude K of $\mathbf{K}$ is given by

$$K = \omega/\mathscr{V} \tag{41}$$

where $\mathscr{V}$ is the velocity of light in the crystal; $\mathscr{V}$ may be expressed as $(\epsilon\mu)^{-1/2}$, where ϵ is the permittivity, or as c/n where c is the velocity of light in free space and n the refractive index. It will be seen that

$$\mathscr{E} = A\omega\mathbf{a}_0 \sin(\mathbf{K}.\mathbf{r} - \omega t) \tag{42}$$

$$\mathscr{H} = -\mu^{-1}A(\mathbf{K} \times \mathbf{a}_0)\sin(\mathbf{K}.\mathbf{r} - \omega t) \tag{43}$$

* J. A. Stratton, *Electromagnetic Theory*. McGraw-Hill (1941), p. 27.

and that $\bar{S}$ the average value of S over a period is given by

$$\bar{S} = \tfrac{1}{2}K\omega A^2/\mu \tag{44}$$

From equation (41) we see that $\bar{S}$ may also be expressed as

$$\bar{S} = \frac{1}{2}\sqrt{\left(\frac{\epsilon}{\mu}\right)}\,\omega^2 A^2 = \tfrac{1}{2}n\epsilon_0\,\omega^2 A^2 c \tag{45}$$

if we assume that $\mu = \mu_0$, the value for free space, so that $n^2 = \epsilon/\epsilon_0$.

The perturbation Hamiltonian H_1 in this case is a differential operator given by

$$H_1 = \frac{ie\hbar}{2m} A\, e^{i(\mathbf{K}.\mathbf{r}-\omega t)}\, \mathbf{a}_0.\nabla + c.c. \tag{46}$$

since we may neglect terms in A^2 in the wave equation for motion in a magnetic field, the fields in the light wave being quite small (see § 11.3.2 equation (109)). Here we have, of course, m and not m_e since we are concerned with the true wave equation and not the effective-mass approximation. The wave function ψ_0 corresponding to the state (0) is the Bloch function

$$\psi_0 = N^{-1/2}\, u_v(\mathbf{r}, \mathbf{k}')\, e^{i\mathbf{k}'.\mathbf{r}} \tag{47}$$

corresponding to the valence band, while the wave function ψ_m for the excited state (m) is the Bloch function

$$\psi_m = N^{-1/2}\, u_c(\mathbf{r}, \mathbf{k}'')\, e^{i\mathbf{k}''.\mathbf{r}} \tag{48}$$

The factor $N^{-1/2}$, where N is the number of unit cells in the crystal, is introduced as a normalising constant so that

$$\int_C u_v^\star u_v\, d\mathbf{r} = \int_C u_c^\star u_c\, d\mathbf{r} = 1 \tag{49}$$

where C indicates that the integrals are taken over a unit cell; these conditions indicate that ψ_0 and ψ_m are normalised for the volume V of the crystal.

The matrix element of the perturbation H_{m0} as defined in equation (28) is therefore given by

$$H_{m0} = \frac{ie\hbar A}{2Nm}\int u_c^\star(\mathbf{r}, \mathbf{k}'')\, e^{-i\mathbf{k}''.\mathbf{r}}(e^{i\mathbf{K}.\mathbf{r}}\, \mathbf{a}_0.\nabla)\, u_v(\mathbf{r}, \mathbf{k}')\, e^{i\mathbf{k}'.\mathbf{r}}\, d\mathbf{r} \tag{50}$$

since only the term in $e^{-i\omega t}$ contributes to H_{m0} (see § 13.2.3). Using equation (40), H_{m0} may be expressed in the form

$$H_{m0} = \frac{ie\hbar A}{2Nm}\int u_c^\star(\mathbf{r}, \mathbf{k}'')[\mathbf{a}_0.\nabla u_v(\mathbf{r}, \mathbf{k}') + i(\mathbf{a}_0.\mathbf{k}')\, u_v(\mathbf{r}, \mathbf{k}')]\, e^{i(\mathbf{k}'+\mathbf{K}-\mathbf{k}'').\mathbf{r}}\, d\mathbf{r} \tag{51}$$

Because of the periodicity of functions u_v, u_c the integral in (51) may be expressed as a sum over the unit cells of the crystal of the form

$$H_{m0} = \int_C \phi(\mathbf{r}, \mathbf{k}', \mathbf{k}'')\, d\mathbf{r} \sum_j e^{i(\mathbf{k}+\mathbf{K}-\mathbf{k}'').\mathbf{R}_j} \tag{52}$$

where $\mathbf{R}_j$ is the lattice vector which determines the jth cell. The sum in equation (52) is equal to zero unless either

$$\mathbf{k}'+\mathbf{K}-\mathbf{k}'' = 0 \tag{53}$$

or

$$\mathbf{k}'+\mathbf{K}-\mathbf{k}'' = 2\pi\mathbf{b}_{j'} \tag{54}$$

where $\mathbf{b}_{j'}$ is a vector of the reciprocal lattice, for in this case

$$(\mathbf{k}'+\mathbf{K}-\mathbf{k}'')\,\mathbf{R}_j = 2\pi(\mathbf{b}_{j'}.\mathbf{R}_j) = 2\pi s$$

where s is an integer (cf. § 11.2.1 equation (21)). Since the wavelength of light is very long compared with the lattice spacing, K is very small and we need only consider condition given by equation (53). We therefore have

$$\mathbf{k}''-\mathbf{k}' = \mathbf{K} \tag{55}$$

If we multiply by $\hbar$ this may be expressed in terms of the crystal momenta $\mathbf{P}'$, $\mathbf{P}''$ as

$$\mathbf{P}''-\mathbf{P}' = \mathbf{p}_p \tag{56}$$

where $\mathbf{p}_p = \hbar\mathbf{K}$ is the momentum of a photon of frequency ν in the crystal. Equation (55) or (56) therefore is an expression of the conservation of crystal momentum and is another justification of the term. Since the wavelength for an electron whose mean energy is of the order of 0·04 eV, for $T = 300°$K, is of the order of 6×10^{-7} cm, and for near infra-red radiation the wavelength is about 10^{-4} cm, we see that $K \ll k'$, or k'' and equation (55) may be approximately expressed as

$$\mathbf{k}'' \simeq \mathbf{k}' \tag{57}$$

Transitions of this kind may therefore only be produced when the wave vector of the excited electron is approximately equal to the wave vector of the electron before excitation, and are usually referred to as 'vertical' transitions or 'direct' transitions (see Fig. 13.1). If this condition is not satisfied the transitions are very strongly forbidden and cannot be produced by the absorption of a photon only; transitions between the top of the valence band at $\mathbf{k} = 0$ and minima in the conduction band *not* at $\mathbf{k} = 0$ cannot be 'vertical' transitions and cannot occur by this direct process. As we shall see later, they require the assistance of a phonon to conserve the crystal momentum. Transitions from states near the top of the valence band at $\mathbf{k} = 0$ to states near a maximum or minimum in the conduction band also at $\mathbf{k} = 0$ can, however, take place as direct transitions, and we may calculate the absorption coefficient by assuming that $\mathbf{k}'' = \mathbf{k}'$.

When the condition (55) is satisfied, the sum in (52) is equal to N; also putting $\mathbf{k}'' = \mathbf{k}'$ in the non-exponential terms of equation (51) we have

$$H_{m0} = \frac{ie\hbar A}{2m}\int_C u_c^\star(\mathbf{k}')[\mathbf{a}_0.\nabla u_v(\mathbf{r},\mathbf{k}')+i(\mathbf{a}_0.\mathbf{k}')\,u_v(\mathbf{r},\mathbf{k}')]\,\mathrm{d}\mathbf{r} \tag{58}$$

Unless the first term in the integral of equation (58) happens to be very small, corresponding to what is known as a 'forbidden' transition, it will generally be much greater (see below) than the second for small values of k'. For the moment we shall omit the second term and write the first part of the integral in terms of the matrix element of the momentum operator $\mathbf{p} = -i\hbar\nabla$, given by

$$\mathbf{p}_{m0} = -i\hbar \int_C u_c^\star(\mathbf{k}')\nabla u_v(\mathbf{k}')\,d\mathbf{r} \tag{59}$$

Expressing H_{m0} in terms of $\mathbf{p}_{m0}$ we have

$$H_{m0} = -\frac{eA}{2m}\mathbf{a}_0 \cdot \mathbf{p}_{m0} \tag{60}$$

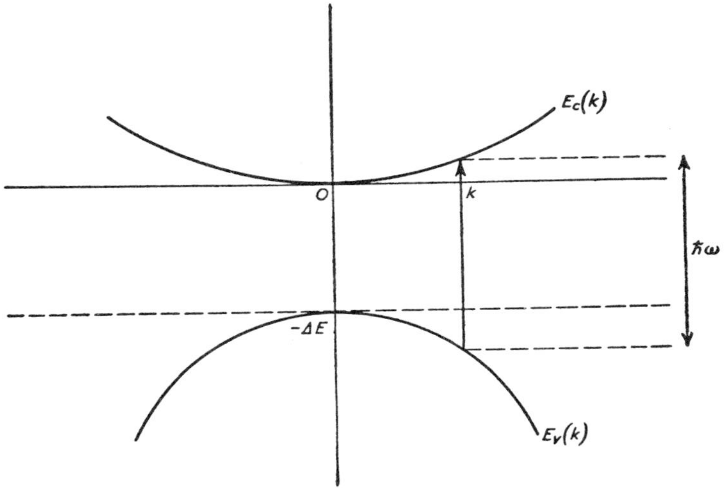

FIG. 13.1. Direct or 'vertical' inter-band transition

Inserting this value in equation (32) we obtain for the transition probability $|a_m|^2$,

$$|a_m|^2 = \frac{e^2 A^2 |\mathbf{a}_0 \cdot \mathbf{p}_{m0}|^2 \sin^2\{\frac{1}{2}(\omega_{m0}-\omega)t\}}{m^2 \hbar^2(\omega_{m0}-\omega)^2} \tag{61}$$

where

$$\hbar\omega_{m0} = E_c(k') - E_v(k') \tag{62}$$

$$= \Delta E + \hbar^2 k'^2/2m_r \tag{62a}$$

and m_r is the reduced effective mass of the electron and hole produced by the transition, given by

$$m_r = m_e m_h/(m_e + m_h) \tag{63}$$

To obtain the total probability of a transition being produced, two alternative procedures are now open to us. We may assume the incident radiation to be strictly monochromatic and sum over all allowed values of $\mathbf{k}'$ or we may assume that we have a continuous spectrum of radiation and determine the probability of transition for a fixed value of $\mathbf{k}'$. In either case we find a probability proportional to t and which varies slowly with the remaining parameter $\boldsymbol{\omega}$ or $\mathbf{k}'$; we shall adopt the former procedure. The transition probability P for monochromatic radiation of frequency $\nu = \omega/2\pi$ is then obtained by summing the expression for $|a_m|^2$ over all allowed values of $\mathbf{k}'$. We shall first of all assume that the valence band is fully occupied and that the density of states (including both values of spin) from which transitions may take place is given as usual by $V/4\pi^3$ where V is the volume of the crystal, replacing the sum by an integral. We therefore have for the transition probability $P(\omega)$ per unit volume and per unit time (we shall verify that $P(\omega)$ is independent of t)

$$P(\omega)t = \frac{\mathbf{e}^2 A^2}{4\pi^3 \boldsymbol{m}^2 \hbar^2} \int \frac{|\mathbf{a}_0 \cdot \mathbf{p}_{m0}|^2 \sin^2\{\frac{1}{2}(\omega_{m0}-\omega)t\}}{(\omega_{m0}-\omega)^2} \mathrm{d}\mathbf{k}' \tag{64}$$

Since ω_{m0} is a function of k' we may perform the angular integration and write

$$4\pi\overline{p_{m0}^2} = \int |\mathbf{a}_0 \cdot \mathbf{p}_{m0}|^2 \mathrm{d}\Omega \tag{65}$$

where $\mathrm{d}\Omega$ is a small solid angle. Thus we have

$$P(\omega)t = \frac{\mathbf{e}^2 A^2}{\pi^2 \boldsymbol{m}^2 \hbar^2} \int \frac{\overline{p_{m0}^2} \sin^2\{\frac{1}{2}(\omega_{m0}-\omega)t\} k'^2 \mathrm{d}k'}{(\omega_{m0}-\omega)^2} \tag{66}$$

As in § 13.2.1 we see that for sufficiently large values of t the integrand is very large only when $\omega_{m0} \simeq \omega$. We may therefore take $k'\overline{p_{m0}^2}$ outside the integral assuming that $k' = k_0$, where k_0 is given by the condition $\omega_{m0} \simeq \omega$, and then evaluate the integral by using a new variable x given by $\frac{1}{2}(\omega_{m0}-\omega)t = x$ and extending the limits of integration to $\pm\infty$. From equation (62a) we have

$$\mathrm{d}x = \hbar t k' \mathrm{d}k'/2m_r \tag{67}$$

so that

$$P(\omega) = \frac{\mathbf{e}^2 A^2 k_0 \overline{p_{m0}^2} m_r}{2\pi^2 \boldsymbol{m}^2 \hbar^3} \int_{-\infty}^{\infty} \frac{\sin^2 x \, \mathrm{d}x}{x^2}$$

$$= \frac{\mathbf{e}^2 A^2 (2m_r)^{3/2} \overline{p_{m0}^2}}{4\pi \boldsymbol{m}^2 \hbar^4} (\hbar\omega - \Delta E)^{1/2} \tag{68}$$

(cf. equation (16)).

We must now relate the transition probability $P(\omega)$ to the absorption coefficient. The number of quanta absorbed per unit time in a thin strip of unit area and thickness δ normal to the light flux is equal to $P(\omega)\delta$, since $P(\omega)$ refers to unit volume; it is also equal to $\alpha F_q \delta$ where F_q is the quantum flux given by $\bar{S}/\hbar\omega$ (equation (45)) and α is the absorption coefficient. We therefore have

$$\alpha = P(\omega)/F_q = \hbar\omega P(\omega)/\bar{S} \tag{69}$$

Expressing the energy flux $\bar{S}$ in terms of the vector potential by means of equation (45) we obtain for α_d, the absorption coefficient for direct transitions

$$\alpha_d = \frac{2\hbar P(\omega)}{\omega n A^2 \epsilon_0 \boldsymbol{c}} \tag{70}$$

Inserting the value of $P(\omega)$ given by equation (68) we obtain, when $h\nu \geqslant \Delta E$,

$$\alpha_d = \frac{\boldsymbol{e}^2(2m_r)^{3/2}\overline{p_{m0}^2}(\hbar\omega - \Delta E)^{1/2}}{2\pi n\omega \boldsymbol{m}^2 \hbar^3 \epsilon_0 \boldsymbol{c}} \tag{71}$$

α_d, is of course zero when $\hbar\omega < \Delta E$. We now introduce a dimensionless quantity f, the average value of a quantity f_{m0} related to the matrix element of the momentum operator p_{m0} by means of the equation

$$f_{m0} = 2|p_{m0}|^2/\boldsymbol{m}\hbar\omega \tag{72}$$

The quantity f_{m0} is known as the 'oscillator strength' and is much used in the quantum theory of dispersion. We may write equation (71) using f, the value of f_{m0} averaged over various directions of the polarisation vector $\mathbf{a}_0$, as

$$\alpha_d = \frac{\boldsymbol{e}^2(2m_r)^{3/2} f(\hbar\omega - \Delta E)^{1/2}}{4\pi n \boldsymbol{m}\hbar^2 \epsilon_0 \boldsymbol{c}} \tag{73}$$

This is equivalent to an expression given by J. Bardeen, F. J. Blatt and L. H. Hall* when the factor $4\pi\epsilon_0$ is removed from the denominator. This factor occurs through our use of M.K.S. units (a factor π given in the denominator of equation (6) of the paper by Bardeen, Blatt and Hall would appear to have been included in error).

To determine the value of the oscillator strength f_{m0} we require to know the wave functions for both bands for small values of $\mathbf{k}$. The use of plane waves gives a poor approximation since in this case f_{m0} is equal to zero. We may estimate the order of magnitude of f_{m0}, however, from a relationship known as the f-sum rule. It may be shown that if we sum the oscillator strengths over transitions from the state (0) to the states with the same value of $\mathbf{k}$ in all the possible bands we have

$$\sum_{m\neq 0} f_{m0}^x = 1 - \frac{\boldsymbol{m}}{\hbar^2}\frac{\partial^2 E}{\partial k_x^2} \tag{74}$$

$$= 1 + \boldsymbol{m}/m_h \tag{75}$$

A proof of this result is given in Appendix 2. For the high-energy bands, f_{m0} will be small both because of the large value of $|E_m - E_0|$ (see Appendix

* J. Bardeen, F. J. Blatt and L. H. Hall, *Atlantic City Photoconductivity Conference* (1954) *Report*. John Wiley, New York, and Chapman & Hall, London (1956), p. 146.

2), and also because the wave functions may be represented by plane waves. For transitions to the lower bands, f_{m0} will be small if the valence band is broad again because $|E_m - E_0|$ will be large. We might therefore expect f_{m0} to be a good deal larger for the transition to the nearest band than for transitions to any of the others unless the wave functions $u_v(\mathbf{r})$ and $u_c(\mathbf{r})$ are such that the matrix element is very small. In this case we have what is known as a forbidden transition in the usual sense, as in atomic transitions. When the transition is not forbidden we shall then have as regards order of magnitude

$$f_{m0} \sim 1 + \boldsymbol{m}/m_h \tag{76}$$

If the hole mass m_h is very small f_{m0} may be quite large but in general will be of the order of magnitude of unity. Assuming that this is so, we may calculate the magnitude of the absorption coefficient. Equation (73) may be put in numerical form as

$$\alpha_d = 2{\cdot}7 \times 10^5 n^{-1}(2m_r/\boldsymbol{m})^{3/2} f(\boldsymbol{h}\nu - \Delta E)^{1/2}\,\text{cm}^{-1} \tag{77}$$

if $\boldsymbol{h}\nu$ and ΔE are expressed in eV. If we take $2m_r = \boldsymbol{m}$ and $n = 4$ we have, assuming $f \sim 1$, $\alpha_d = 6{\cdot}7 \times 10^3$ cm^{-1} when $\boldsymbol{h}\nu - \Delta E = 0{\cdot}01$ eV. Absorption due to direct transitions from the valence band to the conduction band of semiconductors due to such direct transitions have been observed. Of particular interest is the direct transition in Ge which has been carefully studied by G. G. Macfarlane, T. P. McLean, J. E. Quarrington and V. Roberts* who found absorption of the magnitude we have calculated. The variation of absorption as $(\boldsymbol{h}\nu - \Delta E)^{1/2}$ is frequently taken as a criterion for a direct allowed transition. It was shown, however, by the above authors that equation (73) does not hold for very small values of $(\boldsymbol{h}\nu - \Delta E)$. Instead of tending to zero as $\boldsymbol{h}\nu \to \Delta E$, the absorption coefficient tends to a constant value and merges into an absorption spectrum due to formation of excitons (see § 9.3.7). The failure of equation (73) is due to the neglect of the Coulomb interaction between the hole and electron formed through the transition. When this is taken into account, it is found that excellent agreement is obtained with experiment, and that equation (73) holds when $\boldsymbol{h}\nu - \Delta E$ is greater than 10 W_{ex} where W_{ex} is the exciton binding energy. The Coulomb attraction between the hole and electron is, of course, also the cause of the formation of the exciton as discussed in § 9.3.7.

When the matrix element p_{m0} is zero we have a forbidden transition at $\mathbf{k} = 0$ and must now take into account the second term on the right-hand side of equation (58). We then find that H_{m0} is given by

$$H_{m0} = -\frac{\boldsymbol{e}\hbar A}{2\boldsymbol{m}}(\mathbf{a}_0 . \mathbf{k}') \int_C u_c^\star(\mathbf{k}'')\, u_v(\mathbf{k}')\, \mathrm{d}\mathbf{r} \tag{78}$$

* G. G. Macfarlane, T. P. McLean, J. E. Quarrington and V. Roberts, *Proc. Phys. Soc.*, **71**, 863 (1958).

If $\mathbf{k}'$ were strictly equal to $\mathbf{k}''$, H_{m0} would be equal to zero because of the orthogonality of the Bloch wave functions for different bands. If we write

$$f' = \left| \int_C u_c(\mathbf{k}'')\, u_v(\mathbf{k}')\, d\mathbf{r} \right|^2 \tag{79}$$

f' will be a number which we should expect to be somewhat less than 1. If $(a_0 . \mathbf{k}) = \cos\theta$ and we take the average value one-third of $\cos^2\theta$ we obtain instead of equation (68)

$$P(\omega) = \frac{e^2 A^2 (2m_r)^{3/2} f' k'^2 (\hbar\omega - \Delta E)^{1/2}}{12\pi m^2 \hbar^2} \tag{80}$$

$$= \frac{e^2 A^2 (2m_r)^{5/2} f' (\hbar\omega - \Delta E)^{3/2}}{12\pi m^2 \hbar^4} \tag{80a}$$

The absorption coefficient α_d' is therefore given by

$$\alpha_d' = \frac{e^2 (2m_r)^{5/2} f'}{6\pi n c m^2 \hbar^2 \epsilon_0} \left\{ \frac{(\hbar\omega - \Delta E)^{3/2}}{\hbar\omega} \right\} \tag{81}$$

when $\hbar\omega \geqslant \Delta E$. This is equivalent to the expression for α_d' given by Bardeen, Blatt and Hall (their equation (8)) when we take account of the factor $4\pi\epsilon_0$ in the denominator of equation (81) through the use of M.K.S. units. Equation (81) may be written in numerical form as

$$\alpha_d' = 1{\cdot}8 \times 10^5 n^{-1} f' \left(\frac{2m_r}{m} \right)^{5/2} \frac{(\hbar\omega - \Delta E)^{3/2}}{\hbar\omega} \text{ cm}^{-1} \tag{82}$$

if $\hbar\omega$ and ΔE are expressed in eV. We therefore see that for a forbidden transition α_d varies as $(h\nu - \Delta E)^{3/2}$; this is also modified for small values of $(h\nu - \Delta E)$ by the Coulomb attraction between the hole and electron but the modification is less noticeable because of the extra factor $(\hbar\omega - \Delta E)$ in α_d.

The magnitude of the quantity f' is difficult to estimate without a knowledge of the wave functions near $\mathbf{k} = 0$; it may readily be seen, however, that it is likely to be somewhat less than 1. If we take plane waves for the wave functions ψ_0 and ψ_m, so that u_c and u_v are constant, the normalising condition shows that $f'_{m0} = 1$. The functions ψ_0 and ψ_m are then not orthogonal and this is likely to lead to a serious overestimate of f', just as it leads to a serious underestimate of f_{m0}. If we assume that $u_v(\mathbf{r}, \mathbf{k})$ and $u_c(r, \mathbf{k})$ are strictly orthogonal we have, since $\mathbf{k}'' = \mathbf{k}' + \mathbf{K}$,

$$\int u_c^\star(\mathbf{r}, \mathbf{k}'')\, u_v(\mathbf{r}, \mathbf{k}')\, d\mathbf{r} = \int u_v(\mathbf{r}, \mathbf{k}')[1 + \mathbf{K} . \nabla_{k'} + \tfrac{1}{2}(\mathbf{K} . \nabla_{k'})^2 + \ldots]\, u_c^\star(\mathbf{r}, \mathbf{k}')\, d\mathbf{r} \tag{83}$$

The first term in the right-hand side of equation (83) is zero because of the orthogonality of $u_v(r, \mathbf{k}')$ and $u_c(r, k')$; the second term is also zero if the transition is forbidden since it may be shown to be proportional to p_{m0} (Exercise 6 of § 8.8); the third term gives a contribution proportional to

K^2 and so is likely also to be small. If we assume that $f' = 0{\cdot}1$ and $2m_r = \boldsymbol{m}$ we obtain for $\hbar\omega = 1$ eV and $\hbar\omega - \Delta E = 0{\cdot}01$ eV, $\alpha_d = 4{\cdot}5\ \mathrm{cm}^{-1}$; this should be compared with the value $6{\cdot}7 \times 10^3\ \mathrm{cm}^{-1}$ for a direct transition.

When the valence band is degenerate, we have to add the absorption due to the various bands of which it is composed. The formulae for the absorption coefficient also must be modified when the effective mass is not scalar; the most general case is very complex but the results are not significantly different from those we have obtained, the reduced mass m_r being replaced by an averaged effective mass. The rigid selection rule $\mathbf{k}'' = \mathbf{k}' + \mathbf{K}$ is of course not affected, since the form of the bands does not affect the calculation till this result has been established. When the valence band may be represented by a scalar effective mass, and the conduction band has ellipsoidal constant-energy surfaces, the calculation may be readily carried out (see exercise).

In our calculation we have assumed that all the states in the valence band are occupied and that all the accessible states in the conduction band are empty. If this is not so, to an appreciable extent, the absorption coefficient will be modified. Provided we have $\Delta E \gg \boldsymbol{k}T$ either the states in the conduction band or the valence band may be affected, but not both, and the effect will only be marked when we have a strongly degenerate condition. Suppose, for example, that we have a degenerate n-type semiconductor, the Fermi level E_F being at height η *above* the bottom of the conduction band. The energy of the level in the conduction band to which a transition is made, above the bottom of the band, is $m_r[\hbar\omega - \Delta E]/m_e$ and the probability that this level is unoccupied is

$$\frac{1}{1+\exp\{-[m_r(\hbar\omega-\Delta E)-m_e\eta]/m_e\boldsymbol{k}T\}} \tag{84}$$

and the absorption coefficient will be multiplied by this factor. Thus we see that the absorption sets in strongly when

$$\hbar\omega = \Delta E + m_e\eta/m_r \tag{85}$$

instead of when $\hbar\omega = \Delta E$. Similarly, it may be shown that if the Fermi level is at depth η *below* the top of the valence band the absorption coefficient will be multiplied by the factor

$$\frac{1}{1+\exp\{-[m_r(\hbar\omega-\Delta E)-m_h\eta]/m_h\boldsymbol{k}T\}} \tag{86}$$

and the absorption starts strongly when

$$\hbar\omega = \Delta E + m_h\eta/m_r \tag{87}$$

This effect, originally observed for InSb by M. Tanenbaum and H. Briggs* has been discussed by E. Burstein† and by T. S. Moss‡, and has now been

* M. Tanenbaum and H. Briggs, *Phys. Rev.*, **91**, 1561 (1953).
† E. Burstein, *Ibid.*, **93**, 623 (1954).
‡ T. S. Moss, *Proc. Phys. Soc.*, B**67**, 775 (1954).

observed in a number of semiconductors for which the top of the valence band and bottom of the conduction band occur at $\mathbf{k} = 0$, and has been used to estimate the values of the effective masses. The effect is clearly most marked when one of the effective masses is small so that degeneracy may be achieved with not too high a concentration of free electrons or holes.

When the effect of degeneracy is negligible, the equations (73) and (81) show that the absorption coefficient for direct inter-band transitions does not vary with temperature except through the variation of the forbidden energy gap ΔE. The effect of the variation of ΔE is simply to shift the absorption curves along the $\hbar\omega$ axis. It is found that except at low temperatures ΔE may be expressed in the form

$$\Delta E = \Delta E(T_0) + \beta(T - T_0) \tag{88}$$

over quite a large range of temperatures. β is a constant whose magnitude varies from one substance to another but is usually of the order of 10^{-4} eV/°C; the sign of β may be either positive or negative. For small values of T, ΔE varies quadratically with T and may be expressed in the form

$$\Delta E = \Delta E_0 + \alpha T^2 \qquad (89$$

The form of the absorption spectrum for direct inter-band transitions will also be modified when $\hbar\omega$ is so large that the initial and final states occur for values of $\mathbf{k}$ for which the quadratic approximation to the energy no longer holds. In general the absorption will increase less rapidly when this condition is reached and may reach a saturation value of the order of 10^5 cm^{-1} for an allowed transition. It is beyond the scope of this section to discuss in detail the application of infra-red absorption spectroscopy to the study of semiconductors, which, in recent years, has yielded a great deal of information on their properties and is discussed in detail in books on semiconductors*; the absorption of light by metals is discussed in some detail by N. F. Mott and H. Jones†.

The theory of direct transitions may also be applied to transitions between two overlapping bands, such as the valence bands of Ge, and has yielded useful information on these. It cannot be applied to discuss transitions between states in the same band since the selection rule $\mathbf{k}'' = \mathbf{k}' + \mathbf{K}$ makes such transitions forbidden. That such transitions do take place is evidenced by the fact that the free electrons in the conduction band and holes in the valence band cause absorption in the infra-red. The transitions involved require absorption and emission of phonons in order to overcome the selection rule $\mathbf{k}'' = \mathbf{k}' + \mathbf{K}$. These intra-band transitions may be studied by means of classical electromagnetic theory, the

* See, for example, R. A. Smith, *Semiconductors*. Cambridge University Press (1959), Chapter 7.

† M. F. Mott and H. Jones, *The Theory of the Properties of Metals and Alloys*. Oxford University Press (1936), Chapter 3, §§ 7, 8.

absorption being related to the conductivity (see R. A. Smith, *loc. cit.*, Chapter 7). Transitions between the top of the valence band at $\mathbf{k} = 0$ and minima in the conduction band not at $\mathbf{k} = 0$ are also forbidden by the conservation of crystal momentum unless a phonon is either absorbed or emitted. Such transitions correspond to the lowest inter-band transitions in semiconductors such as Si and Ge and give rise to a continuous absorption spectrum which occurs for smaller values of $\hbar\omega$ than the direct transitions. Before discussing transitions in which both electrons and phonons change their states under the influence of a light wave we must discuss transitions in which electrons change their states under the influence of acoustic waves or phonons.

Exercise (1) Show that if the valence band may be represented by a scalar effective mass m_h and the conduction band has ellipsoidal constant-energy surfaces, the effective masses for the principal directions being m_1, m_2, m_3, the factor mass $(2m_r)^{3/2}$ in equation (73) for the absorption coefficient due to direct allowed transitions is replaced by $2^{3/2}m_{r1}^{1/2}m_{r2}^{1/2}m_{r3}^{1/2}$ where

$$\frac{1}{m_{r1}} = \frac{1}{m_1}+\frac{1}{m_h}$$

$$\frac{1}{m_{r2}} = \frac{1}{m_2}+\frac{1}{m_h}$$

$$\frac{1}{m_{r3}} = \frac{1}{m_3}+\frac{1}{m_h}$$

(In performing the integration over allowed values of k' make the transformation $k_x^\star = (m/m_{r1})^{1/2}k_x'$, $k_y^\star = (m/m_{r2})k_y'$, $k_z^\star = (m/m_{r3})^{1/2}k_z'$, so that

$$h\omega_0 = \Delta E+\hbar^2 k^{\star 2}/2m$$

then perform the integration with respect to $k^\star$.)

13.4 Scattering of electrons by lattice vibrations

The analysis which we have used in § 13.2.3 to obtain the transition probabilities for a periodic perturbation may be used to calculate the probabilities of transitions between states having different wave vectors, induced by the lattice vibrations; from these transition probabilities we may then calculate the partial cross section $\sigma(\theta,\phi)$ for scattering and the relaxation time τ. A lattice vibrational mode of frequency $\nu_s = \omega_s/2\pi$ may, as we have seen in § 6.4, be represented by

$$\mathbf{X}_j = \tfrac{1}{2}\mathbf{A}_s \mathrm{e}^{i\mathbf{K}_s.\mathbf{d}_j - i\omega_s t} \tag{90}$$

$\mathbf{X}_j$ being the displacement of the atom at the site whose lattice vector is $\mathbf{d}_j$. If there are more than one atom per unit cell there will be other equations like (90) for the other atoms but, for simplicity of notation we shall omit them, and include all the displacements in (90). As we shall see, displacements of the form (90) produce periodic changes in potential which may be used as the perturbing potential. We must remember, however, that we are dealing with *real* displacements and that to obtain these we should add to the expression for $\mathbf{X}_j$ the complex conjugate of that given in (90), so that we have

$$\mathbf{X}_j = \tfrac{1}{2}\mathbf{A}_s e^{i\mathbf{K}_s.\mathbf{d}_j - i\omega_s t} + \tfrac{1}{2}\mathbf{A}_s^{\star} e^{-i\mathbf{K}_s.\mathbf{d}_j + i\omega_s t} \tag{90a}$$

The amplitude of the vibration of each of the atoms in this mode is therefore $|\mathbf{A}_s|$.

Now let us consider the effect of displacing the atom (j) through a vector distance $\mathbf{X}_j$ keeping all the other atoms at their respective lattice points. We shall take as unit cell a Wigner–Seitz cell (see § 8.5), centred on the atom (j), and not the basic cell of the lattice. If the electrons of the atom move with the nucleus we shall have for the potential in this cell with the atom in the displaced condition

$$V(\mathbf{r}) = V_0(\mathbf{r}-\mathbf{d}_j-\mathbf{X}_j) \tag{91}$$

where $V_0(\mathbf{r})$ is the periodic lattice potential. The potential V may therefore be expressed in the form

$$V(\mathbf{r}) = V_0(\mathbf{r}-\mathbf{d}_j)+U_j(\mathbf{r}) \tag{92}$$

where in the jth cell

$$U_j(\mathbf{r}) = V_0(\mathbf{r}-\mathbf{d}_j-\mathbf{X}_j)-V_0(\mathbf{r}-\mathbf{d}_j) \tag{92a}$$

Outside the jth cell $U(\mathbf{r})$ will rapidly become zero, since all the other atoms are undisplaced, and we shall make the approximation of taking $U_j(\mathbf{r})$ equal to zero outside the jth cell. The assumption that the electrons of the displaced atom move with the nucleus is in any case an approximation, since there is likely to be some deformation of the charge distribution of the atomic electrons during the displacement. A number of other assumptions have been made and the detailed treatment of this problem is extremely complex. The assumption of quasi-rigid motion leads to results in not too serious disagreement with experiment and has the advantage of simplicity. We may now extend the calculation of the change in potential to the situation in which all the atoms of the crystal are displaced by a wave of the form (90a). Since the displacement of each atom is assumed to affect the potential only on its own unit cell, we have the potential given by equation (92) where j refers now to any cell, the value of $U_j(\mathbf{r})$ being given by equation (92a) for the jth cell and equal to zero for other cells.

Since the displacements $\mathbf{X}_j$ are small we write equation (92a) in the form

$$U_j(\mathbf{r}) = \mathbf{X}_j.\nabla V_0(\mathbf{r}-\mathbf{d}_j) \tag{93}$$

Inserting the value of $\mathbf{X}_j$ from equation (90) we have

$$U_j(\mathbf{r}) = \tfrac{1}{2}\mathbf{A}_s.\nabla V_0(\mathbf{r}-\mathbf{d}_j)\, e^{i\mathbf{K}_s.\mathbf{a}_j - i\omega_s t} + c.c. \tag{94}$$

where *c.c.* represents the complex conjugate. We therefore see that $U_j(\mathbf{r})$ is a periodic potential (in time) with period $\omega_s/2\pi$. Some authors have used instead of the potential $U'(\mathbf{r})$ obtained by replacing $\mathbf{a}_j$ by $\mathbf{r}$, so that, in view of the periodicity of $V_0(\mathbf{r})$,

$$U'(\mathbf{r}) = \tfrac{1}{2}\mathbf{A}_s.\nabla V_0(\mathbf{r})\,\mathrm{e}^{i\mathbf{K}_s.\mathbf{r}-i\omega_s t}+c.c. \qquad (95)$$

This is approximately equivalent to assuming that the electron cloud is displaced by the acoustic wave as a continuous medium. This is found in many cases to give results which agree much worse with experiment. The correct form of the perturbing potential probably lies somewhere between the two forms (94) and (95)*; we shall, however, use (94), which we may insert in equation (27) to determine the quantities a_m in the notation of § 13.2.3. In this case the matrix element H_{m0} is given by

$$H_{m0} = \tfrac{1}{2}\sum_j \mathrm{e}^{i\mathbf{K}_s.\mathbf{d}}\int_{c_j}\psi_m^\star\psi_0 A_s.\nabla V(\mathbf{r}-\mathbf{d}_j)\,\mathrm{d}\mathbf{r} \qquad (96)$$

where c_j means that the integral is taken over the jth cell. As before, we take Bloch wave functions for the initial and final states ψ_0, ψ_m but now these correspond to the *same* band which, for the moment, we shall assume to be the conduction band; i.e. we have

$$\psi_0 = N^{-1/2}u_{k_0}(\mathbf{r})\,\mathrm{e}^{i\mathbf{k}_0.\mathbf{r}}$$
$$\psi_m = N^{-1/2}u_{k'}(\mathbf{r})\,\mathrm{e}^{i\mathbf{k}'.\mathbf{r}}$$

the functions $u_{k_0}(\mathbf{r})$, $u_{k'}(\mathbf{r})$ being normalised for a unit cell (cf. equation (49)).

Inserting these wave functions in equation (96) we have

$$H_{m0} = \frac{1}{2N}\sum_j \mathrm{e}^{i\mathbf{K}_s.\mathbf{d}_j}\int_{c_j}\mathrm{e}^{i(\mathbf{k}_0-\mathbf{k}').\mathbf{r}}u_{k'}^\star(\mathbf{r})\,u_{k_0}(\mathbf{r})\,\mathbf{A}_s.\nabla V(\mathbf{r}-\mathbf{d}_j)\,\mathrm{d}\mathbf{r} \qquad (97)$$

In view of the periodicity of the function $u_{k'}^\star(\mathbf{r})$, $u_{k_0}(\mathbf{r})$ and $V(\mathbf{r})$ equation (97) may be written in the form

$$H_{m0} = \frac{1}{2N}\int_c \mathrm{e}^{i(\mathbf{k}_0-\mathbf{k}').\mathbf{r}}u_{k'}^\star(\mathbf{r})\,u_{k_0}(\mathbf{r})\,\mathbf{A}_s.\nabla V(\mathbf{r})\,\mathrm{d}\mathbf{r}\left(\sum_j \mathrm{e}^{i(\mathbf{k}_0+\mathbf{K}-\mathbf{k}').\mathbf{d}_j}\right) \qquad (98)$$

and we see that $H_{m0} = 0$ unless

$$\mathbf{k}_0+\mathbf{K}_s-\mathbf{k}' = 2\pi\mathbf{b}_{j'} \qquad (99)$$

where $\mathbf{b}_{j'}$ is a vector of the reciprocal lattice (cf. equation (54)). When both $\mathbf{k}_0$ and $\mathbf{k}'$ are small we must take $\mathbf{b}_{j'} = 0$ but when they are not small we may have transitions for which $\mathbf{b}_{j'} \neq 0$; we shall discuss these later. When $\mathbf{b}_{j'} = 0$ we have

$$\mathbf{k}'-\mathbf{k}_0 = \mathbf{K}_s \qquad (99a)$$

* A detailed discussion of the form of the perturbing potential has recently been given by J. M. Ziman, *Electrons and Phonons*. Oxford University Press (1960), Chapter 5.

The physical significance of equation (99a) will be more apparent when taken in connection with the condition given by equation (30a) which must hold again for a periodic perturbation, i.e. we must have

$$E(\mathbf{k}') - E(\mathbf{k}_0) = \hbar\omega_s \tag{100}$$

if we assume that $E(\mathbf{k}') > E(\mathbf{k}_0)$. The quantity $\hbar\omega_s$ is just the energy of a phonon of frequency ν_s and we see that the transition corresponds to the *absorption* of a single phonon. Equation (99a) again therefore gives the conservation of momentum, if we multiply by $\hbar$, in the form

$$\hbar\mathbf{k}_0 + \hbar\mathbf{K}_s = \hbar\mathbf{k}' \tag{101}$$

The crystal momentum of the electron in its final state is thus equal to the sum of the crystal momentum of the electron in its initial state plus the momentum of the phonon. This equation justifies the use of the term *phonon momentum* for the quantity $\hbar\mathbf{K}_s$.

Here we must also consider the contribution from the complex conjugate in the perturbing potential as given in equation (94). We replace $\mathbf{K}_s$ by $-\mathbf{K}_s$ and ω_s by $-\omega_s$ obtaining the two conditions

$$\mathbf{k}' - \mathbf{k}_0 = -\mathbf{K}_s \tag{102}$$

$$E(\mathbf{k}_0) - E(\mathbf{k}') = \hbar\omega_s \tag{103}$$

Equation (103) shows that this transition corresponds to *emission* of a phonon of energy $\hbar\omega_s$ in the scattering process, the balance of momentum now being given by equation (102). We therefore see that whenever an electron is scattered by lattice vibrations a phonon is either emitted or absorbed so that the scattering is never strictly elastic. We have already obtained equations (99a), (100), (102), (103) by elementary considerations in § 10.9.2 and have shown that, except for non-degenerate semiconductors at very low temperatures, the energy lost or gained in a collision is a very small fraction of the kinetic energy of the electron; except in these conditions therefore we may assume that $E(\mathbf{k}_0) \simeq E(\mathbf{k}')$, and we need consider only scattering from one point of a constant-energy surface to another.

An interesting and rather subtle point which should be noted is that if the state corresponding to $E(\mathbf{k}')$ lies *below* that corresponding to $E(\mathbf{k}_0)$ we have the possibility of a spontaneous transition to the state of lower energy with emission of a phonon in addition to the transitions stimulated by the lattice vibrations. We shall later see how this may be taken automatically into account; the effect is in any case negligible if $\mathbf{k}T \gg \hbar\omega_s$, a condition which we shall now assume; we shall later discuss briefly what happens when this condition is not satisfied. To obtain the total transition probability when $\boldsymbol{k}T \gg \hbar\omega_s$ we must simply double the transition probability obtained for transitions involving phonon absorption.

When we may take $k' \simeq k_0$ we have seen in § 10.9.2 that we have approximately

$$|\mathbf{k}' - \mathbf{k}_0| = |\mathbf{K}_s| = 2k_0 \sin \theta/2 \tag{104}$$

where θ is the angle of scattering. If k' and k_0 are small compared with the zone-edge values of $\mathbf{K}_s$ so that the relationship between ω_s and K_s is linear, we also have

$$K_s = \omega_s/c_s \tag{105}$$

where c_s is the velocity of sound corresponding to the polarisation vector $\mathbf{A}_s$.

The calculation of the transition probability then proceeds exactly as in § 13.2.1 since the neglect of the energy change $\hbar\omega_s$ at this stage reduces the problem to that of a perturbation independent of the time. We have therefore, using equation (17), for the probability $\mathrm{d}P$ of a transition such that $\mathbf{k}'$ ends in a surface element $\mathrm{d}S$

$$\mathrm{d}P = \frac{4\pi}{\hbar} N(E)|H_{m0}|^2 \,\mathrm{d}S \tag{105a}$$

where we have introduced a factor 2 as compared with equation (17) to take account of phonon emission and absorption. When the condition (99a) is satisfied we have, from equation (98)

$$H_{m0} = \tfrac{1}{2} \int_c \mathrm{e}^{i(\mathbf{k}_0 - \mathbf{k}')} u_{k'}^{\star}(\mathbf{r})\, u_{k_0}(\mathbf{r})\, A_s . \nabla V(\mathbf{r}) \,\mathrm{d}\mathbf{r} \tag{106}$$

For a band with scalar effective mass m_e the probability $\mathrm{d}P$ of scattering into a solid angle $\sin\theta \,\mathrm{d}\theta \,\mathrm{d}\phi$ is given as in equation (24) by

$$\mathrm{d}P = \frac{V k_0 m_e}{2\pi^2 \hbar^3} |H_{m0}|^2 \sin\theta \,\mathrm{d}\theta \,\mathrm{d}\phi \tag{107}$$

We are now faced with the most difficult part of the calculation, namely the evaluation of the matrix element H_{m0} as given by equation (106). An exact evaluation of the matrix element requires a knowledge not only of the functions $u_{k'}(\mathbf{r})$ and $u_{k_0}(\mathbf{r})$ but also of $V(\mathbf{r})$, and these are available in approximate form, only for a few materials. A large variety of approximate methods have been used to obtain H_{m0}; one of the simplest, used by N. F. Mott and H. Jones* for metals, may also be adapted to deal with semiconductors, and is based on the cellular method of obtaining approximate wave functions as discussed in § 8.5. In this we replace the true Wigner–Seitz unit cell by a sphere of radius r_0 and use wave functions of the form

$$\psi_{k'}(\mathbf{r}) = N^{-1/2} a u(r) \,\mathrm{e}^{i\mathbf{k}'.\mathbf{r}} \tag{108}$$

* N. F. Mott and H. Jones, *The Theory of the Properties of Metals and Alloys*. Oxford University Press (1936), p. 252.

where a is a normalising factor. As we saw in § 8.5, over a large fraction of the unit cell $u(r)$ is constant so that it is quite a good approximation to take $a = (N/V)^{1/2}$. The radius r_0 of the equivalent sphere is given by

$$\tfrac{4}{3}\pi r_0^3 = V/N \tag{109}$$

The boundary condition imposed on the function $u(r)$ is (see § 8.5)

$$\left.\frac{du(r)}{dr}\right)_{r=r_0} = 0 \tag{110}$$

We may transform the integral (106) into a form which depends explicitly only on the wave functions $\psi_{k'}^{\star}$, ψ_{k_0} and not on $V(\mathbf{r})$. A knowledge of the wave functions implies a knowledge of $V(\mathbf{r})$ but the calculation is more readily made by use of the transformation. The quantities ψ_{k_0} and $\nabla\psi_{k_0}$ both satisfy Schrödinger's equation with $E = E_{k_0}$; likewise $\psi_{k'}^{\star}$ and $\nabla\psi_{k'}^{\star}$ both satisfy Schrödinger's equation with $E = E_{k'}$. We have seen, however, that $E_{k_0} \simeq E_{k'}$ and we shall assume that $E_{k_0} = E_{k'}$. We have therefore

$$(V - E_k)\left.\begin{matrix} \psi_{k_0} \\ \nabla\psi_{k_0} \\ \psi_{k'}^{\star} \\ \nabla\psi_{k'}^{\star} \end{matrix}\right\} = \frac{\hbar^2\nabla^2}{2m}\left.\begin{matrix} \psi_{k_0} \\ \nabla\psi_{k_0} \\ \psi_{k'}^{\star} \\ \nabla\psi_{k'}^{\star} \end{matrix}\right\} \tag{111}$$

If we multiply the first of these equations by $\psi_{k'}^{\star}$ and operate on it with ∇, we obtain

$$\nabla(V\psi_{k_0}\psi_{k'}^{\star}) - E_k\psi_{k_0}\nabla\psi_{k'}^{\star} - E_k\psi_{k'}^{\star}\nabla\psi_{k_0} - \frac{\hbar^2}{2m}\nabla\psi_{k'}^{\star}\nabla^2\psi_{k_0} - \frac{\hbar^2}{2m}\psi_{k'}^{\star}\nabla^2\nabla\psi_{k_0} = 0 \tag{112}$$

Again, using Schrödinger's equation this may be written in the form

$$\nabla(V\psi_{k_0}\psi_{k'}^{\star}) - V\psi_{k_0}\nabla\psi_{k'}^{\star} - V\nabla\psi_{k_0}\psi_{k'}^{\star} - \frac{\hbar^2}{2m}(\psi_{k'}^{\star}\nabla^2\nabla\psi_{k_0} - \nabla\psi_{k_0}\nabla^2\psi_{k'}^{\star}) = 0 \tag{113}$$

Thus we have

$$\psi_{k_0}\psi_{k'}^{\star}\nabla V = \frac{\hbar^2}{2m}(\psi_{k'}^{\star}\nabla^2\nabla\psi_{k_0} - \nabla\psi_{k_0}\nabla^2\psi_{k'}^{\star}) \tag{114}$$

We may therefore replace $\psi_{k_0}\psi_{k'}^{\star}\nabla V$ in the integral for H_{m0} by the expression on the right-hand side of equation (114). We note that if $u(r)$ were constant and the constant-energy surfaces were spherical so that $k_0 = k'$ then the right-hand side would be equal to zero. Most of the integral giving H_{m0} therefore comes from the variable part of the function $u(r)$. The integral of the right-hand side of equation (114) may also be transformed into an integral over the surface of the unit cell by Green's theorem; this brings about a great simplification and enables us to use the cellular wave functions in a very simple manner. The integral for H_{m0} becomes

$$H_{m0} = \frac{\hbar^2}{4m}\int\limits_S \left(\psi_{k'}^{\star}\frac{\partial}{\partial r}A_s.\nabla\psi_{k_0} - A_s.\nabla\psi_{k_0}\frac{\partial}{\partial r}\psi_{k'}^{\star}\right) \mathrm{d}S \tag{115}$$

where S represents the surface of the equivalent sphere of radius r_0. On the surface S $\partial u/\partial r = 0$, and the only appreciable contribution comes from the term in $\partial^2 u/\partial r^2$; this gives for H_{m0}

$$H_{m0} = \frac{\hbar^2}{4mr_0}\int\limits_S \mathrm{e}^{i(\mathbf{k}_0-\mathbf{k}').\mathbf{r}}(\mathbf{A}_s.\mathbf{r}_d)\, u(r_0)\, u''(r_0)\, \mathrm{d}S \tag{116}$$

where the vector $\mathbf{r}_d$ is directed along the displacement direction given by ∇ or $\mathbf{X}_s$. Now we have, if E_0 is the energy for $k = 0$

$$\nabla^2 u(r) + \frac{2m}{\hbar^2}[E_0 - V(\mathbf{r})]\, u(r) = 0 \tag{117}$$

Since, when $r = r_0$, $u'(r) = 0$ we have if we take E_0, the energy at the bottom of the band equal to zero,

$$u''(r_0) = \frac{2m}{\hbar^2} V(r_0)\, u(r_0) \tag{118}$$

Also near the edge of the zone we may take

$$u(r_0) \simeq (N/V)^{1/2} \tag{119}$$

in order that the functions ψ_{k_0}, $\psi_{k'}^{\star}$ may be normalised. We therefore obtain for H_{m0}

$$H_{m0} = \frac{V_0 N}{2r_0 V}\int (\mathbf{A}_s.\mathbf{r}_d)\, \mathrm{e}^{i(\mathbf{k}_0-\mathbf{k}').\mathbf{r}}\, \mathrm{d}S \tag{120}$$

where we have written V_0 for $V(r_0)$. If we use polar coordinates θ', ϕ' with the polar axis along the direction of $\mathbf{k}_0 - \mathbf{k}'$ we have for the component of H_{m0} along the polar axis

$$H_{m0} = \frac{N\pi r_0^2 V_0 A_s}{V}\int\limits_0^\pi \cos\gamma \cos\theta'\, \mathrm{e}^{i|\mathbf{k}_0-\mathbf{k}'|r_0\cos\theta'} \sin\theta'\, \mathrm{d}\theta' \tag{121}$$

where γ is the angle between the polarisation vector $\mathbf{A}_s$ and the vector $\mathbf{k}_0 - \mathbf{k}$. Now the vector $\mathbf{k}_0 - \mathbf{k}'$ is in the direction of the vector $\mathbf{K}_s$ the propagation vector of the acoustic wave; we therefore see that only longitudinal waves give a contribution to H_{m0}. In general, as we have seen, the acoustic modes cannot be split exactly into longitudinal and transverse waves so that in some instances there will be a contribution from the transverse components; in general, this will be small and we shall neglect it, taking $\cos\gamma = 1$.

13.4.1 Lattice scattering in semiconductors

When $|\mathbf{k}_0-\mathbf{k}'|\,r_0$ is small, which will be so for electrons near the bottom of the conduction band in a semiconductor, we may expand the exponential and retain only the first two terms. We obtain in this way

$$H_{m0} = \frac{2\pi i N r_0^3\, V_0 |\mathbf{k}_0-\mathbf{k}'|\, A_s}{3V} \tag{122}$$

Using equation (109), we obtain

$$H_{m0} = \tfrac{1}{2} i V_0 K_s A_s \tag{123}$$

From equation (107) we therefore obtain for the transition probability dP

$$\mathrm{d}P = \frac{V k_0 m_e K_s^2 V_0^2 A_s^2}{8\pi^2 \hbar^3} \sin\theta \,\mathrm{d}\theta\, \mathrm{d}\phi \tag{124}$$

The amplitude A_s is determined from the mean kinetic energy of the mode which, when $\boldsymbol{k}T \gg \hbar\omega_s$ is equal to $\frac{1}{2}\boldsymbol{k}T$ (see § 7.3). (The mean potential energy is also equal to $\frac{1}{2}\boldsymbol{k}T$ giving a total mean energy of $\boldsymbol{k}T$ per mode.) The mean kinetic energy of the mode is also equal to the sum for all the atoms of the crystal of $\frac{1}{2}M\dot{X}_j^2$ where M is the mass of each atom, and this is equal to $\frac{1}{4}MN\omega_s^2 A_s^2$. We therefore have

$$A_s^2 = 2\boldsymbol{k}T/MN\omega_s^2 \tag{125}$$

Inserting this value in equation (124) we obtain

$$\mathrm{d}P = \frac{V\boldsymbol{k}T k_0 m_e K_s^2 V_0^2 \sin\theta\,\mathrm{d}\theta\,\mathrm{d}\phi}{4\pi^2\hbar^3\omega_s^2 MN} \tag{126}$$

Since, as we have seen, K_s corresponds to a longitudinal acoustic wave having a long wavelength we may take $\omega_s = K_s c_l$ where c_l is the speed of longitudinal sound waves. Inserting this value in equation (126) we obtain

$$\mathrm{d}P = \frac{V\boldsymbol{k}T k_0 m_e V_0^2 \sin\theta\,\mathrm{d}\theta\,\mathrm{d}\phi}{4\pi^2\hbar^3 c_l^2 MN} \tag{127}$$

For a single scattering centre we find that dP is proportional to V^{-1} (cf. equations (20) and (24)) and to obtain a quantity defining the scattering probability per unit time independent of the volume we have introduced the partial scattering cross section $\sigma(\theta,\phi)$ which has the dimensions of an area and is independent of V, being a property of a single scattering centre. The partial cross-section $\sigma(\theta,\phi)$ is related to dP by means of the equation (cf. equation (25))

$$\sigma(\theta,\phi)\sin\theta\,\mathrm{d}\theta\,\mathrm{d}\phi = (V/v)\,\mathrm{d}P \tag{127a}$$

so that $\sigma(\theta,\phi)$ is independent of the volume V. When we have N_s scattering centres in the volume V the probability of scattering dP_t is given by

$$\mathrm{d}P = N_s\mathrm{d}P = (N_s/V)\,\mathrm{d}P_1 = (N_s/V)\,v\sigma(\theta,\phi)\sin\theta\,\mathrm{d}\theta\,\mathrm{d}\phi \tag{128}$$

where dP_1 is the scattering probability per unit volume; dP_t is therefore independent of V, as we should expect.

The probability per unit time dP given by equation (127) is already independent of V, and this is reasonable since we are dealing with scattering by the whole crystal. It is interesting to see how the various factors depending on the volume V arise. We have $V^{-1/2}$ in the normalisation of ψ_0 and ψ_m giving V^{-2} in $|H_{m0}|^2$; we have a factor N coming from the sum over the number of unit cells in the integral for H_{m0} giving a factor N^2 in $|H_{m0}|^2$. We therefore see that the square of the matrix element including the factor $(N/V)^2$ is independent of V. A factor V is introduced in the density of states function $N(E)$ and a factor N^{-1} in $|A_s|^2$ giving a total factor (N/V) in dP. Equation (127) also includes a factor $(V/N)^2$ arising from the integration over the unit cell so a factor (V/N) appears explicitly in this equation. We are now faced with a difficulty if we are to define a scattering cross-section $\sigma(\theta, \phi)$ independent of V. If we now use an equation such as (127) for this definition we obtain a quantity proportional to V^{-1}. The difficulty arises through the uncertainty of what we mean by the number of scattering centres. If this is taken as the number of independent scattering potentials then we have only one for the volume V and we would define a cross-section relating to one scattering centre per unit volume. It is more reasonable to define a new quantity $\sigma_A(\theta, \phi)$ giving the scattering cross-section *per atom* of the crystal; the number of scattering centres in the crystal is then simply N and σ_A is defined by means of equation (128) by taking $N_s = N$. We have, therefore,

$$\sigma_A \sin\theta \, d\theta \, d\phi = V \, dP/vN \tag{129}$$

where dP is given by equation (127); in this case dP is the total scattering probability for all the scattering centres and is equivalent to dP_t in equation (128). We therefore obtain, writing $NM/V = \rho$, where ρ is the density,

$$\sigma_A = \frac{\Omega_0 \boldsymbol{k} T k_0 m_e V_0^2}{4\pi^2 \hbar^3 \rho c_l^2 v} \tag{130}$$

where Ω_0 is the volume of a unit cell. Since $\hbar k_0 = m_e v$ we may also write σ_A in the form

$$\sigma_A = \frac{\Omega_0 \boldsymbol{k} T m_e^2 V_0^2}{4\pi^2 \hbar^4 \rho c_l^2} \tag{131}$$

We therefore see that the scattering cross-section is independent of the scattering angle θ.

The relaxation time τ is given by (see § 10.4)

$$\frac{1}{\tau} = \iint (1-\cos\theta) \, dP = \frac{2\pi}{\Omega_0} \int_0^{\pi} v\sigma_A(\theta)(1-\cos\theta)\sin\theta \, d\theta \tag{132}$$

so that we have

$$\frac{1}{\tau} = \frac{v m_e^2 V_0^2 \boldsymbol{k} T}{\pi \hbar^4 \rho c_l^2} = \frac{(2m_e^3 E_k)^{1/2} V_0^2 \boldsymbol{k} T}{\pi \hbar^4 \rho c_l^2} \tag{133}$$

where E_k is the kinetic energy of the electron. The mean free path $l = \tau v$ is given by

$$l = \frac{\pi \hbar^4 \rho c_l^2}{m_e^2 V_0^2 \boldsymbol{k}T} \tag{134}$$

The mean free path is independent of E_k and is inversely proportional to the absolute temperature; τ varies as T^{-1} and as $E_k^{-1/2}$. We have used a relaxation time with this variation to calculate the mobility when we have pure lattice scattering (see § 10.9.2). The quantity V_0 is not easy to determine; calculations for a number of materials have shown, however, that it is of the order of a few electron volts. Thus we may readily obtain from equation (133) the order of magnitude of the relaxation time. It agrees quite well with values found for pure germanium and silicon although the variation with temperature given by equation (133) is not too well obeyed. We shall later give an alternative method of calculating τ based on the effective-mass approximation. This gives τ in terms of another potential which is more readily interpreted experimentally than V_0 (see § 13.4.4).

An equation similar to (133) may be obtained for the scattering of holes; in this case m_e is replaced by m_h and V_0 by $|V_0' + \Delta E|$ where V_0' is the cellular potential energy associated with the edge of the valence band. This may be seen as follows: the transition of a hole described by means of a wave vector $\mathbf{k}_0$ to a state with wave vector $\mathbf{k}'$ corresponds to the transition of an electron from a state with wave vector $\mathbf{k}'$ to a state with wave vector $\mathbf{k}_0$. This is given by the previous analysis provided the fact that most of the levels in the valence band are already occupied does not change the transition probability. We have seen, however, in § 10.9 that the Pauli Principle has no effect on the transition probability and so we may use the same analysis as before but now applied to the valence band. We shall also see later that a derivation based on the effective-mass equation gives the same result for holes as electrons provided the appropriate effective mass and scattering potential are used.

When the temperature is so low that $\boldsymbol{k}T < \hbar\omega_s$ the value which we have used for A_s^2 no longer applies and we should replace $\boldsymbol{k}T$ in equation (125) by (see § 7.3)

$$\frac{\hbar\omega_s}{e^{\hbar\omega_s/\boldsymbol{k}T} - 1}$$

The maximum value of $\hbar\omega_s$ for phonons which take part in scattering is obtained by giving θ the value π in equation (104). If ω_m is the maximum value of ω_s we have

$$\hbar\omega_m = 2\hbar k_0 c_l \tag{135}$$

so that the maximum value K_m of the magnitude of the phonon wave vector $\mathbf{K}_s$ is given by

$$K_m = 2k_0 \tag{136}$$

The acoustic waves involved in scattering are therefore long waves when k_0 is small. Equation (135) may also be written in the form

$$\hbar\omega_m = 2m_e v c_l \tag{137}$$

where v is the velocity of the electron being scattered. If we take the average value of v as $(2\boldsymbol{k}T/m_e)^{1/2}$ we may write equation (137) in the form

$$\frac{\hbar\omega_m}{\boldsymbol{k}T} = \frac{2(2m_e)^{1/2}c_l}{(\boldsymbol{k}T)^{1/2}} \tag{138}$$

We therefore have $\boldsymbol{k}T < \hbar\omega_m$ when $T < \theta_0$ where

$$\boldsymbol{k}\theta_0 = 16(\tfrac{1}{2}m_e c_l^2) \tag{139}$$

An average velocity $c_l \simeq 10^5$ cm/sec corresponds, for electrons, to a temperature of about 0·03°K (10^7 cm/sec corresponds to 300°K) so that $\theta_0 \simeq 0{\cdot}5$°K. We therefore see that for all practical temperatures it is a good approximation to use the value for τ given by equation (133) and to assume that the scattering is isotropic; as we shall see later this is not true for metals. Lattice scattering, in semiconductors, is in any case unimportant at low temperatures.

We have seen that whenever an electron is scattered by a normal mode of the lattice vibrations it either gains or loses an amount of energy equal to $\hbar\omega_s$. In our treatment of scattering we have assumed that the probabilities of scattering with emission and absorption of a phonon are exactly equal. This is not quite so; when we take into account the spontaneous emission of a phonon the ratio of the two probabilities is

$$\exp(\hbar\omega_s/\boldsymbol{k}T) \simeq 1+\hbar\omega_s/\boldsymbol{k}T$$

when $\hbar\omega_s \ll \boldsymbol{k}T$. A net energy exchange therefore takes place between the conduction electrons and the atoms of the crystal. When a small electric field is applied, an equilibrium is soon established, but for pure semiconductors in strong electric fields the mean energy of the electrons may be appreciably increased, an effect first discussed by W. Shockley* and since observed in a number of semiconductors. (For a detailed discussion, see R. A. Smith†, *Semiconductors* or a review article by J. B. Gunn‡.)

13.4.2 Lattice scattering in metals

For metals with conduction bands about half filled we may no longer assume that $|\mathbf{k}_0-\mathbf{k}'|r_0$ is small. The problem is simplified, however, in that we only require to calculate τ for values of k_0 corresponding to the Fermi surface. In this case the kinetic energy of the electron will be very much greater than the phonon energy at all temperatures. The integral in equation (121) may readily be evaluated without approximation by

* W. Shockley, *Bell Syst. Tech. J.*, **30**, 990 (1951).
† R. A. Smith, *Semiconductors*. Cambridge University Press (1959), § 5.5.
‡ J. B. Gunn, *Progress in Semiconductors*. Heywood, London (1957), Vol. 2, p. 213.

making the transformation $x = \cos\theta'$, and using equation (99a) for $|\mathbf{k}_0 - \mathbf{k}'|$. The integral then becomes (taking $\cos\gamma = 1$)

$$\int_{-1}^{1} x\, e^{iK_s r_0 x}\, dx = \frac{2}{iK_s^2 r_0^2}[\sin K_s r_0 - K_s r_0 \cos K_s r_0] \tag{140}$$

We therefore obtain for H_{m0}

$$H_{m0} = \frac{2\pi V_0 N A_s[\sin K_s r_0 - K_s r_0 \cos K_s r_0]}{iVK_s^2} \tag{141}$$

The transition probability dP is now obtained from equation (107) and is given by

$$dP = \frac{2N^2 k_0 m_e A_s^2 V_0^2[\sin(r_0 K_s) - r_0 K_s \cos(r_0 K_s)]^2 \sin\theta\, d\theta\, d\phi}{\hbar^3 K_s^4 V} \tag{142}$$

We shall now assume that k_0 and k' both correspond to electrons whose wave vector terminates on the Fermi surface. In obtaining equation (141) we have assumed that we may replace the energy band by one having spherical constant-energy surfaces so that we may write

$$\frac{\hbar^2 k_0^2}{2m_e} = E_F \tag{143}$$

For a metal with one atom per unit cell, and one valence electron per atom we have also (cf. § 8.9, equation (86))

$$\frac{N}{2} = \frac{V}{8\pi^3} \cdot \tfrac{4}{3}\pi k_0^3 \tag{144}$$

or

$$k_0 = (3\pi^2/\Omega_0)^{1/3} \tag{144a}$$

Since $\Omega_0 = 4\pi r_0^3/3$ this is equivalent to

$$k_0 r_0 = (9\pi/4)^{1/3} \tag{145}$$

so that from equation (104) we have

$$r_0 K_s = r_0|k_0 - k'| = 2(9\pi/4)^{1/3} \sin\tfrac{1}{2}\theta \tag{146}$$

The validity of equation (146) depends, however, on the assumption that we are able to satisfy the condition $\mathbf{k}' = \mathbf{k}_0 + \mathbf{K}_s$ with values of $\mathbf{k}'$ and $\mathbf{k}_0$ corresponding to points in the first Brillouin zone. The maximum value of K_s is $2k_0$ and a vector of this magnitude may well bring the vector $\mathbf{k}'$ outside the first Brillouin zone when $\mathbf{k}_0$ corresponds to a point on the Fermi surface. In this case we must replace equation (99a) by equation (99), using one of the vectors of the reciprocal lattice b_j to bring $\mathbf{k}'$ back within the first Brillouin zone; this is necessary if we are using a reduced representation of the wave vectors. Transitions for which $b_j \neq 0$ are usually referred to in the literature as '*umklapprozesse*', while transitions for which equation (99a) is satisfied are called 'normal' transitions.

Provided that $\boldsymbol{k}T \gg \hbar\omega_s$ we may replace A_s^2 by $2\boldsymbol{k}T/NM\omega_s^2$ as before. For metals, however, K_s is no longer small compared with the value for the zone edge, and in order to obtain an exact expression for $\mathrm{d}P$ we should know the variation of ω_s with $\mathbf{K}_s$ over most of the zone. This, however, is not generally known for metals and we must have recourse to the Debye approximation as used to calculate the specific heat in § 8.5. In this case $\mathbf{K}_s$ is not restricted precisely to the first Brillouin zone but K_s must have a value less than K_m corresponding to the maximum Debye frequency ν_m. We have seen in § 8.5 that K_m is given by

$$K_m = (6\pi^2/\Omega_0)^{1/3} = (9\pi/2)^{1/3}/r_0 \tag{147}$$

This should be compared with equation (144) for the value of k_0; we see that $K_m = 2^{1/3}k_0$ (the factor 2 comes from the fact that the electron has spin and each state is doubly degenerate). We note from equation (147) that the maximum value which the quantity r_0K_s can take is $(9\pi/2)^{1/3}$; the maximum value θ_m of θ for a normal transition is therefore obtained by inserting this value in equation (146) and is given by

$$\theta_m = 2\sin^{-1}(2^{-2/3}) \simeq 80° \tag{148}$$

In view of the approximate nature of a calculation based on the cellular method we are hardly justified in taking account of the large angle scattering for which $\theta > \theta_m$; it is unlikely to affect the order of magnitude of the scattering cross-section, and we shall neglect *umklapprozesse* in the following calculation. When we express ω_s in terms of K_s by means of the Debye approximation $\omega_s = K_s c_l$, for longitudinal acoustic waves, we obtain the value for A_s^2 we have already used for semiconductors, namely $A_s^2 = 2\boldsymbol{k}T/NMK_s^2c_l^2$, provided $\boldsymbol{k}T \gg \hbar\omega_s$ for all values of ω_s involved in the scattering; we shall later examine more closely the circumstances under which this condition is valid. On substituting for A_s^2 in equation (141), and using equation (146) for r_0K_s we obtain

$$\mathrm{d}P = \frac{4N\boldsymbol{k}Tk_0 m_e V_0^2[f(\theta)]^2 r_0^4 \sin\theta\,\mathrm{d}\theta\,\mathrm{d}\phi}{c_l^2 K_s^2 \hbar^3 VM} \tag{149}$$

where
$$f(\theta) = \frac{\sin(\beta\sin\frac{1}{2}\theta) - \beta\sin\frac{1}{2}\theta\cos(\beta\sin\frac{1}{2}\theta)}{\beta^2\sin^2\frac{1}{2}\theta} \tag{150}$$

and $\beta = 2(9\pi/4)^{1/3} \simeq 3{\cdot}85$. Using equation (145) for k_0r_0, and equation (109) for V/N the expression for $\mathrm{d}P$ reduces to

$$\mathrm{d}P = \left(\frac{4}{9\pi}\right)^{1/3}\frac{3m_e V_0^2\boldsymbol{k}T[f(\theta)]^2 r_0^2\sin\theta\,\mathrm{d}\theta\,\mathrm{d}\phi}{2\pi c_l^2\hbar^3 M(1-\cos\theta)} \tag{149a}$$

The relaxation time τ is therefore given by

$$\frac{1}{\tau} = \iint(1-\cos\theta)\,\mathrm{d}P \tag{151}$$

the integration with respect to θ being taken from 0 to θ_m; this gives

$$\frac{1}{\tau} = \left(\frac{2}{9\pi}\right)^{2/3} \frac{\mathscr{C} m_e V_0^2 \boldsymbol{k} T r_0^2}{M c_l^2 \hbar^3} \tag{152}$$

where $\mathscr{C}$ is a constant given by

$$\mathscr{C} = 6\left(\frac{9\pi}{8}\right)^{1/3} \int_0^{\theta_m} [f(\theta)]^2 \sin\theta \, d\theta \tag{153}$$

whose value is very nearly equal to 1($\mathscr{C} = 0{\cdot}98$ if $\theta_m = 80°$). If we introduce the density ρ we have

$$\frac{1}{\tau} = \left(\frac{2}{9\pi}\right)^{1/3} \frac{3\mathscr{C} m_e V_0^2 \boldsymbol{k} T}{4\pi\rho c_l^2 \hbar^3 r_0} \tag{154}$$

The relaxation time may also be expressed in terms of a Debye temperature θ_l related to c_l by means of the equation

$$c_l = \frac{\boldsymbol{k}\theta_l}{\boldsymbol{h}}\left(\frac{4\pi V}{3N}\right)^{1/3} = \frac{\boldsymbol{k}\theta_l r_0}{\hbar}\left(\frac{2}{9\pi}\right)^{1/3} \tag{155}$$

(cf. § 7.3 equations (8) and (12)). θ_l is equal to the Debye temperature only when the velocity of transverse sound waves c_t is equal to c_l (cf. § 7.3 equation (18)). Substituting for c_l in equation (152) we obtain

$$\frac{1}{\tau} = \mathscr{C}\frac{m_e}{M}\left(\frac{V_0^2}{\boldsymbol{k}^2\theta_l^2}\right)\left(\frac{\boldsymbol{k}T}{\hbar}\right) \tag{156}$$

The mean free path $l = \tau v_0$ is given by

$$l = \mathscr{C}^{-1}\frac{M}{m_e}\left(\frac{\boldsymbol{k}^2\theta_l^2}{V_0^2}\right)\left(\frac{\hbar^2 k_0}{\boldsymbol{k}T m_e}\right) \tag{157}$$

$$= \mathscr{C}^{-1}\left(\frac{4}{9\pi}\right)^{1/3}\left(\frac{2M}{m_e}\right)\left(\frac{\boldsymbol{k}^2\theta_l^2}{V_0^2}\right)\left(\frac{E_F}{\boldsymbol{k}T}\right) r_0 \tag{157a}$$

Alternatively, using equations (145) and (152) we may write

$$l = \frac{1}{2^{1/3}\mathscr{C}}\left(\frac{9\pi}{2}\right)^{2/3}\left(\frac{4\pi}{3}\right)\frac{\rho c_l^2 \hbar^4}{m_e^2 V_0^2 \boldsymbol{k}T} \tag{157b}$$

which differs from the expression given in equation (134) for the mean free path in a non-degenerate semiconductor or only by a numerical factor whose value is approximately 3·2. The atomic scattering cross section $\sigma_A(\theta)$, obtained from equation (149a) is given by

$$\frac{\sigma_A(\theta)}{\pi r_0^2} = \left(\frac{4}{9\pi}\right)^{1/3} \frac{3\mathscr{C} m_e V_0^2 \boldsymbol{k}T[f(\theta)]^2}{4\pi^2 \rho c_l^2 \hbar^3 v_0 \sin^2\theta/2} \tag{158}$$

Since, for small values of θ, $f(\theta) \simeq (\beta \sin\frac{1}{2}\theta)/3$ we see that $\sigma_A(\theta)$ tends to a constant value for small values of θ. We shall later discuss numerical values of τ and l (see § 13.4.5).

We must now examine the condition $\boldsymbol{k}T \gg \hbar\omega_s$, for all appropriate values of ω_s, on which the validity of the expressions we have found for τ depends. The maximum value ω_m of ω_s is related to the temperature θ_l by means of the equation (cf. § 7.3 equation (12)) $\hbar\omega_m = \mathbf{k}\theta_l$. The condition we require is therefore simply $T \gg \theta_l$. When this condition is not satisfied we must replace equation (125) by (see § 13.4.3)

$$A_s^2 = \frac{\hbar}{MN\omega_s} + \frac{2\hbar}{MN(e^{\hbar\omega_s/\mathbf{k}T} - 1)\,\omega_s} \tag{159}$$

We therefore see that A_s^2 will be small for values of θ such that $\hbar\omega_s$ is substantially greater than $\boldsymbol{k}T$, i.e. such that

$$\hbar K_s c_l > \boldsymbol{k}T \tag{160}$$

or

$$2\hbar k_0 \sin\frac{\theta}{2} > \boldsymbol{k}T/c_l \tag{160a}$$

Using equation (155) for c_l and equation (145) for $k_0 r_0$ we see that we shall have appreciable scattering only at angles such that

$$\sin\theta/2 < 2^{-2/3}\,T/\theta_l \tag{161}$$

when $T \ll \theta_l$ this reduces to

$$\theta < 2^{1/3}\,T/\theta_l \tag{162}$$

For values of T such that $T \gg \theta_l$ we see that τ^{-1} and hence the resistivity is proportional to T; as T decreases below the value θ_l, however, the resistivity begins to fall more rapidly since the scattering is restricted to increasingly smaller angles, the magnitude of the small angle scattering decreasing as T. We may obtain an approximate value of the resistivity for metals at low temperatures, such that $T \ll \theta_l$, as follows. We assume that the scattering probability is given by equation (149a) for small values θ such that the inequality (162) is true, and is equal to zero for larger angles. For small values of θ the function $f(\theta)$ has the value $\frac{1}{3}\beta\sin\theta/2$; inserting this value in equation (149a) we obtain on integrating from $\theta = 0$ to $\theta = 2^{1/3}\,T/\theta$

$$\frac{1}{\tau} = \mathscr{C}'\left(\frac{T}{T_h}\right)\left(\frac{T}{\theta_l}\right)^4\frac{1}{\tau_h} \tag{163}$$

where τ_h is the relaxation time for a high temperature T_h such that $T_h \gg \theta_l$ and $\mathscr{C}'$ is a constant. We see that $1/\tau$, and therefore the resistivity, varies as T^5/θ_l^6 at low temperatures, such that $T \ll \theta_l$; while this variation is in reasonable agreement with observations for a number of metals the value of the constant $\mathscr{C}'$ obtained in this way is found to be much too small. This is hardly surprising since the assumption that we have made that $k_0 = k'$ is no longer a valid one at low temperatures. The Fermi function varies rapidly in an energy range of the order of $\boldsymbol{k}T$ and the energy change in scattering $\pm\hbar\omega_s$ is no longer small compared with $\boldsymbol{k}T$.

The assumption that both $\mathbf{k}_0$ and $\mathbf{k}'$ lie on the Fermi surface is therefore no longer a valid one. The exact calculation of the relaxation time at low temperatures is extremely complex. A treatment given by F. Bloch* takes account to some extent of the change in energy on scattering, and equation (163) is obtained with $\mathscr{C}'$ having the value 497·6. The difficulties are discussed in some detail by A. H. Wilson†, who derives a formula which reduces to equation (163) under certain circumstances, the constant $\mathscr{C}'$ also being equal to 497·6.

13.4.3 Quantum theory of phonon absorption and emission

In calculating the transition probability for scattering we have regarded the acoustic wave as a classical displacement wave. In order to calculate the mean energy of the crystal we have seen in §§ 7.1–7.3 that we require to use quantum theory to describe the normal modes in terms of wave functions for simple harmonic oscillators. We have found that when $T \gg \theta_D$, the Debye temperature, a classical treatment gives the same answer as the quantum theoretical treatment and we shall find that this is also so for the transition probabilities. It is interesting and also quite simple to verify this and to derive the modification already quoted for the interaction potential when the temperature is not much greater than the Debye temperature.

We begin by expressing the interaction energy as in equation (93). We then transform to the normal coordinates q_s, defined as in § 3.4.1 but extended by an obvious generalisation to three dimensions. The displacement X_j is expressed in terms of the normal coordinates q_s by the normal orthogonal transformation

$$X_j = \frac{1}{N^{1/2}} \sum_s e^{i\mathbf{K}_s.\mathbf{d}_j} q_s \tag{164}$$

(cf. equation (89)) and by reversing this transformation we have

$$q_s = \frac{1}{N^{1/2}} \sum_j e^{-i\mathbf{K}_s.\mathbf{d}_j} X_j \tag{165}$$

Equation (93) for the perturbing potential may be written in terms of the normal coordinates q_s in the form

$$U_j(\mathbf{r}) = \frac{1}{N^{1/2}} \sum_s q_s e^{i\mathbf{K}_s.\mathbf{d}}\, \mathbf{a}_s.\nabla V \tag{166}$$

where a_s is a unit polarisation vector.

When we quantize the lattice vibrations the wave function describing the motion of the atoms of the lattice and a conduction electron will be of the form

$$\Psi = \prod_s \chi_{n_s}(q_s)\, \psi_{k'}(\mathbf{r}) \tag{167}$$

* F. Bloch, *Handb. Phys.*, **24**, Pt. 2, 499 (1933).
† A. H. Wilson, *Theory of Metals*. Cambridge University Press (1954), p. 279.

where the wave function $\chi_{n_s}(q_s)$ is the harmonic oscillator wave function corresponding to the vibrational state with energy $\hbar\omega_s(n_s+\frac{1}{2})$. When we form the matrix element of the potential (166) we see that because of the orthogonality of the wave functions χ_s only those for the particular co-ordinate q_s occurring in each term will contribute. We may therefore write the matrix element H_{m0} in the form

$$H_{m0} = \frac{H'_{m0}}{N}\sum_s \int \chi^{\star}_{n'_s} q_s \chi_{n_s} \mathrm{d}q_s \tag{168}$$

where H'_{m0} is the matrix element defined in equation (96) replacing the amplitude A_s by unity. Now it is well known that the integrals of the type occurring in equation (168) are zero unless* $n'_s = n_s \pm 1$. When $n'_s = n_s + 1$ we have emission of a phonon of energy $\hbar\omega_s$ and*

$$\int \chi^{\star}_{n_s+1} q_s \chi_{n_s} \mathrm{d}q_s = \left[\frac{\hbar(n_s+1)}{2M\omega_s}\right]^{1/2} \tag{169}$$

when $n'_0 = n_s - 1$ we have absorption of a phonon of energy $\hbar\omega_s$ and

$$\int \chi^{\star}_{n_s-1} q_s \chi_{n_s} \mathrm{d}q_s = \left[\frac{\hbar n_s}{2M\omega_s}\right]^{1/2} \tag{170}$$

When we sum the squares of the two matrix elements giving non-zero transition probabilities for a given value of ω_s we obtain

$$|H^{+}_{m0}|^2 + |H^{-}_{m0}|^2 = \frac{|H'_{m0}|^2}{N^2}\frac{(2n_s+1)\,\hbar\omega_s}{M\omega_s^2} \tag{171}$$

where H^{+}_{m0} and H^{-}_{m0} are, respectively, the matrix elements corresponding to emission and absorption of a phonon. The quantity H'_{m0} we see as before is zero unless equation (99) is satisfied, and this picks out one value of ω_s. Moreover, the integral for H'_{m0} gives the same value when we sum over all values of j, removing one of the factors N^{-1} from (171). The calculation therefore proceeds as before provided we replace A_s^2 by

$$\frac{2}{N}\cdot\frac{(n_s+\frac{1}{2})\,\hbar\omega_s}{M\omega_s^2} \tag{172}$$

Now the average value of $\hbar\omega_s(n+\frac{1}{2})$ is just the mean energy $\bar{E}_s$ per mode which we have seen in § 3.5 is given by

$$\bar{E}_s = \tfrac{1}{2}\hbar\omega_s + \frac{\hbar\omega_s}{\mathrm{e}^{\hbar\omega_s/kT}-1} \tag{173}$$

When $\boldsymbol{k}T \gg \hbar\omega_s$ we see that A_s is replaced by $2\boldsymbol{k}T/NM\omega_s^2$ as in equation (125) and we obtain the same value for the transition probability as given by the classical treatment. When this condition does not hold we obtain for the average value of A_s^2 the value given by equation (159).

* See, for example, L. I. Schiff, *Quantum Mechanics* (2nd Ed.). McGraw-Hill (1955), p. 65.

13.4.4 The deformation potential

Since, as we have seen, we are concerned mainly with scattering by long-wave acoustic modes it seems likely that we could find a slowly varying potential to represent the change in potential due to the wave; this would be expected only in the case of semiconductors where we are concerned with electrons whose wave vectors do not differ much from those corresponding to the minimum of the conduction band or maximum of the valence band. If we were able to represent the effect of the acoustic waves by means of a slowly varying potential we could then use the effective-mass approximation the amplitude wave functions $F(\mathbf{r})$ being plane waves (see § 11.2.4). We cannot use the effective-mass approximation in the method given in § 13.4.1 since the potential ∇V varies rapidly throughout each unit cell. A method of obtaining a slowly varying potential, applicable to semiconductors, and based on the effective-mass approximation has been developed by W. Shockley and J. Bardeen*. The effective-mass wave equation for a perfect semiconductor in which all the atoms are at rest at their equilibrium positions is given by taking $V(\mathbf{r}) = 0$ in equation (10a) of § 11.1 so that we have

$$\nabla_k(-i\nabla)\, F(\mathbf{r}) = EF(\mathbf{r}) \tag{174}$$

We shall suppose first of all that we are dealing with a conduction band for which the energy may be expressed as a function of k in the form

$$E = E_c + \hbar^2 k^2/2m_e \tag{175}$$

In this case equation (174) becomes

$$-\frac{\hbar^2}{2m_e}\nabla^2 F(\mathbf{r}) + E_c F(\mathbf{r}) = EF(\mathbf{r}) \tag{176}$$

The wave functions $F(\mathbf{r})$ which are solutions of equation (176) are of the form

$$F(\mathbf{r}) = V^{-1/2} e^{i\mathbf{k}.\mathbf{r}} \tag{177}$$

the energy E being given by equation (175). Now suppose that we give the crystal a uniform dilatation δ which changes the lattice constant, then the values of all the energy levels will be shifted and we may write

$$E = E'_c + \hbar^2 k^2/2m'_c \tag{178}$$

The change in effective mass $(m'_e - m_e)$ may be shown to be very small and to give only second-order effects (Shockley and Bardeen, *loc. cit.*); we may therefore write equation (176) for the expanded lattice in the form

$$-\frac{\hbar^2}{2m_e}\nabla^2 F(\mathbf{r}) + E_c F(\mathbf{r}) + (E'_c - E_c)\, F(\mathbf{r}) = EF(\mathbf{r}) \tag{179}$$

* W. Shockley and J. Bardeen, *Phys. Rev.*, **77**, 407 (1950); **80**, 72 (1950).

This equation will still be true if the dilatation is a slowly varying function so that we may define a 'local' set of energy levels assuming E_c' to be constant. Strictly, the change $E_c'-E_c$ does not arise only from the change in lattice constant; the dilatation will also affect the carrier density through its effect on the forbidden energy gap and this will introduce a change in electrostatic potential. This may readily be shown to be very small, by a simple argument, when the dilatation is periodic; with a wavelength λ of the order of 10^{-6} cm there will be, even for an electron concentration of 10^{18} cm^{-3}, only one electron per λ^3 of volume and a small change in this value will have a very small effect on the potential; a more rigorous calculation by Shockley and Bardeen (*loc. cit.*) verifies this conclusion, and we shall neglect such effects.

If we write

$$(E_c'-E_c) = \delta E_c(\mathbf{r}) \tag{180}$$

and define the operator H_0 by means of the equation

$$H_0 = -\frac{\hbar^2}{2m_e}\nabla^2+E_c \tag{181}$$

equation (179) may be expressed in the form

$$H_0 F(\mathbf{r})+\delta E_c(\mathbf{r}) = EF(\mathbf{r}) \tag{182}$$

We may then solve equation (182) in terms of the wave functions of the undistorted crystal given by the solutions of equation (176) which may be written in the form

$$H_0 F(\mathbf{r}) = EF(\mathbf{r}) \tag{183}$$

In particular, we may apply the theory of transition probabilities developed in § 13.1 to the equation (182) to determine the probability of transitions between states with different wave vectors $\mathbf{k}_0$ and $\mathbf{k}'$, caused by the potential $\delta E_c(\mathbf{r})$.

It now remains to associate the potential $\delta E_c(\mathbf{r})$ with an acoustic wave of long wavelength. For a crystal with a high degree of symmetry, such as a cubic crystal it may readily be shown that the volume of the unit cell will not be changed to the first order by a displacement in the form of a shear but only by a pure dilatation. We therefore again see that the longitudinal acoustic modes will give the major contribution, though a contribution by the other modes cannot be ruled out when these cannot be resolved exactly into compressional and shear modes. We shall, however, only consider the effect of a dilatation $\delta = \delta V/V$ (see Fig. 13.2). We may express the change in E_c in the form

$$\delta E_c = \frac{\partial E_c}{\partial V}\delta V = V\frac{\partial E_c}{\partial V}.\delta \tag{184}$$

We define the deformation potential V_c by means of the equation

$$V_c = V\frac{\partial E_c}{\partial V}\Big)_{\delta=0} \tag{185}$$

and write δE_c in the form

$$\delta E_c(\mathbf{r}) = V_c \delta(\mathbf{r}) \tag{186}$$

Only if we regard the crystal as a continuous medium may we define a dilatation as a continuous function of a coordinate vector **r**. When δ varies slowly, in the sense for which the effective-mass approximation holds, this will be permissible. If $\delta\mathbf{r}$ represents the displacement vector of a point of the medium the dilatation $\delta(\mathbf{r})$ is given by

$$\delta(\mathbf{r}) = \nabla . \delta\mathbf{r} \tag{186a}$$

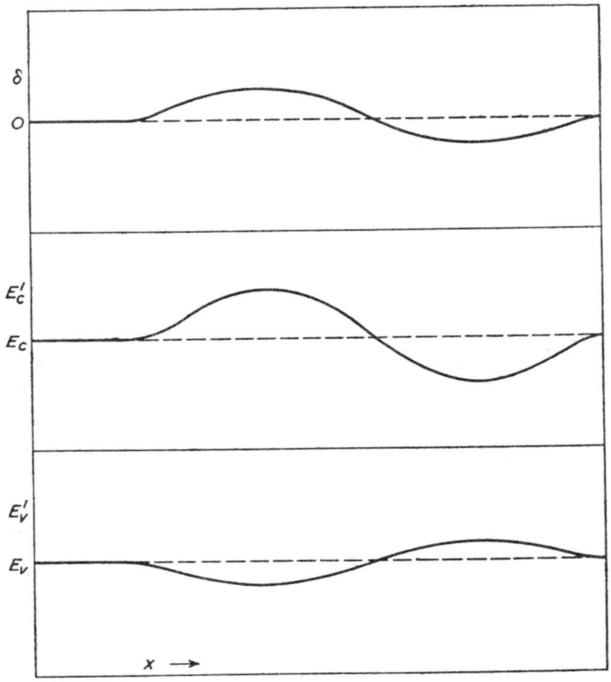

FIG. 13.2. Variation of band edges with dilatation δ

as may readily be seen by calculating the change of volume of a small cube. If $\delta\mathbf{r}_j$ corresponds to a longitudinal acoustic wave it will be given by (cf. equation (90))

$$\delta\mathbf{r}_j = \tfrac{1}{2}\mathbf{A}_s \, e^{i\mathbf{K}_s.\mathbf{d}_j - i\omega_s t} + c.c. \tag{187}$$

where $\mathbf{A}_s$ is a vector parallel to $\mathbf{K}_s$.

To obtain an expression for the dilatation we replace the lattice vectors $\mathbf{d}_j$ in equation (187) by the continuous variable **r** before applying the differential operator ∇; we obtain in this way

$$\delta(\mathbf{r}) = \tfrac{1}{2}i(\mathbf{A}_s . \mathbf{K}_s) \, e^{i\mathbf{K}_s.\mathbf{r} - i\omega_s t} + c.c. \tag{188}$$

Since $\mathbf{K}_s$ and $\mathbf{A}_s$ are parallel we have

$$\delta(\mathbf{r}) = \tfrac{1}{2}iA_s K_s e^{i\mathbf{K}_s.\mathbf{r}-i\omega_s t} + c.c. \tag{189}$$

The potential change δE_c is therefore given by

$$\delta E_c = \tfrac{1}{2}iV_c A_s K_s e^{i\mathbf{K}_s.\mathbf{r}-i\omega_s t} + c.c. \tag{190}$$

We may now apply perturbation theory as in § 13.4, replacing the scattering potential $U_j(\mathbf{r})$ by δE_c and using as initial and final wave functions

$$\psi_0 = V^{-1/2} e^{i\mathbf{k}_0.\mathbf{r}} \tag{191}$$

$$\psi_m = V^{-1/2} e^{i\mathbf{k}'.\mathbf{r}} \tag{192}$$

As before, we use the condition for conservation of energy given by equations (100) and (103) corresponding, respectively, to absorption and emission of a phonon of energy $\hbar\omega_s$. The matrix element H_{m0} is given by

$$H_{m0} = \int \psi_m^{\star} \delta E_c \psi_0 \, d\mathbf{r} \tag{193}$$

$$= \frac{iV_c A_s K_s}{2V} \int e^{i(\mathbf{k}_0+\mathbf{K}_s-\mathbf{k}').\mathbf{r}} \, d\mathbf{r} \tag{194}$$

and a corresponding term with $\mathbf{K}_s$ replaced by $-\mathbf{K}_s$ coming from the complex conjugate in the expression (190) for δE_c. We therefore have as before, since k_0, k' and K_s are small

$$\mathbf{k}' - \mathbf{k}_0 = \pm \mathbf{K}_s \tag{195}$$

since otherwise the integral in (194) is equal to zero. When condition (195) is satisfied we have

$$H_{m0} = \tfrac{1}{2}iV_c K_s A_s \tag{196}$$

which is the same as given by equation (123) but with V_0 replaced by V_c. The two approximations are not quite equivalent but it turns out that V_0 and V_c do not differ very much, each being equal to a few electron volts.

The deformation potential V_c is not readily calculable since, in general, we do not have good theoretical values for the variation of energy with lattice spacing. We may, however, obtain a similar equation for the relaxation time for scattering of holes, the quantity V_c being replaced by a similar quantity V_v, defined in terms of the energy E_v of the top of the valence band so that

$$V_v = V \left.\frac{\partial E_v}{\partial V}\right)_{\delta=0} \tag{197}$$

The effective mass m_e is also replaced by the hole mass m_h. If V_v and V_c have opposite signs we have for the change $\delta\Delta E$ in the forbidden energy gap $\Delta E = E_c - E_v$ (see Fig. 13.2)

$$\delta\Delta E = (|V_c| + |V_v|)\,\delta V/V \tag{198}$$

The quantities V_c and V_v may be determined from measurements of mobility under conditions of pure lattice scattering provided we know m_e and m_h. An independent check may then be obtained since $\delta \Delta E$ may be deduced from the variation of carrier concentration with pressure. These quantities are all known for only a few semiconductors to which the theory is applicable but, in general, a self-consistent set of values is obtained.

The above method of calculating τ has the advantage that it is very easily generalised to deal with semiconductors having non-scalar effective-mass tensors, while that given in § 11.4.1 is not. If the constant-energy surfaces are ellipsoidal, having effective masses m_1, m_2, m_3 in the principal directions, H_0 is given by

$$H_0 = -\frac{\hbar^2}{2}\left(\frac{1}{m_1}\frac{\partial}{\partial x^2}+\frac{1}{m_2}\frac{\partial}{\partial y^2}+\frac{1}{m_3}\frac{\partial}{\partial z^2}\right)+E_c \tag{199}$$

The wave functions ψ_0 and ψ_m given by equations (191), (192) are solutions of the equation $H_0\psi = E\psi$ corresponding to the energy values

$$E_0 = E_c+\frac{\hbar^2}{2}\left(\frac{k_{x0}^2}{m_1}+\frac{k_{y0}^2}{m_2}+\frac{k_{z0}^2}{m_3}\right) \tag{200}$$

$$E_m = E_c+\frac{\hbar^2}{2}\left(\frac{k_x'^2}{m_1}+\frac{k_y'^2}{m_2}+\frac{k_z'^2}{m_3}\right) \tag{201}$$

The transition probability is then given by equation (17) with the appropriate value of the density of states function $N(E)$. Since the matrix element H_{m0} is constant we have

$$\frac{1}{\tau} = \frac{2\pi}{\hbar}|H_{m0}|^2\int(1-\cos\gamma)\,N(E)\,\mathrm{d}S \tag{202}$$

where γ is the angle between the vector $\mathbf{k}_0$ and the vector $\mathbf{k}'$. Since $N(E)$ and the surface element on the constant-energy surface are the same for the vector $-\mathbf{k}'$, with γ replaced by $\pi-\gamma$, as for $\mathbf{k}'$, we see that

$$\int N(E)\cos\gamma\,\mathrm{d}S = 0 \tag{203}$$

and we have

$$\frac{1}{\tau} = \frac{2\pi}{\hbar}|H_{m0}|^2\int_S N(E)\,\mathrm{d}S$$

where S represents the constant-energy surface corresponding to energy E. Now the total number of states with energy between E and $E+\mathrm{d}E$ is given by

$$\mathrm{d}E\int N(E)\,\mathrm{d}S \tag{204}$$

We have seen in § 9.4.2 that for ellipsoidal constant-energy surfaces this is equal to $2\pi V(2m_d)^{3/2}E^{1/2}/\hbar^3$ where m_d is the density of states effective mass given by

$$m_d^3 = m_1 m_2 m_3 \tag{205}$$

This differs from the value for spherical constant energy surfaces only through the replacement of m_e by m_d. We therefore see that the relaxation time τ for lattice scattering, when we have ellipsoidal constant-energy surfaces, is given by the expression (133) with m_e replaced by m_d.

The average value of the relaxation time for a non-degenerate semiconductor with scalar effective mass m_e is obtained from equation (83) of § 10.6 with $s = \frac{1}{2}$ and so is given by

$$\langle\tau\rangle = \frac{2^{3/2}\pi^{1/2}\hbar^4\rho c_l^2}{3m_e^{3/2}V_c^2(\boldsymbol{k}T)^{3/2}} \tag{206}$$

The mobility μ_l is given by (see § 10.6)

$$\mu_l = \boldsymbol{e}\langle\tau\rangle/m_e \tag{207}$$

and so is proportional to $m_e^{-5/2}$ and to $T^{-3/2}$. For ellipsoidal constant-energy surfaces we have

$$\langle\tau\rangle = \frac{2^{3/2}\pi^{1/2}\hbar^4\rho c_l^2}{3m_d^{3/2}V_c^2(\boldsymbol{k}T)^{3/2}} \tag{208}$$

where m_d is the 'density-of-states' effective mass and is equal to $(m_1 m_2 m_3)^{1/3}$.

When we have a degenerate valence band as in Si and Ge the situation is considerably more complex. We may, however, assume, as an approximation, that the valence band consists of two 'spherical' bands with scalar hole masses m_{h1} and m_{h2} the former corresponding to the 'heavy' holes and the latter to the 'light' holes*. We shall have two scattering processes in operation, scattering within a band and scattering between bands. For the former the average relaxation time will be given by equation (206) with m_e replaced by m_{h1} or m_{h2}. If we assume that the bands move together when a dilatation is applied, the deformation potential will be the same for inter-band scattering as for scattering within each band. The only difference in calculating the relaxation time will therefore come from the density of states in the region of the state *to* which a hole is scattered. For scattering of heavy holes to the light hole band we therefore must replace m_e by m_{h2} to obtain the inter-band relaxation time; similarly, for scattering of light holes into the heavy hole band we replace m_e by m_{h1}. If therefore we write τ_0 for the relaxation time for scattering of holes in a non-degenerate 'spherical' band with effective mass m_0 as given by equation (206) we have for τ_1 the relaxation time for the heavy holes

$$\frac{1}{\tau_1} = \frac{1}{\tau_0}\left(\frac{m_{h1}}{m_0}\right)^{3/2} + \frac{1}{\tau_0}\left(\frac{m_{h2}}{m_0}\right)^{3/2} \tag{209}$$

since we have to add the probabilities of scattering or the reciprocals of the relaxation times. Similarly, the relaxation time for scattering of the

* See, for example, R. A. Smith, *Semiconductors*. Cambridge University Press (1959), p. 350.

light holes τ_2 is also given by equation (209). The mobilities of heavy and light holes μ_1 and μ_2 are therefore given by

$$\frac{e}{\mu_1} = \frac{m_{h1}}{\tau_1} = \frac{m_0}{\tau_0}\left\{\left(\frac{m_{h1}}{m_0}\right)^{5/2} + \left(\frac{m_{h2}}{m_0}\right)^{3/2}\left(\frac{m_{h1}}{m_0}\right)\right\} \tag{210}$$

$$\frac{e}{\mu_2} = \frac{m_{h2}}{\tau_2} = \frac{m_0}{\tau_0}\left\{\left(\frac{m_{h2}}{m_0}\right)^{5/2} + \left(\frac{m_{h1}}{m_0}\right)^{3/2}\left(\frac{m_{h2}}{m_0}\right)\right\} \tag{211}$$

For Ge $m_{h2} \ll m_{h1}$ so that the relaxation time for the heavy holes is not very different from that for a single band, and the relaxation time for light holes to change to heavy holes is almost equal to the relaxation time for scattering of the heavy holes. The mean lifetime of an individual light hole is therefore quite short.

For crystals having more than one atom in the basic unit cell, equation (123) for the relaxation time still applies for scattering by the acoustic modes; the mass M is replaced by the sum of the masses in each unit cell as will be seen from the derivation of the expression for the mean amplitude A_s. In this case, however, scattering by the optical modes may also be of importance in determining the relaxation time. For small values of the wave vector $\mathbf{K}_s$ the frequency of the lattice vibrations may be regarded as constant for these modes and equal to the reststrahl frequency ν_0. Emission of an optical phonon will only be possible for scattering of electrons whose energy is greater than $h\nu_0$ so that this process is unlikely to be important for semiconductors except at fairly high temperatures. Also, except at high temperatures, there will be few optical phonons present in the crystal so that scattering with absorption of an optical phonon will also be improbable. For polar semiconductors, on the other hand, because of the strong interaction between electrons and the dipole moment created by the relative movement of the positive and negative ions which make up the crystal, scattering by polar modes is likely to be of importance even at room temperature and has been discussed by H. Fröhlich and N. F. Mott*. For semiconductors such as Ge and Si, which have multiple minima in the conduction band scattering between the different minima will also be of importance. The change in momentum in such scattering is large and the corresponding phonon energy quite large so that the same remarks apply as for optical scattering; this type of scattering, sometimes called 'inter-valley' scattering has been discussed in some detail by C. Herring†.

13.4.5 Some numerical values

We shall now calculate some values of the relaxation time τ and mean free path l in order to compare them with the experimental values. We shall do this for two metals Na and Ag, the former being a typical alkali

* H. Fröhlich and N. F. Mott, *Proc. Roy. Soc.*, **A171**, 496 (1939).
† C. Herring, *Bell Syst. Tech. J.*, **34**, 237 (1955).

metal and the latter a typical noble metal; for other metals more detailed comparison of theory and experiment will be found in books on metals such as those already referred to. We shall assume that in each case we have one conduction electron per atom and that the effective electron mass m_e is equal to the free mass m. Values of various quantities of interest are given in Table 13.1. From the value of the resistivity ρ the

Table 13.1

Various constants for Na and Ag

	N (cm^{-3})	r_0 (cm)	k_0 (cm^{-1})	v_0 (cm/sec)
Na	$2{\cdot}6 \times 10^{22}$	$2{\cdot}1 \times 10^{-8}$	$9{\cdot}2 \times 10^{7}$	$10{\cdot}5 \times 10^{8}$
Ag	$5{\cdot}95 \times 10^{22}$	$1{\cdot}6 \times 10^{-8}$	$1{\cdot}2 \times 10^{8}$	$1{\cdot}4 \times 10^{8}$

N = number of free electrons per unit volume.
r_0 = radius of Wigner–Seitz equivalent sphere.
k_0 = magnitude of the wave vector corresponding to the Fermi surface.
v_0 = velocity of electron corresponding to the Fermi surface.

mobility μ and relaxation time τ may be obtained from the relationship $\rho^{-1} = Ne\mu = Ne^2\tau/m_e$; values obtained in this way are given in Table 13.2, together with the mean free path $l = \tau v_0$.

Table 13.2

Relaxation time and mean free path for Na and Ag for T = 300°K

	ρ (Ω cm)	μ (cm^2/V sec)	τ (sec)	l (cm)
Na	$4{\cdot}3 \times 10^{-6}$	55	$3{\cdot}1 \times 10^{-14}$	$3{\cdot}25 \times 10^{-6}$
Ag	$1{\cdot}45 \times 10^{-6}$	71	$4{\cdot}05 \times 10^{-14}$	$5{\cdot}7 \times 10^{-6}$

We could also calculate τ using equation (156), taking values of V_0 obtained from band-structure calculations; alternatively we may calculate the value of V_0 required to give the values of τ obtained from the resistivity as given in Table 13.2. The values which we take for the quantity θ_l are values of the Debye temperature θ_D and are not strictly equivalent (see § 13.4.2); to the order of approximation for which equation (156) is valid the difference is probably not significant. For Na and Ag the values which we have taken are respectively 150°K and 200°K. For the temperature $T = 300°K$ at which we calculate τ the condition $T \gg \theta$ certainly does not hold; experimentally it is found, however, that the resistivity is fairly accurately proportional to T in this temperature range in accordance with equation (156) which is therefore unlikely to be much in error in this respect. The values of the various quantities used in the calculation are given in Table 13.3, together with values of V_0 required to give the values of the relaxation time obtained from the observed values of the resistivity. Values of V_0 (denoted by $V_0^\star$) obtained from band-structure calculations using the Wigner–Seitz method are also shown in Table 13.3. These are somewhat smaller than the values required to give the observed resistivity, particularly for Ag; some of this discrepancy may be due to our neglect of the large angle scattering

Table 13.3

Comparison of values of V_0 obtained from the resistivity and from Wigner–Seitz type of calculations for Na and Ag

	M/m	θ_D (°K)	V_0 (eV)	$V_0^\star$ (eV)
Na	$4{\cdot}2 \times 10^4$	150	2·3	1·3
Ag	$1{\cdot}95 \times 10^5$	200	5·5	2·3

M = atomic mass.
θ_D = Debye temperature (approximate).
V_0 = potential energy at surface of Wigner–Seitz sphere obtained by comparing calculate and measured value of τ.
$V_0^\star$ = value of V_0 obtained from band-structure calculations (for Na, E. Wigner and F. Seitz, *Phys. Rev.*, **43**, 804 (1933); *Ibid.*, **46**, 509 (1934). For Ag, K. Fuchs, *Proc. Roy. Soc.*, **A151**, 585 (1935).

arising from the *umklapprozesse* but is more likely to be due to the approximations made in calculating the matrix elements and in assuming that $m_e = m$ (see § 13.4.2). A full discussion of these discrepancies is given in a number of books on metals such as those already referred to. In view of the approximate nature of the treatment given in § 13.4.2 the agreement between the calculated and measured values of the resistivity may be regarded as fairly satisfactory and we may conclude that scattering of electrons by lattice vibrations accounts for most of the resistivity of metals, except at very low temperatures at which the effect of impurities and mechanical imperfections becomes of importance.

For semiconductors it is better to use equation (208) giving the averaged relaxation time $\langle\tau\rangle$ in terms of the deformation potential V_c, in order to compare the calculated and experimental values of $\langle\tau\rangle$. The relaxation time for a non-degenerate semiconductor is obtained from the

Table 13.4

Quantities derived from the mobility for Ge, Si and InSb ($T = 300°K$)

	μ_e ($m^2/V\,sec$)	μ_h ($m^2/V\,sec$)	m_{ec}/m	m_h/m	$\bar{v}_e$ ($10^7 cm/sec$)	$\bar{v}_h$ ($10^7 cm/sec$)	$\langle\tau_e\rangle$ (10^{-13} sec)	$\langle\tau_h\rangle$ (10^{-13} sec)	l_e (10^{-6} cm)	l_h (10^{-6} cm)
Ge	0·39	0·18	0·12	0·28	3·4	2·3	2·5	2·8	8·5	6·5
Si	0·14	0·05	0·26	0·49	2·4	1·7	2·1	1·4	5·0	2·4
InSb	8·0	0·08	0·04	0·18	6·0	2·8	18·5	0·82	110	2·2

μ_e = electron mobility in n-type material.
μ_h = hole mobility in p-type material.
m_{ec} = conductivity effective mass for electrons (equal to m_e in InSb).
m_h = effective mass for holes.
$\bar{v}_e, \bar{v}_h$ = 'averaged' velocities for electrons and holes.
$\langle\tau_e\rangle, \langle\tau_h\rangle$ = 'averaged' relaxation times for electrons and holes.
l_e, l_h = 'averaged' mean free paths for electrons and holes.

Table 13.5

Further data for Ge, Si and InSb

	m_{ed}/m	ρc_l^2 (10^{11} N/m^2)	ρ (10^3 kg/m^3)	c_l (10^3 m/sec)	K (10^{-10} m^2/kg)	$d\|\Delta E\|/dp$ (10^{-10} eV m^2/kg)	V_g (eV)
Ge	0·22	1·29	5·33	4·9	1·35	4·6	3·4
Si	0·33	1·66	2·33	8·5	1·0	4·2	4·2
InSb	0·04	0·67	5·79	3·4	2·15	15·5	7·2

m_{ed} = 'density-of-states' effective mass for electrons.
$\rho c_l^2 = c_{11}$ = elastic constant.
ρ = density.
c_l = velocity of longitudinal sound waves.
K = compressibility = $3/(c_{11}+2c_{12})$.
$d\Delta E/dp$ = rate of change of forbidden energy gap with pressure.
$V_g = |d\Delta E/dp| \div K = |V\,d\Delta E/dV|$.

mobility μ by means of the relationship $\langle\tau\rangle = m_c\mu/e$ where m_c is the conductivity effective mass (see § 10.6). We shall calculate $\langle\tau\rangle$ for pure Ge, Si and InSb for which the values of the effective masses are well known*; the values required for subsequent calculation are given in Table 13.4. For the effective mass of electrons in InSb we have used a value $0{\cdot}04\boldsymbol{m}$ rather than the more usual cyclotron-resonance value $0{\cdot}013\boldsymbol{m}$ since there is a good deal of evidence for the higher value at room temperature†.

A comparison made by J. Bardeen and W. Shockley in their original paper on the deformation potential (see § 13.4.4) is invalid since they used the free electron mass $\boldsymbol{m}$, values of the effective masses derived from cyclotron resonance experiments not being then available. It is now known that the effective masses are very different from $\boldsymbol{m}$, so we have repeated the calculations using up-to-date values. We have also used later values of the elastic constants quoted by H. B. Huntington‡. The quantity ρc_l^2 is equal to the elastic constant c_{11} and we give values in Table 13.5; we have also calculated values of c_l as a matter of interest in order to compare them with the average values of the electron velocity.

For a given value of the energy E the velocity will be a function of the direction of motion when the effective mass is not scalar. We have, however, calculated the value of an 'average' velocity $\bar{v}$ defined by means of the equation

$$\bar{v} = (3\boldsymbol{k}T/m_c)^{1/2} \tag{212}$$

and from this have derived an 'average' mean free path $\bar{l}$ given by

$$\bar{l} = \tau\bar{v} \tag{213}$$

These quantities are also given in Table 13.4.

Using the values of ρc_l^2 given in Table 13.5 and equation (208) we may calculate the magnitudes of the deformation potentials V_c and V_v for the conduction and valence bands required to give the values of $\langle\tau_e\rangle$ and $\langle\tau_h\rangle$ quoted in Table 13.4. The 'density-of-states' effective mass m_{ed} is given for electrons in Table 13.5; for holes we have taken $m_{hd} = m_h$, where m_h is the effective mass of the 'heavy' holes. The calculation has been carried out for $T = 300°$K and the results are shown in Table 13.6. There is, unfortunately, no simple method available for obtaining an independent check on the values of the deformation potentials V_c and V_v. The value of a related quantity V_u which is a deformation potential for uniaxial strain has been obtained by G. Weinreich, T. M. Sanders and

* See, for example, R. A. Smith, *Semiconductors*. Cambridge University Press (1959), Chapters 10 and 11; values quoted are taken from this book, unless otherwise stated, where full references to the original papers are also given.

† Private communication from Dr. B. Lax.

‡ H. B. Huntington, *Solid State Phys.*, **7**, 214 (1958).

H. G. W. White* from measurements of the 'electroacoustic' effect—the electromotive force generated by a sound wave; the value found for *n*-type Ge is 16 eV which is of the same order of magnitude, but somewhat larger than the value we have calculated for V_c. If shear modes were to produce no change in the position of the energy bands, the change in forbidden energy gap ΔE under pressure, which has been measured, could be related to the deformation potentials. For cubic crystals such as Ge and Si, however, the strain components cannot be simply resolved into longitudinal and transverse components and the relationship between the change in energy gap under pressure and the deformation potentials is rather complex†. Even if we neglect the shear modes there is still some uncertainty as to whether the conduction band and valence

Table 13.6

Deformation potentials for Gi, Si and InSb derived from mobility

	V_c (eV)	V_v (eV)	(V_c+V_v) (eV)
Ge	9·5	5·8	15·3
Si	9·0	7·5	16·5
InSb	8·3	14·5	22·5

band vary in the same or in opposite directions. If the levels of the bands both increase or decrease we have

$$|\delta\Delta E| = |\delta E_c - \delta E_v| \tag{214}$$

If the level of one decreases while the other increases, on the other hand, we have

$$|\delta E| = |\delta E_c| + |\delta E_v| \tag{215}$$

The change of ΔE with pressure p has been measured for all three semiconductors (see R. A. Smith, *loc. cit.*) and values of $\mathrm{d}|\Delta E|/\mathrm{d}p)$ are given in Table 13.5. If we define a quantity V_g by means of the equation

$$V_g = |V\,\mathrm{d}\Delta E/\mathrm{d}V| \tag{216}$$

it may be obtained from the compressibility K (equal to $-V^{-1}\mathrm{d}V/\mathrm{d}p$) and $\mathrm{d}\Delta E/\mathrm{d}p$, since

$$V_g = |\mathrm{d}\Delta E/\mathrm{d}p) \div K| \tag{217}$$

* H. G. W. White, *Phys. Rev.*, **114**, 33 (1959).

† See, for example, C. Herring and E. Vogt, *Phys. Rev.*, **101**, 944 (1956).

We therefore see that either

$$V_g = V_c + V_v \tag{218}$$

or

$$V_g = |V_c - V_v| \tag{219}$$

On comparison of the values of V_g given in Table 13.5 with the values of V_c and V_v given in Table 13.6 we see that it seems likely that for Si and Ge the valence and conduction bands move in opposite directions but that the values of V_c and V_v obtained from the mobility are about four times too large to agree with the value of V_g. The discrepancy between the value of V_g and the deformation potentials has been discussed by Herring and Vogt (*loc. cit.*) who have shown that the deformation potential for shear modes has opposite sign to that for the longitudinal mode, and this may considerably reduce the value of V_g as compared with the values of the deformation potentials. The calculated values also depend strongly on the values of the effective mass. We have used the low-temperature cyclotron resonance values for Si and Ge and there is some evidence that the effective masses at room temperature may be larger. The large value obtained for V_v for InSb is almost certainly due to the use of too small a value for m_h, the value $0{\cdot}18m$ being somewhat uncertain. A more serious discrepancy between theory and experiment is the fact that for none of these semiconductors is the theoretical prediction, that the mobility should vary is $T^{-1{\cdot}5}$, found to be even approximately true; variations from $T^{-1{\cdot}66}$ to $T^{-2{\cdot}6}$ are found. For n-type Si and Ge this is explained by Herring and Vogt (*loc. cit.*) by the inclusion of some scattering by shear modes, optical modes and inter-valley scattering; it is clear that the lattice scattering in such materials is extremely complex. The treatment using deformation potentials, gives, however, the order of magnitude of the acoustic mode scattering with reasonable values of the potentials.

It is interesting to compare the magnitudes of the mobilities for metals and semiconductors. We may see at once that the large mobilities for the latter are mainly due to the small values of the effective mass. For example, if we take out this factor for Ge the value for μ_m would reduce to about 50 cm^2/V sec and the high value for InSb would come down to 27 cm^2/V sec, values comparable to those found for metals. If we compare equation (157) with equation (134) we see that the energy-independent mean free path l is given by the same expression for a metal as for a semiconductor apart from a numerical factor which makes the value for a metal about six times greater than for a semiconductor with the same parameters; this factor comes from the fact that in a metal the scattering is not isotropic. When we calculate the relaxation time, or mobility, a factor $E^{-1/2}$ must be included. For the metal we have $E = E_F$ and to obtain the order of magnitude of τ for a semiconductor we may take $E = \frac{3}{2}kT$. If we take $E_F = 3$ eV (as for Na) and $T = 300°$K the $E^{-1/2}$ factor is equal to 7·7 in favour of the semiconductor, and just about cancels the factor 6 referred to above. We therefore see, that at room temperature the main difference comes from the effective masses.

13.5 Transitions involving photons and phonons

We have seen in § 13.3 that when a photon is absorbed by an electron in a crystal a rigid selection rule governs the possible transitions which may be made; when we neglect the momentum of the photon, which is generally very small compared with the crystal momentum of the electron, the selection rule is equivalent to conservation of crystal momentum (see § 13.3). If no other process is involved the final and initial states must correspond to the same wave vector $\mathbf{k}$, so that for semiconductors such as Si and Ge transitions between the top of the valence band and bottom of the conduction band would be ruled out since they have different values of $\mathbf{k}$. Such transitions are of great importance since they determine the value of the forbidden energy gap ΔE, and are observed to take place with sufficiently high probability to lead to an absorption coefficient which, although smaller than that for 'vertical' transitions, is still quite appreciable. Since the selection rule referred to is a very rigid one, some additional process must take place which enables crystal momentum to be conserved. This process is the absorption or emission of one or more phonons. As we shall see, multiple phonon processes are much less probable than those involving a single phonon. If the wave vector of the phonon is $\mathbf{K}_s$, momentum of amount $\hbar\mathbf{K}_s$ is lost in the process of absorption and $\hbar\mathbf{K}_s$ gained in emission (see § 13.4). If $\mathbf{k}_0$ and $\mathbf{k}_f$ are the wave vectors of the initial and final states of the electron, and $\mathbf{K}_p$ the wave vector of the photon, strict conservation of crystal momentum is expressed by the condition

$$\mathbf{k}_f \pm \mathbf{K}_s = \mathbf{k}_0 + \mathbf{K}_p \tag{220}$$

the + sign corresponding to emission of a phonon and the − sign to absorption; when we neglect the momentum of the photon, equation (220) reduces to

$$\mathbf{k}_f - \mathbf{k}_0 = \mp \mathbf{K}_s \tag{221}$$

When the wave vector $\mathbf{k}_f$ corresponds to a state in the conduction band and $\mathbf{k}_0$ to a state of the valence band the condition for conservation of energy is then

$$E_c(\mathbf{k}_f) = E_v(\mathbf{k}_0) + \hbar\omega \mp \hbar\omega_s \tag{222}$$

where $\nu = \omega/2\pi$ is the frequency of the photon and $\nu_s = \omega_s/2\pi$ is the frequency of the phonon, the − sign corresponding to emission and the + sign to absorption of a phonon. Transitions of this type may also occur between states in the same band and account for the so-called free-carrier absorption. Since this may be treated by means of a quasi-classical method of analysis we shall restrict our discussion to inter-band transitions which require a treatment based on the quantum theory.

In order to deal with transitions of this kind we shall have to extend the theory of perturbations given in § 13.2, since to the degree of approximation of the former treatment we get zero transition probability. Before doing this, however, we shall consider the physical processes involved in such double transitions in greater detail. The condition that the wave vector **k** should not change appreciably in a transition from one electronic state to another comes from the fact that unless this condition is satisfied the matrix element for the transition is equal to zero (see § 13.3); this rules out *all* electronic transitions except 'vertical' ones. To make a non-vertical transition from a state A in the valence band to a state B in the conduction band with different **k** we may consider the possibility of a transition being made 'vertically' to a state I in the conduction band and subsequently a transition within the conduction band to the state B. Under the conditions which we are considering the energy E_I of the state I will be a good deal higher than the energy E_B of the state B so that if E_A is the energy of the initial state $E_I - A_A$ is a good deal greater than $\hbar\omega$ whereas $E_B - A_A$ differs from $\hbar\omega$ by a small amount corresponding to the energy of a phonon; in other words the transition to the state I is allowed by conservation of momentum but not by conservation of energy. There is no way of avoiding the conservation of momentum, but provided the time spent by the electron in the intermediate state is *very* short, because of the Uncertainty Principle (see § 1.4), the need for conservation of energy may be relaxed. Thus it appears that a transition from state A to state B through state I *is* possible. The transition from state I to state B by emission or absorption of a phonon is possible again with conservation of momentum but not necessary with conservation of energy. Thus we see that conservation of momentum *in each step* is required, the two conditions also giving conservation of momentum between states A and B; conservation of energy is required only between states A and B. That such transitions involving short-lived intermediate or 'virtual' states, as they are sometimes called, can take place is a consequence of the quantum theory of perturbations taken to a second order of approximation which we shall now discuss. This is likely to give a transition probability very much less than for direct transitions; as we shall see, however, a new degree of freedom is introduced through the introduction of the phonon energy, and transitions from many more states may take place for a given value of the frequency, thus compensating to some extent for the very small transition probability.

For a *direct* transition at a *given* frequency ω, transitions are limited to take place from states lying very near to a state in the valence band fixed by the conservation of energy, the final state in each case being fixed by the conservation of momentum; there is thus essentially no freedom of choice of initial and final states. For the indirect transition, however, all states in the valence band with energy in a certain range of values may provide initial states for a transition. Suppose we fix the initial state A; this fixes the intermediate state I from which a transition can be made to a state near a minimum of the conduction band, the wave vector $\mathbf{K}_s$ of the emitted or absorbed phonon being determined so as to conserve energy in the double transition. Thus the energy of the initial state can lie between zero and $\hbar(\omega \pm \omega_s) - \Delta E$ *below* the top of the valence band (see (Fig. 13.3)), and all states in this range may be considered as suitable initial states. The total probability for a transition is then obtained by summing over these states. Correspondingly the states to which transitions may be made have energies lying between

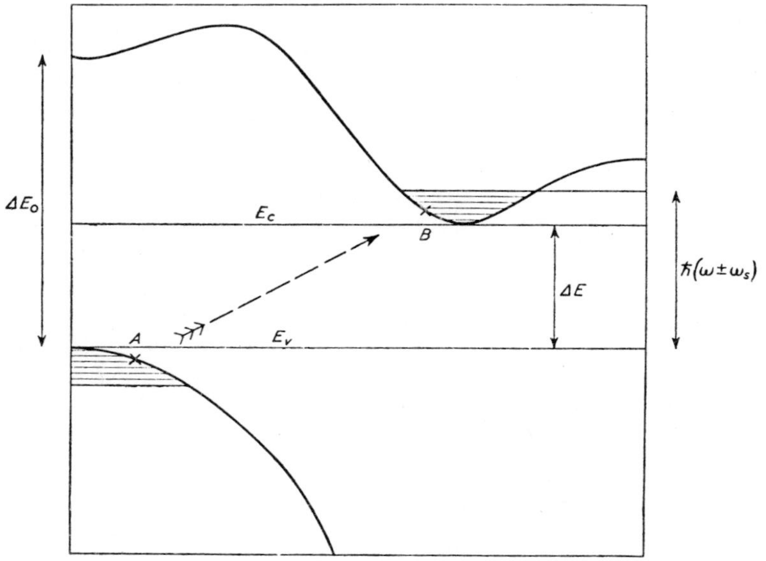

FIG. 13.3. Energy conditions for an indirect transition

$\hbar(\omega \pm \omega_s) - \Delta E$ and zero above the bottom of the conduction band; these regions are shown as shaded in Fig. 13.3. For transitions limited to regions near the top of the valence band and bottom of the conduction band, i.e. for which $(\hbar\omega \pm \hbar\omega_s - \Delta E)$ is small, ω_s may be regarded as constant provided the total change in electron wave vector between the initial and final states is large. In this case the change in phonon wave vector throughout the region of possible transitions is comparatively small giving a small change in the phonon energy $\hbar\omega_s$; this is particularly so in Ge where the minima in the conduction band lie at the band edge. We must now show that indirect transitions of the type we have envisaged may take place with appreciable probability and to do this we consider the probability of transitions induced by the simultaneous action of a light wave and sound wave. As we shall see, we have to extend the theory of perturbations given in § 13.2 to a second-order approximation in order to obtain a transition probability that increases steadily with time.

There is an alternative process to that which we have discussed by means of which an electron may make an indirect transition from a state A near the top of the valence band to a state near the bottom of the conduction band. By absorption or emission of a phonon it may first make a transition to a state I' in the valence band having the same wave vector as the state B in the conduction band. It may then, using I' as an intermediate state make a 'vertical' transition to the state B; both processes are illustrated in Fig. 13.3 and will have to be taken into account in calculating the total transition probability. Moreover, we shall generally have several equivalent minima in the conduction band when the lowest point is not at the zone centre and we must sum over all these minima.

We may apply the theory of § 13.2 to calculate the transition probability of an electron from an initial state having wave function ψ_0 to a state having wave function ψ_m under the simultaneous action of a light wave and a sound wave. Two methods of approach may be adopted corresponding to the treatments of lattice scattering given in §§ 13.4.1, 13.4.2 and 13.4.3, i.e. we may treat the sound wave as a periodic perturbation given by equation (90a) or we may use the quantized normal modes; both give the same results provided we use the appropriate expression for the phonon density and since the former is simpler we shall adopt it. The perturbing Hamiltonian H_1 is then given by

$$H_1 = \tfrac{1}{2}\mathbf{A}_l e^{i(\mathbf{K}.\mathbf{r}-\omega t)} + \tfrac{1}{2}\mathbf{A}_s e^{i(\mathbf{K}_s.\mathbf{r}-\omega_s t)} + c.c. \tag{223}$$

$\mathbf{A}_l$, being an operator given by equations (39) and (46) and $\mathbf{A}_s$ a constant vector given by equation (90a). We shall confine our attention to states associated with two bands only, the conduction band and the valence band. If we start with an initial state near the top of the valence band we must consider transitions to two types of states, (1) other states in the valence band, (2) states in the conduction band. We shall assume that the photon energy $\hbar\omega$ is approximately equal to ΔE, and we see that no transitions of type (1) are possible for which energy is conserved so as to obtain a transition probability which increases steadily with time. Because of the rigid selection rule which demands that inter-band transitions should be vertical, the only non-zero matrix elements are those for which energy cannot be conserved if ΔE_0, the forbidden energy gap at $\mathbf{k} = 0$, is somewhat greater than ΔE (see Fig. 13.3). Thus we have no transitions given by first-order perturbation theory for which the transition probabilities increase steadily with time. Some of the quantities a_n as defined in § 13.3 are not zero—for example, those connecting the initial state with one having the same value of $\mathbf{k}$ but in the conduction band. None of the quantities $\omega_{m0} - \omega$ which occur in the denominator in equation (29) vanish to give a large value of $|a_m|$. The quantities

$|a_m|$ for which the matrix elements do not vanish lead only to small oscillatory probabilities to this order of approximation. When we take the approximation one stage further we shall find that they lead to transition probabilities which increase linearly with time. The general theory of second-order perturbations is given in detail in several books on quantum mechanics*. We shall discuss briefly in the next section those aspects applicable to the problem of indirect transitions.

13.5.1 Second-order perturbation theory

As for the first-order calculation of the probability of a transition between two states by a perturbing Hamiltonian H_1 let us first of all consider a perturbation which does not vary with time, as in § 13.2.1. We shall use the same notation as in § 13.2 to denote the various states and energy levels. We now assume that either the matrix element H_{m0} defined by equation (8) for two states (m) and (0) is equal to zero or that $\omega_{m0} = (E_m - E_0)/\hbar$ is *not* equal to zero so that none of the values of $|a_m|$ given by equation (13) lead to a transition probability which increases with time. We therefore proceed to a higher degree of approximation writing

$$\left.\begin{aligned} a_0 &= 1 + a_0^{(1)} + a_0^{(2)} \\ a_m &= a_m^{(1)} + a_m^{(2)} \quad m \neq 0 \end{aligned}\right\} \tag{224}$$

where $a_0^{(1)}$, $a_m^{(1)}$ represent the first-order approximation given by equation (13) and $a_0^{(2)}$, $a_m^{(2)}$, represent the next degree of approximation; the quantities $a_m^{(2)}$ are assumed to have the order of magnitude of H_1^2 or $a_m^{(1)} H_1$. We substitute the wave function obtained by using the equations (224) to give the coefficients of the stationary-state wave functions in the wave equation and proceed as in § 13.3, now retaining terms up to the second order. We obtain at once an equation for the quantities $a_m^{(2)}$ in the form,

$$i\hbar \frac{\mathrm{d}a_m^{(2)}}{\mathrm{d}t} = \sum_{n \neq 0} a_n^{(1)} H_{mn} \mathrm{e}^{i\omega_{mn}t} \tag{225}$$

if we assume that the average value of the perturbation H_{00} is equal to zero. On substituting for $a_n^{(1)}$ from equation (12) we obtain

$$i\hbar^2 \frac{\mathrm{d}a_m^{(2)}}{\mathrm{d}t} = -\sum_{n \neq 0} \frac{H_{n0} H_{mn}}{\omega_{n0}} (\mathrm{e}^{i\omega_{m0}t} - \mathrm{e}^{i\omega_{mn}t}) \tag{226}$$

since $\omega_{mn} + \omega_{n0} = \omega_{m0}$. We may now integrate equation (226) with the initial condition $a_m^{(2)} = 0$ at $t = 0$ and obtain

$$a_m^{(2)}(t) = \hbar^{-2} \sum_{n \neq 0} \frac{H_{n0} H_{mn}}{\omega_{n0}} \left\{ \frac{\mathrm{e}^{i\omega_{m0}t} - 1}{\omega_{m0}} - \frac{\mathrm{e}^{i\omega_{mn}t} - 1}{\omega_{mn}} \right\} \tag{227}$$

* See, for example, L. I. Schiff, *Quantum Mechanics* (2nd Ed.). McGraw-Hill (1955), p. 201.

We now see that we have a term with ω_{m0} in the denominator and involving matrix elements H_{n0} and H_{mn} of a set of states (n) with the states (0) and (m). We now have the possibility that $\omega_{m0} = 0$, i.e. that energy is conserved between the initial and final states and yet we are not restricted by the condition $H_{m0} = 0$. If there are any of the states (n) for which neither H_{n0} nor H_{mn} is not equal to zero then we may obtain as before a transition probability which increases with time on summing over a set of neighbouring states. The term having ω_{mn} in the denominator is subject to the same restrictions as those in $a_m^{(1)}$ and will not contribute. We may therefore take

$$|a_m^{(2)}|^2 = \sum_{n\neq 0} \frac{4|H_{n0}|^2\,|H_{mn}|^2 \sin^2 \frac{1}{2}\omega_{m0}}{\hbar^4\,\omega_{n0}^2\,\omega_{m0}^2} \tag{228}$$

The quantity $|H_{n0}|^2$ is proportional to the transition probability from the state (0) to the intermediate state (n), while $|H_{mn}|^2$ is proportional to the transition probability from the state (n) to the state (m); each term of equation (228) may therefore be interpreted as a probability of transition from state (0) to state (m) via the intermediate state (n) as already discussed.

The above analysis is readily modified to deal with the case of a perturbation which is periodic in the time, as in § 13.2.3. We shall proceed at once to discuss a perturbation of the form

$$H_1 = \tfrac{1}{2}\mathbf{A}_1\,\mathrm{e}^{i\omega_1 t} + \tfrac{1}{2}\mathbf{A}_2\,\mathrm{e}^{i\omega_2 t} + c.c. \tag{229}$$

$\mathbf{A}_1$ and $\mathbf{A}_2$ being operators or functions of $\mathbf{r}$, with $\omega_1 \gg \omega_2$ so that we only need to discuss terms in which $-\omega_1$ occurs (absorption of photon). If we write

$$H_{mn}^{(1)} = \tfrac{1}{2}\int \psi_m^\star\,\mathbf{A}_1\,\psi_n\,\mathrm{d}\mathbf{r} \tag{230}$$

$$H_{mn}^{(2)} = \tfrac{1}{2}\int \psi_m^\star\,\mathbf{A}_2\,\psi_n\,\mathrm{d}\mathbf{r} \tag{231}$$

we obtain an equation for $a_n^{(1)}$ similar to equation (29) namely

$$\begin{aligned} a_n^{(1)} = & -\frac{H_{n0}^{(1)}}{\hbar}\left\{\frac{\mathrm{e}^{i(\omega_{n0}+\omega_1)t}-1}{\omega_{n0}+\omega_1}+\frac{\mathrm{e}^{i(\omega_{n0}-\omega_1)t}-1}{\omega_{n0}-\omega_1}\right\} \\ & -\frac{H_{n0}^{(2)}}{\hbar}\left\{\frac{\mathrm{e}^{i(\omega_{n0}+\omega_2)t}-1}{\omega_{n0}+\omega_2}+\frac{\mathrm{e}^{i(\omega_{n0}-\omega_2)t}-1}{\omega_{n0}-\omega_2}\right\} \end{aligned} \tag{232}$$

Similarly, we obtain instead of equation (225)

$$\begin{aligned} i\hbar\frac{\mathrm{d}a_m^{(2)}}{\mathrm{d}t} = & \sum_{n\neq 0} a_n^{(1)}\{H_{mn}^{(1)}[\mathrm{e}^{i(\omega_{mn}+\omega_1)t}+\mathrm{e}^{i(\omega_{mn}-\omega_1)t}] \\ & + H_{mn}^{(2)}[\mathrm{e}^{i(\omega_{mn}+\omega_2)t}+\mathrm{e}^{i(\omega_{mn}-\omega_2)t}]\} \end{aligned} \tag{233}$$

On substituting for $a_n^{(1)}$ and integrating we obtain a rather complicated expression containing terms with denominators having various combinations of ω_{n0}, ω_{mn}, ω_1, ω_2. For the particular problem under discussion, namely indirect transitions between a valence band and a conduction band having maxima and minima well separated in **k**-space we shall see that only two of these combinations give an appreciable contribution to the transition probability, namely $\omega_{m0}-\omega_1 \pm \omega_2$; we shall therefore not write down the full expression for $a_m^{(2)}$ which is similar to that given by equation (227) with ω_{m0} replaced by the various possible contributions referred to.

13.5.2 Indirect transitions between bands

We may now apply the above theory to the study of indirect transitions between a valence band and a conduction band. We identify the first term in the perturbation Hamiltonian H_1, given by equation (229) as the incident light wave, so that $\omega_1 = \omega$ and $\mathbf{A}_1$ is given by equations (39) and (46); the second term we identify as the phonon wave so that $\omega_2 = \omega_s$ and $\mathbf{A}_2$ is given by equation (90a). The wave functions ψ_n fall into two sets, the wave functions $\psi_c(\mathbf{k},\mathbf{r})$ for the conduction band and the wave functions $\psi_v(\mathbf{k},\mathbf{r})$ for the valence band so that when we perform a sum over n we now mean two separate sums over the allowed values of **k**. For the initial state we shall choose one state of the valence band whose wave function is $\psi_v(\mathbf{k}_0,\mathbf{r})$, $\mathbf{k}_0$ having a value near $\mathbf{k} = 0$; for the final state for which we wish to calculate the transition probability P_{f0} we shall take the state of the conduction band whose wave function ψ_m is $\psi_c(\mathbf{k}_f,\mathbf{r})$, $\mathbf{k}_f$ having a value near $\mathbf{k}_m$, the wave vector corresponding to one of the minima of the conduction band. We must now consider the various types of matrix element which can be formed; they are given by

$$H_{cc}^{(1)}(\mathbf{k},\mathbf{k}') = \tfrac{1}{2}\int \psi_c^\star(\mathbf{k},\mathbf{r})\,\mathbf{A}_1\,\psi_c(\mathbf{k}',\mathbf{r})\,d\mathbf{r}$$

$$H_{cv}^{(1)}(\mathbf{k},\mathbf{k}') = \tfrac{1}{2}\int \psi_c^\star(\mathbf{k},\mathbf{r})\,\mathbf{A}_1\,\psi_v(\mathbf{k}',\mathbf{r})\,d\mathbf{r}$$

$$H_{vv}^{(1)}(\mathbf{k},\mathbf{k}') = \tfrac{1}{2}\int \psi_v^\star(\mathbf{k},\mathbf{r})\,\mathbf{A}_1\,\psi_v(\mathbf{k}',\mathbf{r})\,d\mathbf{r}$$

$$H_{cc}^{(2)}(\mathbf{k},\mathbf{k}') = \tfrac{1}{2}\int \psi_c^\star(\mathbf{k},\mathbf{r})\,\mathbf{A}_2\,\psi_c(\mathbf{k},\mathbf{r})\,d\mathbf{r}$$

$$H_{cv}^{(2)}(\mathbf{k},\mathbf{k}') = \tfrac{1}{2}\int \psi_c^\star(\mathbf{k},\mathbf{r})\,\mathbf{A}_2\,\psi_v(\mathbf{k},\mathbf{r})\,d\mathbf{r}$$

$$H_{vv}^{(2)}(\mathbf{k},\mathbf{k}') = \tfrac{1}{2}\int \psi_v^\star(\mathbf{k},\mathbf{r})\,\mathbf{A}_2\,\psi_v(\mathbf{k},\mathbf{r})\,d\mathbf{r}$$

Of these $H_{cc}^{(1)}$ and $H_{vv}^{(1)}$ are zero when we neglect the photon momentum and give no contribution; $H_{cv}^{(2)}$ corresponds to inter-band transitions induced by phonons and may also be neglected. $H_{cv}^{(1)}$ is zero unless $\mathbf{k} = \mathbf{k}'$, i.e. only 'vertical' transitions are allowed. We are therefore left with the matrix elements $H_{cv}^{(1)}$, $H_{cc}^{(2)}$, $H_{vv}^{(2)}$ corresponding, respectively, to vertical inter-band transitions induced by a photon, and phonon-induced transitions within the conduction and valence bands. From equation (228) we see that only matrix elements of the form H_{m0} and H_{nm} appear in the terms giving an appreciable contribution and it will readily be seen that the same applies here, since only in this combination can we have a denominator which can be zero. The fact that H_{cv} is zero unless $\mathbf{k} = \mathbf{k}'$ then indicates that we need consider only two intermediate states having the wave functions $\psi_c(\mathbf{k}_0, \mathbf{r})$ and $\psi_v(\mathbf{k}_f, \mathbf{r})$, i.e. corresponding to a 'vertical' transition from the initial state or to a 'vertical' transition to the final state. This reduces the matrix elements $H_{cv}^{(1)}$ which we require to $H_{cv}^{(1)}(\mathbf{k}_0, \mathbf{k}_0)$, which we shall write as H_0^p, and $H_{cv}^{(1)}(\mathbf{k}_f, \mathbf{k}_f)$, which we shall write as H_f^p. The intermediate states being reduced to two, we require only two matrix elements of type $H_{cc}^{(2)}$ and $H_{cv}^{(2)}$, namely $H_{cc}^{(2)}(\mathbf{k}_f, \mathbf{k}_0)$ and $H_{vv}^{(2)}(\mathbf{k}_0, \mathbf{k}_f)$ which we shall write as H_c^s and H_v^s, respectively. If we include only those terms with denominators which may be zero we obtain from equations (232) and (233), on integrating

$$a_f^{(2)}(t) = \frac{1}{\hbar^2}\left\{\frac{H_0^p H_c^s}{\omega_{i0}-\omega} + \frac{H_f^p H_v^s}{\omega_{i'0}-\omega}\right\}\left\{\frac{e^{i(\omega_{f0}-\omega-\omega_s)t}-1}{\omega_{f0}-\omega-\omega_s} + \frac{e^{i(\omega_{f0}-\omega+\omega_s)t}}{\omega_{f0}-\omega+\omega_s}\right\} \tag{234}$$

where

$$\hbar\omega_{f0} = E_c(\mathbf{k}_f) - E_v(\mathbf{k}_0)$$

$$\hbar\omega_{i0} = E_c(\mathbf{k}_0) - E_v(\mathbf{k}_0)$$

$$\hbar\omega_{i'0} = E_c(\mathbf{k}_f) - E_v(\mathbf{k}_f)$$

On forming the square of the modulus of $a_f^{(2)}$ we may neglect cross products, since from our previous analysis, we see that they do not lead to probabilities which increase with time. We may therefore write

$$|a_f^{(2)}|^2 = \frac{4|H_0^p|^2\,|H_c^s|^2\sin^2\frac{1}{2}(\omega_{f0}-\omega-\omega_s)}{\hbar^4(\omega_{i0}-\omega)^2(\omega_{f0}-\omega-\omega_s)^2} + \frac{4|H_0^h|^2\,|H_c^s|^2\sin^2\frac{1}{2}(\omega_{f0}-\omega+\omega_s)}{4\hbar^4(\omega_{i0}-\omega)^2(\omega_{f0}-\omega+\omega_s)^2}$$
$$+ \frac{4|H_f^h|^2\,|H_v^s|^2\sin^2\frac{1}{2}(\omega_{f0}-\omega-\omega_s)}{\hbar^4(\omega_{i'0}-\omega)^2(\omega_{f0}-\omega-\omega_s)^2} + \frac{4|H_f^p|^2\,|H_v^s|^2\sin^2\frac{1}{2}(\omega_{f0}-\omega+\omega_s)}{\hbar^4(\omega_{i'0}-\omega)^2(\omega_{f0}-\omega+\omega_s)^2} \tag{235}$$

The first two terms in equation (235) correspond to transitions from state A to state B through the intermediate state I of Fig. 13.4, while the third and fourth terms correspond to transitions through the state I'. The conditions that the denominators $(\omega_{f0} - \omega \pm \omega_s)$ be zero show that the first term involves a transition from state I to state B with phonon absorption, the second from state I to state B with phonon emission, the third from state A to state I' with phonon absorption and the fourth from state A to state I' with phonon emission. We may calculate the transition probabilities and hence the optical absorption for each of these terms separately and later add them to obtain the total absorption. Let us consider the first term; to obtain the transition probability we must sum over all allowed values of the final wave vector $\mathbf{k}_f$ keeping $\mathbf{k}_0$ fixed. This is somewhat different from the procedure for direct transitions in which $\mathbf{k}_0$ and $\mathbf{k}_f$ varied together, and is similar to the summation (integration) carried out in § 13.2.1. The terms other than

$$\sin^2 \tfrac{1}{2}(\omega_{f0}-\omega-\omega_s)/(\omega_{f0}-\omega-\omega_s)^2$$

may be given the value for which $\omega_{f0} = \omega+\omega_s$ and the resulting integration gives as before (cf. equation (17)) a factor $\frac{1}{2}\pi t N_c(E)$ where $N_c(E)$ is the density of states at energy E above the minimum of the conduction band. To obtain the contribution P_{ia} to the transition probability per

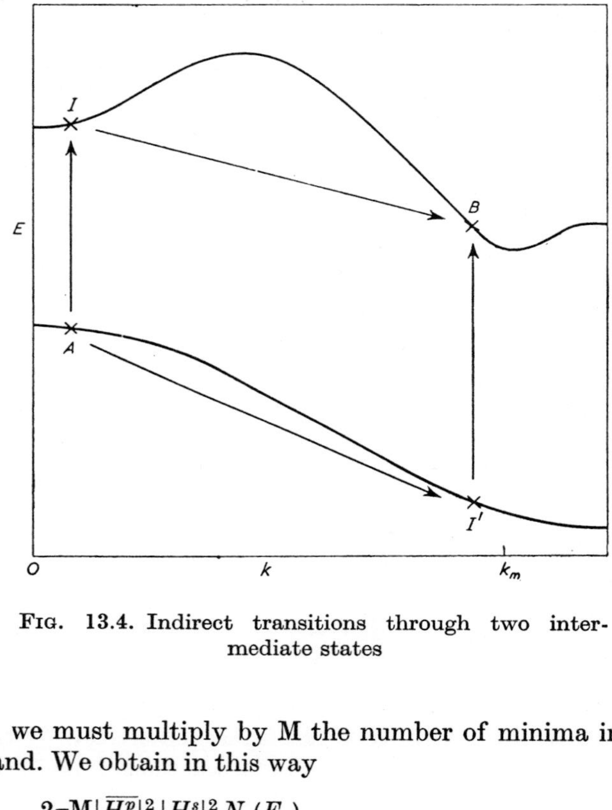

FIG. 13.4. Indirect transitions through two intermediate states

unit time, we must multiply by M the number of minima in the conduction band. We obtain in this way

$$P_{ia} = \frac{2\pi \mathrm{M}|\overline{H_0^p}|^2\,|H_c^s|^2 N_c(E_1)}{\hbar^4(\omega_{i0}-\omega)^2}\,\delta(\omega_{f0}-\omega-\omega_s) \tag{236}$$

$$= \frac{2\pi \mathrm{M}|\overline{H_0^p}|^2\,|H_c^s|^2 N_c(E_1)}{\hbar^2[E_c(k_0)-E_v(k_0)-\hbar\omega]^2}\,\delta[E_c(\mathbf{k}_f)-E_v(\mathbf{k}_0)-\hbar\omega-\hbar\omega_s] \tag{237}$$

where $|\overline{H_0^p}|$ is an averaged value as in equation (65) and

$$E_1 = E_c(\mathbf{k}_f)-E_c(\mathbf{k}_m)$$

The matrix element H_c^s is equal to zero unless (cf. equation (99a))

$$\mathbf{k}_f = \mathbf{k}_0 + \mathbf{K}_s \tag{238}$$

and this determines the energy $\hbar\omega_s$. We may, as a good approximation, take $\mathbf{K}_s = \mathbf{k}_m$ the value of the wave vector at the minimum of the conduction band so that $\hbar\omega_s$ is equal to $\hbar\omega_m$ the phonon energy corresponding to the wave vector $\mathbf{k}_m$. The value of $\mathbf{k}_f$ is then determined by the δ-function in equation (237) giving

$$E_c(\mathbf{k}_f) = \hbar\omega + \hbar\omega_m + E_v(\mathbf{k}_0) \tag{239}$$

This corresponds to conservation of energy between the initial and final states and confirms that we are dealing with phonon absorption. Let us write $E_c(\mathbf{k}_m) = E_c$ and $E_v(0) = E_v$ so that

$$E_1 = E_c(\mathbf{k}_f) - E_c \tag{240}$$

and let us also define E_2 by means of the equation

$$E_2 = E_v - E_v(\mathbf{k}_0) \tag{241}$$

E_2 is therefore the depth in energy of the state corresponding to $\mathbf{k}_0$ in the valence band *below* the top of the band. From equation (239) we may write equation (240) in the form

$$E_1 = \hbar\omega + \hbar\omega_m - E_2 - \Delta E \tag{242}$$

since $\Delta E = E_c - E_v$. We may therefore omit the δ-function from equation (237) provided E_1 is given by equation (242).

From equation (237) we see that we obtain a finite value of P_{ia} for any allowed value of $\mathbf{k}_0$ for which equation (242) can be satisfied with *positive* values of E_1 and E_2, i.e. for values of E_2 lying between 0 and $\hbar\omega + \hbar\omega_s - \Delta E$. To obtain the total transition probability P_{ca} due to this type of process we must sum over all allowed values of $\mathbf{k}_0$ for which E_2 lies between these limits. Since, apart from the quantity $N_c(E_1)$, the other factors in equation (327) do not vary much with $\mathbf{k}_0$ we may give them their values at the $\mathbf{k} = 0$ for the valence band and $\mathbf{k} = \mathbf{k}_m$ for the conduction band. We obtain in this way

$$P_{ca} = \frac{4\pi \mathrm{M}|\overline{H_0^p}|^2\,|H_c^s|^2}{\hbar^2\,(\Delta E_0 - \hbar\omega)^2} \int_0^{E_m} N_v(E)\,N_c(E_m - E)\,\mathrm{d}E \tag{243}$$

where $$E_m = \hbar\omega + \hbar\omega_m - \Delta E$$

A factor 2 has been introduced into equation (243) to take account of the fact that an electron in its initial state may have either spin, $N_c(E)$ and $N_v(E)$ being the density of states functions, apart from spin. (We did not include a factor 2 in summing over the final states since it is assumed that the spin will not change in a 'vertical' transition.)

We shall consider first a non-degenerate valence band so that $N_v(E)$ is given by (§ 2.4 equation (24))

$$N_v(E) = V(2m_h)^{3/2} E^{1/2}/4\pi^2 \hbar^3 \tag{244}$$

$N_c(E)$ is given by (§ 9.4.2 equation (38))

$$N_c(E) = V(2m_d)^{3/2} E^{1/2}/4\pi^2 \hbar^3 \tag{245}$$

m_d being the 'density-of-states' effective mass for the conduction band. Inserting these values of $N_v(E)$ and $N_c(E)$ in equation (243) we obtain

$$P_{ca} = \frac{2MV^2\,|\overline{H_0^p}|^2\,|H_c^s|^2\,m_h^{3/2}\,m_d^{3/2}}{\pi^3\,\hbar^8[\Delta E_0 - \hbar\omega]^2} \int_0^{E_m} E^{1/2}(E - E_m)^{1/2}\,\mathrm{d}E \tag{246}$$

The integral may readily be evaluated and we have

$$\int_0^{E_m} E^{1/2}(E_m - E)^{1/2}\,\mathrm{d}E = \frac{\pi E_m^2}{8}$$

so that

$$P_{ca} = \frac{MV^2\,|\overline{H_0^p}|^2\,|H_c^s|^2\,m_h^{3/2}\,m_d^{3/2}[\hbar\omega + \hbar\omega_m - \Delta E]^2}{4\pi^2\,\hbar^8[\Delta E_0 - \hbar\omega]^2}$$

provided

$$\hbar\omega > \Delta E - \hbar\omega_m,$$
$$= 0, \quad \hbar\omega \leqslant \Delta E - \hbar\omega_m \tag{247}$$

Let us first of all suppose that vertical transitions from states near $\mathbf{k} = 0$ are allowed. The matrix element H_0^p is then approximately constant near $\mathbf{k} = 0$ and may be expressed in terms of the oscillator strength f_c defined by means of equation (72). We have from equations (72), (60) and (65)

$$f_c = 8\boldsymbol{m}|\overline{H_0^p}|^2/\boldsymbol{e}^2 A^2\,\hbar\omega_{i0} \simeq 8\boldsymbol{m}|\overline{H_0^p}|^2/\boldsymbol{e}^2 A^2 \Delta E_0 \tag{248}$$

where A is the amplitude of vector potential for the incident radiation.

The matrix element H_c^s may be calculated approximately in terms of the deformation potential as in § 13.4.4. This is rather a crude approximation since the effective-mass approximation with a quadratic energy function certainly does not hold over the whole range of wave vectors from $\mathbf{k} = 0$ to $\mathbf{k} = \mathbf{k}_m$. We obtain in this way (cf. equation (196))

$$H_c^s = \tfrac{1}{2} i V_c K_s A_s \tag{249}$$

where V_c is the deformation potential associated with the conduction band. For elevated temperatures A_s is given by equation (125) but since we are frequently concerned with low temperatures in absorption measurements we should replace $\mathbf{k}T$ but by $\hbar\omega_s/[\exp(\hbar\omega_s/\boldsymbol{k}T) - 1]$ as required by the quantum theoretical treatment of § 13.4.3. Thus we have

$$|H_c^s|^2 = \left(\frac{V_c^2}{\rho c_l^2}\right) \frac{\hbar\omega_s}{2V(e^{\hbar\omega_s/\boldsymbol{k}T} - 1)} \tag{250}$$

The quantity $V_c^2/\rho c_l^2$ may be expressed in terms of l the high temperature value of the mean free path for lattice scattering. If T_0 is a temperature for which equation (134) for l holds we have

$$V_c^2/\rho c_l^2 = \pi\hbar^4/m_d^2 l_{0c} \boldsymbol{k} T_0 \tag{251}$$

where l_{0c} is the mean free path for acoustic mode scattering in the conduction band at temperature T_0. We have replaced m_e by m_d to take account of the 'non-spherical' nature of the minima in the conduction band, and we note that the quantity $l_{c0}\boldsymbol{k}T_0$ is a constant. On substituting in equation (250) we obtain

$$|H_c^s|^2 = \frac{\pi\hbar^5\omega_s}{2Vm_d^2(l_{c0}\boldsymbol{k}T_0)(e^{\hbar\omega_s/\boldsymbol{k}T}-1)} \tag{252}$$

On substituting for $|H_0^p|^2$ and $|H_c^s|^2$ in equation (247) and taking $\omega_s = \omega_m$ we obtain

$$P_{ca} = \frac{\mathrm{M}Ve^2A^2f_c\omega_m m_h^{3/2}\Delta E_0(\hbar\omega+\hbar\omega_m-\Delta E)^2}{64\pi \boldsymbol{m} m_d^{1/2}(l_{c0}\boldsymbol{k}T_0)\hbar^3(\Delta E_0-\hbar\omega)^2(e^{\hbar\omega_m/\boldsymbol{k}T}-1)} \tag{253}$$

If the valence band is degenerate and we have two effective hole masses m_{h1} and m_{h2} we must add the contributions from the two bands replacing $m_h^{3/2}$ by $(m_{h1}^{3/2}+m_{h2}^{3/2})$.

To obtain the contribution α_{ca} to the absorption coefficient from P_{ca} we must take $V=1$ and multiply P_{ca} by $2\hbar/\omega nA^2\epsilon_0\boldsymbol{c}$ where n is the refractive index (cf. equation (70)); this gives

$$\alpha_{ca} = \frac{\mathrm{M}e^2m_h^{3/2}f_cE_p\Delta E_0(\hbar\omega+E_p-\Delta E)^2}{32\pi\epsilon_0\boldsymbol{m}m_d^{1/2}\hbar^2cn\omega(l_{c0}\boldsymbol{k}T_0)(\Delta E_0-\hbar\omega)^2(e^{E_p/\boldsymbol{k}T}-1)}, \quad \hbar\omega > \Delta E - E_p,$$
$$= 0, \quad \hbar\omega < \Delta E - E_p \tag{254}$$

where $E_p = \hbar\omega_m$ is the phonon energy.

Similarly we have from the second term of equation (235) corresponding to phonon emission

$$\alpha_{ce} = \frac{\mathrm{M}e^2m_h^{3/2}f_cE_p\Delta E_0(\hbar\omega-E_p-\Delta E)^2e^{E_p/\boldsymbol{k}T}}{32\pi\epsilon_0\boldsymbol{m}m_d^{1/2}\hbar^2cn\omega(l_{c0}\boldsymbol{k}T_0)(\Delta E_0-\hbar\omega)^2(e^{E_p/\boldsymbol{k}T}-1)}, \quad \hbar\omega > \Delta E + E_p,$$
$$= 0, \quad \hbar\omega < \Delta E + E_p \tag{255}$$

the factor $N_p = [\exp(E_p/\boldsymbol{k}T)-1]^{-1}$ being replaced by N_p+1 as shown in § 13.4.3. An exactly similar treatment may be given for the third and fourth terms of equation (235) corresponding to transitions through the state I' (Fig. 13.4). The absorption coefficients corresponding to absorption and emission of a phonon, α_{va}, α_{ve} are given by

$$\alpha_{va} = \frac{\mathrm{M}e^2m_d^{3/2}f_vE_p\Delta E_1(\hbar\omega+E_p-\Delta E)^2}{32\pi\epsilon_0\boldsymbol{m}m_h^{1/2}\hbar^2cn\omega(l_{v0}\boldsymbol{k}T_0)(\Delta E_1-\hbar\omega)^2(e^{E_p/\boldsymbol{k}T}-1)}, \quad \hbar\omega > \Delta E - E_p,$$
$$= 0, \quad \hbar\omega < \Delta E - E_p \tag{256}$$

$$\alpha_{ve} = \frac{\mathbf{M}\boldsymbol{e}^2 m_d^{3/2} f_v E_p \Delta E_1(\hbar\omega - E_p - \Delta E)^2 \mathrm{e}^{E_p/kT}}{32\pi\epsilon_0 \boldsymbol{m} m_h^{1/2} \hbar^2 \boldsymbol{c} n\omega(l_{v0}\boldsymbol{k}T_0)(\Delta E_1 - \hbar\omega)^2(\mathrm{e}^{E_p/kT}-1)}$$

$$\hbar\omega > \Delta E + E_p,$$

$$= 0, \quad \hbar\omega < \Delta E + E_p \qquad (257)$$

f_v is the oscillator strength for the transition from state I' to state B (assuming it is allowed) and ΔE_1 is the energy separation between the bottom of the conduction band and the point 'vertically beneath' in the valence band; l_{v0} is the mean free path for acoustic mode scattering at an elevated temperature T_0 of holes in the valence band. To obtain the total absorption we must add the contributions from $\alpha_{ca}, \alpha_{ce}, \alpha_{va}, \alpha_{ve}$. We see that the contributions start at $\hbar\omega = \Delta E \pm E_p$ and are approximately quadratic functions of the excess energy of the incident quanta over these starting values. This quadratic dependence will only be found if the transitions between states A and I and between I' and B are allowed transitions. If we neglect the energy E_p and add the four contributions taking $\Delta E_1 = \Delta E_0$ and $m_h = m_d$ we obtain the expression first obtained for the absorption due to indirect transitions in Ge by J. Bardeen, F. J. Blatt and L. H. Hall*.

When one or other of the transitions between the states A and I or between I' and B are forbidden we must use the expression given by equation (78) for H_0^p instead of equation (60), so that we have

$$|\overline{H_0^p}|^2 = \frac{\boldsymbol{e}^2 A^2 \hbar^2 f_c' k^2}{12\boldsymbol{m}^2} \qquad (258)$$

where f_c' is the number defined for the transition from state A to state I at $k = 0$ by means of equation (79). Expressing k in terms of E_1 we have

$$|\overline{H_0^p}|^2 = \frac{\boldsymbol{e}^2 A^2 m_h f_c' E_1}{6\boldsymbol{m}^2} \qquad (259)$$

The factor E_1 must be taken inside the integration sign in equation (246); we have now

$$\int_0^{E_m} E^{3/2}(E - E_m)\,\mathrm{d}E = \frac{\pi a^3}{16}$$

so that we replace f_c in equations (254), etc., by

$$2m_h f_c'(\hbar\omega + E_p - \Delta E)/3\boldsymbol{m}\Delta E_0$$

and in equation (255), by

$$2m_e f_c'(\hbar\omega - E_p - \Delta E)/3\boldsymbol{m}\Delta E_1$$

* J. Bardeen, F. J. Blatt and L. H. Hall, *Proceedings of Atlantic City Photoconductivity Conference*. John Wiley, New York, and Chapman & Hall, London (1956), p. 146.

Similarly, for the transition from state I' to state B we replace f_v in equation (256) by

$$2m_d f_v'(\hbar\omega - E_p - \Delta E)/3m\Delta E_1$$

and in equation (257) by

$$2m_d f_v'(\hbar\omega + E_p - \Delta E)/3m\Delta E_1$$

Thus we now have

$$\alpha_{ca} = \frac{Me^2 m_h^{5/2} f_c' E_p(\hbar\omega + E_p - \Delta E)^3}{48\pi\epsilon_0 m^2 m_d^{1/2} \hbar^2 cn\omega(l_{c0}\, kT_0)(\Delta E_0 - \hbar\omega)^2(e^{E_p/kT} - 1)} \tag{260}$$

and similar expressions for α_{ce}, α_{ve}, α_{va}. We see that near each limit the absorption is now proportional to the cube of the excess of the energy of the photon above the limiting value.

For direct transitions one may estimate the numerical value of the total absorption coefficient α for Ge as about 10 cm^{-1} when the excess photon energy is about 0·05 eV, in good agreement with values found by G. G. Macfarlane, T. P. McLean, J. E. Quarrington and V. Roberts* who have made a careful study of the indirect transitions in Ge and Si. In order to explain their absorption curves photons due to transverse modes as well as the longitudinal modes had to be taken into account. The absorption curves are quite complex and require careful interpretation. When this is done, however, they yield valuable information on the band structure and the phonon energies. In contrast to the absorption due to direct transitions the absorption due to indirect transitions varies rapidly with the temperature, especially at low temperatures. Apart from the variation due to the change in the value of ΔE this comes about mainly from the variation of phonon density, at low temperatures the absorption being mainly due to transitions involving emission of a phonon. The reverse of these transitions namely emission of radiation in indirect transitions has also been observed by J. R. Haynes, M. Lax and W. F. Flood†. Absorption due to exciton formation together with phonon emission and absorption has also been observed by Macfarlane *et al.* (*loc. cit.*). It will thus be seen that the optical absorption of semiconductors for values of the quantum energy near the energy of the forbidden gap is quite complex. The subject has been discussed in some detail by various authors including R. A. Smith‡, H. Y. Fan§ and T. P. McLean**.

* G. G. Macfarlane, T. P. McLean, J. E. Quarrington and V. Roberts, *Phys. Rev.*, **108**, 377 (1957); *Ibid.*, **111**, 1245 (1958).
† J. R. Haynes, M. Lax and W. F. Flood, *J. Phys. Chem. Solids*, **8**, 392 (1959).
‡ R. A. Smith, *Semiconductors*. Cambridge University Press (1959), Chapter 7.
§ H. Y. Fan, *Rep. Progr. Phys.*, **19**, 107 (1956).
* T. P. McLean, *Progress in Semiconductors*. Heywood, London, **5**, 53 (1960).

Appendix 1

EQUATION OF MOTION FOR AN ELECTRON IN A CRYSTAL IN A MAGNETIC FIELD

In § 8.8 we proved the result that for a force **F** due to an electric field, or to any source which does not produce a force at right angles to the velocity vector **v**, we have the rate of change of crystal momentum **P** equal to the force **F**; we now wish to prove this when the force acting on an electron is due to a magnetic field. In this case the instantaneous force is equal to $-e(\mathbf{v}\times\mathbf{B})$ and is always at right angles to **v**. There seems to be no simple way of proving this result. Various proofs have been proposed and that given here is based on one by A. H. Wilson*; all involve rather complicated analysis which is not difficult in principle but is tedious to carry out. The result is in fact only an approximation, true for not too strong magnetic fields, and this may be why no simple and elegant demonstration has been given.

In the absence of a magnetic field we shall assume that the motion of an electron may be represented by a wave packet of the form

$$\Psi_0 = \int F(\mathbf{k})\,\psi_{\mathbf{k}}(\mathbf{r})\,d\mathbf{k} \tag{1}$$

where $\psi_{\mathbf{k}}(\mathbf{r})$ represents a Bloch-type wave function, i.e.

$$\psi_{\mathbf{k}}(\mathbf{r}) = \exp\{i(\mathbf{k}.\mathbf{r}.)-i(E/\hbar)t\}\,U_{\mathbf{k}}(\mathbf{r}) \tag{2}$$

where $U_{\mathbf{k}}(\mathbf{r})$ is a periodic function with periodicity of the lattice. The wave function $\psi_{\mathbf{k}}(\mathbf{r})$, apart from its time factor, satisfies the Schrödinger equation

$$H_0\psi = -\frac{\hbar^2}{2m}\nabla^2\psi+V\psi = E(\mathbf{k})\,\psi \tag{3}$$

or with the time factor included, the time-dependent wave equation

$$i\hbar\frac{\partial\psi}{\partial t} = H_0\psi \tag{4}$$

When we have a magnetic field present the wave equation becomes

$$i\hbar\frac{\partial\Psi}{\partial t} = H_0\Psi-\frac{ie\hbar}{m}\mathbf{A}.\nabla\Psi-e\phi\Psi \tag{5}$$

where **A** and ϕ are the vector and scalar potentials, and we have assumed, as is usual, that div $\mathbf{A} = 0$. We have also neglected quadratic terms in A^2

* A. H. Wilson, *Theory of Metals* (2nd Ed.). Cambridge University Press (1954), p. 50.

which limits the discussion to moderate values of the magnetic field and does not take account of effects due to the quantization of angular momentum which takes place at high fields*.

When we have a magnetic field, the expressions

$$\frac{\hbar i}{2m}\int(\Psi\nabla\Psi^{\star}-\Psi^{\star}\nabla\Psi)\,d\mathbf{r} \tag{6}$$

or

$$-\frac{\hbar i}{m}\int\Psi^{\star}\nabla\Psi\,d\mathbf{r} \tag{6a}$$

for the average velocity do not hold, and in general and we have to add a term*

$$\frac{e}{m}\int\mathbf{A}\Psi^{\star}\Psi\,d\mathbf{r} \tag{7}$$

As is well known, the vector potential is not uniquely defined, and since it is convenient to use the simpler expressions (6) or (6a) we shall choose **A** so that the extra term (7) is equal to zero, as suggested by H. Jones and C. Zener† who have also given a detailed discussion of this problem.

If we have a constant magnetic flux **B** it is usual to take

$$\left.\begin{aligned}\mathbf{A} &= \tfrac{1}{2}\mathbf{B}\times\mathbf{r}\\ \phi &= 0\end{aligned}\right\} \tag{8}$$

If we take

$$\mathbf{A} = \tfrac{1}{2}\mathbf{B}\times(\mathbf{r}-\bar{\mathbf{r}}) \tag{9}$$

where $\bar{\mathbf{r}}$ is the quantum mechanical average of the position vector $\mathbf{r}$ given by

$$\bar{\mathbf{r}} = \int\Psi^{\star}\mathbf{r}\Psi\,d\mathbf{r} \tag{10}$$

we may readily verify that we obtain the same magnetic field as before and the expression (7) is equal to zero (it being assumed that Ψ is normalised to unity). The form (9) for **A**, however, gives rise to an electric field given by

$$\mathscr{E} = -\frac{\partial\mathbf{A}}{\partial t} = \tfrac{1}{2}(\mathbf{B}\times\bar{\mathbf{v}}) \tag{11}$$

since $d\bar{r}/dt = \bar{v}$ for a localised wave packet.

We must therefore also take

$$\nabla\phi = \tfrac{1}{2}(\mathbf{B}\times\bar{\mathbf{v}}) \tag{12}$$

in order to obtain $\mathscr{E} = 0$. If the magnetic flux has strength B and is directed along the z-axis we may take

$$\phi = -\frac{B}{2}(x\bar{v}_y-y\bar{v}_x) \tag{13}$$

* See, for example, L. I. Schiff, *Quantum Mechanics* (5th Ed.). McGraw-Hill (1955), p. 137. (N.B. We take $-e$ for the electronic charge.)

† H. Jones and C. Zener, *Proc. Roy. Soc.*, **A144**, 101 (1934).

Inserting these values of **A** and ϕ into equation (5) we obtain

$$i\hbar\frac{\partial\Psi}{\partial t} - H_0\Psi = -\frac{ie\hbar B}{2m}\left[(x-\bar{x})\frac{\partial}{\partial y} - (y-\bar{y})\frac{\partial}{\partial x}\right]\Psi + \frac{e\mathbf{B}}{2}[x\bar{v}_y - y\bar{v}_x]\Psi \quad (14)$$

We now substitute the wave function (1) in (14) and use equation (3) to obtain

$$i\hbar\int \dot{F}(\mathbf{k})\,\psi_{\mathbf{k}}(\mathbf{r})\,d\mathbf{k} = \int F(\mathbf{k})\,H_1\,\psi_{\mathbf{k}}(\mathbf{r})\,d\mathbf{k} \quad (15)$$

where we have written $H_1(\mathbf{k})$ for the term on the right-hand side of equation (14). We now multiply equation (15) by $\psi^{\star}_{\mathbf{k}'}(\mathbf{r})$ and integrate over the space for which the wave functions are normalised; we thus obtain

$$i\hbar\iint \dot{F}(\mathbf{k})\,\psi^{\star}_{\mathbf{k}'}(\mathbf{r})\,\psi_{\mathbf{k}}(\mathbf{r})\,d\mathbf{k}\,d\mathbf{r} = \iint F(\mathbf{k})\,\psi^{\star}_{\mathbf{k}'}(\mathbf{r})\,H_1\,\psi_{\mathbf{k}}(\mathbf{r})\,d\mathbf{k}\,d\mathbf{r} \quad (16)$$

Now $F(\mathbf{k})$ is very small except for values of **k** near $\mathbf{k} = \mathbf{k}_0$ corresponding to the mean value of the wave vector, and we seek to perform the integration over **k** first of all.

There is a rather general result of which we can make use which is derived as follows. To begin with we may replace the integral over **k** with a sum; this will in fact be what we have in practice since only the discrete values of **k** permitted by the boundary conditions are involved. The density of allowed states is $1/8\pi^3$ per unit volume of **k** space so we replace

$$\frac{1}{8\pi^3}\int d\mathbf{k} \quad \text{by} \quad \sum_{\mathbf{k}}$$

Let $\phi(\mathbf{k}, \mathbf{r})$ be a function periodic in **r**, with the periodicity of the lattice. We have then, for an integral I defined as follows,

$$I = \frac{1}{8\pi^3}\iint e^{i(\mathbf{k}-\mathbf{k}').\mathbf{r}}\,\phi(\mathbf{k}, \mathbf{r})\,d\mathbf{k}\,d\mathbf{r} \quad (17)$$

on replacing the integration over **k** by a sum over the allowed values

$$I = \sum_{\mathbf{k}}\int e^{i(\mathbf{k}-\mathbf{k}').\mathbf{r}}\,\phi(\mathbf{k}, \mathbf{r})\,d\mathbf{r} \quad (18)$$

Because of the periodicity of $\phi(k, r)$ we may write this integral as a sum over unit cells of the lattice in the form

$$I = \sum_{\mathbf{k}}\sum_{\mathbf{n}} e^{i(\mathbf{k}-\mathbf{k}').\mathbf{d_n}}\int_1 e^{i(\mathbf{k}-\mathbf{k}').\mathbf{r}}\,\phi(\mathbf{k}, \mathbf{r})\,d\mathbf{r} \quad (19)$$

where the integral is now taken over a unit cell, and $\mathbf{d_n}$ is a lattice vector. Now it is well known that a sum of the form

$$\sum_{\mathbf{n}} e^{i(\mathbf{k}-\mathbf{k}').\mathbf{d_n}}$$

is zero unless $\mathbf{k} = \mathbf{k}'$, and when $\mathbf{k} = \mathbf{k}'$ the sum is equal to the number of unit cells. If the normalisation is carried out per unit volume this is equal to V_d^{-1} where V_d is the volume of a unit cell of the lattice. Thus we have

$$I = \frac{1}{V_d}\int_1 \phi(\mathbf{k}', \mathbf{r})\,\mathrm{d}\mathbf{r} \tag{20}$$

Now the integral on the left-hand side of equation (16) is of this form and may be written as

$$\frac{8\pi^3}{V_d}\dot{F}(\mathbf{k}')\int_1 \psi^\star_{\mathbf{k}'}\psi_{\mathbf{k}'}\,\mathrm{d}\mathbf{r} = 8\pi^3\dot{F}(\mathbf{k}') \tag{21}$$

The integrals on the right-hand side are of the types

$$I_1 = \iint F(\mathbf{k})\,\psi^\star_{\mathbf{k}'}\frac{\partial\psi_{\mathbf{k}}}{\partial x}\,\mathrm{d}\mathbf{r}\,\mathrm{d}\mathbf{k} \tag{22}$$

$$I_2 = \iint F(\mathbf{k})\,\psi^\star_{\mathbf{k}'}x\psi_{\mathbf{k}}\,\mathrm{d}\mathbf{r}\,\mathrm{d}\mathbf{k} \tag{23}$$

$$I_3 = \iint F(\mathbf{k})\,\psi^\star_{\mathbf{k}'}x\frac{\partial\psi_{\mathbf{k}}}{\partial y}\,\mathrm{d}\mathbf{r}\,\mathrm{d}\mathbf{k} \tag{24}$$

I_1 is of the required form and may be written as,

$$I_1 = \frac{8\pi^3 F(\mathbf{k}')}{V_d}\int_1 \psi^\star_{\mathbf{k}'}\frac{\partial\psi}{\partial x}\,\mathrm{d}\mathbf{r}$$

Using equation (6a) we have then

$$I_1 = 8\pi^3 miF(\mathbf{k}')\,\bar{v}_x/\hbar \tag{25}$$

I_2 is not obviously of the required form, since x is not periodic. It may, however, be expressed in the required form by noting that

$$x\,\mathrm{e}^{i\mathbf{k}.\mathbf{r}} = -i\frac{\partial}{\partial k_x}(\mathrm{e}^{i\mathbf{k}.\mathbf{r}})$$

Inserting this in I_2 and integrating partially with respect to k_x, the integrated terms being zero because of periodicity, we obtain

$$I_2 = i\int \mathrm{e}^{-i\mathbf{k}'.\mathbf{r}}\,\mathrm{d}\mathbf{r}\int \mathrm{e}^{i\mathbf{k}.\mathbf{r}}\frac{\partial}{\partial k_x}[F(\mathbf{k})\,U^\star_{\mathbf{k}'}U_{\mathbf{k}}]\,\mathrm{d}\mathbf{k} \tag{26}$$

This is of the required form and may be written using (20) in the form

$$\left.\begin{aligned} I_2 &= \frac{8\pi^3 i}{V_d}\cdot\frac{\partial F(\mathbf{k}')}{\partial k'_x}\int_1 U^\star_{\mathbf{k}'}U_{\mathbf{k}'}\,\mathrm{d}\mathbf{r} + \frac{8\pi^3 iF(\mathbf{k}')}{V_d}\int_1 U^\star_{\mathbf{k}'}\frac{\partial U_{\mathbf{k}'}}{\partial k'_x}\,\mathrm{d}\mathbf{r} \\ &= 8\pi^3 i\frac{\partial F(\mathbf{k}')}{\partial k'_x} + \frac{8\pi^3 iF(\mathbf{k}')}{V_d}\int_1 U^\star_{\mathbf{k}'}\frac{\partial U_{\mathbf{k}'}}{\partial k'_x}\,\mathrm{d}\mathbf{r} \end{aligned}\right\} \tag{27}$$

2 H

To evaluate I_3 we note that

$$\frac{\partial \psi_{\mathbf{k}}}{\partial y} = U_{\mathbf{k}} \frac{\partial}{\partial y}(e^{i\mathbf{k}.\mathbf{r}}) + e^{i\mathbf{k}.\mathbf{r}} \frac{\partial U_{\mathbf{k}}}{\partial y}$$

$$= i\, e^{i\mathbf{k}.\mathbf{r}}\left(k_y U_{\mathbf{k}} - i \frac{\partial U_{\mathbf{k}}}{\partial y}\right)$$

so that

$$x \frac{\partial \psi_{\mathbf{k}}}{\partial y} = \left[\frac{\partial}{\partial k_x}(e^{i\mathbf{k}\mathbf{r}})\right]\left(k_y U_{\mathbf{k}} - i \frac{\partial U_{\mathbf{k}}}{\partial y}\right) \tag{28}$$

Integrating by parts as before we obtain

$$I_3 = \iint i \frac{\partial F(\mathbf{k})}{\partial k_x} \psi^\star_{\mathbf{k}'} \frac{\partial \psi_{\mathbf{k}}}{\partial y} \mathbf{dr}\, \mathbf{dk}$$

$$- \iint F(\mathbf{k})\, U^\star_{\mathbf{k}'} k_y \frac{\partial U_{\mathbf{k}}}{\partial k_x} \mathbf{dr}\, \mathbf{dk}$$

$$+ i \iint F(\mathbf{k})\, U^\star_{\mathbf{k}'} \frac{\partial^2 U_{\mathbf{k}}}{\partial k_x\, \partial y} \mathbf{dr}\, \mathbf{dk}$$

Integration by parts with respect to y transforms the last of these integrals into

$$-i \iint F(\mathbf{k}) \frac{\partial U^\star_{\mathbf{k}'}}{\partial y} \cdot \frac{\partial U_{\mathbf{k}}}{\partial k_x} \mathbf{dr}\, \mathbf{dk}$$

These integrals are now all in the required form and we may write

$$I_3 = \frac{8\pi^3 i}{V_d} \frac{\partial F(\mathbf{k}')}{\partial k'_x} \int_1 \psi^\star_{\mathbf{k}'} \frac{\partial \psi_{\mathbf{k}'}}{\partial y} \mathbf{dr} - \frac{8\pi^3 F(\mathbf{k}')}{V_d} k'_y \int U^\star_{\mathbf{k}'} \frac{\partial U_{\mathbf{k}'}}{\partial k'_x} \mathbf{dr}$$

$$- \frac{8\pi^3 i F(\mathbf{k}')}{V_d} \int_1 \frac{\partial U^\star_{\mathbf{k}'}}{\partial x} \frac{\partial U_{\mathbf{k}'}}{\partial k'_y} \mathbf{dr} \tag{29}$$

The first term in I_3 is just equal to

$$-8\pi^3 m \frac{F(\mathbf{k}')}{\partial k'_x} \frac{\bar{v}_y}{\hbar}$$

Inserting the value of H_1 in equation (16) we may now obtain an expression for $\dot{F}(\mathbf{k})$ in terms of the integrals I_1, I_2, I_3; the resulting expression is somewhat complex but a great simplification takes place if we form the expression for

$$F^\star(\mathbf{k})\, \dot{F}(\mathbf{k}) + F(\mathbf{k})\, \dot{F}^\star(\mathbf{k}) \tag{30}$$

This may be seen from the following considerations: the terms given by I_1 cancel because of the factor i when (30) is formed: also we note that the second terms in I_2 and in I_3 go out when (30) is formed since

$$\int U^\star_{\mathbf{k}'} \frac{\partial U_{\mathbf{k}'}}{\partial k'_x} \mathbf{dr} = - \int U_{\mathbf{k}'} \frac{\partial U^\star_{\mathbf{k}'}}{\partial k'_x} \mathbf{dr}$$

which follows from differentiating the normalising condition with respect to k'_x, under the integral sign. The first term of the integral I_2 gives the same result as the first terms of the integrals I_3, and together give the coefficients of $F^\star \partial F/\partial k_x + F\, \partial F^\star/\partial k_x$, etc. The remaining terms also vanish since they may be expressed in terms of the effective mass tensor m_{rs}^{-1}, and on subtracting the complex conjugate they cancel. We are therefore left with the simple equation

$$\frac{\partial}{\partial t}|F(\mathbf{k})|^2 = \frac{eB}{\hbar}\left\{\bar{v}_y \frac{\partial}{\partial k_x}|F(\mathbf{k})|^2 - \bar{v}_x \frac{\partial}{\partial k_y}|F(\mathbf{k})|^2\right\} \tag{31}$$

This may be written in vector form

$$\frac{\partial}{\partial t}|F(\mathbf{k})|^2 = -\frac{e}{\hbar}[\bar{\mathbf{v}} \times \nabla_k . \mathbf{B}]|F(\mathbf{k})|^2 \tag{32}$$

which is the same as

$$\frac{\partial}{\partial t}|F(\mathbf{k})|^2 = \frac{e}{\hbar}[\bar{\mathbf{v}} \times \mathbf{B} . \nabla_k]\,|F(\mathbf{k})|^2 \tag{33}$$

The equation

$$\frac{\partial Q}{\partial t} = \mathbf{a} . \nabla_k Q \tag{34}$$

where $\mathbf{a}$ as a constant vector, has general solution

$$Q = \phi(\mathbf{k} + \mathbf{a}t) \tag{35}$$

as may be seen by taking the k_x direction along the vector $\mathbf{a}$. We have then

$$\frac{\partial Q}{\partial t} = |\mathbf{a}|\,\frac{\partial Q}{\partial k_x} \tag{36}$$

The solution is then clearly

$$Q = \phi(k_x + |\mathbf{a}|\, t) \tag{37}$$

so that we have

$$|F(\mathbf{k})|^2 = \phi\left[\mathbf{k} + \frac{et}{\hbar}(\bar{\mathbf{v}} \times \mathbf{B})\right]$$

Now the function $F(k)$ is only appreciable at $t = 0$ when $\mathbf{k} = \mathbf{k}_0$ so the function ϕ is also only appreciable when its argument is equal to $\mathbf{k}_0$. Thus we must have

$$\mathbf{k} = \mathbf{k}_0 - \frac{et}{\hbar}(\bar{\mathbf{v}} \times \mathbf{B}) \tag{38}$$

so that

$$\frac{\mathrm{d}\mathbf{P}}{\mathrm{d}t} = \hbar\frac{\mathrm{d}\mathbf{k}}{\mathrm{d}t} = -e[\bar{\mathbf{v}} \times \mathbf{B}] \tag{39}$$

This is the important result which we set out to prove. We note that the velocity which occurs in equation (39) is the average velocity and that the term on the right-hand side of equation (39) is the force acting on an electron, averaged in the same way.

Appendix 2

THE f-SUM RULE

THE f-sum rule expressed in the form

$$\sum_{m \neq n} f_{mn}^{x} = 1 - \frac{m}{\hbar^2} \frac{\partial^2 E(\mathbf{k})}{\partial k_x^2} \tag{1}$$

where the quantities f_{mn} are the oscillator strengths defined in terms of the momentum matrix elements p_{mn} given by

$$p_{mn} = -i\hbar \int \psi_m^{\star} \nabla \psi_n \, d\mathbf{r} \tag{2}$$

as in § 13.3 equation (72), appears first to have been derived by H. Bethe*. The proof which we shall give follows closely one given by A. H. Wilson†.

When ψ_m and ψ_n are Bloch functions of the form

$$\left.\begin{aligned} \psi_m &= N^{-1/2} U_m(\mathbf{r}, \mathbf{k}) \, e^{i\mathbf{k}.\mathbf{r}} \\ \psi_n &= N^{-1/2} U_n(\mathbf{r}, \mathbf{k}') \, e^{i\mathbf{k}'.\mathbf{r}} \end{aligned}\right\} \tag{3}$$

it may readily be shown that $p_{nm} = 0$ unless $\mathbf{k} = \mathbf{k}'$, provided ψ_m and ψ_n refer to different bands (see exercise (4) of § 8.7). We shall therefore take the same value of $\mathbf{k}$ in ψ_m and ψ_n. It may also be shown that, in this case, p_{nm} may be expressed in the two alternative forms

$$p_{mn} = -i\hbar \int_C U_m^{\star} \nabla U_n \, d\mathbf{r} \tag{4}$$

$$\doteq m \frac{(E_n - E_m)}{\hbar} \int_C U_m^{\star} \nabla_k U_n \, d\mathbf{r} \tag{5}$$

where C denotes integration over a unit cell. $E_n(\mathbf{k})$ and $E_m(\mathbf{k})$ are the energies associated with the two bands (Exercises (5) and (6) of § 8.7). The complex conjugate $p_{nm}^{\star}$ may then be expressed in the two alternative forms

$$p_{mn}^{\star} = \hbar \int_C U_m \nabla U_n^{\star} \, d\mathbf{r} \tag{6}$$

$$= \frac{m(E_n - E_m)}{\hbar} \int_C U_m \nabla_k U_n^{\star} \, d\mathbf{r} \tag{7}$$

* H. Bethe, *Handb. der. Phys.*, **24**, (2), 378 (1933).

† A. H. Wilson, *Theory of Metals* (2nd Ed.). Cambridge University Press (1954), p. 47.

The oscillator strength f^x_{mn} for the direction of the operator ∇, which we shall indicate by the symbol x, is defined as

$$f^x_{mn} = \frac{2|p_{mn}|^2}{m(E_m - E_n)} \tag{8}$$

We may express $|p_{mn}|^2 = p_{mn}\, p^{\star}_{mn}$ in the alternative forms

$$|p_{mn}|^2 = im(E_m - E_n) \int_C U^{\star}_m(\mathbf{r})\, \nabla U_n(\mathbf{r})\, d\mathbf{r} \int_C U_m(\mathbf{r}')\, \nabla_k U^{\star}_n(\mathbf{r}')\, d\mathbf{r}' \tag{9}$$

$$= im(E_m - E_n) \int_C U_m(\mathbf{r})\, \nabla U^{\star}_n(\mathbf{r})\, d\mathbf{r} \int_C U^{\star}_m(\mathbf{r}')\, \nabla_k U_n(\mathbf{r}')\, d\mathbf{r}' \tag{10}$$

or as half the sum of these two expressions. We take the latter expression for $|p_{mn}|^2$ and add to it a corresponding term with $m = n$, which is zero, as may readily be seen by differentiating the normalising condition with respect to $\mathbf{k}$. We may then form the sum (1), using (8), by summing over *all* values of m for a fixed value of n. In this way we obtain

$$\sum_{m \neq n} f^x_{mn} = i \sum_{\substack{\text{all}\\ m}} \left\{ \int_C U^{\star}_m(\mathbf{r})\, \nabla U_n(\mathbf{r})\, d\mathbf{r} \int_C U_m(\mathbf{r}')\, \nabla_k U^{\star}_n(\mathbf{r}')\, d\mathbf{r}' \right.$$

$$\left. - \int_C U_m(\mathbf{r})\, \nabla U^{\star}_n(\mathbf{r})\, d\mathbf{r} \int_C U^{\star}_m(\mathbf{r}')\, \nabla_k U_n(\mathbf{r}')\, d\mathbf{r}' \right\} \tag{11}$$

Let us now consider the sum

$$\phi(\mathbf{r}) = \sum_m U^{\star}_m(\mathbf{r}) \int_C U_m(\mathbf{r}')\, \nabla_k U^{\star}_n(\mathbf{r}')\, d\mathbf{r}' \tag{12}$$

If we multiply by $U_m(\mathbf{r})$ and integrate over a unit cell, we obtain from the normal and orthogonal property of the Bloch functions, and hence of the functions U_n for a given value of $\mathbf{k}$,

$$\int_C \phi(\mathbf{r}')\, U_n(\mathbf{r}')\, d\mathbf{r}' = \int_C U_n(\mathbf{r}')\, \nabla_k U^{\star}_n(\mathbf{r}')\, d\mathbf{r}' \tag{13}$$

This equation is satisfied by taking

$$\phi(\mathbf{r}) = \nabla_k U^{\star}_n(\mathbf{r}) \tag{14}$$

That this solution is unique is a consequence of the fact that the functions $U_n(\mathbf{r})$ from a complete orthogonal set for a given value of $\mathbf{k}$ and provide a unique expansion of a function $\phi(\mathbf{r})$ provided it has the periodicity of the lattice. We may similarly evaluate the sum over m in the second term of equation (12) and obtain

$$\sum_m U^{\star}_m(\mathbf{r}) \int_C U_m(\mathbf{r}')\, \nabla_k U^{\star}_n(\mathbf{r}')\, d\mathbf{r}' = \nabla_k U^{\star}_n(\mathbf{r}) \tag{15}$$

We therefore have

$$\sum_{m \neq n} f_{mn}^{x} = i \int (\nabla U_n(\mathbf{r}) \nabla_k U_n^{\star}(\mathbf{r}) - \nabla U_n^{\star}(\mathbf{r}) \nabla_k U_n(\mathbf{r}))\, d\mathbf{r} \tag{16}$$

We may express the integral in equation (16) in terms of the diagonal elements of the effective mass tensor m_{rr}^{-1} (see exercise at end of § 8.8); the integral is equal to

$$1 - (m/m_{rr}) \tag{17}$$

Since

$$m_{rr}^{-1} = \frac{1}{\hbar^2} \frac{\partial^2 E}{\partial k_r^2}$$

we obtain the sum given in equation (1) on writing k_x for k_r.

Author Index

Subject Index